Biology for Life

GRAMPIAN REGIONAL COUNCIL
BANCHORY ACADEM

D1081210

To Philip and Anna

Biology for Life

M.B.V. ROBERTS
M.A. Ph.D.

Head of Biology, Cheltenham College

Nelson

Thomas Nelson and Sons Ltd
Nelson House   Mayfield Road
Walton-on-Thames   Surrey KT12 5PL

P.O. Box 18123   Nairobi   Kenya

116-D JTC Factory Building
Lorong 3   Geylang Square   Singapore 1438

Thomas Nelson Australia Pty Ltd
19–39 Jeffcott Street   West Melbourne   Victoria 3003

Nelson Canada Ltd
1120 Birchmount Road   Scarborough   Ontario M1K 5G4

Thomas Nelson (Hong Kong) Ltd
Watson Estate   Block A   13 Floor
Watson Road   Causeway Bay   Hong Kong

Thomas Nelson (Nigeria) Ltd
8 Ilupeju Bypass   PMB 21303   Ikeja   Lagos

Note to the teacher

Experiments involving the use of micro-organisms, the drawing of blood, or where the pupil acts as a subject, may be potentially hazardous. Such experiments should be carried out under close supervision. Detailed advice is given in the Association for Science Education booklet entitled *Safeguards in the School Laboratory*. Copies are available from:

The Association for Science Education
College Lane
Hatfield
Herts   AL10 9AA

© M.B.V. Roberts 1981

First published 1981
Reprinted 1982
ISBN 0-17-448081-4

NCN 210-3077-1

All rights reserved. No part of this publication may be reproduced, stored in a retrieval system, or transmitted, in any form or by any means, electronic, mechanical, photocopying, recording or otherwise, without the prior permission of the publishers.

Illustrations by Milne Stebbing Illustration
Picture Research by Michael Spillard
Designed by The New Book Factory, London
Composition by Filmtype Services Limited, Scarborough, England
Origination by Parkway Illustrated Press, London and Abingdon
Printed in Great Britain by Butler & Tanner Ltd., Frome and London
Cover design by Colin Lewis

Preface

For most people, O-level or CSE is the last biology course they will ever do, and naturally one hopes that it may be of some use to them afterwards. With this in mind I have written a book for 13 – 16 year olds in which the topics taught in O-level and CSE courses are related to everyday life: hence the title *Biology for Life*. Much of the human physiology is related to medicine and health, and the various organisms which are dealt with are looked at from the point of view of their application to man.

In the last few years a number of surveys have been carried out which suggest that many science text books at this level present students with serious reading difficulties. I have therefore endeavoured to make *Biology for Life* easy to read. This has meant keeping the language and sentence construction as simple as possible. I hope this will make the book equally usable by O-level and CSE pupils, and will help to bridge the gap between these two traditionally separated groups.

Writing a text book which is readable and yet sufficiently rigorous academically has not been easy. If I have had any success in solving the readability problem it is largely because of the help I have received from Drs Nikolas and Justine Coupland of the University of Wales Institute of Science and Technology. Specialists in English, the Couplands have made a special study of the linguistic aspects of communication in science teaching. They kindly read the entire manuscript of *Biology for Life* and made many valuable suggestions as to how the text might be simplified and its clarity improved. I am most grateful to them for their shrewd counsel.

On the question of the language, I obtained much useful information from a trial which was carried out in a selection of schools during 1978. The schools were deliberately chosen for the widely differing ability ranges of their pupils. They are listed below, and I would like to record here my thanks to the heads of the biology departments and their colleagues for their help and cooperation.

The book is divided up into a large number of short topics. Most of the topics are self-contained so that, within reason, they can be studied in any order. I hope this will enable the individual teacher to weave the topics together in a sequence which suits his or her particular course.

Each topic is normally followed by one or more investigations and about six homework assignments. Some of the investigations can be carried out by students, either individually or in groups; others are best done as demonstrations by the teacher. I have deliberately not incorporated the instructions into the text because this would have interrupted the flow and made it difficult to see the overall picture.

The assignments are all questions of the kind which occur in O-level or CSE examinations. They are of graded difficulty with simple ones coming first and more difficult ones later. The simple ones involve mainly factual recall, whereas the harder ones test more advanced skills such as formulating hypotheses and designing experiments.

Many people, in addition to those involved in the trial, have assisted in the production of this book. I am particularly grateful to Dr James Parkyn and my father, Dr Llywelyn Roberts, for advising me on medical matters; to Mrs Gill Williams for her advice on topics relating to health education; to Mr Malcolm Ashby for help with the dental section; to Mr David Alford for his constructive comments on the first draft; and to Dr James Parkyn for obtaining some of the more specialised medical photographs. In addition I would like to thank Mr John Barker, Mr Peter Fry and Dr John Land for their useful comments. Despite the help which I have received, the book is bound to contain some errors. These are entirely my fault and I hope that teachers and students will not hesitate to point them out to me as they use the book.

Biology for Life was written while I was holding a Research Associateship at Chelsea College, University of London. I would like to thank Professor Paul Black, Director of the Centre for Science Education, for providing me with facilities, and Professor Peter Kelly for his continual help and encouragement. I am also grateful to Miss Pat Stevens who typed the manuscript with patience and good humour.

Finally I must thank my publishers, Thomas Nelson and Sons Ltd. It was they who first suggested that I should write this book, and their support has been a constant source of encouragement. I owe particular thanks to Mrs Donna Evans for her unfailing patience and efficiency, and to all those who have been involved with the design and production of the book.

M. B. V. Roberts
Kensington, August 1980

Schools which took part in the trial.

Rydens School, Walton-on-Thames, Surrey.
Isleworth Grammar School, Isleworth, Middlesex.
Hounslow Manor School, Middlesex.
Willesden High School, London, N.W.10.
Woodberry Down School, London, N4.
Kidbrooke School, London, S.E.3.
Marlborough College, Marlborough, Wiltshire.
Winchester College, Winchester, Hampshire.
The Holt School, Wokingham, Berkshire.
Theale Green School, Theale, Berkshire.
Earls High School, Halesowen, West Midlands.

Riland-Bedford School, Sutton Coldfield, West Midlands
North Bromsgrove High School, Bromsgrove, Hereford and Worcester.
JordanThorpe School, Sheffield.
Valley Comprehensive School, Worksop, Nottinghamshire.
Burnage High School, Manchester.
Parrswood High School, East Didsbury, Manchester.
Central High School for Girls, Manchester.
Spurley Hey High School, Gorton, Manchester.
Stewards Comprehensive School, Harlow, Essex.
St. Bernard's Convent High School, Westcliff-on-sea, Essex.
Lucton County Boys' School, Loughton, Essex.
Exmouth School, Devon
Mayfield School, Putney, London S.W.15.

Contents

The continuation of life

Living things and their environment

Some basic principles

*Biology is the
study of life and living
things. In the first few Topics
we shall introduce the subject
and look at some of its
basic principles.*

Introducing biology

Biology is the study of life and living things. We call living things organisms. One of the most complicated organisms is the human being, and studying man is an important part of biology.

Figure 1 Biology comes into almost everything we do. Growing crops is an important aspect of biology. The photograph is of farmland in Iowa, U.S.A.

The different branches of biology

Here are some of the main branches of biology:

Zoology: the study of animals
Botany: the study of plants
Anatomy: the study of the structure of living things
Physiology: the study of how the body works
Nutrition: the study of food and how living things feed
Heredity: the study of how characteristics are passed from parents to offspring
Ecology: the study of where organisms live

What are living things made of?

All living things, including humans, are made of chemical substances. These are composed of **elements**. The main elements found in organisms are carbon, hydrogen and nitrogen. Other elements occur in smaller amounts: these include phosphorus, calcium and sodium.

Some of the substances which make up the body are very complex and contain lots of carbon atoms. We call them **organic substances**; they are a very important ingredient of all living things. The other substances are simpler, and usually lack carbon. They are called **inorganic substances**.

Here is a summary of the main substances found in the body:

Organic substances: Carbohydrates
Fats
Proteins
Inorganic substances: Salts
Water

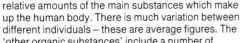

Figure 2 This diagram, called a pie chart, shows the relative amounts of the main substances which make up the human body. There is much variation between different individuals – these are average figures. The 'other organic substances' include a number of important substances derived from vitamins. These and all the other substances in this chart come from the food we eat.

inorganic substances 1%
other organic substances 1%
carbohydrate 5%
protein 18%
fat 10%
water 65%

Each of these substances has certain jobs to do. These are their main jobs:

Carbohydrates give us energy; one of the best known carbohydrates is sugar.
Fats also give us energy; there is a lot of fat under the skin, it helps to keep us warm.
Proteins help to build the body; they form important structures like bones and muscles.

Salts help to make the various structures in the body work properly.
Water provides a liquid in which other substances can be moved about inside the body.

Figure 2 shows the proportions of each of these substances in the human body. You may be surprised to see that our bodies contain far more water than any other substance. The fact is that we are really very wet.

All these substances are made up of **molecules**. The molecules, in turn, are composed of **atoms**. Take water, for example. A molecule of water consists of two hydrogen atoms and one oxygen atom. We usually represent it as H_2O. This is its **chemical formula**.

Water is one of the simplest substances found in living things. A more complicated substance is glucose which is a type of sugar. Its formula is $C_6H_{12}O_6$: this shows that a molecule of it contains six carbon atoms, twelve hydrogen, and six oxygen atoms.

Chemical reactions in organisms

All manner of chemical reactions take place inside living organisms. These reactions are essential for life. If they stop, the organism dies.

We call these reactions **metabolism**. Some of the reactions build things up: they make important substances which the organism needs. These reactions require energy. Other reactions break things down: they give out energy, mainly in the form of heat. The difference between these two kinds of chemical reactions is illustrated in Figure 3.

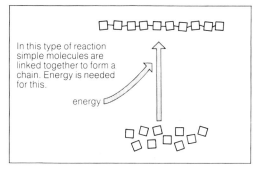

In this type of reaction simple molecules are linked together to form a chain. Energy is needed for this.

energy

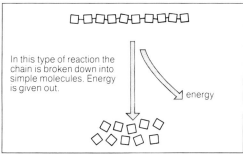

In this type of reaction the chain is broken down into simple molecules. Energy is given out.

energy

Figure 3 This diagram illustrates the two main types of chemical reactions which take place in living organisms.

Naming living things

Every organism is given two names. The first is the name of the **genus** to which the organism belongs. It shares this name with a number of other closely related organisms. We can liken the genus name to a person's surname. The organism shares this name with other members of its genus, just as we share our surname with other members of our family.

The organism's second name is the name of the **species** to which it belongs. This name is possessed by only one kind of organism: it does not share it with any other organisms in the genus.

It is customary to start the genus name with a capital letter, and the species name with a small letter, and to print both names in *italics*.

Now for an example. The domestic cat's full name is *Felis catus*: *Felis* is the genus name, and *catus* is its species name. This name applies to the family pet which spends so much of its time curled up on the hearth-rug. However, the genus *Felis* also includes several animals which would be less welcome in our houses, the lion and tiger for example (Figure 4). The lion's proper name is *Felis leo*, and the tiger's is *Felis tigris*. Other wild cats include the leopard, jaguar and cheetah.

This system of naming organisms was developed by the 18th Century naturalist Carl Linnaeus (1707 – 78). Because it involves giving organisms *two* names, it is known as the **binomial system**.

The names just described are called **proper names**. They are usually Latin names, or have Latin endings. The trouble is that they are often long and difficult to remember. For example, there is a certain kind of worm which is called *Haploscoloplos bustorus*! To make things easier we often call animals and plants by simpler **common names**: cat, lion, tiger, and so on. These common names may be likened to a person's nickname.

People sometimes ask why we bother with proper names. Why don't we just use common names? The reason is that common names are often not precise enough. An organism may have several different common names, and sometimes the same one is used for several different organisms. Common names can therefore be confusing.

Figure 4 This is a lion. It belongs to the genus *Felis*.

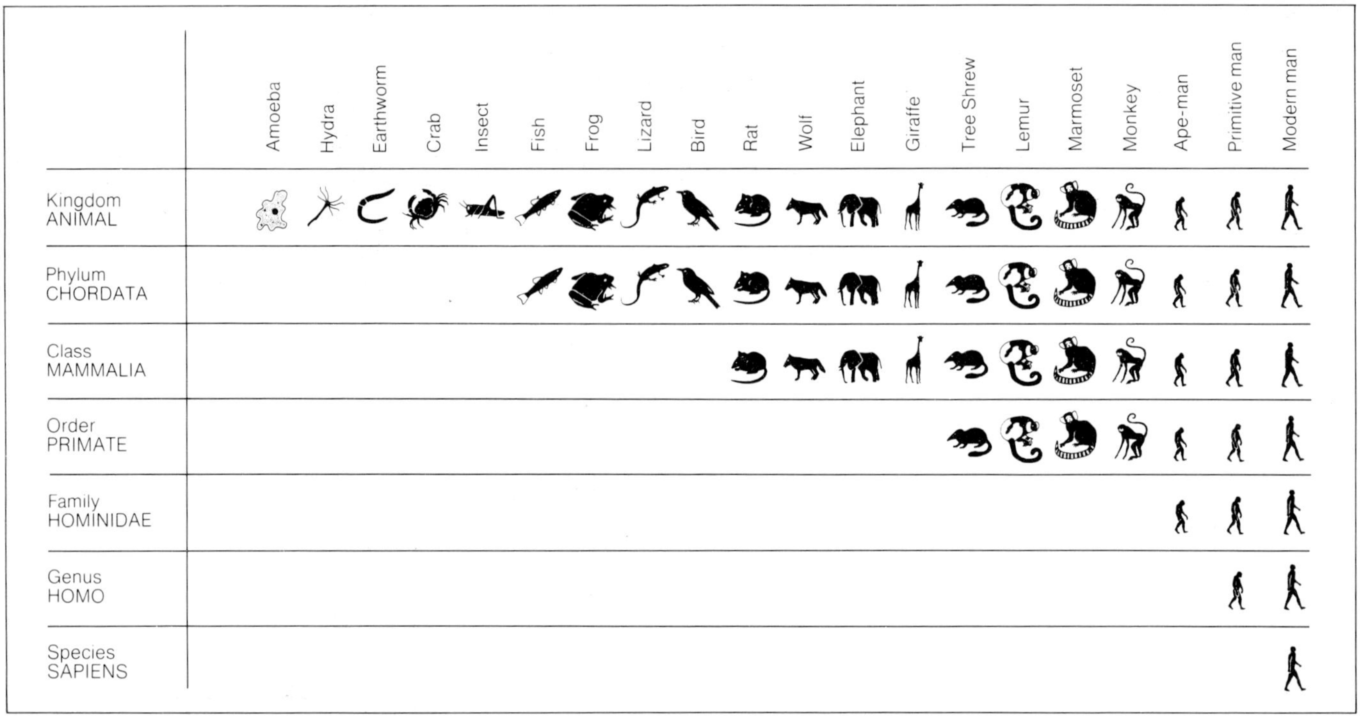

	Amoeba	Hydra	Earthworm	Crab	Insect	Fish	Frog	Lizard	Bird	Rat	Wolf	Elephant	Giraffe	Tree Shrew	Lemur	Marmoset	Monkey	Ape-man	Primitive man	Modern man
Kingdom ANIMAL	●	✦	⬤	🦀	🦗	🐟	🐸	🦎	🐦	🐀	🐺	🐘	🦒	⬤	⬤	⬤	🐒	𝆑	𝆑	𝆑
Phylum CHORDATA						🐟	🐸	🦎	🐦	🐀	🐺	🐘	🦒	⬤	⬤	⬤	🐒	𝆑	𝆑	𝆑
Class MAMMALIA										🐀	🐺	🐘	🦒	⬤	⬤	⬤	🐒	𝆑	𝆑	𝆑
Order PRIMATE														⬤	⬤	⬤	🐒	𝆑	𝆑	𝆑
Family HOMINIDAE																		𝆑	𝆑	𝆑
Genus HOMO																			𝆑	𝆑
Species SAPIENS																				𝆑

Figure 5 This picture shows how man is classified. As we go downwards from top to bottom the number of organisms in each group decreases and the similarities between them increases. Ape-man and primitive man are, of course, extinct and are known only from their fossil remains.

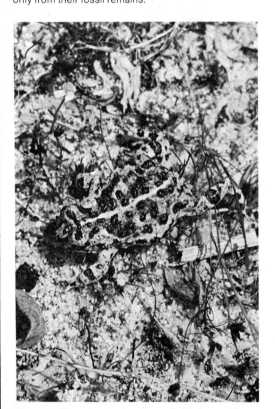

Figure 6 There is an animal here. Can you see it? What kind of animal is it?

How do we classify living things?

Scientists have described about one and a half million kinds of organisms and new ones are constantly being discovered. With so many organisms in existence, we must have some way of classifying them. This is done by arranging them into groups. Each group is then split into smaller groups, and these groups into even smaller groups and so on. The members of each group have certain features in common which distinguish them from other groups.

Living things are first split into **kingdoms**, such as the animal and plant kingdoms. These kingdoms are then split up into a large number of smaller groups called **phyla** (singular: phylum). All the members of a phylum have certain things in common. Each phylum is broken down into **classes**, classes into **orders**, orders into **families**, families into **genera** (singular: genus), and genera into **species**. Each of these groups contain progressively fewer and fewer kinds of organisms. Thus a phylum contains a wide variety of organisms: they all have certain basic features in common, but there are a lot of differences between them. However, the organisms belonging to a genus are all very similar. This is illustrated in Figure 5, which shows how man is classified.

Where do organisms live?

Living things are found almost everywhere in the world: on land and in the air, in water and underground. You find them in the most unlikely places, such as hot dry deserts and the freezing cold polar regions. You even find them in salt lakes and hot springs where the water is almost boiling. Many organisms live on or in the bodies of other organisms.

The place where an organism lives is called its **habitat**. The conditions which exist in its habitat make up the **environment**. Every organism is suited, or **adapted**, to live in its particular habitat. For example, some animals are wonderfully camouflaged so they cannot be seen by their enemies (Figure 6). Organisms can survive only if they are suitably adapted. This is one of the most important principles in biology and we will meet it frequently in later Topics.

The world of living things

We will now look at the main groups of living things. This will give you a glimpse of the variety that's found amongst animals and plants. Before getting down to details, look at Figure 7. This shows at a glance the main groups into which living things are divided. This is not the only way of splitting them up, but it is one that is commonly used. On the next three pages we will look briefly at each of the groups of living things shown in Figure 7.

Figure 7 The main groups of living things at a glance. (Viruses and bacteria are not included in this classification. They are put into separate kingdoms of their own.)

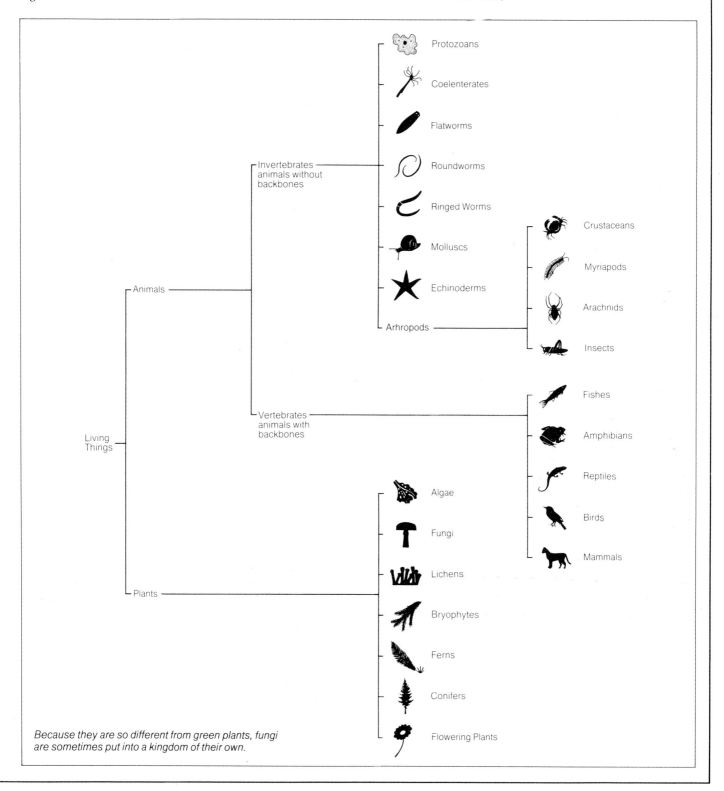

Because they are so different from green plants, fungi are sometimes put into a kingdom of their own.

Animals without backbones (invertebrates)

Protozoans

Microscopic animals whose body consists of only one cell. Live in water, or as parasites inside other organisms.

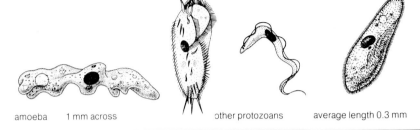

amoeba   1 mm across   other protozoans   average length 0.3 mm

Coelenterates

Many-celled animals with tentacles and sting cells. Most of them live in the sea.

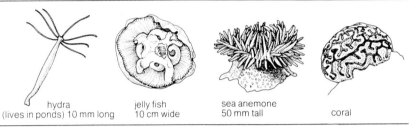

hydra
(lives in ponds) 10 mm long

jelly fish
10 cm wide

sea anemone
50 mm tall

coral

Flatworms

Body elongated and flat. Some of them live in ponds and streams, but most are parasites causing diseases.

10 mm long
fresh-water flatworm

blood fluke   15 mm long

tapeworm   5 m long

Roundworms

Body elongated and thread-like, round in cross-section. This group includes some harmful parasites.

threadworms   10 mm long

Annelids

Body divided up by rings into a series of segments.

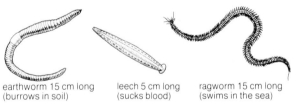

earthworm 15 cm long
(burrows in soil)

leech 5 cm long
(sucks blood)

ragworm 15 cm long
(swims in the sea)

tube worm   10 cm long
(lives in a tube in the sea)

Molluscs

Have a soft body usually protected by a shell. In some the shell is greatly reduced.

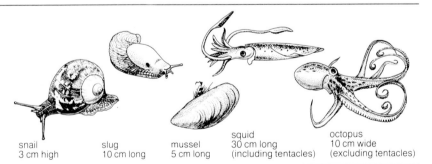

snail
3 cm high

slug
10 cm long

mussel
5 cm long

squid
30 cm long
(including tentacles)

octopus
10 cm wide
(excluding tentacles)

Echinoderms

Have a tough spiny skin. Most of them are star-shaped. They all live in the sea.

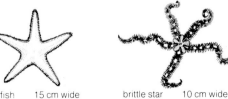

starfish   15 cm wide

brittle star   10 cm wide

sea urchin   10 cm wide

Arthropods

Have a hard cuticle and jointed limbs. Divided into four groups mainly on the basis of the number of legs.

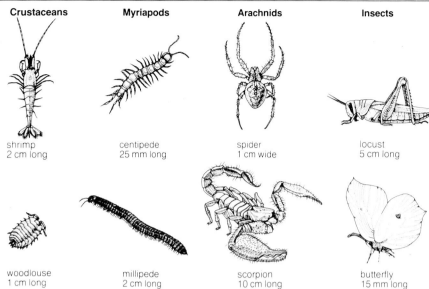

Crustaceans

shrimp
2 cm long

woodlouse
1 cm long

Myriapods

centipede
25 mm long

millipede
2 cm long

Arachnids

spider
1 cm wide

scorpion
10 cm long

Insects

locust
5 cm long

butterfly
15 mm long

Animals with backbones (vertebrates)

Fishes

Live in water. Have gills for breathing, scales on their skin, and fins for movement.

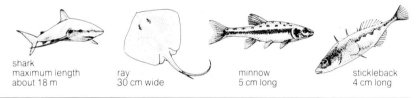

shark
maximum length
about 18 m

ray
30 cm wide

minnow
5 cm long

stickleback
4 cm long

Amphibians

Have moist skin without scales. Live on land but lay eggs in water. Have fish-like tadpole larva which changes into the adult.

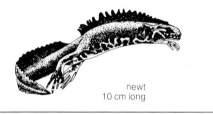

newt
10 cm long

frog   6 cm long
(excluding legs)

Reptiles

Have dry waterproof skin with scales. Eggs have a leathery shell and are laid on land.

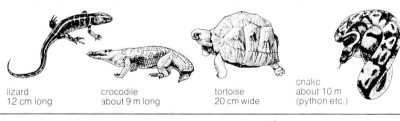

lizard
12 cm long

crocodile
about 9 m long

tortoise
20 cm wide

snake
about 10 m
(python etc.)

Birds

Have feathers. Eggs have hard shells. Wings for flying, and a beak for feeding.

sparrow
15 cm long

ostrich   2.5 m tall
(does not fly)

Mammals

Have hair. The young develop inside the mother and after birth are fed on her milk.

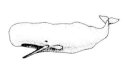

kangaroo 2 m high

lion   2 m long

whale about 33 m long

man
2 m high

Plants

Algae

Simple plants which do not have roots, stems and leaves. Usually green, but sometimes brown or red. Live in water.

single-celled algae 10 μm wide

spirogyra
(a thread-like alga)

sea-weed 50 cm long

Fungi

Simple non-green plants which do not photosynthesise. Some are parasites and cause serious diseases of plants.

pin mould
(grows on bread etc.)

yeast each cell 5 μm wide
(single cells or short chains)

mushroom
5 cm wide

Lichens

Consist of an alga and fungus combined together. Grow on rocks and tree trunks. Very hardy.

shrubby
10 mm high

lichens

leafy
(flat)

Bryophytes

Have simple leaves or leaf-like form. Found mainly in damp places.

moss 10 mm high

liverworts 5 mm wide

Ferns (Pteridophytes)

Have proper roots and stems, and frond-like leaves. Found mainly in damp places. Reproductive spores can be seen on the undersides of the fronds in the autumn.

common fern
(has unbranched fronds)

40 cm high

bracken
(has branched fronds)

Conifers (Gymnosperms)

Large plants with cones for reproduction. Good at surviving in dry or cold climates. Most of them keep their leaves through the winter.

pine tree   30 m high

Flowering plants (Angiosperms)

Wide range of plants with flowers for reproduction. Range from small herbs to massive trees. Many of them drop their leaves in the winter.

foxglove   45 cm high

grass   30 cm high

oak tree   25 m high

Note: Fungi are sometimes put in a separate kingdom of their own.

Investigation 1

What are you made of and how much are you worth?

1 Weigh yourself.

2 Write down your mass in kilograms.

3 Study Figure 2. This tells you the percentage of different substances in an average human being.

4 Assuming that you contain these substances in the same proportions, work out the mass of each substance in your body.

5 Find out the cost of each of the substances in the shops. Assume that all the carbohydrate is sugar, all the protein is meat, and the salts are all common table salt. Ignore the 'other organic substances'.

6 Work out the value of your body in pounds and pence.

Do you think this is a valid way of expressing the value of a human being?

Investigation 2

Putting some familiar animals and plants into groups

1 Examine various organisms, or pictures of organisms, provided by your teacher. All of them are featured in the classification on pages 6–8.

2 Write down the name of the group to which each organism belongs. Use the classification on pages 6–8 to help you.

3 Look carefully at each organism.

From its structure, what can you say about the sort of place where it lives, and the kind of life it leads?

Investigation 3

Putting some unfamiliar animals and plants into groups

1 Examine various organisms which are *not* illustrated in the classification on pages 6–8.

2 Write down the name of the group to which you think each organism belongs. Do this by relating the characteristics of the organism to the information given on pages 6–8.

3 Which specific animal or plant illustrated on pages 6–8 does each organism resemble most closely?

Assignments

1 Which branch or branches of biology listed at the begining of this Topic must each of the following people know about:

a farmer, a gardener, a nurse, a family doctor, a game warden, a person who breeds dogs, a PE teacher, a forester, a surgeon, a keeper in a zoo.

2 Which chemical substances mentioned in this Topic play the most important part in:

a) helping us to run fast,
b) making us strong,
c) keeping us warm,
d) making our blood 'runny',
e) mending a broken leg?

3 What is the connection between each of the following pairs:

a) carbohydrates and fat,
b) insect and crustacean,
c) genus and species,
d) leopard and cheetah,
e) molecules and atoms?

4 What group does each of these organisms belong to: moss, jelly fish, tortoise, tapeworm, whale, mushroom, mould, tube worm, sea weed, newt?

5 What would be the easiest way of telling the difference between:

a) an arthropod and a vertebrate,
b) an insect and an arachnid,
c) an amphibian and a reptile,
d) an alga and a fungus,
e) a conifer and a flowering plant?

6 From books, try to find out the largest member of each of the following plant groups: Algae, Ferns, Conifers, Flowering plants.

In each case give the proper name and common names of the organism, and state its approximate size.

7 Give the name of an animal which:

a) is shaped like an umbrella and has sting cells,

b) lays eggs with a leathery shell,
c) has a pouch in which the young develop,
d) is shaped like a star,
e) has two pairs of wings,
f) consists of only one cell,
g) lives on land but lays its eggs in water,
h) has a long flat body,
i) has tentacles and belongs to the same group as snails,
j) has four pairs of legs.

8 Give the name of a plant which:

a) reproduces by means of flowers,
b) has frond-like leaves,
c) consists of only one cell and is coloured green,
d) causes a disease,
e) has no chlorophyll.

The characteristics of living things

If we examine the things organisms do, and the processes which take place inside them, we find certain features that are common to them all.

Figure 1 This gymnast is showing one of the basic properties of life: movement.

Living things move

This is obvious in the case of a human being (Figure 1). We move our arms and legs by means of muscles. Most animals can move in this kind of way, at least at some period of their lives.

However movement is not so obvious in a plant. To see movement in a plant you must look inside it, under a microscope. Then you may see things moving about, though it is not always easy (Investigation 1).

Living things respond to stimuli

If you sit on a drawing pin, you jump up quickly. The pricking of your bottom is called the **stimulus** (plural: stimuli). Your jumping is called the **response**.

Living things respond to different kinds of stimuli. The main ones are touch, chemicals, heat, light and sound. For example, when we see something we are responding to light entering our eyes, and when we taste things we are responding to chemicals in the mouth.

At first sight you might think that plants are an exception to the rule that all organisms respond to stimuli. After all, if you hit a tree, it doesn't move away. However, plants *do* respond to certain stimuli, but much more slowly than animals. They do not have muscles. Instead they respond by *growing* in a particular direction. For example, most plants grow towards light (Figure 2).

There are a few plants which respond quickly to touch, like animals do. For example, the leaves of the mimosa plant close up when you touch them (Investigation 2). However, there are no sense organs, nerves or muscles in the leaves: the response is brought about in a quite different way.

Living things grow

As an animal or plant develops, it gets larger and heavier. In other words, it **grows**.

Growth takes place by substances being taken into the organism from outside. These substances are then built up into the structures of the body: they become part of the organism.

Figure 2 This plant is growing towards the light.

Figure 3 All organisms feed. This lizard is eating a cricket.

A plant, such as a tree, goes on growing throughout its life. Animals usually stop growing when they reach a certain age. For example, humans stop growing at about the age of eighteen.

Even when growth stops, the materials of the body are constantly replaced by new substances coming in from outside. This process of renewal goes on throughout life. It has been worked out that in about seven years all the chemicals in the human body are replaced by new ones.

Living things feed

We have just seen that in order to grow an organism must take substances into its body. This is achieved by **feeding** (**nutrition**). All organisms need food.

Animals and plants feed in quite different ways. Animals feed on complex organic substances which are often in solid form (Figure 3). In man the food is taken into the mouth. It is then broken down into a liquid: this process is called **digestion** and it is carried out in the gut.

Any food which cannot be digested passes out of the body through the anus. Meanwhile the digested food is absorbed and used.

In contrast to animals, plants make their own food. They take in simple things like carbon dioxide and water and build them up into complex organic substances. Energy is needed for this: it comes from sunlight. The green pigment **chlorophyll** enables the plant to use sunlight in this way: this is why plants are usually green. The process by which plants make food is called **photosynthesis** (Figure 4).

Living things produce energy

Living things need energy to move, grow, replace worn-out structures, and so on. They obtain this energy by burning food. The food is not really burned, but it comes to the same thing chemically: the food is broken down into carbon dioxide gas and water. This process is called **respiration**.

Respiration normally requires oxygen. Organisms get this vital gas from the air or water around them. We call this process **breathing**. For example, in man air is sucked into a pair of lungs (Figure 5). In fishes water flows over the surface of gills. As well as taking up oxygen into the body, lungs and gills get rid of carbon dioxide.

Not all organisms have lungs or gills. Some just let oxygen 'seep' into the body across the surface. This is called diffusion. It's a slow process, but it's good enough for small or inactive organisms which don't need much energy.

Many animals are able to carry oxygen quickly round the body. In man this job is done by the **blood system**: the blood is pumped by the heart through a system of blood vessels. The blood system is also used for transporting dissolved food substances.

Living things get rid of poisonous waste

In many ways an organism is like a chemical factory. Substances are constantly being broken down to produce energy, or built up to make structures, such as bones and muscles.

Some of the by-products of metabolism are poisonous. They must not be allowed to pile up in the body, or they will kill the organism. So the body must get rid of them. This is called **excretion**.

In animals one of the most poisonous waste substances is **ammonia**. Ammonia contains nitrogen. It is so poisonous that most animals quickly turn it into a less poisonous substance. This is then expelled from the body, along with water in the form of a liquid called **urine**.

Plants get rid of waste substances in a different way. They turn them into harmless crystals which they store within their bodies out of harm's way.

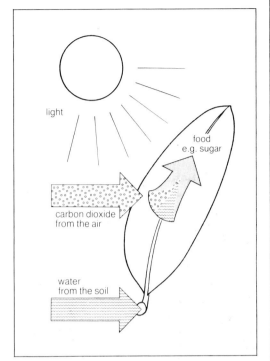

Figure 4 Plants feed by photosynthesis. The plant takes up carbon dioxide and water and turns them into complex food substances such as sugar. This process is carried out mainly in the leaves.

Figure 5 An athlete breathing heavily after a race. His muscles have done a lot of work and they need plenty of oxygen.

Living things produce offspring

Organisms produce offspring. This is known as reproduction. Usually it involves the union of two individuals, a male and a female. We call this **sexual reproduction** (Figure 6).

The male produces sperms, and the female produces eggs. Sperms are much smaller than eggs.

The male and female usually come into close contact, and the male's sperms are put into the female's body. The sperms then unite with the eggs, one sperm per egg. This is called **fertilisation**. The fertilised egg then develops into a new individual.

Some organisms can reproduce on their own without the help of another individual. This is called **asexual reproduction**. At its simplest, the organism merely splits in two. In good conditions asexual reproduction may take place very quickly and sometimes a very large number of offspring are produced at the same time.

The series of events which take place from the time an organism reproduces to the time its offspring reproduce is called the **life cycle**.

Living things die

The life of an individual animal or plant does not go on for ever. Eventually all the processes which have been described in this Topic stop and the organism dies.

Some organisms live much longer than others. One of the longest living animals is man: there are many cases of people living to over 100 years. The only other animals to come anywhere near this are tortoises, whales and elephants, and perhaps certain fish. At the other extreme, certain moths live only a few hours: just long enough to find a mate and reproduce.

With plants it is a different story altogether: they can live much longer than animals. The record is held by certain pine trees in California: some of them are estimated to be well over 4000 years old.

Figure 6 Reproduction is one of the basic features of all living things.

Investigation 1

Detecting movement inside a plant

Movement is difficult to see in most plants, but here is an exception.

1 Obtain a sprig of the water plant Canadian pondweed (*Elodea*) which has been kept in the light for several hours.

2 Cut off one of the leaves and put it in a drop of water on a microscope slide.

3 Cover the leaf with a coverslip (a thin square of glass) so as to keep the leaf flat. Try not to get any air bubbles trapped under the coverslip.

4 Look at the leaf under a microscope. (If you don't know how to use a microscope, look up page 400).

Can you see lots of small green objects inside the leaf?

These are chloroplasts (see page 15).

If they are moving, describe their movement as fully as you can.

The chloroplasts of Canadian pondweed don't *always* move, so don't be disappointed if you see nothing happening.

Investigation 2

Getting a plant to respond to touch

Few plants respond quickly when you touch them, but one that does is the sensitive mimosa plant *Mimosa pudica*.

1 Obtain a potted specimen of *Mimosa pudica*.

2 Gently touch the top side of a leaf with a needle. What happens?

3 Gently touch other parts of the plant, including the lower side of the leaves, and the stem.

Describe what happens in each case.

4 Pipette a drop of water onto one of the leaves. What happens?

Of what use do you think this response is to the plant?

How do you think the response might be brought about?

Investigation 3

Recognising the characteristics of life in organisms

1 Look at the list of characteristics of life given in this Topic. (The eight headings to the sections.)

2 Examine various organisms or pictures of organisms, provided by your teacher.

3 For each organism write down the particular characteristics of life which you can see it possesses. Do not write down the characteristic unless you can actually see it.

After doing this, you will realise that some of the characteristics of life are difficult to see in organisms.

How could you find out if an organism possesses a characteristic of life which you cannot actually see?

Assignments

1 Name three different activities for which we need food.

2 Of all the characteristics of living things mentioned in this Topic, which ones are most important in each of the following?

The number of characteristics which you should mention in each case is given in brackets.

a) a person watching television (1)
b) a footballer kicking a ball (2)
c) a lion stalking a zebra (2)
d) germs spreading through your body when you are ill (1)
e) a plant bending towards the light (1)
f) a person panting after a race (1)
g) a bean plant climbing up a bamboo cane (2)

3 Which of the following activities are shown by all animals and plants?

Respiration, eating, sexual reproduction, growth, escaping from enemies.

4 Explain in your own words what is meant by a stimulus.

What sort of stimuli does a pot plant such as a geranium respond to?

5 If you blow up a balloon and then hold it in front of the fire, it increases in size.

Is the balloon growing in a biological sense?

Give reasons for your answer.

6 What is the difference between an object which is dead and one which is non-living?

7 A visitor to our planet from outer space says he thinks motor cars are alive.

In order to put him straight, make a list of ways in which a motor car is similar to living organisms, and another list of ways in which it is different.

8 The table below shows the main differences between a typical animal such as you, and a typical plant such as an oak tree:

Animal	Plant
Moves around	Does not move around
Has a mouth	Has no mouth
Has eyes and ears	Has no eyes or ears
Has no green leaves	Has green leaves

Explain the reason for each of these differences.

Cells, the bricks of the body

An organism is made of cells in much the same way as a house is made of bricks. This Topic is about cells: how we study them, what they look like, and what goes on inside them.

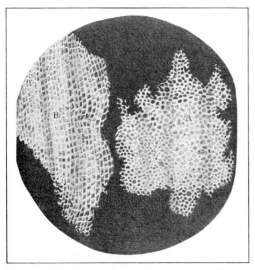

Figure 1 The first drawing of cells ever made. These cells were observed in a piece of cork by Robert Hooke. This drawing was published in Hooke's famous *Micrographia* in 1665.

Figure 2 This kind of light microscope is used in many schools and colleges.

How were cells discovered?

Cells were discovered in 1665 by the English inventor and scientist, Robert Hooke. Hooke examined a piece of bark which he stripped from a tree. Near the surface of bark is a layer of cork: Hooke cut a thin slice of the cork and placed it under a microscope which he had made himself. Hooke described the cork as being made up of hundreds of little boxes, giving a kind of honeycomb appearance (Figure 1). He called these little boxes **cells**.

As more and more organisms were examined under the microscope, it became clear to scientists that virtually all living things are made of cells. And so cells came to be regarded as the basic unit of which organisms are made.

How can we see cells?

The human body consists of about one hundred million million cells, and each one is very small. Because they are so small, we usually express their size in **micrometres**. A micrometre is one thousandth of a millimetre, and is given the symbol μm. A typical cell is about 20 μm wide.

Objects this size are too small to be seen with the naked eye, or even with a magnifying glass. To see them you must use a microscope such as the one shown in Figure 2. This is called a **light microscope**. When you look at a specimen down this kind of microscope, light rays pass from the specimen to your eye. On their way, the light rays pass through a series of glass lenses which magnify the specimen. The light microscope in Figure 2 can magnify things up to about 400 times. Enlarged to this extent, a pinhead would be bigger than the wheel of a car.

For cells to be seen under the microscope, they must be spread out flat. They must not be on top of each other, otherwise light cannot get through them and they will be invisible.

One of the easiest places to get cells from is inside your cheek. If you scrape the inside of your cheek, the cells will come away and you can put them on a glass slide. Adding a drop of dye will stain the cells and make them show up under the microscope (Investigations 1 and 2).

Sometimes scientists want to look at the cells in the middle of a thick solid organ such as the liver or kidney. To do this it is necessary to cut thin slices of the organ. This is done with an instrument which works rather like a bacon-slicer. It is called a microtome. The slices, or **sections** as they are called, are then stained with a dye which makes the cells show up.

Inside a typical animal cell

Figure 3 shows the structure of a typical animal cell. The cell is bounded by a thin **cell membrane**. In the centre is a tiny ball, the **nucleus**. This is surrounded by a material called the **cytoplasm**.

The nucleus

It is possible to take the nucleus out of certain cells. If this is done, the cell dies. From this experiment we conclude that the nucleus is essential for the life of the cell. It controls the various processes which go on inside it.

The nucleus contains a number of thread-like bodies called **chromosomes**. However, these can only be seen clearly when the cell is about to split in two (see page 336). The chromosomes determine the organism's characteristics such as the colour of the eyes.

The cytoplasm

The cytoplasm is told what to do by the nucleus. The cytoplasm produces energy, makes things, and stores food. Hundreds of chemical reactions take place inside it. Together, these reactions make up **metabolism** (see page 3).

Scattered about in the cytoplasm are small granules. Seen under the light microscope, these look like little dots. The larger ones are **mitochondria** (singular: mitochondrion). The mitochondria have been described as the

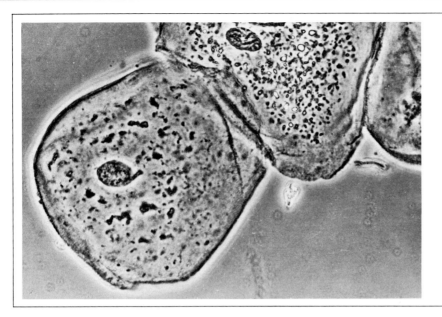

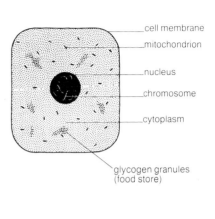

'powerhouse of the cell': their job is to produce energy.

The smaller granules in the cytoplasm are tiny bits of stored food. Many of them consist of a substance called **glycogen**.

Figure 3 This diagram shows a typical animal cell. On the left are some cheek cells as they actually appear under the microscope.

Inside a typical plant cell

A typical plant cell is shown in Figure 4. It differs from animal cells in the following ways:

1 In addition to the cell membrane, the plant cell has a **cell wall**. It is made of **cellulose**, a rubbery material which helps to make plants tough.

Instructions for using the microscope are on pages 400–401.

2 In the centre of the cell there is a large cavity called the **vacuole**, which is filled with fluid. This means that the cytoplasm is pushed towards the edge of the cell. The nucleus is usually found in this layer of cytoplasm. However, in some plant cells the nucleus is suspended in the middle of the vacuole by fine strands of cytoplasm.

3 The cytoplasm contains **starch grains**. This is how plants store food. The starch grains are equivalent to the glycogen granules in animal cells.

4 Many plant cells possess **chloroplasts**. These are located in the cytoplasm, and they contain the green pigment **chlorophyll** which is used in **photosynthesis**. Chloroplasts only occur in the green parts of the plant which are exposed to the light. Roots and other underground structures lack them

Figure 4 This diagram shows a typical plant cell. On the left are some leaf cells as they appear under the microscope.

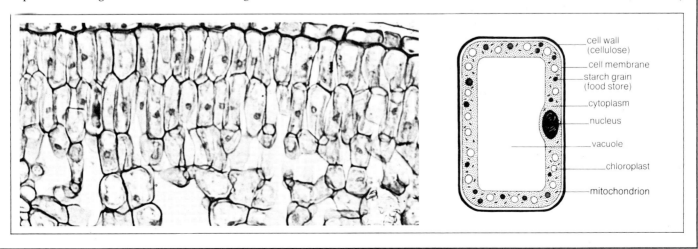

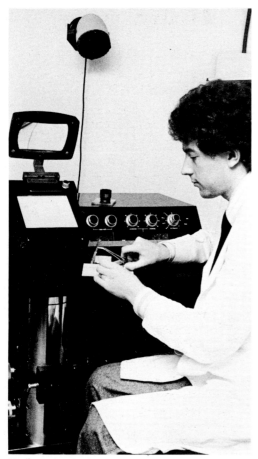

Figure 5 A scientist about to view a specimen in an electron microscope.

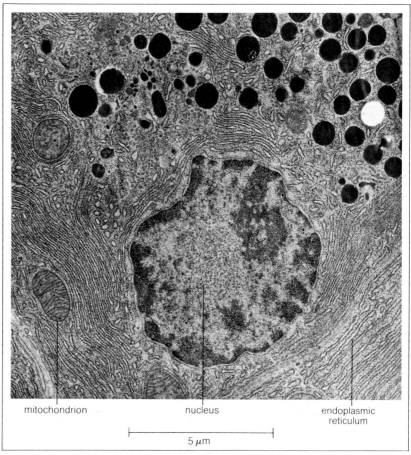

mitochondrion   nucleus   endoplasmic reticulum

5 μm

Figure 6 This is a cell from the pancreas as seen in the electron microscope. The black blobs are substances which the cell produces.

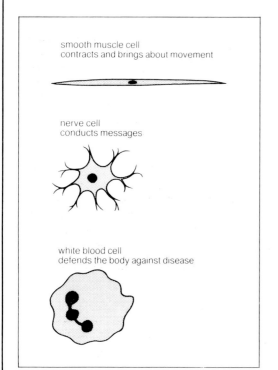

smooth muscle cell
contracts and brings about movement

nerve cell
conducts messages

white blood cell
defends the body against disease

Figure 7 Three types of cell found in the human body.

A new look at the cell

In the late 1930s a new kind of microscope was invented: the **electron microscope** (Figure 5). It uses a beam of electrons instead of light rays and is much more powerful than the light microscope. It is able to magnify things as much as *half a million times*. Enlarged to this extent, a pinhead would cover ten football pitches side by side.

Figure 6 shows part of an animal cell as it appears in the electron microscope. We can now see much more detail. For example, the cytoplasm appears to consist of a network of membranes and channels. This is called the **endoplasmic reticulum**. Scientists think that it helps to transport material inside the cell.

Stuck to the membranes of the endoplasmic reticulum are minute granules called **ribosomes**. They play an important part in the process by which cells make proteins. The mitochondria show up as sausage-shaped bodies. These and many other intriguing structures have been revealed by the electron microscope.

Different cells for different jobs

Practically all cells contain a nucleus and cytoplasm. However, they vary tremendously in their shape and form. In the human body there are at least twenty different types of cell, each with a particular job to do. Three are shown in Figure 7.

There is thus a **division of labour** between cells. It's rather like a factory or an office in which each person has his or her own job to do. This is more efficient than if each individual tried to do everything.

Investigation 1

Looking at cheek cells

1 Obtain a blunt instrument such as a spatula. It must be clean.

2 Gently scrape the inside of your cheek with the instrument.

3 Put the scrapings onto the surface of a microscope slide.

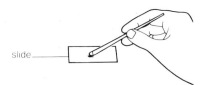

slide

4 Add a drop of methylene blue to the scrapings on the slide. This will stain the cells and help you to see them.

5 Cover with a coverslip. Lower it carefully onto the slide. The stain will spread out beneath it.

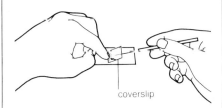

coverslip

6 Examine the slide under the microscope. First use low power to find some of the scrapings, then look at one of the cells under high power.

Which of the structures shown in Figure 3 can you see?

7 Draw the cheek cell and label it as fully as you can.

Investigation 2

Looking at plant cells

1 Slice an onion in two longways.

2 Take out one of the thick 'leaves' from inside it.

3 With forceps pull away the thin lining from the inner surface of the 'leaf'.

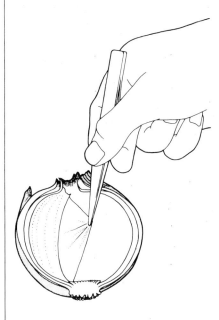

4 With scissors cut out a small piece of the lining, about 5 mm square.

5 Place the piece of lining on a slide and add a drop of dilute iodine.

This will stain the cells and make their nuclei easier to see.

6 Put on a coverslip as shown in Investigation 1.

7 Examine the slide under the microscope, first under low power, then high power.

Which of the structures shown in Figure 4 can you see?

8 Draw one of the onion cells and label it as fully as you can.

9 Place a single leaf of moss in a drop of water on a slide and put on a coverslip.
What structures, absent in the onion cells, can you see inside the moss cells?

Assignments

1 A typical cell is 20 micrometres wide.

Suppose that cells of this size were placed side by side.

How many would there be in a row that was the same length as the second line of this question?

2 Each word in the left hand column below is related to one of the words in the right hand column.

Write them down in the correct pairs.

glycogen	inheritance
chloroplast	energy
mitochondrion	sunlight
chromosomes	elastic
cellulose	storage

3 Which of the structures listed below are found (a) in animal cells only, (b) in plant cells only, and (c) in both animal and plant cells?

cytoplasm
chloroplasts
starch grains
nucleus
vacuole
glycogen granules
cell wall
chromosomes
cell membrane
mitochondria

4 Which of the following structures can be seen with a light microscope and which ones can only be seen with an electron microscope?

nucleus
endoplasmic reticulum
ribosomes
chromosomes
chloroplasts

5 The picture below shows a group of cells from a certain organ of the human body. The cells were obtained by cutting a very thin slice (section) of the organ with a microtome.

Why do some of the cells appear not to have a nucleus?

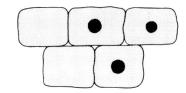

How are living things constructed?

In an organism like man the different kinds of cells are arranged in a precise way. The health and well-being of the individual depend on this.

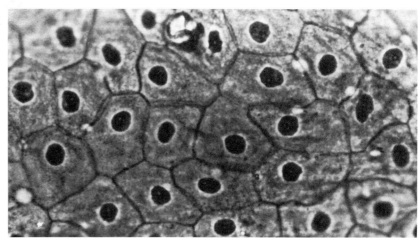

Figure 1 A simple type of epithelium seen under the microscope. It comes from one of the thin membranes inside the body.

Cells are grouped into tissues

Cells don't normally float around on their own. Usually large numbers of them are massed together into a **tissue** (Investigations 1 and 2).

One of the simplest tissues is shown in Figure 1. It is called **epithelium**. It consists of a sheet of cells. The cells fit neatly together, like paving stones. This kind of tissue forms the lining of spaces and tubes inside the body and is also found on the surface of the skin.

In Figure 2, an epithelial tissue is compared with two other kinds of tissue: smooth muscle and nerve tissue. **Smooth muscle** consists of lots of slender muscle cells packed together. You find this tissue in the wall of the gut, amongst other places. Its job is to squeeze the food along. **Nerve tissue** consists of a network of nerve cells connected with one another. This kind of

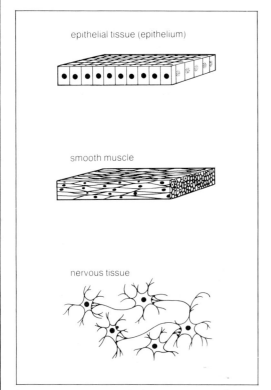

Figure 2 Three types of tissue found in the human body.

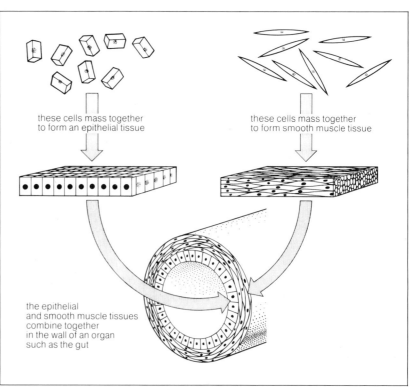

Figure 3 This diagram shows in a simplified way how cells combine to form tissues and how tissues combine to form organs.

tissue occurs in the brain. It carries messages from place to place, as in a complicated telephone system.

The main tissues of animals and plants are summarised in Tables 1 and 2. Some of them consist of just one type of cell. However, most of them contain two or three types of cells mixed together.

Tissues are combined into organs

In most animals, including man, tissues are combined together to form **organs** (Figure 3). An organ is a complex structure which has a particular job to do. The main organs in the human body are shown in Figure 4. Look at some organs obtained from the butcher (Investigation 3), and examine thin sections of them under a microscope (Investigation 4).

Some organs do just one job. For example, the only job the heart does is pump blood round the body. Other organs do more than one job. For example, the kidneys get rid of poisonous waste substances and control the amount of water in the body. The organ with the greatest number of jobs is the liver: scientists have worked out that it does about 500 jobs altogether.

Organs are grouped into systems

In the human body certain tasks are carried out by several different organs working together. These organs all belong to a **system**. An example is the digestive system. This consists of the gut, together with the liver, pancreas and gall bladder. Its job is to digest and absorb food.

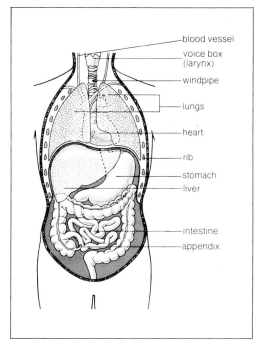

Figure 4 This diagram shows some of the main organs in the human body.

Animal tissues (based on human)

Name of Tissue	What it consists of	Main Functions
Epithelial tissue	Sheets of cells	To line tubes and spaces and form the skin
Connective tissue	Tough flexible fibres	To bind other tissues together
Skeletal tissue	Hard material	To support the body and permit movement
Blood tissue	Runny fluid	To carry oxygen, and food round the body
Nerve tissue	Network of threads with long cable-like extensions	To conduct and coordinate messages
Muscle tissue	Bundles of elongated cells	To bring about movement

Table 1 Summary of the main tissues found in animals.

Plant tissues (based on flowering plant)

Name of tissue	What it consists of	Main Functions
Epidermal tissue	Sheets of cells	To line the surface of plants
Photosynthetic tissue	Cells with chloroplasts	To feed the plant
Packing tissue	Round balloon-like cells	To fill in spaces inside the plant
Vascular tissue	Long tubes	To transport water and food substances
Strengthening tissue	Bundles of tough fibres	To support the plant

Table 2 Summary of the main tissues found in plants.

Systems in the human body

Name of system	Main organs in the system	Main Functions
Digestive system	Gut, liver and pancreas	To digest and absorb food
Respiratory system	Windpipe and lungs	To take in oxygen and get rid of carbon dioxide
Blood (circulatory) system	Heart, blood vessels	To carry oxygen and food round the body
Excretory system	Kidneys, bladder, liver	To get rid of poisonous waste substances
Sensory system	Eyes, ears, nose	To detect stimuli
Nervous system	Brain and spinal cord	To conduct messages from one part of the body to another
Musculo-skeletal system	Muscles and skeleton	To support and move the body
Reproductive system	Testes and ovaries	To produce offspring

Table 3 Summary of the systems in the human body.

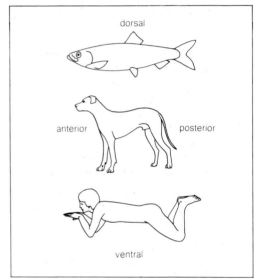

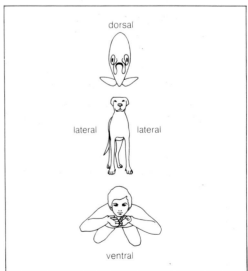

Figure 5 Most animals have an anterior and posterior end and dorsal and ventral sides. They are also bilaterally symmetrical: the two sides of the body are mirror images.

The various systems found in the human body are summarised in Table 3. Some organs belong to more than one system. The liver, for example, belongs to the digestive and excretory systems.

Division of labour

In most organisms we see a division of labour between different kinds of cells. We also see a division of labour between different tissues, and between different organs. They all work in harmony. This is essential for the smooth running of a complex animal like man.

Division of labour is only possible in an organism whose body is made up of many cells. Such organisms are described as **multicellular**.

There are a few multicellular plants whose cells are all identical. And there are some simple animals and plants which consist of only one cell. In these simple organisms there is obviously no division of labour between cells: all jobs have to be carried out within the one cell.

The shapes of living things

A plant such as a tree has a rather irregular shape, with branches sticking out all over the place. Animals, on the other hand, are more regular and compact.

Most animals move with one end of the body in the lead. This is the front or **anterior end**. The other end is the **posterior end**. In most animals there is some kind of head at the anterior end.

The lower side of the body, that is the side closest to the ground, is known as the **ventral side**. The upper side is known as the **dorsal side** (Figure 5).

Most animals, man included, have symmetrical right and left sides. In other words the two sides of the body are mirror images of one another. We call this **bilateral symmetry**. Other types of symmetry are also found amongst animals, but bilateral symmetry is the most common type.

Investigation 1

Looking at epithelium

1 Put a drop of stain on the surface of a microscope slide.

2 Your teacher will provide you with a piece of frog's skin.

3 With a knife or scalpel gently scrape the surface of the frog's skin. This will remove the epithelium.

4 Dip the knife in the drop of stain. The epithelium will come off the knife and float in the stain.

5 Cover it with a coverslip.

6 Examine your slide under the microscope: low power first, then high power.

How does the tissue compare with the one shown in Figure 1?

Draw a small group of cells, showing how they fit together.

What job does this tissue do?

Investigation 2

Looking at plant packing tissue

1 Obtain a fruit such as a tomato or grape.

2 Cut the fruit in half.

3 With a knife remove a small piece of the soft pulpy material from inside. This is packing tissue.

4 Put the tissue on a slide and spread it out.

5 With a pipette add a drop of water to the tissue.

6 Cover it with a coverslip.

7 Examine your slide under the microscope: low power first, then high power.

How does the tissue compare in appearance with the epithelial tissue in Investigation 1?

Draw a small group of the cells.

What job does this tissue do?

Assignments

1 Explain each of these words: tissue, organ, muscle, epithelium, multicellular.

2 Each of the tissues listed in the left hand column is related to one of the words in the right hand column. Write them down in the correct pairs.

photosynthetic tissue   transport
epithelial tissue   protection
connective tissue   messages
blood tissue   feeding
nervous tissue   strength

3 What kind of tissue:

  a) fills spaces inside a plant stem,
  b) carries oxygen round the human body,
  c) brings about movement in animals,
  d) supplies a plant with food,
  e) transports water in a plant,
  f) supports your body,
  g) lines the surface of a leaf,
  h) conducts messages from one part of your body to another,
  i) binds other tissues together,
  j) lines an animal's body cavity?

4 Why is it an advantage to have a division of labour between different organs in the body?

5 Why is photosynthetic tissue not found in animals, and why is muscle tissue not found in plants?

6 Name two organs in Figure 4 which occur in pairs and two which occur singly.

Why is it an advantage to have pairs of organs rather than single ones?

7 Make a list of all the functions you can think of which are performed by your head.

Why is it an advantage to an animal to have its head at the anterior end of its body?

Investigation 3

Looking at organs

1 Examine some or all of the following organs obtained from a butcher: lungs, stomach, intestine, tongue, liver, pancreas, heart, kidney, muscle, brain and eye.

2 Try tearing, or cutting, each organ to see how tough it is.

3 Cut open each organ with a sharp knife in order to see its inside.

4 Describe what each organ looks and feels like.

5 Find out what each organ has to do in the body.

In what ways does the structure of each organ suit it to its job?

6 For each organ write down the main tissues which are found in it.

How would you relate the presence of the particular tissue to the job which the organ has to do?

Investigation 4

Examining the inside of an organ

Your teacher will give you a thin section of an organ which has been cut with a microtome, stained to show up the cells, and mounted on a microscope slide.

1 Look at your section under the low power of the microscope.

2 Firstly try to see the cells of which the organ is composed.

You should be able to recognise them from their nuclei which will be darkly stained.

3 Make a list of particular kinds of tissue which you think you can see in your section.

4 Find out the main function of the organ from which your section was obtained.

Make a list of the ways the internal structure of the organ seems to be suited to carrying out its function.

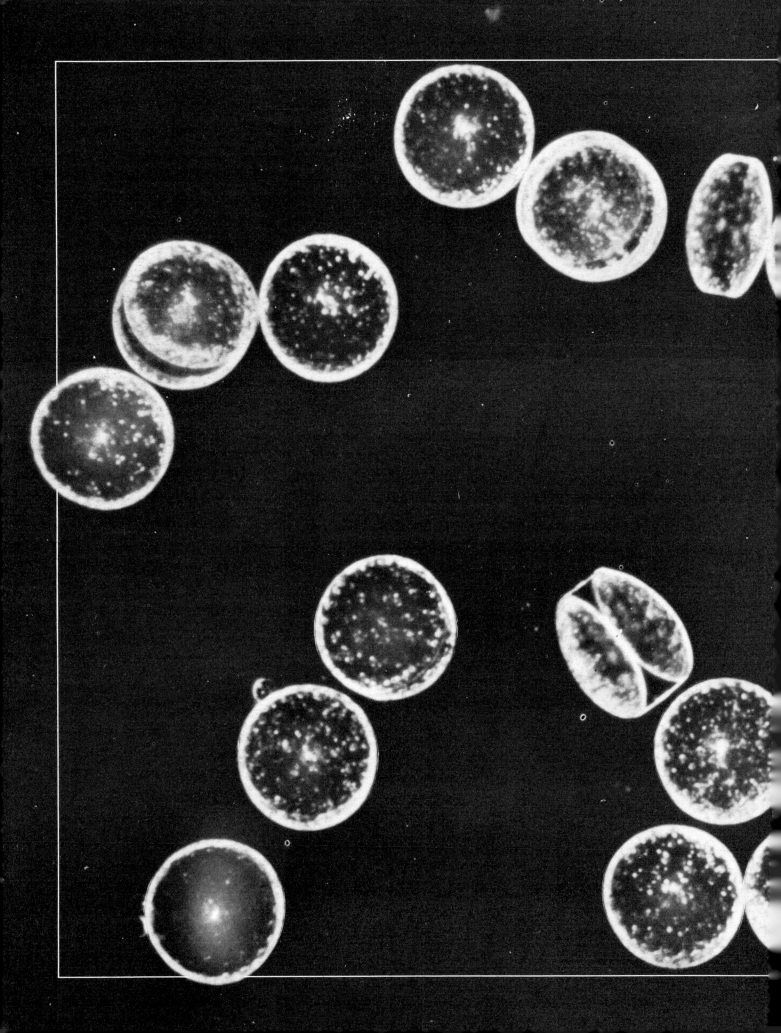

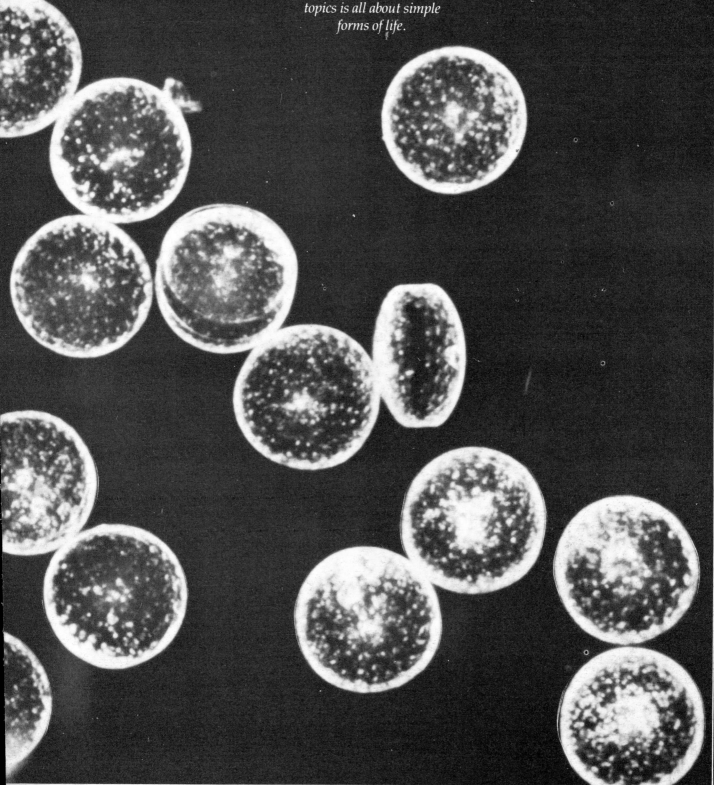

Simple life forms

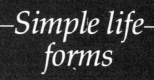

This photograph,
taken down a microscope,
shows some simple living organisms
called diatoms. The next series of
topics is all about simple
forms of life.

Amoeba and other Protozoans

If you take a drop of water from a pond or ditch and look at it under a microscope, you will see little organisms which consist of only one cell. These one-celled organisms are protozoans.

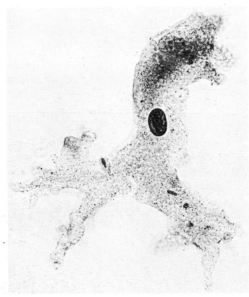

Figure 2 Photograph of an amoeba taken through a light microscope.

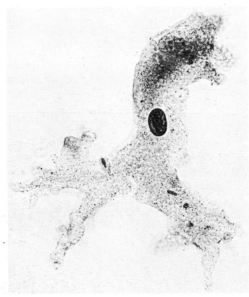

Figure 3 Diagram of an amoeba.

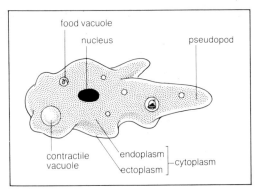

Figure 4 The amoeba moves by the runny endoplasm flowing forward. At the front end the fluid endoplasm changes into the more solid ectoplasm, and at the rear end the reverse takes place.

Figure 1 A boy looks for protozoans and other small animals in a pond.

The amoeba

This is one of the largest protozoans. It can be the size of a pinhead. This makes it easy to see under the microscope (Figure 2).

The amoeba lives in ponds and slow-flowing streams where it moves around in the mud and on the surfaces of rocks and stones. It also lives in damp soil.

With a microscope you can look either at live specimens, or at dead specimens which have been stained so as to show up the structures inside the cell (Investigation 1). You really need to do both if you are to build up a complete picture of this little animal.

The structure of the amoeba

The amoeba is a single cell (Figure 3). It has a **nucleus** and **cytoplasm**. The cytoplasm is bounded by a very thin **cell membrane**, and is divided into two parts: the outer part is clear and jelly-like and is called the **ectoplasm**; the inner part is granular and runny and is called the **endoplasm**. Various structures float around in the endoplasm. These will be explained presently.

The animal constantly changes shape. This is because the cell membrane is thin and elastic, and the fluid endoplasm flows around inside it. If you examine a live amoeba under the microscope, you will probably think its insides look chaotic. Linnaeus, the famous eighteenth century naturalist, thought so too. When he first saw an amoeba, he christened it *Chaos chaos*!

How does the amoeba move?

If you watch a live amoeba moving under the microscope, you will see that the runny endoplasm flows towards one end of the cell. As a result a bulge grows out. This is a **pseudopod** or 'false foot'.

Watching an amoeba, it looks as if the soft endoplasm is being squeezed into the pseudopod, rather like toothpaste being squeezed along its tube. In this way the animal creeps slowly from place to place (Figure 4).

How does the amoeba feed?

The amoeba feeds on tiny organisms in the water, mainly bacteria and single-celled plants. Suppose a little organism of this kind comes close to an amoeba. The amoeba's cytoplasm flows round it, forming a cup (Figure 5).

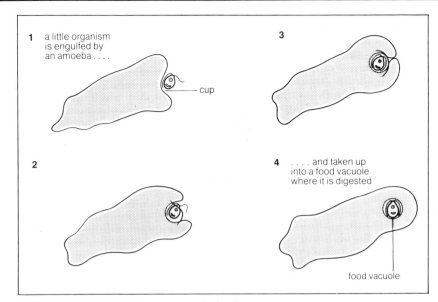

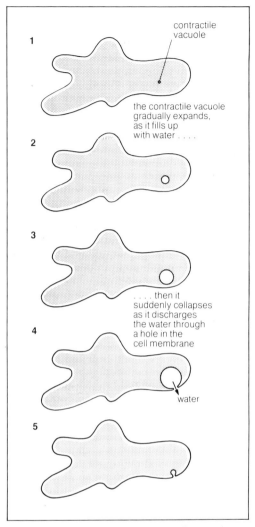

1 a little organism is engulfed by an amoeba . . . .

cup

3

2

4 . . . . and taken up into a food vacuole where it is digested

food vacuole

Figure 5 These diagrams show an amoeba feeding.

Eventually the prey becomes completely trapped in the cavity. We call this a **food vacuole**. Here the prey is killed and digested, and its goodness is absorbed into the surrounding cytoplasm. Any bits and pieces that cannot be digested are got rid of across the cell membrane.

How does the amoeba breathe and excrete?

The amoeba breathes in a simple way. It takes up oxygen, and gets rid of carbon dioxide, by **diffusion** across the thin cell membrane.

Excretion takes place in a similar way. Poisonous waste substances, such as ammonia, diffuse passively across the cell membrane into the surrounding water.

How does the amoeba control its water content?

Water constantly enters the amoeba's body from outside. This is because of **osmosis** (see page 109).

What would happen if the amoeba did not control this intake of water? It would gradually get bigger and bigger, and eventually it would burst.

So the animal must get rid of the water as quickly as it comes in. It's rather like a boat with a leak in the bottom: if the boat is to be prevented from sinking, the water must be bailed out as fast as it enters.

The amoeba gets rid of water by means of its **contractile vacuole** (Figure 6). This is a tiny sac situated in the cytoplasm. The sac gradually fills up with water, getting larger and larger like a balloon. When it is full, it empties its contents to the outside. The contractile vacuole then becomes tiny again, and the process is repeated.

The contractile vacuole empties once every few minutes. If you're lucky you may see it happening.

How does the amoeba reproduce?

Most organisms reproduce sexually (see page 12). However, the amoeba does not reproduce this way, at any rate nobody has ever seen it do so. Instead it simply splits into two (Figure 7). First the nucleus splits, and then the rest of the body. The two little amoebas then feed and grow. When fully grown, each splits again. In good conditions this may happen about once a day.

We call this process **binary fission**. Binary means 'two', and fission means

1 contractile vacuole

the contractile vacuole gradually expands, as it fills up with water . . . .

2

3

. . . . then it suddenly collapses as it discharges the water through a hole in the cell membrane

4

water

5

Figure 6 The amoeba uses its contractile vacuole to get rid of water.

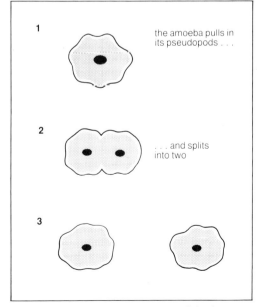

1 the amoeba pulls in its pseudopods . . .

2 . . . and splits into two

3

Figure 7 An amoeba reproduces by splitting into two (binary fission).

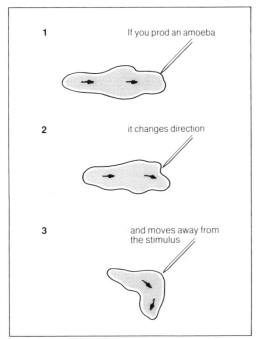

Figure 8 This amoeba responds to an unpleasant stimulus by moving away.

Figure 9 Four common single-celled organisms which you might see under the microscope. They are all found in ponds and ditches. The black object in each one is the nucleus.

'splitting', so the term literally means 'splitting in two'. Because sex is not involved, it is a form of **asexual reproduction**.

How does the amoeba react to stimuli?

It's rather difficult to prod an animal as small as an amoeba, but scientists have managed to do this. The animal reacts by moving away (Figure 8).

Touch is one of many stimuli to which amoebas react. Other stimuli include temperature (hot or cold), and chemical substances in the water. The animal moves either towards or away from the stimulus, depending on whether it is pleasant or unpleasant. What sort of stimuli do you think it moves towards, and which ones do you think it moves away from?

The amoeba's ability to react in this way is very important. It has the effect of guiding it away from harmful situations into places where it is most likely to survive.

Other harmless protozoans

Many other one-celled organisms besides the amoeba live in ponds and streams. Figure 9 shows some of them. They vary in size and shape, and in the way they move and feed (Investigation 2).

Some of them swim by means of tiny hairs called **cilia** which beat like the oars of a boat. Others use a whip-like **flagellum** which lashes to and fro. A few of them don't swim at all, but are attached to weeds or stones by a stalk.

Many of them feed by sweeping tiny organisms into a **gullet** by means of beating cilia. Others feed like plants: they contain the green pigment chlorophyll and feed by photosynthesis.

Single-celled organisms of this sort are also included in the plant kingdom as members of the Algae. They show that there is no hard and fast line between animals and plants.

Euglena
Up to 100 μm long. Cigar-shaped. Bright green because it possesses chloroplasts for feeding by photosynthesis. It is included in the plant kingdom as a member of the Algae. It has a whip-like flagellum for swimming through water.

Paramecium
About 200 μm long. Slipper-shaped. Covered with beating cilia which 'row' the animal through the water. Feed on tiny organisms which are swept into the gullet by cilia.

Stylonichia
About 100 μm long. Looks spiky. Groups of cilia are stuck together to form little 'legs' which are used for a rather jerky kind of movement.

Vorticella
Can be 1 mm long. Shaped like a bell. Attached to pieces of weed and so on by flexible stalk which can contract like a spring if the animal is disturbed. The cilia are used for sweeping food into the gullet.

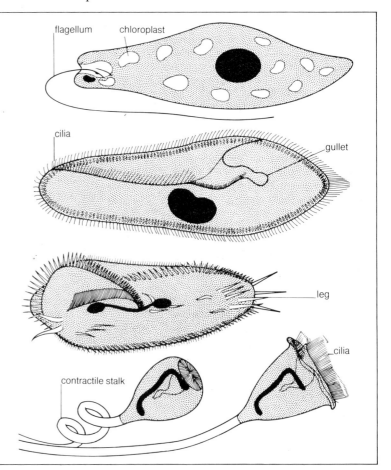

Investigation 1

Looking at an amoeba

1 Examine a prepared slide of an amoeba under the microscope.

2 Which of the structures shown in Figure 3 can you see?

Draw a specimen, and label it as fully as you can.

3 Your teacher will give you a slide with a live amoeba on it.

4 Examine it under the microscope: using low power first, then high power.

5 Watch the amoeba moving.

6 Make outline drawings of it at one minute intervals to show its changes in shape.

7 Put arrows into your drawing to show the direction in which the endoplasm is flowing.

How would you describe the process by which movement takes place?

Investigation 2

Looking at other protozoans

1 Obtain a jar of dirty pond water which has been standing in the laboratory for some weeks.

2 Suck up a little of the water into a pipette and put a drop onto a slide.

3 Cover the drop with a coverslip.

4 Examine your slide under the microscope.

5 Look out for protozoans in the water.

You may see the ones illustrated in Figure 9, and possibly some others as well.

6 Make sketches of the different protozoans which you observe.

How does each one move?

7 If you see *Paramecium*, look for a contractile vacuole: you may see it filling up and collapsing every now and again.

Assignments

1 Make a list of those structures shown in the diagram of the amoeba in Figure 3 which are not found in a typical animal cell.

2 Each word in the left hand column below is related to one of the words in the right hand column. Write them down in the correct pairs.

contractile vacuole	runny
pseudopod	reproduction
food vacuole	digestion
endoplasm	water
binary fission	movement

3 If an amoeba divides once a day, how many amoebas would be formed from a single one after 14 days?

4 What do each of the following pairs of structures have in common:

a) pseudopod and cilia
b) ectoplasm and endoplasm
c) food vacuole and gullet?

5 Explain clearly and briefly:

a) what happens if an amoeba hits a large piece of grit as it moves along,
b) how an amoeba gets rid of water which enters its body,
c) how an amoeba ingests a small piece of food,
d) how *Paramecium* swims,
e) what happens if you touch *Vorticella* with a needle.

6 What structures in an amoeba do the same job as:

a) your mouth,
b) your skin,
c) your kidneys,
d) your legs,
e) your intestine?

Write down five ways in which the amoeba is simpler than a human, either in its structure or in the way it carries out its life processes.

7 A schoolboy did a project to find out if two species of protozoans, A and B, preferred to be in the light or in the dark. He made a special microscope slide which was brightly lit at one end and gradually got darker as one went towards the other end. He then placed an equal number of the two species of protozoan on the slide so that they were spread out evenly. He then left the organisms for an hour. At the end of this period he examined the slide under the microscope, and estimated the density of each species in different parts of the slide. His results are shown in the graph below.

a) What conclusions would you draw from the results?
b) What criticisms, if any, do you have of the way the boy did this experiment?
c) If you were doing it, how would your method differ from his?

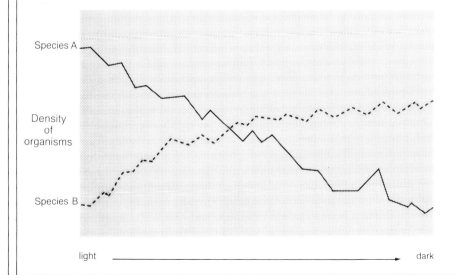

Some harmful protozoans

Many protozoans live inside other animals. Some of them live in the human bloodstream. Others live in the gut or amongst the cells. They are parasites, and can do a lot of harm.

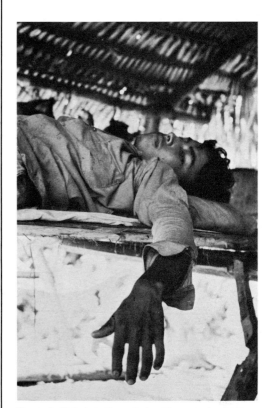

Figure 1 Malaria is characterised by recurring bouts of fever.

The malarial parasite

Every year about 200 million people get malaria, and about two million die of it. It is carried from person to person by a certain type of mosquito called *Anopheles*. This mosquito is found mainly in hot countries. Malaria only occurs in places where the mosquito is found.

What happens when you get malaria?

John is camping in the bush somewhere in the middle of Africa. During the night he is bitten by an *Anopheles* mosquito. During the next two weeks John feels poorly, but he doesn't realise he has malaria.

One night he wakes up with a terrible fever (Figure 1). His temperature soars up. He sweats, shivers and becomes delirious. Then suddenly the fever dies down and he feels better. Exhausted, he falls asleep.

Several days later he has another attack of fever. He goes on having attacks every few days for many weeks.

What has been happening in John's body? To understand this we must study the life cycle of the parasite (Figure 2).

Life cycle of the malarial parasite

When the mosquito bit John, it injected a drop of saliva (or spit) into his bloodstream (see page 64). The saliva contained tiny worm-like parasites.

Once in the bloodstream, the parasites made for John's liver. They stayed in the liver for the next two weeks, feeding and multiplying. This was when John felt poorly.

After two weeks, the parasites left the liver and got into John's bloodstream. They then invaded his red blood cells. Each little worm-like parasite bored its way into a red blood cell. Once inside it changed its shape. It became like a little amoeba, and it fed like an amoeba on the contents of the cell. Gradually it grew, until it just about filled the cell. Then it split into lots of tiny offspring. Finally the red blood cell burst, releasing the new batch of parasites (Investigation 1).

This grisly procedure was undergone not just by one parasite, but by thousands of them all at the same time. John's temperature gradually went up while the parasites were inside his red blood cells. His fever reached its height when his blood cells burst and the parasites were released.

After being set free, the new parasites invaded more red blood cells, and the cycle was repeated. This is why the attacks of fever kept coming back.

What happens inside the mosquito?

When a mosquito bites you, it sucks up your blood. If malarial parasites are present in your bloodstream, the mosquito takes these up too.

Inside the mosquito's stomach the parasites multiply. They then make their way to the salivary glands. Here they wait until the mosquito bites another person. They are then injected into that person's bloodstream with the mosquito's saliva.

So the mosquito carries the malarial parasite from one person to another. We call it a **vector**. The word vector means 'carrier'. It is applied to any animal which transmits parasites or germs from one individual to another. Vectors play an important part in spreading disease.

How does the malarial parasite reproduce?

The malarial parasite reproduces at three stages in its life cycle: in the liver, in the red blood cells, and in the stomach of the mosquito.

The main method is **asexual**. The parasite grows and then splits into lots of offspring. The nucleus divides first, then the rest of the cell. This process is called **multiple fission**.

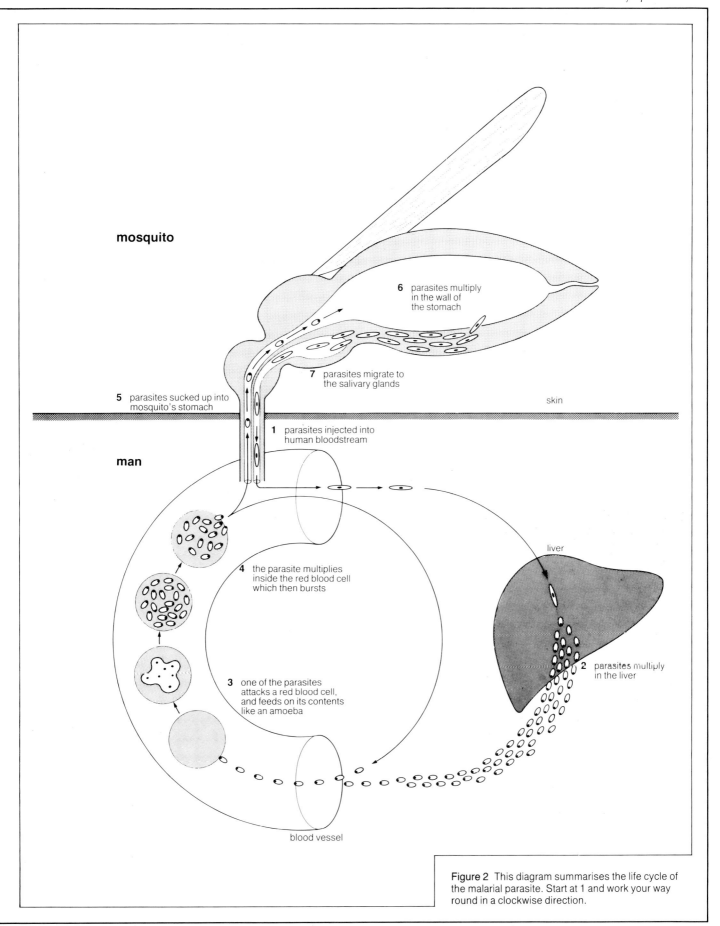

mosquito

6 parasites multiply in the wall of the stomach

7 parasites migrate to the salivary glands

5 parasites sucked up into mosquito's stomach

skin

1 parasites injected into human bloodstream

man

4 the parasite multiplies inside the red blood cell which then bursts

liver

3 one of the parasites attacks a red blood cell, and feeds on its contents like an amoeba

2 parasites multiply in the liver

blood vessel

Figure 2 This diagram summarises the life cycle of the malarial parasite. Start at 1 and work your way round in a clockwise direction.

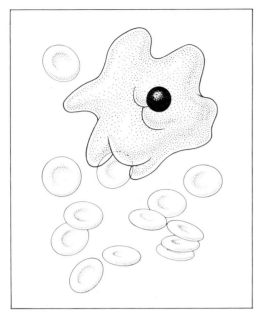

Figure 3 The dysentery amoeba is here seen feeding on a red blood cell in the lining of the large intestine.

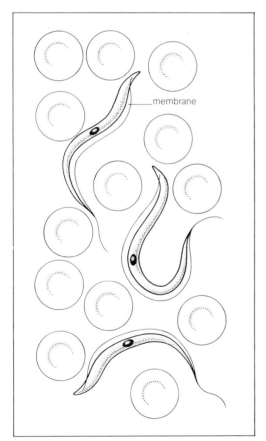

membrane

Figure 4 The sleeping sickness parasite has a worm-like shape with a membrane down one side. It moves by flapping the membrane.

Vast numbers of offspring are produced quickly by this method. In the liver, for example, as many as a thousand offspring may be formed every time one of the parasites undergoes multiple fission.

The malarial parasite also reproduces **sexually**. Eggs and sperm are formed, and these unite with each other. This takes place in the stomach of the mosquito.

How can malaria be controlled?

John was eventually cured of malaria by being treated with certain **drugs**. These killed the parasites in his body.

Such drugs can save lives, but obviously it is better not to get the disease in the first place. If you are going into an infected area, you can protect yourself by taking **anti-malarial tablets** beforehand. If John had done this he would have saved himself a lot of trouble. He should also have slept under **mosquito netting**.

The best way of conquering malaria would be to get rid of the parasite altogether. How could this be done? We know that malaria is spread by mosquitoes, so if we could get rid of mosquitoes we would get rid of malaria. Of course this is easier said than done. It is discussed on page 77.

Other harmful protozoans

Here are two other parasitic protozoans that cause unpleasant diseases. One lives in the human gut, the other in the bloodstream.

The dysentery amoeba

This is like the ordinary amoeba, but instead of living in ponds and streams it lives in the human large intestine (Figure 3). It feeds on the lining, and causes bleeding and diarrhoea. Occasionally the parasites pass out with the person's faeces. If they get into food or drinking water, other people can become infected.

The dysentery amoeba is common in countries where people have a poor standard of hygiene. In many places it has been brought under control by improvements in community health and personal cleanliness.

The sleeping sickness parasite

Sleeping sickness afflicts many people in tropical Africa. It is caused by a little worm-like parasite called a trypanosome. This lives in the bloodstream of human beings and wild animals (Investigation 2). The parasite moves around by flapping a membrane which sticks out from the side of the body (Figure 4).

Unlike the malarial parasite, the sleeping sickness parasites do not attack the blood cells. Instead they wriggle around in the fluid part of the blood (the plasma), soaking up its goodness. They release poisonous substances which get to the brain and cause the person to become unconscious, hence the name of the disease.

The sleeping sickness parasite is passed from one individual to another by the blood-sucking tsetse fly (pronounced 'tetsy fly'). A lot of progress has been made in controlling this disease by getting rid of tsetse flies. The trouble is that the parasites live in wild animals such as buffalo without causing them any ill effects. These animals serve as a kind of 'reservoir' from which the parasites are carried to humans by the tsetse fly.

Investigation 1

Looking at the malarial parasite

1 Obtain a prepared slide of blood taken from an individual suffering from malaria.

2 Examine the slide under the high power of the microscope.

3 Observe normal red blood cells first, so you know what they look like (see page 164).

4 Now look for red blood cells which appear to have something unusual inside them.

Can you see anything which might correspond to stages 3 and 4 in the life cycle in Figure 2?

Investigation 2

Looking at the sleeping sickness parasite

1 Obtain a prepared slide of blood taken from an individual suffering from sleeping sickness.

2 Examine the slide under the microscope: low power first, then high power.

3 Find some sleeping sickness parasites (Trypanosomes) amongst the red blood cells.

4 Make a simple outline drawing of a parasite and a red blood cell side by side to show how they compare in size.

What do the parasites feed on?

Why does this parasite cause such a serious disease?

Investigation 3

Looking at a live parasitic protozoan

Your teacher will give you a watch glass containing the contents of the rectum of a frog, mixed with dilute salt solution.

1 With a pipette, place a drop of the material on a microscope slide.

2 Cover it with a coverslip.

3 Examine it under the microscope: low power first, then high power.

4 Look out for a small protozoan covered with beating cilia.

What can you say about its shape and the way it moves?

This animal is called *Opalina*. It lives in the rectum of frogs, but it's uncertain how much harm it does there.

Assignments

1 The following words are used in this Topic. What does each one mean?

parasite
fever
delirious
life cycle
fission

2 Why does malaria occur mainly in hot countries?

Occasional cases occur in colder countries like Britain: how would you explain this?

3 Why do you think the body temperature of a person suffering from malaria goes up during the disease?

4 What do the malarial parasite, the sleeping sickness parasite, and the dysentery amoeba feed on? Be as exact as you can.

5 What precautions could John have taken to prevent himself getting malaria?

6 The malarial parasite has been likened to 'Dr Jekyll and Mr Hyde', the characters in the famous horror story by Robert Louis Stevenson.

In what respect does the malarial parasite resemble them?

7 The graph below shows the body temperature of a person suffering from a certain kind of malaria.

Explain what the malarial parasite is doing in the person's body at point A on the graph.

Why does the temperature keep going up and down?

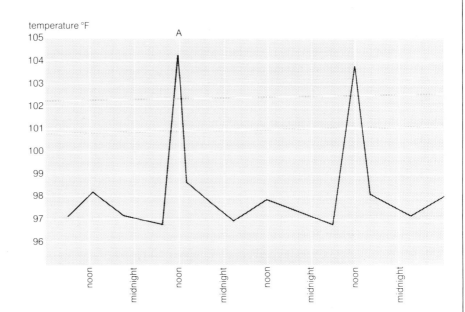

Bacteria

*Bacteria cause diseases,
but they also play a vital part
in the balance of nature. Without
them, life as we know it would
come to an end.*

What are bacteria?

Bacteria are amongst the smallest organisms. They were discovered in the seventeenth century by a Dutch draper, Antonie van Leeuwenhoek. He was a skilful amateur biologist, and he saw them under a microscope which he had made himself. Some bacteria cause very serious diseases. However, most are harmless, and some are extremely useful.

Bacteria occur almost everywhere: in air, water, soil, and inside other organisms. They like warmth, but some can survive at the tops of high mountains such as the Alps where it is very cold. Others live in hot springs at near-boiling temperatures.

They often occur in vast numbers. A teaspoonful of soil may contain several hundred million.

How can we grow bacteria?

Scientists must be able to grow bacteria in the laboratory. This is necessary if we are to investigate them, and find ways of fighting the diseases which they cause.

To grow them, bacteria must be given moisture, warmth and plenty of food. Many years ago it was discovered that they will grow on the surface of a jelly-like material obtained from sea weed. This is called **agar**. Various food substances are added to the agar: this makes it an ideal **nutrient medium** in which to grow, or **culture**, bacteria.

The agar is usually put in a shallow **petri dish**. This must be sterilised beforehand and kept covered, otherwise moulds may grow on the agar. To speed up their growth the bacteria should be kept warm: this is best achieved by putting the petri dish in an **incubator**, a warm box in which the temperature can be kept constant.

Now suppose you put some bacteria on the surface of some nutrient agar. In the course of the next day or two the bacteria multiply into **colonies**. Each colony consists of thousands of bacteria clumped together. The individual bacteria are too small to be seen with the naked eye, but the colonies are clearly visible (Figure 1).

Bacterial colonies vary in size, shape and colour, according to the type of bacteria which gives rise to them. How many types of colony can you see in Figure 1?

Culturing bacteria is something you can do yourself. You can use the technique to find out what kind of bacteria occur in different places (Investigation 1).

How can bacteria be isolated?

The petri dish in Figure 1 contains at least 3 types of bacteria. Scientists often

Figure 1 Bacterial colonies growing on an agar plate. The colonies differ in size, shape and colour.

Figure 2 On the left are streptococcal bacteria as seen with a light microscope. On the right are the same kind of bacteria as seen with an electron microscope.

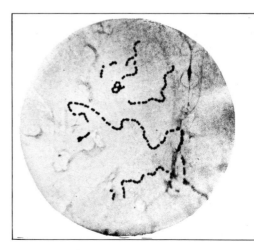
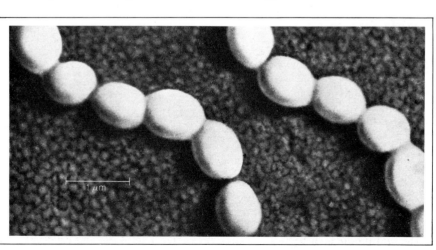

want to separate the different types of bacteria from such a mixture, and grow each on its own. This is also something you can do yourself (Investigation 2).

This is how it is done. You pick up a little bit of *one* colony on the end of a wire loop, and transfer it to another petri dish. You try to make sure that you don't pick up any bacteria from other colonies at the same time. With luck only one type of colony develops in this second petri dish.

If however your petri dish is contaminated with other kinds of bacteria, you transfer a small bit of the new colony to a third petri dish. If the resulting colony is still contaminated, you transfer it to a fourth petri dish. You go on doing this until you have nothing but the one type of colony. This is called a **pure culture**.

Why is it important to be able to make pure cultures of bacteria? There are three main reasons. Firstly, it enables scientists to look at individual kinds of bacteria under the microscope (Investigation 3). Secondly, it means that experiments can be carried out on individual types of bacteria. Thirdly, it enables individual types of bacteria to be produced for use in medicine and industry.

What size are bacteria?

A typical bacterium is about a thousandth of a millimetre wide. This is far too small to be seen with the naked eye. Even with a light microscope you can't see bacteria very well. However, with the much more powerful electron microscope you can see them better.

Look at Figure 2. Both pictures show streptococcal bacteria, which can cause sore throats. The left hand picture shows the bacteria as you would see them under a good light microscope. They are enlarged about 500 times.

The right hand picture shows the same kind of bacteria as they appear in the electron microscope. Here they are enlarged about 12 000 times. If a pinhead was magnified to the same extent, it would be larger than the dome of St. Paul's cathedral in London.

The structure of bacteria

Bacteria vary in their shapes (Figure 3). In some cases they are linked together in chains or small groups. Some have whip-like **flagella** which lash from side to side, propelling the body along.

Scientists can identify bacteria from their shapes, as well as from other characteristics, such as the way they can be stained with certain dyes.

Thanks to very careful work using the electron microscope, we now know a lot about the structure of bacteria (Figure 4). They are single cells, but the cell is simpler than those of most other organisms.

There is no proper nucleus, just a long thread-like **chromosome** coiled up in the middle.

The cytoplasm consists of a dense mass of **granules**. There are few other structures inside it.

The body is surrounded by a thin **cell membrane**. Beyond this is a protective **cell wall**. In addition some bacteria are surrounded by a slimy **capsule**. This gives them extra protection and prevents them drying out.

How do bacteria respire?

Most bacteria need oxygen for producing energy. They absorb oxygen by **diffusion** across their body surface. Carbon dioxide diffuses out.

Certain bacteria can live without oxygen, i.e. they are **anaerobic** (see page 160). Such bacteria live in places where there is a shortage of oxygen, such as inside an animal's gut, or in mud at the bottom of a lake. Some of these bacteria prefer using oxygen if it is available, but can go over to anaerobic respiration if necessary. Others can *only* respire anaerobically, and may indeed be killed by oxygen.

The end products of bacterial respiration vary. Some of the end products

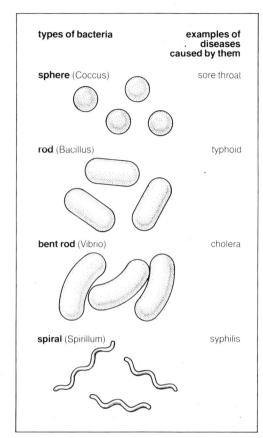

Figure 3 Different types of bacteria have different shapes. This is one way scientists can tell them apart.

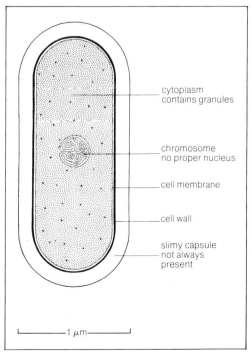

Figure 4 Diagram of a typical bacterial cell.

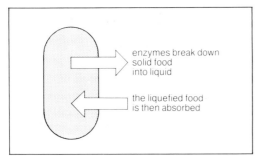

Figure 5 Some bacteria feed like animals. Complex food substances which may be in solid form, are broken down into a liquid and then absorbed.

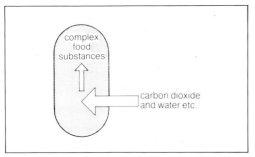

Figure 6 Some bacteria feed like plants. Simple substances such as carbon dioxide gas and water are built up into complex food substances.

are useful to man, and bacteria can be used in the manufacture of various foods such as butter, cheese and vinegar.

How do bacteria feed?

Bacteria have two quite different ways of feeding: some feed like animals, others like plants. Through their feeding activities these bacteria play an important part in the cycling of elements in nature.

Bacteria that feed like animals
These bacteria feed on ready-made organic food. They get their food from the bodies of living or dead animals and plants. Those that feed on dead material are **saprophytes**, they help to bring about decay. Those that feed on living organisms are generally **parasites** and cause disease.

To feed on solid material, bacteria must break it down into liquid form first. They do this by shedding **digestive enzymes** through their body surface. These enzymes break the food down outside the bacteria's body. They then soak it up (Figure 5). A large population of bacteria can eventually turn a solid object into a liquid. This is what happens when things decay.

Bacteria that feed like plants
These bacteria make their own organic food from simple substances like carbon dioxide and water (Figure 6). Many of them get the necessary energy from sunlight, just as green plants do, and they possess a special kind of **chlorophyll**. This is a type of **photosynthesis**.

Some bacteria which can make their own food do not get energy from sunlight. Instead they produce energy by special chemical reactions which take place inside their bodies. We call this **chemosynthesis**. Many of these bacteria put useful nitrogen compounds into the soil which plants need.

How do bacteria reproduce?

Bacteria reproduce very quickly, that's why there are so many of them. They do this by splitting. One cell splits into two, the two daughter cells grow then split again like this.

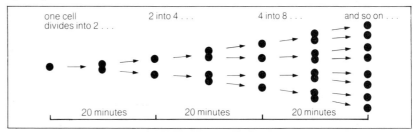

This is a type of **binary fission** and it is **asexual**.

In suitable conditions some bacteria can split once every twenty minutes. This may not seem very fast, but try working out how many bacteria would be formed from one original cell after six hours. It's a large number!

Bacteria can also reproduce **sexually**. In this process part of the chromosome from one bacterial cell passes into another. It is the simplest form of sexual reproduction known.

How do bacteria survive bad conditions?

Many bacteria are very good at surviving bad conditions such as poisons, drought and heat. They do this by forming a thick protective coat round themselves. They are then known as **spores**. Inside the spore the bacterial cell becomes **dormant** and may remain so for a long time.

When conditions become satisfactory again, the spore bursts open and the bacterial cell is released (Figure 7). It then resumes its normal life. The spores of some bacteria can survive for more than 50 years.

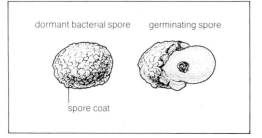

Figure 7 On the left is a bacterial spore with a thick protective coat. On the right the coat has split open and the bacterial cell is coming out.

Investigation 1

To find out what sort of bacteria occur in different places

1 Wash your hands thoroughly.

2 Obtain a petri dish containing sterile nutrient agar. Keep the lid on, whenever possible.

3 With scissors cut out a piece of filter paper, approximately 2 cm × 5 cm.

4 Rub the piece of filter paper on a dirty surface, such as a dusty shelf or water pipe.

5 Remove the lid of the petri dish. Lay the strip of filter paper, dirty side downwards, on the nutrient agar. Leave it there for 10 seconds.

6 With forceps remove the piece of filter paper from the agar.

7 Replace the lid on the petri dish and fasten it down with Sellotape. Label it 'dirty filter paper.'

8 Repeat steps 2 – 6 with a piece of filter paper which has not been rubbed on a dirty surface. Label this petri dish 'clean filter paper'.

9 Put the two petri dishes in an incubator at 35°C.

10 After a day or two examine the agar in each petri dish.
DO NOT REMOVE THE LID AS THE BACTERIA MAY BE HARMFUL

Does the agar in the two petri dishes appear the same?

Are the bacterial colonies all the same?

If not, how do they differ from each other?

Why did you need to wash your hands at the beginning of this experiment?

What was the purpose of setting up the second petri dish, with clean filter paper?

CARE Work with bacteria can be dangerous and should be carried out under strict supervision by the teacher.

Always wash your hands after working with bacteria.

Investigation 2

Making a pure culture of bacteria

1 Obtain a petri dish containing sterile nutrient agar.

2 Your teacher will give you a petri dish containing colonies of bacteria.

3 Sterilise a wire loop by passing it quickly through a small bunsen flame.

4 Scoop up a small piece of the colony on the end of the wire loop.

5 With the wire loop make a zig-zag streak on the surface of the agar in the new petri dish.

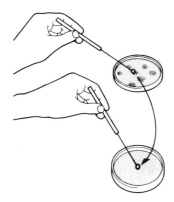

6 Incubate the petri dish at 35°C.

7 After a day or two examine the agar in the dish.

Are the colonies all of one type?

If they are, you now have a pure culture.

If they aren't, what should you do in order to make the culture pure?

Why should scientists want to make pure cultures of bacteria?

Investigation 3

Looking at bacteria under the microscope

1 Your teacher will give you a petri dish containing colonies of bacteria.

2 Pick up part of a colony on the end of a wire loop, and smear the bacteria on a microscope slide.

3 With a pipette put a drop of methylene blue on the smear. Leave it there for one minute.

4 Wash the stain off the smear with a *gentle* stream of water from a tap.

5 Dry the slide by blotting it *very gently* with filter paper.

6 Examine the smear under the microscope (high power).

Although the bacteria will be very small you may just be able to see them.

Can you make a rough estimate of their size?

Assignments

1 List five ways in which the bacterial cell in Figure 4 differs from a typical animal cell.

2 How many bacteria could be fitted side by side in a row the same length as the second line of this question? (Assume that the bacteria are spherical with a diameter of one micrometre.)

3 Give three ways by which a scientist can tell the difference between different kinds of bacteria.

4 In good conditions a bacterial cell splits every 20 minutes.

How many would be formed from a single original one after 10 hours?

5 A politician makes a speech in which he urges scientists to find a way of getting rid of all bacteria from the world.

Write a letter to a newspaper giving your view on this idea and explaining its probable consequences if it was carried out.

Viruses

It used to be thought that bacteria were the smallest organisms. We now know that there are even smaller ones. These are called viruses. Viruses cause many diseases such as colds, influenza, smallpox and poliomyelitis.

Figure 1 This tobacco plant is suffering from mosaic disease. Notice the spots on the lower leaves. Studies carried out on this disease led to the discovery of viruses.

Figure 2 These viruses, seen here in the electron microscope, cause mosaic disease in tobacco plants. In this picture they are magnified 85 000 times.

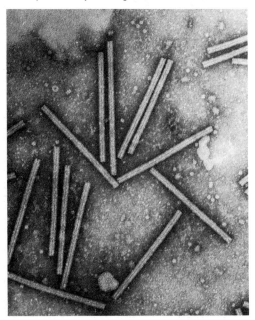

How were viruses discovered?

Towards the end of the nineteenth century a way was found of getting rid of bacteria from a liquid. You filter the liquid by passing it through a very fine sieve made of unglazed porcelain. The sieve holds back the bacteria, so the liquid is freed of them.

In 1900 a Dutch professor called Beijerinck did an experiment on some tobacco plants. These plants had a disease in which the leaves became spotted: it is called mosaic disease (Figure 1).

Beijerinck got some juice out of the leaves. He filtered the juice so as to remove any bacteria present. He then rubbed the juice onto the leaves of a healthy plant. Although the bacteria had been removed, the leaves soon went spotty and developed the disease.

How can we explain this? One possible explanation is that the disease is caused by organisms which are smaller than bacteria. Being smaller, they pass through the sieve which holds back bacteria.

We now know that this is the correct explanation. But at the time no-one could actually *see* these organisms: the electron microscope had not yet been invented, and there were no other microscopes powerful enough. In fact Professor Beijerinck did not think the infection was caused by an organism at all, but by an infectious fluid. He called this fluid *virus*, a Latin word which means 'poison'. However, later experiments showed that the filtered fluid definitely contained organisms of some kind.

In the early 1940s, the electron microscope started to be used. Juice obtained from diseased tobacco plants was examined in this powerful microscope. The fluid turned out to contain tiny objects like those in Figure 2. For the first time viruses could actually be seen.

Structure of viruses

Viruses are so small that we have to express their size in a unit called the **nanometre** (nm). A nanometre is one millionth of a millimetre. A typical virus is about 100 nm wide. It is difficult to imagine anything so small, but you can look at it this way: if you lined them up in a row across this page, there would be over two million of them.

Viruses have various shapes. Some are rod-shaped. Others look spherical, though on close examination they turn out to be many sided (Figure 3). Most viruses have a simple shape. However, the last one in Figure 3 has quite a complex body.

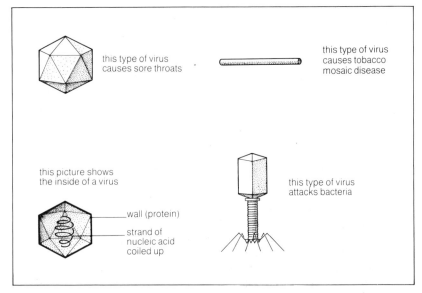

this type of virus causes sore throats

this type of virus causes tobacco mosaic disease

this picture shows the inside of a virus

wall (protein)

strand of nucleic acid coiled up

this type of virus attacks bacteria

Figure 3 These diagrams show the structure of viruses. They are not drawn to the same scale.

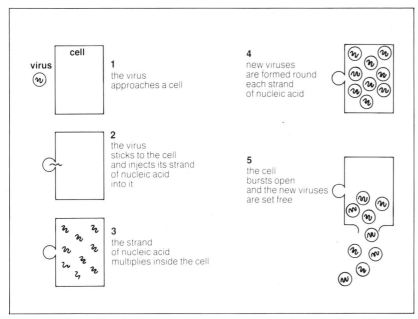

Figure 4 Viruses can only reproduce inside a living cell. Here a virus attacks and destroys a cell.

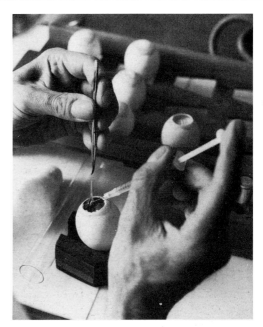

Figure 5 Influenza viruses being injected into a chicken embryo for growing in the laboratory.

Viruses are simpler than any other organisms, including bacteria. There is no nucleus or cytoplasm, so we cannot call them cells. There is a **wall** which is made of protein; inside is a coiled up strand of **nucleic acid** (see page 338).

How do viruses reproduce?

The way viruses reproduce is remarkable, and it explains why they are so harmful. They can only reproduce inside the cells of a living organism. The cells may belong to an animal or plant or bacterium.

First the virus attaches itself to the surface of the cell (Figure 4). Then it injects its strand of nucleic acid into the inside, rather like a doctor injects a person with a hypodermic needle. The nucleic acid then multiplies into lots of separate strands. Round each strand a new virus is formed. The materials for making the new viruses come from inside the cell. So the virus is a thief, robbing the cell of its contents. Eventually, the cell bursts open, and the new viruses are set free.

The whole process takes about half an hour. Thousands of new viruses may be released from a single cell.

Different viruses attack different cells. For example, the common cold virus attacks cells in the nose and throat. The much more serious poliomyelitis virus attacks nerve cells in the spinal cord, which is why the disease often leaves people paralysed.

Growing viruses

Scientists need to be able to grow (or culture) viruses in the laboratory. This is necessary for understanding them, and for developing ways of protecting people against them (see page 233).

Unfortunately, you cannot grow viruses on agar jelly as you can bacteria. This is because they need living cells in order to multiply. So you have to grow them on living tissue. Fertile hens eggs are sometimes used for this purpose. The virus which you want to cultivate is injected into the eggs, where it proceeds to multiply (Figure 5).

Nowadays it is possible to take a few cells out of an animal or a plant, and grow them on their own in the laboratory. This procedure is called **tissue culture**. It provides a convenient source of cells for growing viruses.

Assignments

1 Name four diseases of humans, and one disease of plants, which are caused by viruses.

2 What fraction of a millimetre is (a) a micrometre, and (b) a nanometre?

   Why is it better to express the size of viruses in nanometres rather than micrometres?

3 What part did (a) tobacco plants and (b) unglazed porcelain play in the way viruses were discovered?

4 Why do you think viruses are always harmful?

5 The virus shown at the top of Figure 3 (the one which causes sore throats) is approximately 70 nanometres wide. How many could be fitted side by side in a row the same length as the fifth line of this question?

Hydra and its relatives

The hydra is one of the simplest animals that is made of many cells. In many respects its body works like ours but more simply.

The hydra

The hydra lives in ponds and streams (Figure 1). There are two main types, the green hydra and the brown hydra. You can discover a lot about these animals by looking at them under a hand lens or low-powered microscope (Investigation 1).

The structure of the hydra

The structure of the hydra is shown in Figure 2. Its body consists of a cylindrical **column**. At the top of the column is the **mouth**, surrounded by about eight **tentacles**. At the bottom is the foot. The foot is sticky and is used for attaching the hydra to stones or weeds.

The mouth opens into a large **digestive cavity**. Here food is broken down. The mouth is the only opening into this cavity.

The body wall is made of two layers of cells: the **ectoderm** towards the outside, and the endoderm towards the inside. In between is a thin layer of jelly-like material called the **mesogloea**. The endoderm lines the digestive cavity.

The body wall is composed of seven different types of cells. Each has a particular job to do. Here is a summary of them:

In the outer layer
1 **Epithelial cells** fit together to form a protective skin (epithelium). The inner ends of these cells are drawn out into slender muscle tails which enable the hydra to move.
2 **Sting cells** poison and paralyse other animals.
3 **Sensory cells** detect stimuli such as touch.
4 **Nerve cells** transmit 'messages' from one part of the body to another.

In the inner layer
5 **Gland cells** shed digestive juices (enzymes) into the digestive cavity.
6 **Absorptive cells** take in bits of food, and stir up the digestive fluid by means of flagella. These cells also have muscle tails.

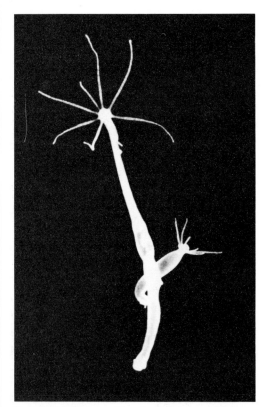

Figure 1 Hydra, one of the simplest many-celled animals.

Figure 2 This diagram shows the structure of the hydra. On the left is the whole animal. On the right is part of the body wall highly magnified to show the different kinds of cells.

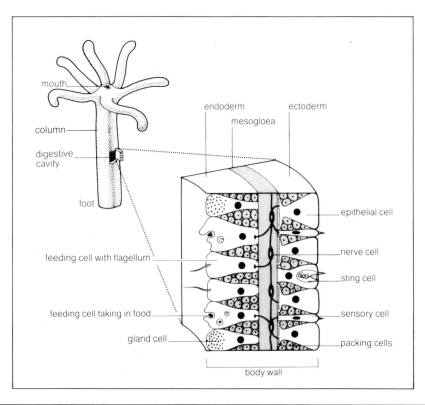

In both layers

7 **Packing cells** fill in spaces between other cells, and develop into other kinds of cells whenever they are needed.

In a one-celled animal like the amoeba, all jobs have to be carried out by one cell. But in the hydra different jobs are carried out by different types of cells. Each type of cell is specialised to perform a particular task. In other words, the hydra shows **division of labour** between its cells.

Why is the green hydra green?

The green hydra's inner layer of cells contains hundreds of tiny one-celled plants (algae). They are called *Zoochlorella*. They possess the green pigment chlorophyll, and they feed by photosynthesis. There are so many of them that they make the whole hydra look green.

These little plants make their home inside the hydra's cells. They do not harm the hydra. Instead they help it, by giving it some of the sugar which they make by photosynthesis. In return the hydra gives the plant shelter, and carbon dioxide. The carbon dioxide comes from the hydra's respiration, and the plant uses it for photosynthesis.

The green hydra and the plant live in a partnership in which both organisms help each other. This is an example of **symbiosis** (see page 362).

How does the hydra breathe and excrete?

The hydra breathes by diffusion. Oxygen diffuses passively into its cells from the surrounding water. Carbon dioxide diffuses out. There are no special organs to aid this. Nor is there a circulatory system for transporting things within the body.

Excretion is carried out in the same way. Poisonous waste substances, such as ammonia, diffuse out of the cells into the surrounding water.

How does the hydra feed?

The hydra feeds on small animals like water fleas. When the animal bumps into the hydra's tentacles, it is in for trouble. You can watch what happens by putting a water flea in a dish with a hungry hydra (Investigation 2).

The moment the water flea touches the tentacle, the **sting cells** leap into action. Each one shoots out a **thread**, like a harpoon (Figure 3). The thread has a pointed tip which pierces the prey's skin. A drop of **poison** then oozes out of the tip: this paralyses the water flea, stopping it moving. Other similar cells have a long coiled thread which wraps itself round the water flea's bristles, holding it firm.

Meanwhile the tentacles close round the unfortunate animal, and pull it towards the mouth. It is then forced into the **digestive cavity**.

Inside the digestive cavity, the prey is broken down into pieces by **digestive enzymes** produced by the gland cells. The pieces are then taken up by the absorptive cells: which engulf them rather like the amoeba does. When the cell has done this, it develops a flagellum. This waves about and stirs up the digestive fluid.

Finally bits which cannot be broken down are forced out of the mouth.

How does the hydra move?

The hydra spends most of its time attached to stones or pieces of weed by its foot. But if the animal is hungry, or if there is something in the water which it does not like, it will go for a 'walk'. It does this by a strange looping or somersaulting motion (Figure 4). The body alternately stretches out, then closes up. This is brought about by contraction of its **muscle tails**.

The muscle tails are also responsible for the writhing movements of the tentacles which help the animal to catch prey.

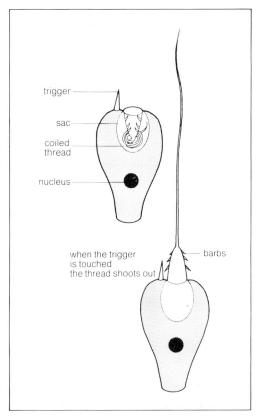

Figure 3 The sting cell of a hydra before and after its thread is shot out.

Figure 4 The hydra moves by somersaulting or looping.

Figure 5 The hydra reproduces asexually by budding.

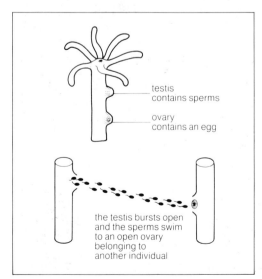

Figure 6 Sexual reproduction in the hydra.

How does the hydra react to stimuli?

If you poke a hydra with a needle, it closes up. The stimulus is detected by the **sensory cells** in the ectoderm. Messages are then sent through the nerve cells to the muscle tails. The muscle tails then contract. This is a simple reflex which helps to protect the hydra from danger.

The hydra reacts much more slowly than higher animals. This is mainly because its nervous system is very primitive. The nerve cells are linked together into a network rather like a string bag. This is called a **nerve net**. Messages travel slowly through the nerve net in all directions. This is nothing like as efficient as the human nervous system. We have long cable-like nerves which can transmit messages very quickly from one end of the body to the other.

How does the hydra reproduce?

Look again at Figure 1. In this picture the hydra has another one growing out of the side. The hydra can reproduce this way. The process is called **budding** (Figure 5). It takes place in the spring and summer when there is plenty of food and conditions are good for growth. This method of reproduction is asexual.

The hydra also reproduces sexually. This occurs in the autumn when food is getting scarce and conditions are not so good. If you look at hydras at this time of year, you will see bumps sticking out from the side of the body. These are **testes** and **ovaries** (Figure 6).

Each testis contains numerous **sperms**, and each ovary contains an **egg**. The sperms and eggs are formed from packing cells in the ectoderm.

When the testes and ovaries are mature, they burst open. The sperms swim to an open ovary of another hydra, and one of them fertilises the egg. The fertilised egg then splits up into a little ball of cells, the **embryo**. A hard wall is formed round it: this is called a **cyst**. The cyst drops out of the ovary, and sinks to the bottom of the pond. When the winter comes the parent hydra dies, but the cyst survives until the next spring. It then bursts open, and a young hydra emerges. So sexual reproduction enables the hydra to survive the winter.

The hydra in Figure 6 has a testis *and* an ovary. It can function as both a male and a female, producing sperms and egg. Organisms which can do this are called **hermaphrodites**. Some hydras are hermaphrodites, others have separate sexes.

The eggs of a hermaphrodite may be fertilised by its own sperms: this is known as **self-fertilisation**. But self-fertilisation has disadvantages, and most hermaphrodites have ways of preventing it.

The hydra prevents self-fertilisation by producing its sperms and eggs at different times. In an individual hydra the testes mature before the ovary, so the sperms are released before the egg is ready to be fertilised. The sperms are therefore forced to swim to an egg in another individual. We call this **cross-fertilisation**.

The hydra's powers of regeneration

The hydra gets its name from a legendary monster which had many heads: every time its heads were cut off, new ones grew in their place. The hydra is similar: if you cut it into pieces, each piece can grow into a new hydra.

This process is called **regeneration**. Hydras are good at regenerating because of their **packing cells**. Whenever a hydra is damaged, the packing cells multiply and develop into whatever cells are needed.

Many lowly animals have good powers of regeneration. This is in sharp contrast to man whose ability to regenerate lost parts is very poor. How useful it would be if a person could form a new leg or arm after an accident!

The hydra's packing cells are also used for forming new sting cells. Once a sting cell has sent out its thread, it cannot be used again. It is replaced by a neighbouring packing cell.

Investigation 1

Looking at hydra

1 Obtain a watch glass containing a hydra. Wait for the hydra to open out fully.

2 Look at the hydra through a hand lens.

Can you see the parts of the body shown in Figure 2?

Does the hydra change its shape?

3 Gently poke one of the tentacles with a needle.

How does the hydra respond?

What kind of muscles bring about the response?

4 Put the watch glass on the stage of a microscope, supporting it underneath with a glass slide.

5 Look at it under *low power*. (Don't try using high power.)

Can you see the difference between the two layers of cells which make up the body wall?

Why is the outer layer much more transparent than the inner layer?

Can you see any sting cells in the outer layer?

Investigation 2

Watching hydra feeding

1 Obtain a watch glass containing a hydra which has been starved for several days.

2 Obtain a jar containing water fleas.

3 With a pipette transfer one or two water fleas from the jar to the watch glass.

4 If a water flea bumps into the tentacles, the hydra may catch it and eat it.

If this happens, watch the hydra's feeding behaviour.

5 Make drawings every now and again to show what happens.

Someone has suggested that hydra must suffer from severe indigestion. Why should this be?

Assignments

1 What job do the hydra's epithelial cells do besides forming its 'skin'?

2 The hydra's packing cells are sometimes called 'reserve cells'.

Why do you think they are given this name?

3 What is the difference between self-fertilisation and cross-fertilisation?

How does the hydra prevent self-fertilisation?

4 Although most of the hydra's sensory cells are in its outer layer of cells, there are a few in the inner layer as well. What do you think their job is there?

5 Describe one experiment which you might do to find out if the algae which live in the green hydra's cells are essential to the hydra.

How could you find out if the hydra is essential to the algae?

6 When a water flea bumps into a hydra's tentacles, the sting cells send out their threads. However, when the hydra touches the ground with its tentacles while it is moving (Figure 4), the sting cells do not send out their threads.

Put forward explanations of why the sting cells respond differently to the water flea and to the ground.

7 The hydra belongs to a group of animals called the coelenterates. Two other members of this group are shown in the photographs below.

a) Name the two animals
b) In what ways are these animals similar to the hydra and
c) in what ways are they different?
Use the classification on page 6 to help you.

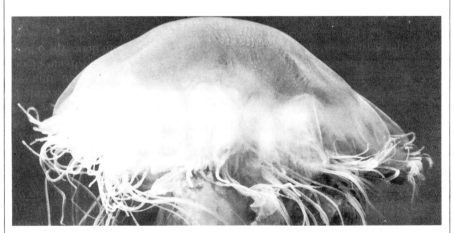

Algae are simple green plants. They are very important to man because they provide food for fish. Our entire fishing industry depends on them. This is the main reason why scientists study them.

Euglena, a single-celled alga

Countless millions of single-celled algae live in the surface waters of lakes, seas and ponds where there is plenty of light for photosynthesis. These organisms are too tiny to be seen individually, but sometimes there are so many of them that the water looks green.

One such organism is *Euglena* which lives in fresh-water ponds. The largest specimens are not more than a tenth of a millimetre long (100 μm). With a microscope you can see them swimming about in the water (Investigation 1).

Euglena's structure

The structure of *Euglena* is shown in Figure 1. The cell is bounded by a tough, elastic 'skin' called the **pellicle**. Inside there is a **nucleus** and **cytoplasm**, just as in any other normal cell.

One of the most noticeable features of this little organism is its bright green colour. This is caused by the presence of **chloroplasts** in the cytoplasm. The cytoplasm also contains **food storage granules** similar to starch.

At the front end there is a little **reservoir** into which opens the **contractile vacuole**. The contractile vacuole does the same job as the amoeba's (see page 25). It collects water from inside the cell and every now and again discharges it to the outside via the reservoir.

From the reservoir springs a long whip-like **flagellum**. This is used for swimming. Waves pass along the flagellum, driving the organism through the water in a kind of corkscrew motion.

At the base of the flagellum there is a **light-sensitive swelling**, and to one side of it a red **pigment spot**. These structures guide the organism towards light when it is swimming.

If the water dries up, *Euglena* stops using its flagellum and wriggles around like a little worm.

How does Euglena reproduce?

Euglena reproduces asexually by splitting into two (**binary fission**). No-one has ever seen it reproduce sexually. However, other single-celled algae reproduce sexually.

On occasions *Euglena* pulls in its flagellum and produces a thick coat round

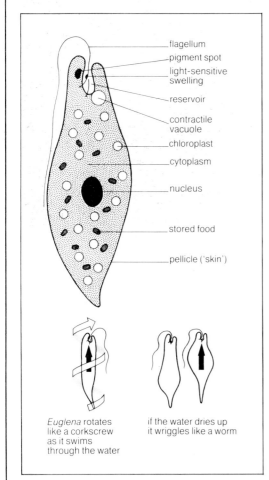

Euglena rotates like a corkscrew as it swims through the water

if the water dries up it wriggles like a worm

Figure 1 *Euglena*, showing its structure and how it moves.

Figure 2 This diagram shows how the *Euglena's* light-sensitive swelling and pigment spot guide it towards light. A response in which an organism moves towards, or away from, a stimulus is called a taxis.

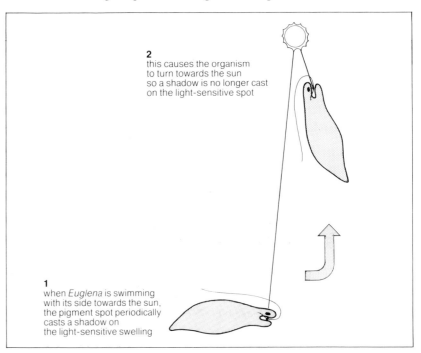

2 this causes the organism to turn towards the sun so a shadow is no longer cast on the light-sensitive spot

1 when *Euglena* is swimming with its side towards the sun, the pigment spot periodically casts a shadow on the light-sensitive swelling itself. It thus becomes a **cyst**. Within the cyst it may split up into several individuals. Eventually the cyst opens and several little *Euglenas* are set free.

How does Euglena feed?

When the sun is shining and there is plenty of light, *Euglena* feeds like a plant. Using its chloroplasts, it makes its own food by photosynthesis.

To help it do this, *Euglena* always swims towards well-lit places (Investigation 2). The mechanism which guides it towards light depends on the light-sensitive swelling and red pigment spot mentioned earlier. Figure 2 shows how it works.

An amazing thing about *Euglena* is that when it's dark, or the light is dim, it can feed like an animal. It takes in organic substances from the surrounding water. It cannot take in solid food, but it can absorb dissolved substances across its surface.

In having these two alternative methods of feeding, *Euglena* seems to be partly a plant and partly an animal – a sort of combination of the two. It shows us that the dividing line between animals and plants is not always very clear.

Spirogyra, a filamentous alga

If you look into a pond, you sometimes see slender green threads floating about. If you lift them out of the water, they feel slimy. It is likely that they belong to *Spirogyra*, one of the simplest plants to be made up of many cells.

The threads are long unbranched filaments, cylindrical in shape. They are coated with a thin layer of **slime** which prevents them getting tangled. Each filament is made up of a chain of cells. The cells are all identical, so this is an example of an organism in which there is *no* division of labour between cells (see page 20). Figure 3 shows some filaments under the microscope.

The structure of one of the cells is shown in Figure 4. It is bounded by a cellulose **cell wall**. There is a large **vacuole**, and the **nucleus** is suspended in the centre by slender strands of cytoplasm. *Spirogyra* feeds entirely by photosynthesis, for which purpose it has a **chloroplast**. This is shaped like a ribbon, and runs in a spiral round the edge of the cell. **Starch grains**, produced by photosynthesis, are stored in the chloroplast in special bodies called **pyrenoids**.

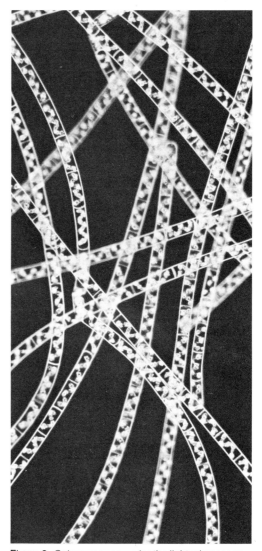

Figure 3 *Spirogyra* seen under the light microscope.

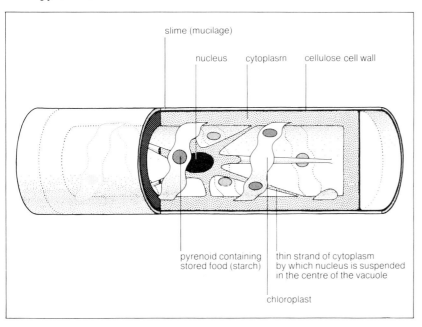

slime (mucilage)

nucleus   cytoplasm   cellulose cell wall

pyrenoid containing
stored food (starch)

thin strand of cytoplasm
by which nucleus is suspended
in the centre of the vacuole

chloroplast

Figure 4 Diagram showing the structure of a cell of *Spirogyra*.

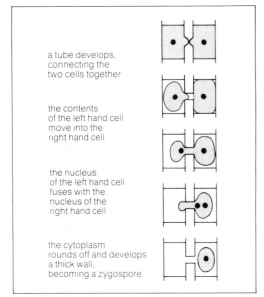

Figure 5   The filaments of *Spirogyra* grow by the cells dividing in two as shown in these diagrams.

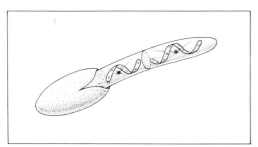

Figure 6   These diagrams show what happens when two cells of *Spirogyra* conjugate.

Figure 7   A zygospore of *Spirogyra* splits open and a new filament grows out.

How does Spirogyra reproduce?

The filaments of *Spirogyra* often get broken up into fragments. In good conditions these can grow into new filaments: the cells divide across the middle as shown in Figure 5. In this way the filaments gradually increase in length. This process is called **fragmentation** and it is a form of asexual reproduction.

Spirogyra also reproduces sexually. This takes place by **conjugation**. Two filaments lie side by side. Short tubes grow out from each of the filaments, connecting next-door cells (Figure 6). Through the tubes the contents of one cell move into the other, and their nuclei combine together. The cytoplasm then rounds off to form a **zygote**. This develops a thick wall, thus becoming a **zygospore**. The process just described takes place between an entire row of cells all at the same time. As a result, a zygospore develops in each cell. Eventually the cell walls break open and the zygospores sink to the bottom of the pond. Here they survive the winter. Next spring they burst open and new filaments grow out (Figure 7).

Sea weed

Sea weeds are the most complex algae. There are many different kinds and most of them are brown. This is because they contain a brown pigment in addition to the green pigment chlorophyll. Both are used for photosynthesis.

Many sea weeds live on the sea shore between the high and low tide marks. They have many adaptations which enable them to withstand the buffeting of the waves.

The body consists of one or more **fronds** which are large and flat, rather like a leaf. This is where most of the photosynthesis takes place.

As you probably know, the fronds are slippery. This is because they are covered with **slime**. The slime helps to protect them from being torn and also from drying out at low tide.

The inside of a sea weed is more complex than *Spirogyra*. There are several different kinds of cell, each responsible for doing a different job. Some of the cells contain chloroplasts and carry out photosynthesis. Others provide strength. In this way there is **division of labour**, though nothing like as much as in more advanced plants.

Some sea weeds are not attached to rocks; they float freely in the open sea. Some of them are very large, reaching lengths of 60 metres or more, ten times longer than an average-sized sailing dinghy. In parts of the Atlantic these giant weeds form a tangled mass which can be a real hazard to ships.

Why are algae important to man?

Algae are important because they help to make up **plankton**. This consists of innumerable single-celled algae and other small organisms which swim or float in the surface waters of seas and lakes. Plankton provides food for fish, indeed our entire fishing industry depends on it.

Algae can, however, be a nuisance in reservoirs. They can make the water smell and taste unpleasant, and may be a danger to health. The algae must therefore be removed from the water before it is fit for drinking. This is normally done by filtering the water. However, in the summer there may be so many algae present that it is necessary to put chemicals in the water to kill them.

Sea weeds contain certain useful substances, for example, iodine. Brown sea weeds give us a jelly-like substance which is used for thickening ice cream, sauces and ointments. It is also used for making agar, on which bacteria are grown in laboratories (see page 32).

Sea weed is sometimes used as a fertiliser. In some parts of the world, particularly the Far East, it is also used as food for livestock and even humans. However, it is difficult to digest and its food value is not very high.

Investigation 1

Looking at Euglena

1 Obtain a jar of water containing a large number of *Euglenas*.

2 Suck up a little of the water into a pipette; put a drop of it onto a slide.

3 Cover it with a coverslip.

4 Look at your slide under the microscope.

5 Look out for *Euglena*. Notice its bright green colour.

Which structures shown in Figure 1 can you see?

6 Watch *Euglena* swimming.

How does it move through the water?

7 Let the water under the coverslip dry up a little.

How does *Euglena* move now?

Investigation 2

To see if Euglena is attracted towards light

1 Obtain three specimen tubes containing large numbers of *Euglena*.

2 Set the three tubes up like this:

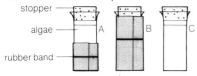

A Wrap black paper around the bottom half of tube.
B Wrap black paper around the entire tube.
C Leave uncovered.

3 Place the three tubes under a lamp (not too bright) for about 24 hours.

4 After 24 hours remove the black paper from the tubes, and examine the distribution of algae in each tube.

Whereabouts are the algae?
What is the reason for setting up tubes B and C?

From the results of this experiment can you suggest how plankton is distributed in lakes and seas?

How would you find out if your suggestion is true?

Investigation 3

Looking at Spirogyra

1 Obtain a jar of water containing *Spirogyra*.

What does *Spirogyra* look like?

2 With forceps lift a little *Spirogyra* out of the jar, and put it on a slide with a little water.

3 Cover it with a coverslip, then look at it under the microscope: low power first, then high power.

4 Notice that *Spirogyra* consists of slender filaments which are made up of cells.

Are all the cells identical?

5 Look at one cell in detail.

Which of the structures shown in Figure 4 can you see?

How many chloroplasts are there in each cell?

6 With a pipette put a drop of iodine to one side of the coverslip. It will immediately flow under it.

7 Draw the iodine across by pulling water from the other side of the coverslip with a piece of filter paper.

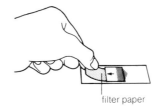

filter paper

8 As the iodine moves across, it will stain any starch blue-black.

Where is starch located in *Spirogyra*?

How do you think the starch gets there?

Assignments

1 Why are single-celled algae important to man?

2 List two ways in which *Euglena* is like a plant, and two ways in which it is like an animal.

3 A large number of *Euglenas* are put into a dish of water. Half the dish is placed in the light, the other half is kept dark. After ten minutes all the *Euglenas* are found to be in the lighted half of the dish.

Explain how they got there.

4 What do each of the following pairs have in common:

a) fission and conjugation,
b) agar and iodine,
c) zygospore and cyst,
d) light-sensitive swelling and pigment spot?

5 In what ways is sea weed adapted to living on the sea shore between the high and low tide marks?

6 Scientists have discovered that algae are very good at taking up salts from the water around them. As a result certain salts may be several million times more concentrated inside the alga than in the surrounding water.

Explain why this makes sea weed a particularly useful fertiliser.

Pin mould and other Fungi

*Fungi are plants
which feed on other
organisms, living or dead.
They include the well-known
moulds which grow on food. Fungi
are very important to man because
of their peculiar way of feeding,
which is quite different from
that of most plants.*

Because of their unplantlike method of feeding,
Fungi are sometimes put in a kingdom of their own.

How do Fungi feed?

Most plants are green because they contain chlorophyll which they use for making complex organic food by photosynthesis.

Fungi are usually white because they lack chlorophyll. They do not make their own food by photosynthesis. Instead they feed on ready-made organic food. In this respect Fungi are more like animals than plants.

Fungi get their food from two sources. Some of them feed on dead material: these are **saprophytes**. They make food go mouldy (Investigation 1). They help to bring about decay, and are useful to man in many ways.

Other Fungi feed on living organisms: they are **parasites**. Some of them do a lot of harm, damaging crops and killing trees.

In this topic we will study pin mould, which is a typical saprophytic fungus. We will then look briefly at some other important Fungi.

Pin mould

Look at Figure 2. This is what a piece of stale bread looks like after it has been left lying around for about a week. It looks as if it's covered with cotton wool: this is pin mould. To understand how the bread got like this we must study the life cycle of the mould.

Life cycle

Fungi produce **spores**. These are small and light, like specks of dust, and they float through the air. For a spore of pin mould to develop, it must land on a damp surface (Investigation 2).

After the spore has landed, it bursts open and a thread grows out (Figure 3). The thread grows over the surface of the bread, branching this way and

Figure 1 A fungus growing on a tree trunk.

Figure 2 Pin mould growing on a piece of stale bread.

that. Eventually the bread becomes covered by a tangled mass of fine silvery threads. The threads are called **hyphae**, and the whole mass of threads is called a **mycelium**. You can see the way the threads branch by looking at them under the microscope (Investigation 3).

After a time short branches grow upwards. The tip of each branch swells up into a little knob. These knobs are **spore cases**. They are known as **sporangia** (singular: sporangium). Inside each one hundreds of spores are formed. Eventually the spore case opens, and the spores are released. They are then carried away by wind or in some cases by people's fingers or small animals such as insects. If one of them lands on a suitable surface, a new mould develops and the cycle is repeated.

Structure of the threads

Scientists have looked at the insides of the threads under the microscope. They contain **cytoplasm** and many **nuclei**, but they are not divided up into separate cells (Figure 4). There is a **vacuole** in the centre.

How does pin mould feed and respire?

The mould feeds on the bread, soaking up its goodness. The threads produce digestive enzymes which break down the solid starch into liquid sugar. This is then absorbed by the threads.

Pin mould can live on many other things besides bread: jam is a great favourite, and even old football boots will do.

The threads always stay near the surface. This is because they need oxygen for respiration. Oxygen diffuses into them, and carbon dioxide diffuses out.

How does pin mould reproduce?

We have seen one way pin mould reproduces: by producing spores. This is its asexual method of reproduction. It is very rapid and enables the fungus to spread quickly to new places.

Pin mould can also reproduce sexually by **conjugating** (Figure 5). Two threads from different moulds grow towards each other. Their tips meet and swell up. The walls separating them break down, and the nuclei from the two threads fuse together in pairs. A round ball-like **zygote** is formed. This develops a thick wall and becomes a dormant **zygospore**. The zygospore can survive for up to a year, even in bad conditions.

When conditions are suitable, the zygospore bursts open and sends out a thread which grows upwards. A spore case is formed at the end of this thread. Inside the spore case, spores are formed in the usual way. These are released and can develop into new moulds.

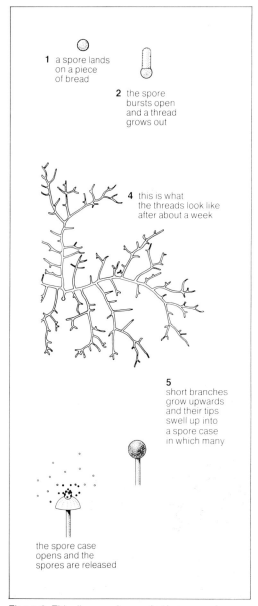

1 a spore lands on a piece of bread

2 the spore bursts open and a thread grows out

4 this is what the threads look like after about a week

5 short branches grow upwards and their tips swell up into a spore case in which many

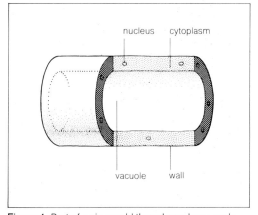

the spore case opens and the spores are released

Figure 3 This diagram shows what happens when a spore of pin mould lands on a piece of damp bread.

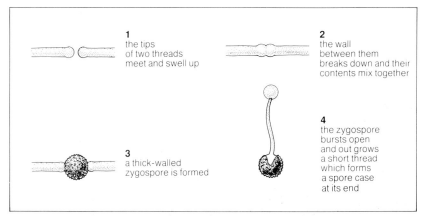

1 the tips of two threads meet and swell up

2 the wall between them breaks down and their contents mix together

3 a thick-walled zygospore is formed

4 the zygospore bursts open and out grows a short thread which forms a spore case at its end

Figure 5 Pin mould can reproduce sexually. Threads from the two neighbouring moulds conjugate.

nucleus   cytoplasm

vacuole   wall

Figure 4 Part of a pin mould thread on a large scale.

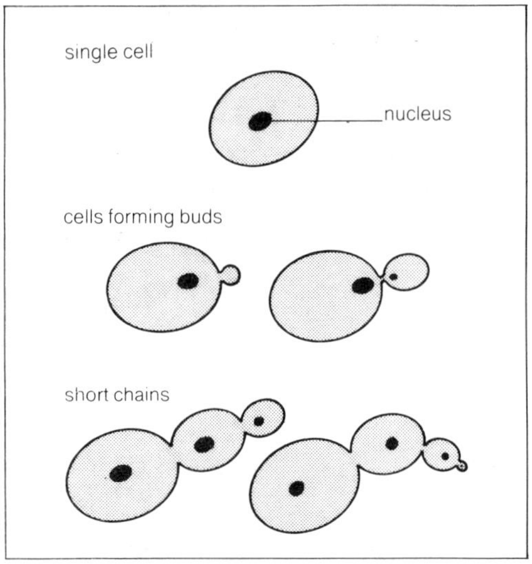

single cell

nucleus

cells forming buds

short chains

Figure 6 Diagram of yeast cells budding.

Some other Fungi

All Fungi have certain features in common. They all feed on ready-made organic food. Their bodies usually consist of thread-like hyphae, and they generally reproduce by means of spores.

Mushrooms

A mushroom doesn't look much like pin mould. However, in the soil there are lots of fine threads, similar to those of pin mould. The mushroom itself is the spore-forming body, and it is the only part of the fungus to grow above the ground.

If you look on the underside of a mushroom, you will see delicate membranes radiating from the stalk. They look like the pages of a book. This is where the spores are formed (Investigation 4).

In the course of a few days, one mushroom may produce ten thousand million spores. Even so the chances of one of them landing in a suitable place and giving rise to a new mushroom are slim; that's why it produces so many.

Some kinds of mushroom are good to eat. The common mushroom grows in fields and on lawns, but it does not need light, so it can grow just as well indoors. Large numbers are grown in cellars, sheds, caves and even disused mines where it is cool and moist. These mushrooms are sold in shops. But although mushrooms taste good, they don't have much food value.

Yeast

Yeast differs from most Fungi in that it usually consists of single cells. It thrives in places where there is plenty of sugar, on the surface of fruit for example.

Each cell contains a nucleus and cytoplasm. The cells multiply by **budding**: the cell sends out a small outgrowth which gets larger and eventually breaks off as a new cell (Figure 6). Sometimes the new cell starts budding before it has broken away from the old cell: this can give rise to chains or clumps of cells (Investigation 5).

For centuries man has used yeast for making wine and other alcoholic drinks, and for baking bread (see page 161).

Dutch elm fungus

The sight of bare elm trees is all too common in Britain these days. The present epidemic of Dutch elm disease started in the late 1960s. By the end of 1975 six million elms had been killed.

The fungus lives in the woody part of the tree. It causes blockage of the water pipes (xylem vessels) in the branches, so water and minerals cannot reach the leaves. As a result the leaves go brown and die.

The fungus forms spores under the bark and the spores are carried from one elm tree to another by a type of beetle.

Attempts have been made to control Dutch elm disease by means of fungicides. The fungicide is introduced into the xylem vessels at the base of the trunk by means of tubes (Figure 7). Despite these measures, by 1980 about half the elms in Britain were dead.

Ringworm

Despite its name, this is a fungus – a parasite of man and other animals. One kind attacks the skin of the head, causing the hair to fall out in clumps so bald patches develop. Another kind lives in between the toes where it causes itching: this is known as **athlete's foot**.

As with other Fungi, ringworm forms spores. These quickly spread from person to person. Athlete's foot can be picked up by walking in bare feet on a changing-room floor which has been contaminated by the feet of an infected person.

Figure 7 This elm tree is being treated with a fungicide to protect it against Dutch elm disease.

Investigation 1

Watching moulds growing on different kinds of food

1 Obtain five or six different kinds of food, e.g. bread, cheese, jam, orange skin and meat.

2 Put them side by side (but not touching) in a dish.

3 Sprinkle a little water into the dish to keep the food moist.

4 Place the dish in a warm place in the laboratory, and leave it open for 24 h.

5 Next day cover the dish with a loose-fitting lid.

6 Examine the dish at regular intervals during the next week.

Which kind of food has the most mould growing on it?

Do all the moulds look alike or do there seem to be more than one type?

How many types can you see? Describe each one.

Investigation 2

To find out if pin mould needs moisture to grow

1 Obtain two slices of white bread.

2 Put one slice in a dish. Sprinkle a little water on it. Put a loose-fitting lid on the dish.

3 Put the other slice in another dish. Leave it dry, do not put a lid on this dish.

4 Place the two dishes side by side in a warm cupboard.

5 Look at the dishes a week later.

Compare the appearance of the two pieces of bread.

Which piece of bread has the greatest amount of mould growing on it?

Is moisture needed for pin mould to grow?

Investigation 3

Looking at pin mould

1 Examine pin mould at various stages in its growth.

Can you see the structures shown in Figure 3?

2 With forceps pick up a few threads of pin mould.

3 Put them on a slide with a drop of water.

4 Cover it with a coverslip.

5 Look at the threads under the microscope: low power first, then high power.

6 Make a drawing of a few threads to show how they branch.

Investigation 4

To see if a mushroom produces spores

1 Obtain a mature mushroom and cut off its stalk.

2 Place the cap, lower surface downwards, on a sheet of paper.

3 Cover it with an inverted dish.

4 After a day or two, remove the dish and carefully lift up the mushroom cap.

What does the paper look like now?

Explain what you see.

5 Place a few spores on a slide and look at them under the microscope.

Estimate their approximate size (see page 401).

Why do you think they are so small?

Investigation 5

Looking at yeast

1 Obtain a jar of fermenting yeast.

2 With a pipette put a drop of the yeast on a slide.

3 Add a drop of a stain such as lactophenol.

4 Cover it with a coverslip.

5 Look at your slide under the microscope: low power first, then high power.

Can you see the yeast cells clearly?

Are any of them budding?

Assignments

1 Why is pin mould white instead of green like most plants?

2 What is the difference between a parasite and a saprophyte?

3 Explain each of the following:
   a) If a piece of bread is kept dry it will not go mouldy.
   b) In the Middle Ages soldiers used to rub mould on their wounds to prevent them going septic.

4 Fungi usually produce very large numbers of spores. Why is this necessary?

5 An elm tree dies. What would you do to find out if it was killed by the Dutch elm fungus?

6 Give the name of a fungus which:
   a) tastes good,
   b) is the source of an antibiotic,
   c) bakers use,
   d) causes 'athlete's foot'.

The tapeworm and other parasitic worms

Many people are infected with 'worms' sometime during their lives. These worms are parasites which make their home inside human beings and other organisms. Some of them make people very ill. Others cause only mild irritation.

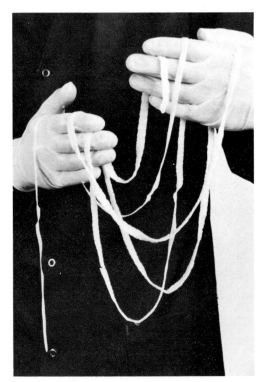

Figure 1 A beef tapeworm from the intestine of a human being.

Figure 2 The pork tapeworm. The beef tapeworm is similar except that its head has no hooks.

The tapeworm

The animal in Figure 1 is the beef tapeworm. It is a type of flatworm (see page 6). You can get it by eating infected beef, hence its name. It has a close relative which can be got by eating infected pork: the pork tapeworm.

These worms live in man's small intestine where they soak up his digested food. Although they look alarming, they don't actually do much harm.

The beef and pork tapeworms can be about 5 metres long. However, one of their close relatives, the fish tapeworm, can reach a length of 20 metres. This is a more harmful worm because it can block the intestine.

The structure of the tapeworm is shown in Figure 2. It is flat, like a long piece of ribbon. It hangs onto the wall of the intestine by its head. To give it a firm grip, the head has four **suckers**, and, in the pork tapeworm, **hooks** as well (Investigation 1).

The body is divided up into a series of segments, about a thousand in all. The youngest segments are at the head end, the oldest ones at the back.

The worm constantly produces new segments just behind the head. As new ones are produced, the older ones get pushed further and further back, enlarging as they do so. The largest ones are about 2 cm long and 1.5 cm wide.

Life cycle of the tapeworm

The life cycle of the beef tapeworm is summarised in Figure 3. Each mature segment contains a full set of sex organs. Two segments can mate with each other by the worm doubling back on itself. By the time the segments reach the rear end of the worm, they are full of eggs.

The segments at the extreme back end drop off and pass out with the host's faeces, taking their eggs with them.

A person with a single worm may pass eight or nine segments a day, releasing a total of three quarters of a million eggs.

To continue their development the eggs must be eaten by a cow. The cow is called the **intermediate host.** In the case of the pork tapeworm the intermediate host is a pig.

The tapeworm's eggs get into the intermediate host if it eats food which is contaminated by human faeces. In the animal's gut the egg shell dissolves, and a tiny **embryo** emerges. This bores through the gut wall and gets into the muscles. Here it forms a **bladder**, about the size of pea.

No further development takes place unless the bladders are eaten by man. This can happen if a person eats infected meat which hasn't been cooked properly. In the person's intestine, the bladder turns inside out and a young tapeworm pops out. This buries its head in the wall of the intestine, and grows to full size.

How can we get rid of tapeworms?

To avoid getting these tapeworms people should make sure that they don't

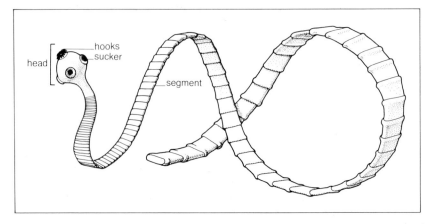

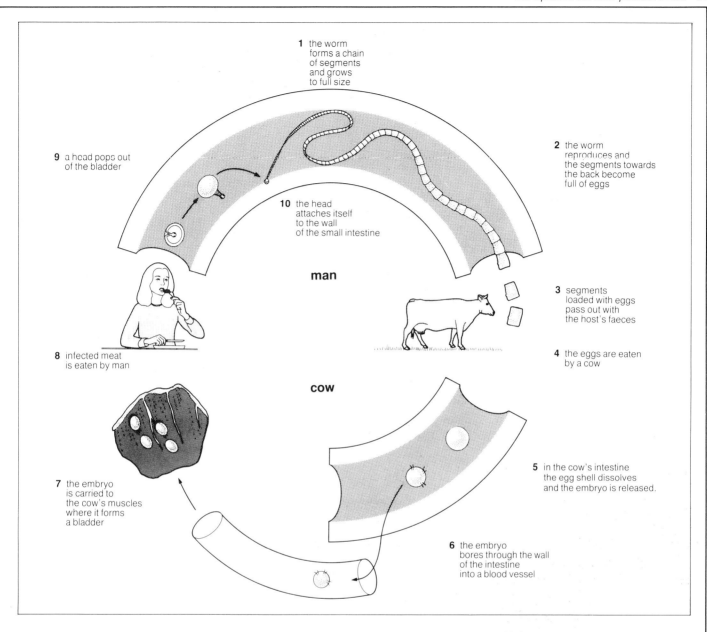

1 the worm forms a chain of segments and grows to full size

2 the worm reproduces and the segments towards the back become full of eggs

9 a head pops out of the bladder

10 the head attaches itself to the wall of the small intestine

man

3 segments loaded with eggs pass out with the host's faeces

8 infected meat is eaten by man

cow

4 the eggs are eaten by a cow

7 the embryo is carried to the cow's muscles where it forms a bladder

5 in the cow's intestine the egg shell dissolves and the embryo is released.

6 the embryo bores through the wall of the intestine into a blood vessel

Figure 3 This diagram summarises the life cycle of the beef tapeworm. The life cycle of the pork tapeworm is similar, except that the intermediate host is a pig.

eat infected meat. In Britain and many other countries meat is inspected to make sure it does not contain tapeworm bladders. In such countries tapeworms are rare, though it is said that in Britain the tapeworm is on the increase at the moment. Proper disposal of sewage is also important in preventing these worms from spreading. To some extent infection can be avoided by cooking meat thoroughly: prolonged heating destroys the tapeworm bladders.

If a person does get a tapeworm he can be given doses of medicine which cause the worm to let go of the wall of the intestine. The worm is then flushed out with the faeces. It is essential that the head doesn't get left behind. If it does, a new worm will grow from it.

The blood fluke

This is another kind of flatworm. It occurs in warm parts of Africa, South America and the Far East where it causes '**snail fever**' (the proper name is **bilharzia** or **schistosomiasis**). People with this disease suffer from sickness, diarrhoea and loss of blood. Sometimes the body swells up with fluid. They become weak, and if they are not treated they usually die.

Figure 4 This diagram summarises the life cycle of the blood fluke.

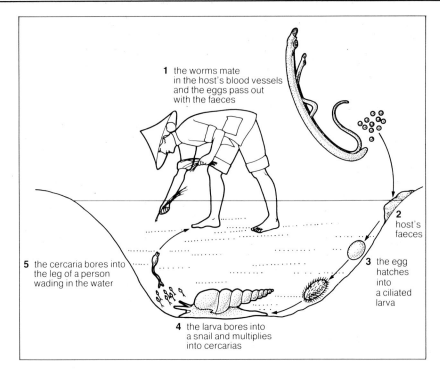

1 the worms mate in the host's blood vessels and the eggs pass out with the faeces

2 host's faeces

3 the egg hatches into a ciliated larva

4 the larva bores into a snail and multiplies into cercarias

5 the cercaria bores into the leg of a person wading in the water

There are several kinds of blood fluke. The one to be described here lives in the blood vessels in the wall of the intestine. The worms are one or two centimetres long. There may be so many of them that they block the blood vessels.

They have a flat body, rather like a curled up leaf. There is a **sucker** near the front for holding onto the sides of the blood vessels. They feed by sucking blood through the mouth.

Life cycle of the blood fluke

The blood fluke's life cycle is summarised in Figure 4. The worms mate in the host's blood vessels. A single female may lay over 3000 eggs a day. Eventually the blood vessels burst, and the eggs are released into the intestine. They then pass out of the host with the faeces.

The eggs will only hatch in water. The egg opens, and a tiny creature called a **larva** comes out. The larva is covered with beating cilia which enables it to swim through the water in search of a particular kind of water snail, the **intermediate host.** If it finds one, it bores into its soft body. Inside the snail the larva reproduces asexually to form thousands of little organisms called **cercarias** – these move by means of a muscular tail which bends from side to side.

The cercarias creep out of the snail into the water. If someone is bathing or paddling in the water, the cercarias attach themselves to the skin and bore through it into the bloodstream. They are then taken to the blood vessels of the intestine. Here they feed and grow into adult flukes. The cycle is then repeated.

How can we get rid of the blood fluke?

Drugs can be used to cure people of bilharzia. However, they are not very successful, and some of them have unpleasant side effects.

A better approach is to get rid of the parasite altogether. This can be done by killing the snails. People have tried putting chemicals in the water to kill the snails, but this has not been very successful.

The best solution is to stop people drinking, or paddling in, water which contains human faeces. In countries where sewage is got rid of properly,

Figure 5 Rice seedlings being planted by a bare-footed farm worker in Thailand. When human dung is used as manure, this is a sure way to get bilharzia.

there is no problem. But in some parts of the world, particularly the Far East, human dung is used as manure in the rice fields. The fields are flooded, and the rice seedlings are planted by farm workers who wade through the water in bare feet (Figure 5). In such places the chances of infection are very high.

Threadworms (a type of roundworm)

Many children get infected with these worms, even in advanced countries. They look like little bits of white thread, about a centimetre long. They live in the rectum, and can cause itching of the anus.

The worms mate in the host's rectum, and the eggs pass out with the faeces. It is very easy for other people to get infected. This is the sort of thing that happens: the child's bottom itches, he scratches it, and gets some eggs on his fingers; later his mother holds his hands, then puts her finger in her mouth; she swallows some eggs and they hatch in her gut. It's quite common for an entire family to become infected, even in the cleanest of households.

This sounds alarming but these worms do little harm, apart from making your bottom itch. Various medicines can be taken to get rid of them: these are available in tablet form, or as a pleasantly flavoured drink. The itching can be eased by washing one's bottom regularly and putting on an ointment.

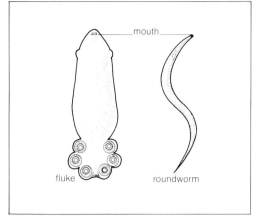

Figure 6 Diagram of a fluke and a roundworm found in the lungs of frogs.

Investigation 1

The front and back of a tapeworm

1 Look at a prepared slide of the head of a tapeworm under the microscope.

What structures help it to cling to the wall of the host's intestine?

2 Now look at a prepared slide of a mature segment from the back end of the tapeworm.

It is full of small round objects: what are they?

3 Make a sketch of the head and a mature segment. Label them.

Investigation 2

Looking at live flukes and roundworms

Your teacher will dissect a frog and take out the lungs.

1 With scissors cut out the lungs, and put them in a watch glass with some saline (1 per cent salt solution).

2 With needles pull the lungs to pieces.

3 With luck, two kinds of worms may be released from the lungs: a fluke related to the blood fluke described in this Topic, and a type of roundworm. They are illustrated in Figure 6. Both can be as much as a centimetre long, so they should be visible to the naked eye.

4 With a pipette transfer a few of these parasites from the watch glass to a slide.

5 Add a little salt solution and cover them with a coverslip.

6 Examine them under the microscope.

In what ways do these two worms appear to be adapted to a parasitic life?

What might they feed on?

How could you find out what they feed on?

Do you think they harm the frog?

How could you find out if they harm the frog?

Assignments

1 Imagine you are a tapeworm living in the intestine of a human being.

What difficulties might you encounter living in such a place?

2 Tapeworms have no gut and no sense organs.

How do you think they manage without them?

3 What advice would you give to people to prevent them becoming infected with (a) the beef tapeworm, and (b) bilharzia?

4 Parasites usually produce very large numbers of eggs. Why is this necessary?

5 Make a list of the ways that either the tapeworm or the blood fluke are adapted to a parasitic life.

6 In Africa, bilharzia is common in lowland areas, but absent from mountainous regions.

Suggest two reasons why bilharzia does not occur in mountainous regions.

More complex life forms

Swans in flight.
Flight is only possible in
animals which have a complex
structure. The next group of
topics is about complex
animals and plants.

The earthworm

You often come across earthworms if you are digging in the garden. They are useful animals, improving the soil and making it better for plants to grow in.

Figure 1 An earthworm.

External structure

A full grown earthworm is about 15 cm long. Its external structure suits it well to burrowing through the soil. The body is elongated and has a streamlined shape (Figure 1). The skin is soft and slimy. There are no sense organs or other structures sticking out which might get in the way as it pushes through the soil.

The outside of the body is marked by a series of **rings** (Figure 2). These divide the body up into **segments.** Sticking out of each segment are four pairs of stiff bristles called **chaetae.** The worm can push these out, or pull them in. What do you think they are for?

There is a **mouth** at the front end, and an **anus** at the back end. The **head** is barely distinguishable from the rest of the body, except that it is rather darker in colour. About a third of the way back there is a region where the skin is thicker than in other places. This is called the **saddle** or **clitellum.** It plays an important part in reproduction.

Inside the worm

Figure 3 shows the internal structure of the earthworm in simplified form. Much of the inside is taken up by a large **body cavity.** This contains a watery fluid. The body cavity is divided up into segments by a series of partitions, called **septa** (singular: septum). They correspond to the rings on the outside of the worm.

The body cavity is surrounded by the **body wall.** This is made of muscle:

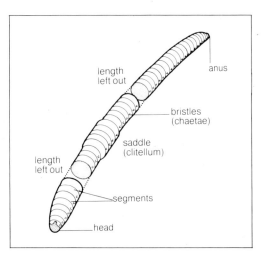
Figure 2 This diagram shows the external structure of the earthworm.

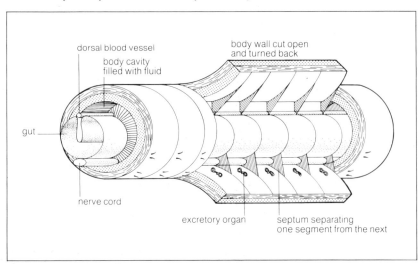
Figure 3 This diagram shows the main structures and organs inside the earthworm.

when the muscles contract the worm changes its shape.

Running down the centre of the body cavity is the **gut**. Beneath the gut is the **nerve cord**: this is continuous with a very small **brain** at the extreme front end. From the nerve cord, slender **nerves** pass out to the body wall muscles in each segment.

The earthworm has a **blood system** which contains red blood similar to man's. The main blood vessel lies above the gut: because of its position on the top side of the body, it is called the **dorsal blood vessel**.

Nearly all the segments contain a pair of **excretory organs**. Each one is a little tube which runs from the body cavity to the outside. It gets rid of poisonous nitrogenous waste.

How does the earthworm move?

You can learn a lot about how worms move by watching one crawling on a piece of paper (Investigation 1). Bulges pass backwards along the body like waves (Figure 4). Where a bulge occurs, the bristles stick out. They enable the worm to get a firm grip on the sides of its burrow as it moves forward.

If you watch an earthworm moving, you will see that it constantly changes its shape: one moment it is long and thin, the next moment it is short and fat. These changes in shape are brought about by the muscles in the body wall which press against the fluid inside the body cavity.

The muscles in the body wall consist of numerous slender muscle fibres packed close together. They are arranged in two layers. In the outer layer the muscle fibres run round the worm in a circular direction: these are called **circular muscles**. In the inner layer the muscle fibres run longways: they are called **longitudinal muscles**.

To make the worm long and thin, the circular muscles contract and the longitudinal muscles relax. To make it short and fat, the longitudinal muscles contract and the circular muscles relax.

How do earthworms feed?

Worms feed on dead leaves and other bits of plants which are lying on the surface of the ground. The worm grasps hold of the leaf with its mouth and pulls it down into its burrow. Worms do this at night when they come up to the surface (Figure 5).

Worms also eat the soil itself as they tunnel through it. They grind the soil up in a special part of the gut called the **gizzard**. Useful substances are absorbed, and undigested matter passes out through the anus. The worm deposits this on the surface of the ground as **worm castings**.

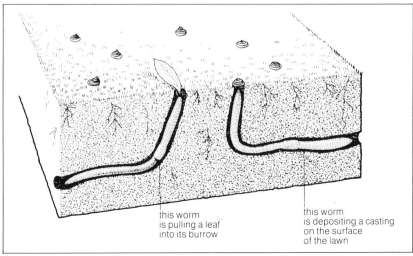

Figure 5 Earthworms feeding and defaecating.

this worm is pulling a leaf into its burrow

this worm is depositing a casting on the surface of the lawn

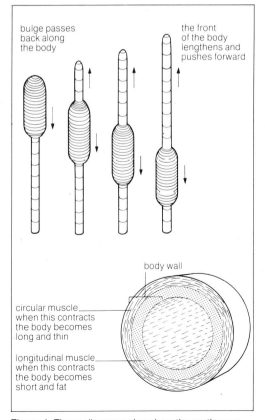

bulge passes back along the body

the front of the body lengthens and pushes forward

body wall

circular muscle when this contracts the body becomes long and thin

longitudinal muscle when this contracts the body becomes short and fat

Figure 4 These diagrams show how the earthworm moves by bulges passing back along the body.

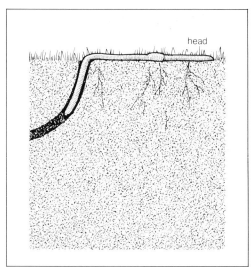

Figure 6 When a worm comes out of its burrow at night, it keeps its back end in its burrow.

How do earthworms respire?

The worm breathes through its moist skin by simple **diffusion**. Oxygen diffuses in, and carbon dioxide diffuses out. There are no special breathing organs such as lungs or gills.

Once the oxygen has diffused into the skin, it is carried by the blood to the various organs inside the body. The blood contains the red pigment **haemoglobin** just as ours does: this carries the oxygen (see page 164).

How do earthworms respond to stimuli?

If you go out into the garden with a torch on a warm wet night, you will see earthworms lying on the ground. During the day they stay underground, but at night when it is dark they come up onto the surface.

When a worm comes out of the soil it keeps its hind end in its burrow (Figure 6). If you touch the worm, or if a bird pecks at it, the worm retreats quickly into its burrow. It does this by pushing out its bristles: this gives it a firm grip on the sides of the burrow. At the same time the longitudinal muscles throughout the body contract rapidly. All this happens very quickly, so the worm pulls back into its burrow before it is harmed. This rapid **escape response** is brought about by messages being sent at high speed down the nerve cord. These messages are sent off as soon as the skin is touched.

Despite this useful response, worms often get damaged. However, they are remarkably good at **regenerating**: broken pieces grow into new worms.

Earthworms respond to other stimuli besides touch. For example, they move away from light: this is why they burrow into the soil and only come out at night. They also move away from unpleasant chemicals in the soil.

How do earthworms reproduce?

Earthworms reproduce sexually by **copulating** with one another. They do this at night on the surface of the ground. The worms come together in pairs as shown in Figure 7. They become glued together by slime (mucus) which is produced by their saddles.

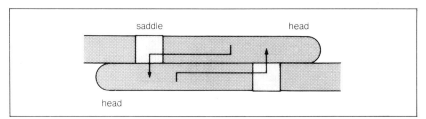

Figure 7 This diagram shows what happens when two earthworms copulate. Sperms pass from each worm into the other as indicated by the arrows.

Worms are hermaphrodites: each individual has both male and female organs and can produce sperms and eggs. When worms copulate, sperms pass from each individual into the other. In this way the eggs in both worms become fertilised. The eggs are then laid in the soil. Eventually they hatch into new worms.

Investigation 1

Looking at the earthworm

For Investigations 1 and 2 you will be given a live earthworm in a dish.

Handle it as little as possible, otherwise it will get tired and you will not be able to see much movement.

1 Observe the structure of the worm.

Can you see the structures shown in Figure 2?

2 Put the worm on a piece of rough paper and watch it moving.

What happens to its shape as it moves?

3 Put your ear very close to the worm and listen carefully.

What can you hear?

Explain the sounds.

4 Repeat steps 2 and 3 with the worm on a white tile or sheet of glass.

Does it move as quickly now?

Can you hear the same sounds?

Explain your observations.

What sort of muscles would be needed to bring about the changes in shape which you have observed?

Investigation 2

To see how the earthworm responds to stimuli

1 Put a live worm on a piece of paper. Let it crawl forward.

2 Put a penny in its way.

What does the worm do when its head touches the penny?

What might it be about the penny that the worm does not like?

3 Tap the worm's head with a blunt instrument such as a pencil.

What does the worm do?

Explain its response as fully as you can.

4 With a pipette place a drop of vinegar (acetic acid) on the worm's head.

What does the worm do this time?

Explain the response.

What part do the responses which you have observed in this investigation play in the normal life of the worm?

Assignments

1 What does the earthworm use the following structures for:

bristles (chaetae),
saddle (clitellum),
gizzard,
haemoglobin,
circular muscle?

2 Explain how an earthworm escapes quickly into its burrow when disturbed.

3 List five ways the earthworm is adapted for burrowing through the soil.

4 For the earthworm to move in the way it does (Figure 4), the body must be divided up into a series of separate watertight segments.

Why is this necessary?

5 If you put an earthworm on the surface of some soil, it soon burrows into it.

Here are two possible reasons why it does this.

a) It dislikes light.
b) It likes to have something solid touching all sides of its body.

Describe an experiment you could do to find out if (a) is correct, and another experiment to find out if (b) is correct.

Can you suggest any other reasons why earthworms burrow into the soil?

6 A student observes an earthworm in the laboratory. In order to find out how it responds to stimuli, he taps its head with a pencil. The worm immediately changes its shape from that shown in diagram A below to that shown in diagram B.

a) Explain how this change in shape is brought about.
b) What use is this response in the worm's natural environment?
c) Name another response which occurs at the same time as the change in shape.

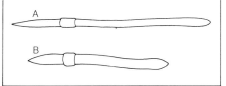

The insect body plan

Over a million different species of animals have been discovered. Of these, more than three quarters are insects. Insects are one of the most successful groups of animals. In this Topic we will look at their basic structure.

Figure 1 This dragonfly illustrates the basic design of insects.

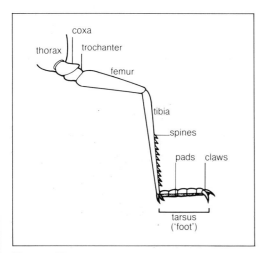

Figure 3 The leg of an insect. This diagram is based on the hind leg of a locust.

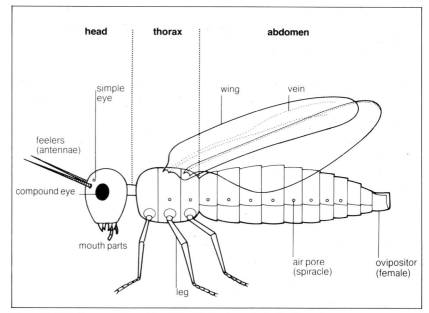

Figure 2 This diagram shows the external structure of a typical insect.

Where do insects belong in the animal kingdom?

Insects belong to a major group of animals called **Arthropods**. This group includes crabs, lobsters, spiders and many other similar animals.

It is characteristic of all these animals that they have **jointed legs** and a hard 'skin' or **cuticle**.

Insects share these features with their arthropod relatives. They also have certain unique characteristics of their own.

The external structure of insects

You can study the external structure of a typical insect by examining a locust or cockroach (Investigation 1).

The body consists of three main regions: the **head**, **thorax** and **abdomen** (Figure 2). Each is divided into **segments**, but these are not as clear as in the earthworm. However, you can see the individual segments clearly in the abdomen.

Insects are made up of twenty segments altogether: the abdomen has eleven, the thorax three, and the head six. The head segments are all fused together, so it is hard to see where one ends and the next begins. It is also hard to make out the last few segments in the abdomen because they fit into each other like a collapsible telescope.

All insects have three pairs of legs; this is one of the easiest ways of recognising them. Each leg has a series of **joints** where they bend. One pair of legs is attached to each of the three segments of the thorax.

The end part of the leg is the **foot** or **tarsus**, and at the end of the foot there is a pair of **claws** (Figure 3). These enable the insect to grip rough surfaces. On the underside of the foot there are sticky **pads** which enable insects like flies to walk on a smooth wall, or even upside down on a ceiling. Some insects, including the locust, have legs which bear rows of curved **spines** which are useful for defence.

Most insects have two pairs of **wings** which are attached to the second and third segments of the thorax. These wings are thin and delicate, and are supported by a network of tough **veins**, rather like a leaf.

When the insect is flying, the wings stick out sideways and flap up and down. But when the insect is walking or resting, they are usually folded together over the abdomen, the forewings on top of the hindwings.

On each side of the thorax and abdomen there is a row of tiny holes called

spiracles. These let air into the body and are important in breathing.

The head is equipped with various **sense organs**. A pair of jointed **feelers**, called **antennae**, stick out in front. The insect sees with a pair of large **compound eyes**. There are also several smaller **simple eyes** which register changes in light intensity.

Round the mouth are feeding structures. These are called **mouth parts**. The mouth parts of different insects are adapted for feeding on different kinds of food.

At the rear end of the abdomen is the **reproductive opening**. In the male this region forms a device by which sperms are put into the female. In the female the reproductive opening is flanked by various plates and valves which form an **egg tube** (**ovipositor**). The eggs pass down this tube when they are laid.

The cuticle

The insect's cuticle is made of a tough material called **chitin**. It consists of two main layers (Figure 4). The outer layer is hard and rigid; the inner layer is soft and flexible. On the surface is a thin layer of **wax** which makes the cuticle waterproof.

In certain parts of the body the hard outer layer of the cuticle is absent, leaving only the flexible inner layer. At these points the cuticle can bend. This happens at the joints in the legs and mouth parts, and between the segments in the abdomen. These joints enable the insect to move: they work like the joints in a suit of armour.

The most obvious function of the insect's cuticle is protection. However, it is also important in movement. Muscles are attached to its inner side. In this way the cuticle serves as a skeleton. As it is on the outside of the body, external to the muscles which are attached to it, we call it an **exoskeleton**.

The insect's cuticle is extremely useful, but it has one major disadvantage: it prevents growth. In order to grow, the insect must cast off its hard cuticle. This is called **moulting** or **ecdysis**.

Variations on insect structure

Some insects show interesting deviations from the basic plan just described (Investigation 2). For example, in beetles the first pair of wings have thickened to form a hard **shield** (Figure 5). They are not used for flying. Instead they protect the delicate hind wings and abdomen underneath.

In flies the second pair of wings is missing altogether: they have been turned into a pair of tiny structures called **halteres** (Figure 6). These are sense organs which help the fly to keep its balance while it is flying.

Another interesting departure from the basic plan is shown by female bees and wasps. They have turned their egg tube into a **sting**. This is a needle-like weapon through which a drop of poison is injected into the victim.

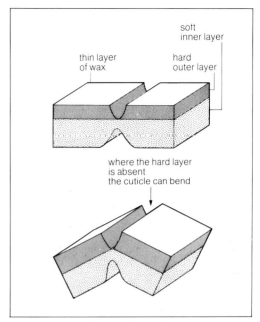

Figure 4 The insect's cuticle provides protection but is able to bend.

Figure 5 This giant Hercules Beetle, from South America, is over 15 cm long. It is the largest insect known.

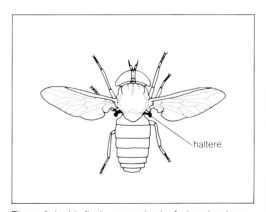

Figure 6 In this fly the second pair of wings has been turned into delicate balancing organs called halteres.

Figure 7 The false 'eyes' on the wings of this bull's eye moth help to protect it from its enemies.

The shape and colour of insects

Many insects, particularly butterflies, are brightly coloured. Usually the female is more colourful than the male: she uses her colours to attract a mate.

Some insects are wonderfully camouflaged. Many of them are the same colour as their background. Others have the same colour and shape as objects like leaves, twigs and thorns. In this way they avoid being seen by other animals.

Some moths have false 'eyes' on their wings (Figure 7). These markings frighten other animals away. Do you think they have any other function?

Some insects with distinctive markings taste unpleasant. Animals such as birds learn to recognise these markings, and they avoid eating these particular insects. Some insects which taste nice have the same markings as those that do not. So predators will avoid *these* insects too. This is known as **mimicry**, and it is a good means of defence.

The size of insects

Insects are generally not more than several centimetres long, and many are much smaller than this.

Why are insects small? One reason is their cuticle. If insects got larger, the cuticle would be so heavy that they would have difficulty holding themselves up and moving. The beetle illustrated in Figure 5 is one of the largest insects; it is a cumbersome beast, rather like a miniature tank.

Another thing that limits the size of insects is their method of breathing (see page 64). This depends mainly on **diffusion** which is a slow way of getting oxygen to the tissues, and is effective only over short distances.

Inside the insect

The internal structure of a typical insect is shown in Figure 8 and is explored in Investigation 3.

There is a **gut** running from the mouth to the anus. A pair of **salivary glands** are connected to the mouth by a tube.

The gut is situated in the large **body cavity**. This is filled with blood called the **blood space**. The blood is pumped by a long tubular **heart**, situated above the gut. Beneath the gut is a **nerve cord**, which is connected to a small **brain** in the head.

Connected to the hind end of the gut is a bunch of fine **excretory tubules**. Their job is to get rid of poisonous waste substances from the insect's blood. The **reproductive organs** consist of **testes** in the male and **ovaries** in the female, and these are connected by a tube to the rear end.

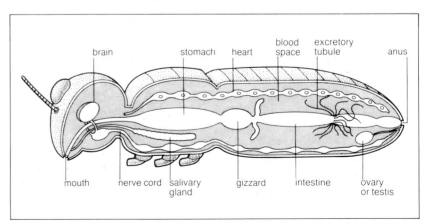

Figure 8 Looking inside an insect. This diagram shows the insect's main internal organs.

Investigation 1

Looking at the external structure of the locust

1 Obtain a preserved locust.

2 Look at its external structure.

Can you see all the structures shown in Figure 2?

3 Examine the hind leg of the locust in detail.

Do your observations agree with the diagram in Figure 3?

4 Draw a side view of your locust.

Label its parts as fully as you can, using Figures 2 and 3 to help you.

Investigation 2

Looking at other insects

1 Obtain preserved specimens of the following insects: beetle, housefly, wasp, ant.

2 Examine them carefully.

3 In each case write down the main ways the insect differs from the locust.

Investigation 3

Examining the inside of an insect

Your teacher will give you a locust or a cockroach in a dissecting dish.

1 With pins, fix the insect to the floor of the dissecting dish.

The insect's dorsal (upper) side should be towards you.

2 Using forceps and scissors, remove the hard covering from the dorsal side of the thorax and abdomen.

3 Put water into the dissecting dish, until it just covers the insect.

This causes the organs to float, and makes them easier to see.

Which of the structures shown in Figure 8 can you see?

The heart should have come away from the dorsal covering of the thorax and abdomen: you will see it adhering to its underside.

To see the nerve cord you will need to deflect the intestine to one side.

Assignments

1 Which of the following features are possessed only by insects, and which ones belong to all arthropods:

a) cuticle (exoskeleton),
b) joints,
c) 6 legs,
d) feelers (antennae),
e) 2 pairs of wings?

2 Why is an insect's cuticle described as an exoskeleton?

3 How does a fly manage to walk upside down on a ceiling?

4 Why is it an advantage to an insect to taste unpleasant?

5 Hover flies have yellow stripes and look very like wasps, but they are not wasps at all, they are flies and do not sting.

Why is it useful to hover flies to look like wasps?

6 The insect in Figure 5 has a large horn on its head.

Suggest three functions which this horn might perform.

How could you investigate which of your suggestions, if any, is correct?

7 Three hundred million years ago there were some very large insects. For example a certain dragonfly had a wing span of nearly a metre.
Why do you think these large insects died out?

8 Give the common name, and briefly explain the function, of each of the following:

a) ovipositor,
b) antenna,
c) tarsus,
d) ecdysis,
e) spiracle.

9 From your knowledge of the structure of the human body, write down five ways that an insect's body differs from yours.

10 Suggest two reasons why it is an advantage to an insect to be small.

11 The picture below is of a small insect which lives in the soil. Name two important structures, typical of most insects, which it lacks. Why do you think this insect does not need these particular structures?

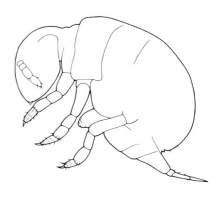

The insect way of life

In this Topic we shall see how insects carry out basic life processes such as feeding respiration and movement. In many respects they are very different from human beings, and yet they are just as efficient in their own way.

How do insects feed?

For feeding, insects have special structures round the mouth. These are called **mouth parts.** Different insects have mouth parts which are adapted for feeding on different foods (Investigations 1 and 2). Here are four examples.

The locust chews plants

The locust's mouth parts include a pair of powerful jaw-like **mandibles** (Figure 1). These cut off pieces of leaf, and grind them up. Behind the mandibles are structures which push the food into the mouth. Sensitive finger-like **palps** hang down on either side: with these the insect tastes the food to see if it is suitable for eating. While the food is being chewed, it is moistened with saliva, which comes from a pair of salivary glands in the thorax.

The mosquito sucks blood

The mosquito's mouth parts take the form of a needle-like **proboscis** (Figure 2A). When a mosquito lands on a person's body, it pushes its proboscis through the skin. The proboscis is protected by a sheath which holds it in place when it is being driven into the skin. It then injects a drop of saliva into the wound. This stops the blood congealing, otherwise it might block the proboscis. The mosquito then sucks blood through its proboscis.

The butterfly sucks nectar

The butterfly has a long proboscis like a tongue (Figure 2B). It pushes this into flowers in order to suck up the nectar. It coils its proboscis up when not in use.

The housefly sucks solids

The housefly has a proboscis which acts like a vacuum cleaner (Figure 2C). It has a pair of swollen pads at the end. The pads are covered with narrow grooves which are connected with a tube that runs up the middle of the proboscis to the gut. Flies can feed on solid things like lumps of sugar. The fly puts its pads in contact with the sugar. A drop of saliva flows down the proboscis. This dissolves the sugar and then the fly sucks it up (Figure 3).

How do insects breathe?

In man oxygen is carried from the lungs to the tissues by the blood. Insects have a very different system. They have hundreds of breathing tubes through which oxygen passes to all parts of the body. This is called the **tracheal system** (Figure 4) (Investigation 3).

Air enters the tracheal system through the **spiracles** which are tiny holes in

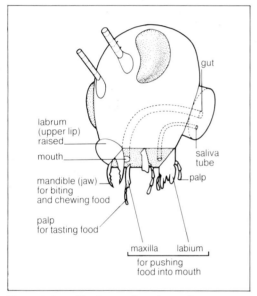

labrum (upper lip) raised

mouth

mandible (jaw) for biting and chewing food

palp for tasting food

gut

saliva tube

palp

maxilla   labium
for pushing food into mouth

Figure 1 In reality, the mouth parts are closer together than shown here, but they have been separated to show them clearly.

Figure 2 The mouth parts of various insects. Each is suited to feeding on a particular kind of food.

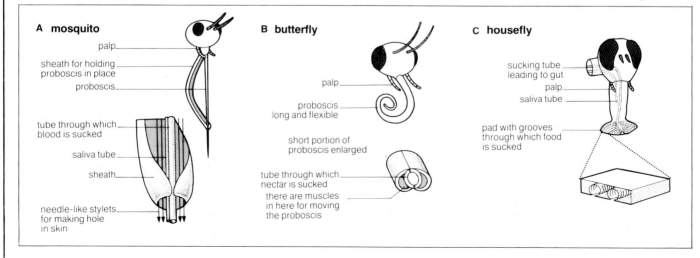

A mosquito

palp

sheath for holding proboscis in place

proboscis

tube through which blood is sucked

saliva tube

sheath

needle-like stylets for making hole in skin

B butterfly

palp

proboscis long and flexible

short portion of proboscis enlarged

tube through which nectar is sucked

there are muscles in here for moving the proboscis

C housefly

sucking tube leading to gut

palp

saliva tube

pad with grooves through which food is sucked

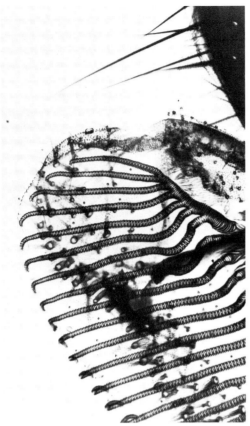

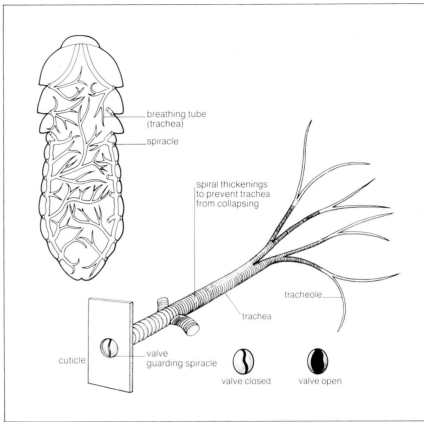

Figure 3 Photograph of one of the pads at the end of the proboscis of a blue-bottle.

Figure 4 The tracheal system consists of breathing tubes which carry oxygen to the organs and tissues.

the cuticle. In some insects the spiracles are fitted with **valves** which can open and close like sliding doors.

The spiracles open into tubes called **tracheae** (singular: trachea). These are lined with hard cuticle which is thickened into a spiral. This prevents their walls from caving in.

The tracheae branch like a tree. The ends of the branches are fine tubes called **tracheoles**. These have thin walls which are not lined with cuticle. They contain a watery fluid. The tracheoles envelop and penetrate all the organs and tissues, bringing oxygen to them.

In most insects oxygen passes through the tracheal system by diffusion. However, in large and active insects like the locust, air is pumped through the tracheal system by muscular movements of the abdomen. If you watch a locust, you can sometimes see these movements. The spiracles open and close in such a way that air is sucked in through the front spiracles and expelled through the back ones. When the insect is active, the spiracles open more frequently, and the pumping movements are more vigorous than when the insect is resting.

How are things transported in insects?

Insects have a **blood system** but it is very different from ours. The blood is a colourless fluid. It does not contain a red pigment, and it plays no part in carrying oxygen round the body. Its main job is to transport food substances and waste matter.

The insect does not have blood vessels like we do. Instead the blood is contained in an open **blood space**. It is kept moving by the tubular heart above the gut. Blood is sucked into the heart through little holes in its sides. It is then pumped forward and expelled into the blood space at the front end (Figure 5).

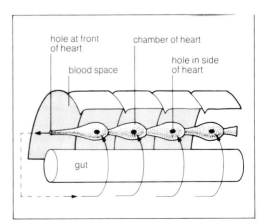

Figure 5 Blood is kept moving in the blood space by the heart. The heart sucks the blood in through the holes in the sides and pumps it out through the hole at the front.

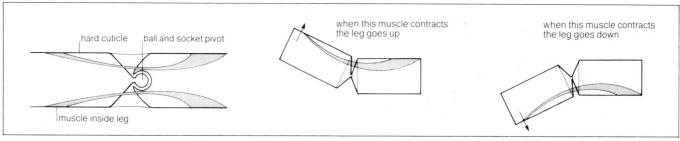

Figure 6 This diagram shows how the muscles are arranged inside the leg of an insect and how they make the leg move.

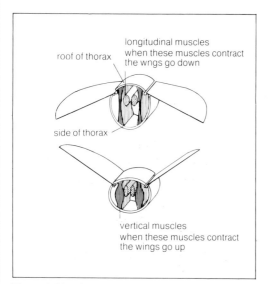

Figure 7 Muscles inside the thorax make the wings go up and down.

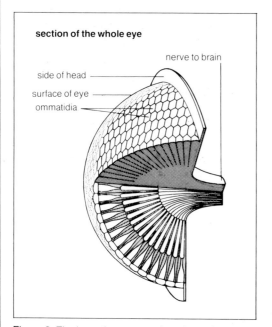

Figure 8 The insect's compound eye is made up of numerous ommatidia.

How do insects get rid of poisonous waste and save water?

Liquid waste matter is released from the tissues into the blood. It is then taken up by the **excretory tubules** (see page 62). Inside the tubules the waste matter is turned into a solid substance called **uric acid**. Water is removed from it and absorbed back into the blood. Meanwhile the solid waste passes out of the body through the anus.

Insects get rid of their excretory waste in solid form so as not to lose water. It enables them to live in hot, dry places without drying out. Insects have other ways of preventing water being lost. For example, the cuticle is waterproof. This prevents water evaporating from the surface of the body. The cuticle is made waterproof by a thin layer of wax on its surface.

The only problem is that water may evaporate through the spiracles. To keep this to a minimum, the spiracles are kept closed as much as possible. So, we see that insects are very good at saving water.

How do insects move?

Figure 6 shows how the muscles move the leg. At the joint there is a ball and socket **pivot** which works like a see-saw.

The muscles which straighten the leg are called **extensors**. Those which bend it are called **flexors**. These two muscles produce opposite effects: when one contracts, the other must relax, otherwise the leg will simply stay still.

Some insects, such as grasshoppers and locusts, can hop. Hopping is achieved by a third pair of legs. These are longer, stouter and more powerful than the others, and the extensor muscles inside them are particularly well developed.

The **wings** are worked by muscles inside the thorax. When the muscles contract, the wings go up and down (Figure 7). The locust beats its wings about 20 times a second, but certain midges can beat them over 1000 times a second.

How do insects react to stimuli?

Whenever you try to swat a fly, the beast gets away! This is because it has very good sense organs and quick reflexes.

The insect's **compound eyes** are particularly useful (Figures 8 and 9). The compound eye is composed of numerous little eyes called **ommatidia.**

What do insects actually see with their compound eyes? Each ommatidium forms an image of the world immediately in front of it. The whole eye forms an image which looks rather like a picture made out of lots of dots. The insect's view of the world is probably blurred and indistinct. However, it has a very large field of vision. This is because the eyes cover a large area of the head, extending round the sides as well as the front. This makes them very good at detecting movement.

Insects react to the slightest touch. The sense organs responsible for this are tiny **bristles** located mainly on the feelers. The base of the bristle is connected to a nerve fibre. When the bristle is moved, messages are sent to the brain. In insects like the housefly these bristles are so sensitive that they are stimulated by air movements, and even by sound. So when you try to swat a fly, even though it may not be able to see you, it can *feel* and *hear* you.

Figure 9 This picture taken with a scanning electron microscope, shows the compound eyes and feelers on the head of a fly.

Investigation 1

Looking at live locusts

1 Look at live adult locusts in a cage.

2 Watch them moving: observe the action of the hind legs.

3 Look at a locust hanging on a twig.

How does it do it?

4 Watch a locust feeding.

What are its mouth parts doing?

Do the legs play any part in feeding?

5 Look at the locust's abdomen.

Can you see any pulsating movements of the abdomen?

What might be the function of these movements?

Make brief notes, with sketches, on your observations.

Investigation 2

Examining the mouth parts of insects

1 Obtain a preserved locust or cockroach.

2 With a needle lift up the labrum ('upper lip').

Can you now see the mandibles (jaws)?

3 Now try to find the other feeding structures shown in Figure 1.

How are the mouth parts suited to dealing with the kind of food which this particular insect eats?

4 Look at the mouth parts of other insects, such as the housefly, under a microscope.

Assignments

1 List three reasons why it is an advantage to an insect to have valves guarding its spiracles.

2 Insects have no red pigment in their blood.

What does this mean that the blood *cannot* do?

What are the main functions of an insect's blood?

3 List three ways insects prevent water being lost from their bodies.

4 Why is it difficult to swat a fly?

5 Describe an experiment which could be done to find out if bees can tell the difference between red and blue.

6 Can you think of an experiment which could be done to test the suggestion that the locust sucks air into its body through its thoracic spiracles and expels air through its abdominal spiracles?

7 Look at Figure 9. Why is it an advantage to an insect to have eyes covering such a large area of its head?

Investigation 3

Examining the tracheal system

1 Cut open an insect (see Investigation 3, page 63).

2 Observe the breathing tubes (tracheae).

3 With forceps pull out a small piece of muscle from the thorax.

4 Mount the muscle in a drop of water on a slide, and put on a coverslip.

5 Look at your slide under the microscope: low power first, then high power.

6 Draw part of the tracheal system, showing how the breathing tubes branch.

The insect life cycle

In this Topic we shall see how insects reproduce, grow and develop. These three processes make up the life cycle. It is important to study the life cycle of insects. By doing so we can learn to control insect pests.

How do insects reproduce?

All insects reproduce sexually. The male is attracted to the female by her smell, bright colours, or by sounds which she makes. After they have met, the male and female go through a short period of **courtship**. For example, the male locust stalks the female for a short time and may make chirping noises by rubbing his hind legs against the hardened edge of his wings. This makes the female receptive to him.

Courtship is followed by **mating**. During this process the male puts his sperms into the female's body. Different insects do this in different ways. In the locust the male jumps on the female's back and grips her thorax with his front legs (Figure 1). He curves his abdomen round hers, so the tip can reach her reproductive opening. Some insects, bees for example, are so agile that they can copulate while they are flying.

The sperms do not fertilise the female's eggs straight away but are taken up into **sperm sacs** in her body. There they wait until the eggs are ready. As the ripe eggs pass out of her body, each one is fertilised by a sperm. In many insects the fertilised eggs are enclosed in a horny **egg case** before they leave the female's body.

Some insects take great care where they lay their eggs, so that the young have a supply of food when they hatch. For example, a certain kind of wasp lays its eggs in the body of a caterpillar. When the eggs hatch, the youngsters (larvae) feed on the caterpillar's tissues.

Many insects bury their eggs to protect them from the sun and enemies. For example, the locust lays her eggs in a hole in the sand. She digs the hole with her long flexible abdomen, the valves at the tip serving as little 'trowels'. The eggs are laid at the bottom. As she pulls her abdomen out, she produces a frothy liquid which hardens and forms a protective case round the eggs.

Figure 1 Two locusts mating. The male is on top, his abdomen twisted around the female's.

How do insects develop?

Insects are divided into two groups according to the way they develop during their life cycle. Some develop gradually, while others go through a complete change from one kind of animal to another. When an animal changes its form, we say it has undergone **metamorphosis**. This is made up of two Greek words: *meta* means 'change' and *morph* means 'form' or 'shape'.

Insects which develop gradually

This kind of development is shown by locusts, grasshoppers, cockroaches and many other insects (Figure 2 and Investigation 1).

The egg hatches into a creature which looks like a miniature version of the adult, except that it has no wings. We call this a **nymph**. The nymph has a hard cuticle, and it cannot grow unless it sheds it. So after a while it moults, and grows a bit bigger. This happens about five times altogether, the nymph getting bigger each time.

In the later stages small **wing buds** appear on either side of the thorax. At the final moult into the **adult**, the wings expand. You can see this happening in Figure 5. By this time the sex organs have developed, so the adult insect can reproduce.

The period of time between one moult and the next is called an **instar**. In the locust each instar lasts about a week, and there are five or six of them altogether. So it takes the locust about six weeks to complete its development.

The change from a newly-hatched nymph to the adult takes place gradually, step by step. For this reason it is sometimes called **gradual** or **incomplete metamorphosis**.

Figure 2 Life cycle of the locust.
This is an example of an insect with gradual development. After mating the female lays her eggs in the sand. The eggs hatch after about two weeks. The young nymphs crawl out of the sand. They moult five times, growing a little each time. They hang from twigs and branches as they moult. As they have no wings yet they cannot fly – they can only hop or walk. They feed on plants, using their jaws (mandibles) for biting and chewing the leaves. It takes between four and seven weeks for the adult stage to be reached. At the final moult the wings expand by blood being forced into them. Like the nymphs, the adults feed on plants.

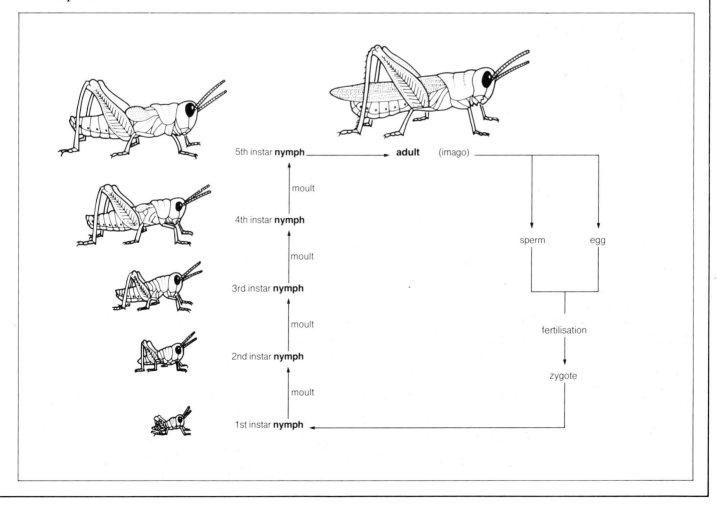

Figure 3 Life cycle of the cabbage white butterfly. This is an example of an insect with complete development. Cabbage white butterflies can be seen flying around during the summer. The adults fly from flower to flower, feeding on the nectar. They live for about three weeks. After mating, the female lays her eggs in batches on the leaves of cabbages and other similar plants. Each egg is shaped like a little barrel and has a sculptured shell. After a week or two the eggs hatch into caterpillars. The caterpillars feed on the cabbage leaves, using their jaws (mandibles) for cutting off pieces. They moult several times, and grow. After about a month the caterpillar crawls to a wall, fence or tree trunk and turns into a pupa. The pupa is anchored by a silk thread which runs around its body and a sticky pad at the back end. Pupae formed during the summer may give rise to adults after only a few weeks. Pupae formed in the autumn survive the winter before the adults emerge the following spring.

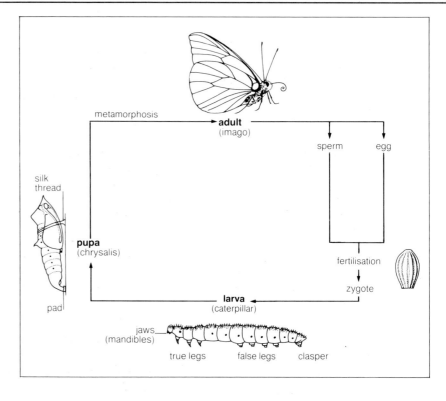

Figure 4 This cabbage white caterpillar has just started eating this cabbage leaf.

Insects which undergo a complete change

This kind of development is shown by insects such as butterflies, moths and flies (Figure 3 and Investigation 2).

The egg hatches into a **larva** or **grub**. The caterpillars of butterflies and moths, and the maggots of flies, are larvae.

The larva is quite different from the adult. Its cuticle is thin and flexible, enabling it to crawl or wriggle around. As it lacks the protective armour of the adult, and is unable to move quickly, it is liable to be eaten by birds and other predators. So many of them have special ways of protecting themselves. For example, caterpillars are often covered by hairs which make them unpleasant to eat, and some are cleverly camouflaged.

Larvae have no sex organs, so they cannot reproduce. Their job is to feed and grow (Figure 4). In this way they build up a store of energy which enables them to develop into the adult. For example, the caterpillars of the cabbage white butterfly feed almost non-stop on cabbage leaves.

After a time, the larva settles down and changes into a **pupa** or **chrysalis**. Sometimes the larva spins a cover of fine threads round itself. This forms a **cocoon** which surrounds and protects the pupa.

Although the pupa looks lifeless, the inside is the scene of much activity. The larval tissues break down into a kind of creamy soup. Out of this the adult is formed. Eventually, when conditions are suitable, the wall of the pupa opens and the adult clambers out. The wings then expand and it flies away.

The adult possesses sex organs, and its main job is to reproduce. Once it has done this, it has fulfilled its purpose and can die. Sometimes the adult lives for only a day or so, just long enough to find a mate and reproduce.

In the kind of insect just described, the life cycle has two distinct types of animals: the larva and the adult. Not only do they look different, but they behave differently too. Each feeds on its own kind of food, and each lives in its own particular habitat. For example, in the cabbage white butterfly the larva moves like a worm and chews up cabbage leaves, whereas the adult flies and sucks up the nectar of flowers.

Because the larva and adult are so different, the change from one to the other is described as **complete metamorphosis**.

How do insects grow?

An insect can only grow if it sheds its cuticle first. This process is called **moulting** or **ecdysis**.

The first thing that happens during moulting is that a fluid is formed underneath the cuticle. This fluid dissolves all but the outermost layer of the cuticle. The insect then expands, usually by swallowing air, and blows up like a balloon. This causes the old cuticle to split, and the insect struggles free (Figure 5).

Meanwhile a new cuticle has been formed under the old one. This is soft at first and as the body expands, it stretches. When the old cuticle has been cast off, the new one starts to harden. Once the new cuticle has hardened, the insect cannot expand any further until the cuticle is cast off again. An insect such as the locust may moult five or six times before it is fully grown. So insects do not grow smoothly as humans do, but in a series of spurts.

There is one great disadvantage with this kind of growth. While moulting, the insect is defenceless. It cannot escape if an animal such as a bird or lizard attacks it. To overcome this difficulty, insects often hide when they are moulting. When locusts moult they hang by their legs from the branches of trees and bushes.

Figure 5 This locust has just moulted. The old cast off cuticle is hanging from a twig. The new one has not yet hardened and the body and wings are expanding.

Investigation 1

The life cycle of the locust

1 Observe locusts mating in a cage.

   Which is the male and which is the female?

   How does the male put his sperms into the female?

2 Look at live locust nymphs ('hoppers') in a cage. You may be lucky enough to see them moulting. If you happen to see the final moult into the adult you may see the wings expanding.

3 Examine preserved nymphs, from the first to last stages. Draw them, using Figure 2 to help you.

   How do they differ from each other?

   How does the largest nymph differ from the adult?

4 With a ruler measure the length of each nymph, and the adult, from the front of the head to the tip of the abdomen.

5 Draw a bar chart to compare their lengths.

The locust is an example of an insect with 'gradual development'.

What does this mean?

Investigation 2

The life cycle of the cabbage white butterfly

1 Obtain preserved specimens of the following stages in the life cycle of the cabbage white butterfly: adult, eggs, caterpillar (larva) and pupa.

2 Examine them carefully. Draw each one, using Figure 3 to help you.

   How do they differ from each other?

   How is each adapted to its way of life?

3 Watch, or see a film of, cabbage white caterpillars moving and feeding.

   What do they feed on?

   What sort of mouth parts do they have?

   What are the functions of the legs?

4 Watch, or see a film of cabbage white butterflies visiting flowers.

   How do they feed?

   What sort of mouth parts do they have?

The butterfly is an example of an insect with 'complete development'.
What does this mean?

Assignments

1 Many insects bury their eggs to protect them from the sun.

   What harm might the sun do to them?

2 Explain the following words, all of which are used in this Topic:

   moulting
   mating
   metamorphosis
   instar
   cocoon

3 What is the difference between a nymph and a larva?

4 What happens inside a pupa?

5 Explain how an insect like the locust grows from the time the egg hatches until it becomes an adult.

6 Before they copulate, the male and female of a certain insect face one another and the female waves its feelers.

   Someone has suggested that the male will only mount the female after it has seen her waving her feelers.

   How could you find out if this is true?

Bees and other social insects

Certain insects, live together in an organised society or colony. We call them social insects. Man is a social animal too, but an insect's society is very different from ours. There are several kinds of social insects. The one we know most about is the honey bee.

Figure 1 Worker bees in a hive.

Figure 2 The three types of individual found in a bee colony. Each has its own job to do.

The honey bee

Wild bees live in a nest in a hollow tree or some other suitable place. Bee-keepers rear them in specially constructed hives so as to get honey from them.

Who's who in the beehive?

A large hive may contain more than 50 000 bees. They are all descended from one individual, the **queen** (Figure 2). So a colony of bees is really one enormous family. The queen is the head of the colony. Her only job is to lay eggs.

Several hundred male bees may be present in the hive. These are called **drones**. Their only job is to mate with a queen.

The remaining bees are **workers**. They are sterile females which cannot reproduce. Their job is to look after the hive, feed the queen and the drones, rear the young and perform sundry other tasks; in short, to do all the work.

Each of these three types of individuals has its own particular job to do. No-one steps out of line; the idea of worker bees going on strike in unheard of! There is thus a strict division of labour within the hive (Investigation 1).

The structure of the hive

Inside the hive, the workers make **combs** out of **wax**. The wax is produced by glands on the abdomen. The worker takes wax from these glands, moistens it with saliva, and moulds it into shape with its mandibles.

A comb consists of numerous chambers called **cells**. The cells are about 2 cm deep, and usually they slope so that the contents do not fall out. They fit together neatly as shown in Figure 3.

In the cells towards the top of the comb, the bees store honey. In the cells lower down, pollen is stored. In the bottom-most cells the young are reared: workers are reared in the smallest cells, drones in slightly larger cells, and queens in specially large cone-shaped cells near the edge of the comb.

What is honey?

Honey is the main food that bees live on. It is made by the workers from **nectar,** the sugary fluid found in flowers (see page 318). When a worker visits a flower, it sucks up the nectar with its tongue-like proboscis, and stores it in its stomach. In the stomach the nectar is turned into honey.

When the bee returns to the hive, it regurgitates the honey into one of the cells of the comb. The worker then closes the cell with a wax lid. Honey is stored in the comb for use during the winter.

Man can take honey from bees because they normally make far more than they need. In a man-made hive, the bees construct their combs inside wooden frames. The top ones, where the honey is stored, can be taken out easily. In a good summer a hive can produce enough honey to fill about 100 pots.

Producing new individuals

In the small cells towards the bottom of the comb, the queen lays eggs. These hatch into larvae which will give rise to workers (Figure 4).

For the first few days the workers feed the larva on a special substance from their mouths, called **royal jelly.** Then they switch the diet to a mixture of pollen and honey.

By the end of a week, the larva is fully grown. The worker now puts a lid on the cell. Inside, the larva pupates and two weeks later an adult worker emerges.

Drones and new queens are produced in a similar way except that they are reared in large cells and new queens are fed on nothing but royal jelly.

The worker bee

The workers live for about two months, during which time they move from one job to another. To begin with they clean out the cells and feed the larvae. Then they build new combs. Later they defend the hive against intruders. In the last few weeks they collect nectar and pollen from flowers.

The structure of the worker bee suits it for these jobs (Figure 5). Its mouth parts are adapted for sucking nectar from flowers, and its legs are adapted for collecting pollen. It has special structures for making the wax combs, and the egg-laying tube has been turned into a sting for defence (Investigation 2).

Defending the hive

A worker on guard duty stands at the entrance of the hive, looking out for wasps, mice and other animals which might come to the hive to steal honey. Some bees are particularly fierce and will make a mass attack on anyone who comes too near their nest.

The worker's main weapon is its sting. This is a sharp needle-like tube with a poison sac at the base. The sting has tiny barbs sticking out of it, like a fish hook, so when the bee has used its sting it cannot pull it out. The result is that when the bee flies away, the sting gets left behind. Unfortunately, for the bee, part of its gut gets left behind too, so it dies soon afterwards. Obviously a worker bee can only use its sting once.

The queen bee has no barbs on her sting, so it does not stay in the victim's body and can be used again. The same applies to wasp stings.

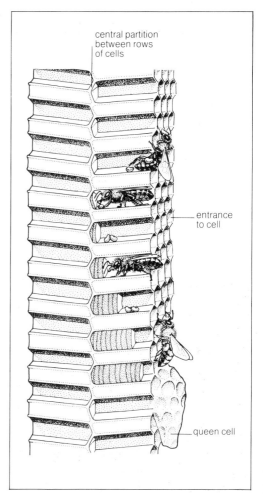

Figure 3 Part of a comb. The picture shows worker bees putting pollen into the cells.

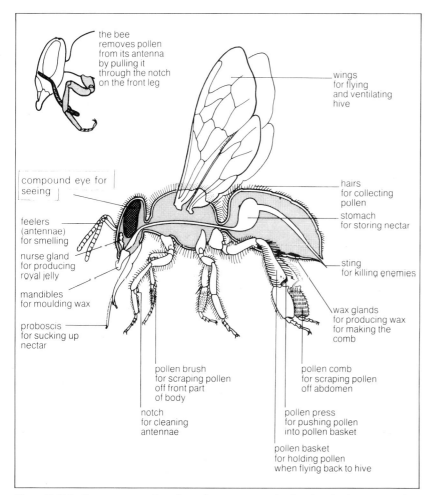

Figure 5 This diagram summarises the various ways a worker bee is suited to carry out its different jobs.

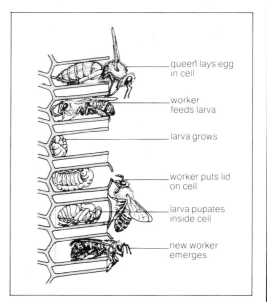

Figure 4 This picture shows how worker bees are produced in a hive.

Figure 6 A swarm of bees hanging from a tree branch.

How is a new colony started?

If the number of bees becomes too great, the reigning queen leaves the hive and starts a new colony somewhere else. Accompanied by about half the workers, she flies off in a **swarm**. Often the swarm hangs from a branch of a tree, while some of the workers go off in search of a suitable nesting site (Figure 6). A bee-keeper who wants to start a new hive will capture a swarm in a basket and put them in his hive.

Meanwhile, back in the old hive, a new queen takes over. A few weeks before the swarm took off, the workers reared several new queens. The first queen to emerge immediately stings the other ones and kills them, thereby establishing herself as head of the colony.

After a short while, the new queen flies away from the hive with all the drones in hot pursuit. This is called her **marriage flight**. The first drone to catch her mates with her in mid-air. In the process his reproductive organs are ripped out of his body, and soon afterwards he dies. From this one drone the queen receives enough sperms to last her entire life.

After mating, the queen returns to the hive. She now moves from cell to cell, laying eggs. In the summer she may lay several thousand a day, but she stops during the winter.

How do worker bees communicate?

On returning to the hive from a food-gathering trip, a worker can tell other workers where to find the food. This was discovered by the German scientist Karl von Frisch.

When the worker gets back to the hive, it does a dance (Figure 7). If the food is close to the hive the bee dances in a circle. This is called the **round dance**. The closer the food, the faster the dance.

If the food is more than about 100 metres away, the worker dances in a figure of eight, waggling its abdomen as it does the straight run in the middle. This **waggle dance** tells other bees not only how far away the food is, but also in what direction it is. The distance is given by the speed of the dance: the faster the dance, the closer the food. Von Frisch claims that its direction is given by the position of the dance relative to the sun. While a worker is dancing, other workers gather round in an excited manner and after a few seconds they fly off in the exact direction of the food.

While the bees are dancing they make various piping noises. Some scientists feel that these noises, rather than the dances, tell the other bees where the food is.

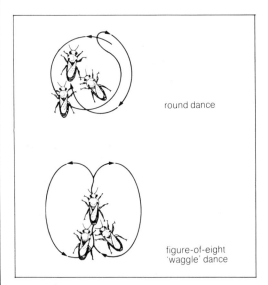

round dance

figure-of-eight 'waggle' dance

Figure 7 The two kinds of dance which the worker bees perform.

Figure 8 Termites live in large mounds which stick up from the surface of the ground. These termite nests are in Queensland, Australia.

Other social insects

Wasps, ants and termites have a social system which is similar to the bee's, though the details are different.

Wasps build nests out of a papery material which they make by chewing wood to a pulp. Their nests hang from the branches of trees or the eaves of buildings. They feed mainly on nectar and any other sugary substances they can find. They have different kinds of individuals like bees, but they don't communicate with each other by dances.

Termites ('white ants') live on wood, grass and leaves. They do much damage, particularly in the tropics where they weaken and even destroy wooden buildings. Their nests consist of large mounds (Figure 8). Unlike other social insects, termites develop gradually like the locust. The workers are nymphs which never grow up into adults, so termites can be said to use 'child labour'. Termites have individuals called **soldiers** which defend the nest. For fighting, these soldiers have large jaws, or in some cases a 'squirt gun' on the head (Figure 9). This shoots out an unpleasant liquid which drives the enemy away.

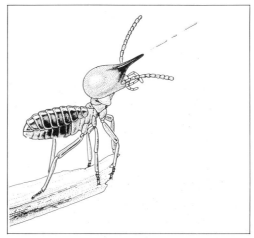

Figure 9 This soldier termite squirts a sticky liquid at any animal that tries to enter its nest.

Investigation 1

Looking at the three types of bee

1 Obtain preserved specimens of a queen bee, drone and worker.

2 Use Figure 2 to help you decide which is which.

Why do you think they differ in size?

3 Draw the three bees in outline, making clear their differences in size and shape.

What are the main jobs performed by each type of bee?

Investigation 2

To see how the worker bee is adapted to do its jobs

1 Watch, or see a film of, worker bees visiting flowers and doing their other duties.

2 Now examine a preserved specimen of a worker bee. Use a hand lens to help you see it in more detail.

3 Make a list of the adaptations shown in Figure 5 which you can see in your own specimen.

4 Examine prepared slides of the mouth parts of a worker bee, under the microscope. Observe the proboscis and the mandibles.

What is each used for?

5 Examine prepared slides of the front and hind legs of a worker bee under the microscope.

Observe in detail the adaptations shown in Figure 5 for collecting pollen.

Assignments

1 Explain each of the following, all of which are mentioned in this Topic.

drone,
royal jelly,
queen substance,
marriage flight,
round dance.

2 Why does a worker bee die after it has stung someone, and why does a drone bee die after it has mated with a queen?

3 It is claimed that the worker bees' figure-of-eight dance (waggle dance) tells other bees where food is.

How could you find out if this is true?

Put forward your *own* suggestion as to how this might work.

4 How does the organisation of a bee hive compare with the organisation of a human society such as a town?

5 Suggest a reason for each of the following:

a) the cells in a bee comb are hexagonal in shape;
b) in a poor summer bees make less honey than usual;
c) when a bee colony gets above a certain size, half the bees leave in a swarm;

d) an individual bee will often visit only one kind of flower, ignoring others;
e) worker bees often sit on the cells which contain developing youngsters.

6 Drone bees have been described as 'lazy, stupid, fat and greedy'.

Do you think this is a fair description?

Insects and man

The locust is one of the world's most serious pests. It is just one of many insects which are harmful to man. In this Topic we will summarise the various ways insects affect man, and how he copes with them.

Figure 1 A swarm of locusts in North Africa. A swarm like this may contain ten thousand million locusts.

The locust

Locusts thrive in warm parts of the world such as Africa, the Middle East and South America. There are several types of locust but they all lead the same kind of life. Much of the time they live singly or in small groups, feeding on grass and leaves. But sometimes their numbers build up, and then they do a great deal of damage to man's crops. Locusts have enormous appetites and a few of them can strip a plant very quickly (Figure 2).

The female locust lays her eggs in the sand. The eggs hatch into nymphs (see page 69). These are called **hoppers.** They have no wings, and cannot fly. As their numbers build up, they crowd together. Food begins to run out, and this causes them to start 'marching' in bands.

They march during the day, eating the leaves of plants as they go. They move about a kilometre a day. At night they rest in shrubs and small trees. Every week or so they moult and grow. After about six weeks they undergo their final moult, their wings expand and they became **adults.**

They now start to fly. They move across the country in a vast **swarm**, like the one in Figure 1. A single swarm may contain ten thousand million locusts, covering an area larger than Greater London. With the aid of the wind, the locusts fly about 80 km a day. They may travel several thousand kilometres before settling down to breed.

The swarming locusts will strip a vast area of all its vegetation. A large swarm may eat 160 000 tonnes of food each day. This amount of corn would feed 800 000 people for a whole year.

At one time, locust swarms occurred regularly in many parts of the world. They caused widespread famine and did millions of pounds-worth of damage. Fortunately man is now learning how to control them.

How are locusts controlled?

In the old days farmers tried to drive locusts away by lighting fires or beating drums. The hoppers were driven into trenches, then buried or burned, or they were killed by putting poisoned bait in their path. Where possible, the eggs were dug up and burned.

Nowadays crops are sprayed with powerful insecticides which kill the locusts. The insecticide is sprayed from vehicles or aeroplanes. Spray from an aeroplane on one flight can kill as many as 180 million locusts.

Constant watch is kept on locusts and in this way scientists can forecast when and where swarming is likely to occur, so the crops can be sprayed in good time. This requires co-operation between different countries and much of this work is co-ordinated by the United Nations.

Figure 2 This corn cob is being eaten by locusts.

Figure 3 A mosquito sucking blood from a person's arm. This type of mosquito carries malaria.

The mosquito

There are many different kinds of mosquitoes, or gnats as they are called. Those found in hot countries carry serious diseases, such as malaria and yellow fever (see pages 28 and 226). The female mosquito sucks blood, and carries the parasites which cause these diseases from one person to another (Figure 3). Fortunately British mosquitoes do not normally carry diseases.

Mosquitoes need water for breeding: ponds, lakes, water-tanks – any place where the water is still. The swamp shown in Figure 5 is ideal.

The female mosquito lays her eggs on the surface of water. The eggs hatch into small wriggling larvae. The larva lives in the water; it has a **breathing tube** at the back end, by which it hangs onto the surface film. The end of the breathing tube is open, so air can get in.

After several weeks the larva pupates. The pupa hangs onto the surface film by a pair of breathing tubes on the head. After a few days, the pupa splits open and the adult mosquito emerges (Figure 4).

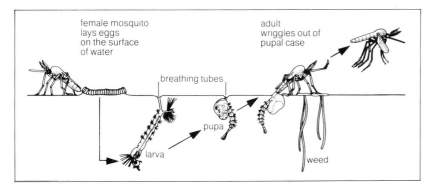

Figure 4 In the life cycle of the mosquito, the larva and pupa live in water as shown on the left.

How can we get rid of mosquitoes?

The adult mosquito can be killed by spraying with insecticides such as DDT. However, DDT is thought to be a hazard to health, so it is better to use other methods.

The larvae can be destroyed by spraying oil onto the water (Investigation 1). The oil lowers the surface tension of the water, causing the larvae to let go. Water then enters their breathing tubes and they drown. Pupae are destroyed in the same way. Usually the oil is mixed with an insecticide so as to make absolutely sure they die.

Another way of getting rid of mosquitoes is to stock up lakes and ponds with fish that eat the larvae or pupae. Or one can drain swamps, so as to get rid of the mosquito's breeding areas.

None of these methods on its own is much good, but together they are quite effective. However, the mosquito is still a major pest in many parts of the world, and the diseases which it carries have not yet been wiped out. In the tropics windows are usually covered with a fine-mesh screen to keep them out, and campers should always use tents with mosquito netting.

Figure 5 This swamp in East Africa is just the kind of place where mosquitoes breed. The men are spraying oil onto the surface of the water to kill the larvae and pupae.

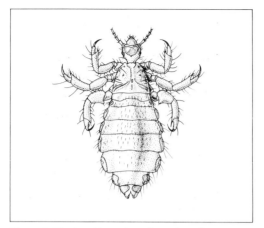

Figure 6 The head louse has sharp claws for clinging onto the skin while it sucks blood. Outbreaks of this insect pest occur in schools from time to time.

Summary of how insects affect man

Here are some ways that insects are harmful:

1 *They eat and destroy crops.*
Insects like the locust will eat just about anything with leaves on it. Other insects are more fussy; for example, caterpillars of the cabbage white butterfly eat nothing but cabbage leaves and closely related plants. They can be killed by spraying or dusting the leaves with an insecticide before the caterpillars start eating them.

2 *They spread diseases.*
Many insects, besides the mosquito, spread diseases from person to person. Flies, fleas and lice are all guilty (see page 228). Some insects spread diseases amongst plants. One reason why green flies and black flies are such a nuisance to gardeners is that they carry harmful viruses from one plant to to another.

3 *They spoil food.*
The housefly and blue-bottle lay their eggs on food, particularly meat. Because of their dirty habits, these insects also bring germs onto our food, thereby causing diseases such as cholera. Fortunately outbreaks of such diseases are now rare in Britain.

4 *They destroy buildings and furniture.*
A major culprit here is the famous death watch beetle, whose larva bores into wood, weakening it and making it rot. Some insects, such as termites, eat wood. In Africa, they can completely destroy a wooden hut.

5 *They ruin clothes.*
The larva of the clothes-moth eats the fibres of woollen garments, making holes in them. You can protect clothes from this insect by means of 'moth balls'. They contain a chemical substance which kills the larvae.

6 *They are irritating.*
The head louse sometimes occurs in schools and other places where people live or work close together (Figure 6). It clings to the skin and sucks blood, and can cause intense itching. The eggs, known as nits, stick to the person's hair. Head lice can be got rid of by rubbing an insecticide preparation into the hair, and washing it later. Any remaining nits are then combed out.

Here are some ways that insects are helpful:

1 *They pollinate plants.*
Insects such as butterflies and bees carry pollen from one flower to another (Figure 7). A bee may visit hundreds of flowers in one day.

2 *They kill harmful insects.*
You have probably heard that ladybirds are useful insects. This is because their larvae eat greenflies, thus helping to get rid of this garden pest. Ichneumon wasps kill cabbage white caterpillars and lay their eggs in them, so they are useful too. In hot countries, the praying mantis eats several insects which destroy crops.

3 *Bees make honey.*
This is discussed on page 72. Nowadays honey is a luxury food, but before sugar was discovered it was the only way of sweetening things.

4 *Some insects produce silk.*
The caterpillar of the silkworm moth makes its cocoon from a single strand of silk. This may be over a kilometre long. At one time these insects were cultivated to obtain silk, but nowadays its place has largely been taken by synthetic fibres.

Figure 7 Insects play an important part in pollinating flowers. In this picture a bee visits a fleabane flower.

Investigation 1

Getting rid of mosquitoes

1 Obtain a dish containing mosquito larvae and/or pupae.

2 Notice that the larvae and pupae hang on the surface film. The slightest disturbance causes them to let go, and dive down into the water.

3 Wait till the larvae and pupae have settled at the surface. Then very gently run some oil or paraffin onto the surface of the water.

What happens to the larvae and pupae?

Do you think this is a good way of getting rid of mosquitoes from the world?

Can you think of any disadvantages?

Why do you think oil and paraffin have this effect on the larvae and pupae?

Suggest two other ways the larvae and pupae of mosquitoes might be destroyed.

4 Obtain a preserved adult mosquito, and put it on a piece of white paper.

5 Measure its length, and width with wings outstretched.

What advice would you give to the manufacturers of mosquito netting to make sure their product is effective?

Investigation 2

A look at some harmful insects

1 Examine the head of a female mosquito under the microscope.

What does the mosquito feed on?

Why is it harmful?

What adaptations can you see which enable it to live in the way it does?

2 Examine a louse under the microscope.

Where does the louse live?

What adaptations can you see which enable it to live in such a place?

3 Examine the head of a cabbage white caterpillar under the microscope.

What does this caterpillar feed on?

What structures can you see which enable it to feed efficiently?

Assignments

1 Why are the following insects regarded as pests:

Locust,
Mosquito,
Louse,
Cabbage white butterfly,
Housefly?

2 Which is best: to spray locusts with an insecticide from an aeroplane or from a vehicle on the ground?

3 Suggest reasons why locust hoppers get excited as their numbers increase.

What investigations could you carry out to find out which of your suggestions is correct?

4 Each word in the left hand column is related to one of the words in the right hand column.

Write them down in correct pairs.

locust	pollination
death watch beetle	insecticide
ladybirds	swarms
bees	greenflies
DDT	furniture

5 Why do head lice occur particularly in cold, overcrowded places?

6 An insect pest may be controlled either by spraying it with an insecticide or by bringing in another insect which eats it.

Put forward arguments for and against each method.

7 Below is a map of Northern Africa. The crosses show the occurrence of swarms of the desert locust during a particular year.

a) From an up-to-date map of Africa, list the names of the countries in which swarming was observed.

b) Suggest reasons why swarming occurred in these particular parts of Africa and not elsewhere.

c) The information shown on the map was obtained some years ago, and swarms of locusts in these areas are less common now. Why do you think this is?

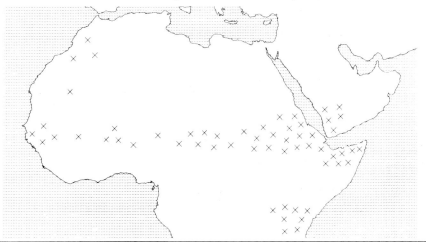

Fish: life under water

Many organisms live in water, but the greatest experts at this are fishes. In order to survive under water, they have many special features. In this Topic we shall be mainly concerned with these features.

Figure 1 The grayling is a relative of the salmon.

Figure 2 These diagrams show the structure of a typical bony fish.

The structure of fish

Fish are vertebrates, that is they possess a backbone. You can study the structure of a typical fish by examining a herring (Investigation 1). Figure 2 will help you to identify its parts.

The skin is covered with **scales** which overlap each other like the tiles on the roof of a house. They protect the fish, and prevent water passing through the skin.

The body is streamlined, enabling the fish to move quickly and smoothly through the water.

On either side of the head there is a flap of skin, which is stiffened by bones. This is called the **operculum**. Underneath the operculum there are four feathery structures side by side. These are **gills** – they are used for breathing.

At the posterior end of the body there is a **tail**. Underneath the skin there are muscles, which are arranged in a series of W-shaped blocks. These muscles play an important part in swimming, and they are the part of the fish that people eat.

At various points there are **fins**. Each fin consists of a thin flap of skin supported by slender spines. Some of the fins are arranged in pairs and stick out from the sides of the body. We refer to them as **paired fins**: they include the pectoral fins just behind the head, and the pelvic fins a little further back.

The other fins are single and are attached to the mid-line of the fish. We call them the **median fins**. The median fins include the dorsal, ventral and tail fins. Some fish have more than one dorsal fin, and there is a good deal of variation in their exact positions.

Fish have good sense organs. The **nostrils** are used for smelling, but they play no part in breathing. The **eyes** can see very well under water. Running along each side of the body is a faint **lateral line**. This is a canal just beneath the surface of the skin; it contains sense organs which detect movements of the water.

Inside the mouth are rows of identical teeth. They are constantly falling out throughout life and are replaced by new ones which grow in their place. The anus is on the ventral side of the body, about two thirds of the way back.

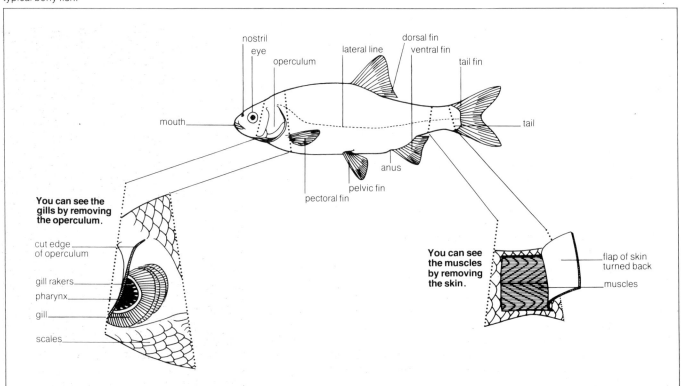

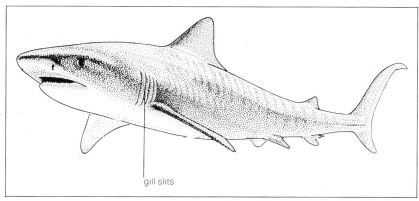

Figure 3 Cartilagenous fish lack an operculum. In this drawing of a tiger shark, notice the five separate gill slits.

Different types of fish

The fish just described belongs to a group called **bony fish**: that is, their skeleton is made of bone. These include fresh water fish such as minnows, sticklebacks and carps, and sea water fish such as cod, mackerel and herring.

There is another group of fish called **cartilaginous fish**: their skeleton is made of cartilage or gristle, a material which is much softer than bone. This group includes sharks and rays which live in the sea.

One of the main differences between bony and cartilaginous fishes concerns their gills. In bony fish the gills are covered by the operculum shown in Figure 2. However, in cartilaginous fish there is no operculum; instead a series of **gill slits**, five in all, open separately on either side of the head (Figure 3).

There is a lot of variation in the shape of fish. Some are long and thin, like the eel. Others are flat, like the plaice: it is flattened from side to side and lies on the sand at the bottom of the sea.

Fish also vary greatly in size. The largest fish in the world is the whale shark which can be twice as long as a double decker bus. However, it feeds on plankton and is quite harmless.

The largest flesh-eating fish is the great white shark which lives in some of the warmer seas of the world. It can be over 11 metres long and is a menace to swimmers.

At the other end of the scale, certain tropical fish are a mere centimetre or two in length.

How do fish feed?

Most fish feed on small organisms such as worms, crabs and plankton (see page 44. As the food passes through the pharynx, it is prevented from getting between the gills by the **gill rakers**: these are slender bars which stick out from the bases of the gills (Figure 2).

Some fish eat much larger prey, which they bite into pieces or swallow whole (Figure 4). Scientists have opened up the stomachs of sharks and found such surprising objects as buckets and old petrol cans. The stomach of a large specimen, caught in the 18th Century, is said to have contained an entire suit of armour!

For hunting down their prey, sharks use their excellent sense organs. Take the great white shark for example. This huge fish is extremely sensitive to movements of the water: it can detect a moving object, such as a swimmer, over a kilometre away. As the shark moves closer to its prey, its sense of smell takes over. As it closes in for the kill, it depends mainly on its eyes. Finally, when it is almost on top of its prey, it closes its eyes and opens its mouth: it now relies on an 'electrical' sense which works rather like radar. As soon as the shark feels the pressure of the prey on its teeth, its jaws automatically snap shut.

Figure 4 The huge dagger-like teeth of this Great White shark are used for biting the prey into pieces.

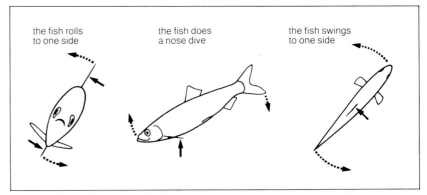

Figure 6 These diagrams explain how a fish is kept on an even keel when it is swimming. If the fish is displaced, the water presses against the fins (solid arrows). The fish then returns to its original position (dotted arrows).

How do fish swim?

You can discover much about how fish swim by watching one swimming in an aquarium (Investigation 2). You will see that it swings its tail from side to side As the tail sweeps through the water, it drives the fish forward (Figure 5). The **tail fin**, with its large surface area, increases the forward thrust.

The movements of the tail are brought about by contraction of the **muscle blocks**, first on one side and then on the other. Some fish can swim quite fast: speeds of over 60 kph are quite common, and certain fish can move at over 100 kph.

When a boat moves through the water it tends to rock about. In fish this is prevented by the fins, which keep the body steady, rather like the feathers at the back of an arrow or dart (Figure 6).

If you watch a fish swimming in an aquarium, you will see that it constantly changes direction, and sometimes slows down and stops. These steering and braking movements are achieved mainly by movements of the pectoral fins.

Most fish possess a swim bladder which keeps the body up in the water. This is a sausage-shaped bag full of air, rather like a balloon, which is situated towards the upper side of the body cavity (Figure 7).

Sharks and rays do not have a swim bladder, so their only way of staying up in the water is by swimming. If they stop swimming they sink to the bottom.

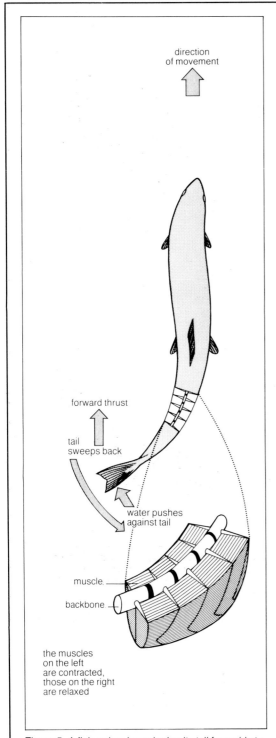

Figure 5 A fish swims by swinging its tail from side to side as shown in this diagram.

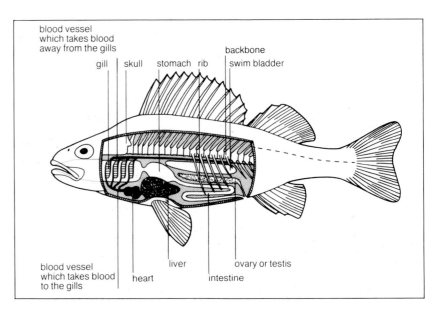

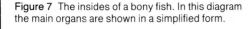

Figure 7 The insides of a bony fish. In this diagram the main organs are shown in a simplified form.

How do fish breathe?

Fish use their gills for breathing. Figure 8 shows how the gills of a bony fish are arranged. Notice that there are holes in between them: these connect the pharynx with the outside (Investigation 3).

The fish sucks water in through its mouth. The water flows between the gills and passes out of the fish by the opercular opening as indicated by the arrows in Figure 8. In this way a continuous stream of water is kept flowing past the gills; this stream is maintained by flapping movements of the operculum and the opening and closing of the mouth.

The gills have a large surface area and blood continuously passes through them. As water flows over the gills, oxygen is taken up by the blood and carbon dioxide diffuses out.

How do fish reproduce?

Most bony fish reproduce by the male and female releasing their sperms and eggs into the surrounding water. The sperms then swim to the eggs and fertilise them. Some fish produce vast numbers of eggs. For example, a cod can produce as many as eight million at one time.

The chances of the eggs being fertilised are greatly increased if the male releases his sperms close to the female's eggs. The **stickleback** or 'tiddler' shows how this can be achieved. In the spring, the male stickleback develops a red breast and builds an underwater nest out of pieces of weed which he glues together with a substance made by his kidneys. He then lures a ripe female to his nest by showing her his red breast. The female enters the nest and lays her eggs (Figure 9). She leaves the nest through the other side. The male may persuade several other females to lay eggs in the same nest. When there are between 50 and 100 eggs in the nest, he enters and releases his sperms on top of them.

The male now looks after the fertilised eggs, guarding them against other fish which might eat them, and fanning them with his tail. This stirs up the water and helps to get oxygen to the eggs. After they have hatched, the male looks after the young sticklebacks for a few days until they are able to fend for themselves.

Fish which migrate

Some fish only reproduce after they have migrated to special breeding grounds. Such is the case with the **eel** and the **salmon**.

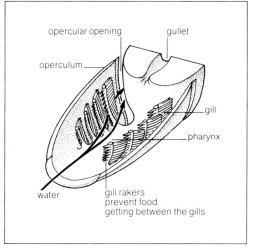

Figure 8 The head of this fish has been sliced horizontally so we can see into its pharynx. The gills lie on either side. Water flows between the gills as indicated by the arrows.

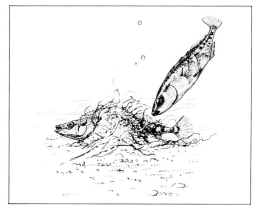

Figure 9 This drawing shows a female stickleback inside the nest which has been made by the male. The male prods her tail and this makes her lay her eggs.

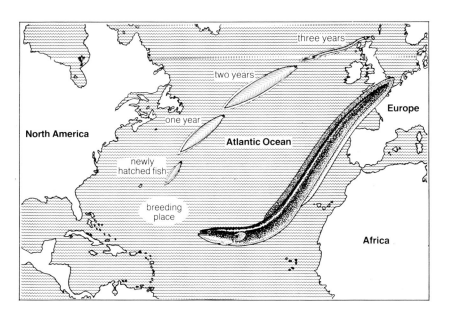

Figure 10 The European eel swims to the other side of the Atlantic Ocean to breed. The young then return to Europe.

Eels live in lakes and rivers in Britain and other parts of Europe. When they are 7 – 12 years old they develop an instinctive urge to move downstream to the sea. After reaching the sea, the eels swim across the Atlantic Ocean to an area south of Bermuda called the Sargasso Sea (Figure 10). Here they breed. The eggs hatch into tiny youngsters, which swim back across the Atlantic, growing as they do so. It takes them about three years to reach Europe. By this time they have grown into miniature eels, called **elvers**. The elvers swim up the rivers and mature into adult eels.

The salmon lives in the sea. When mature it migrates up rivers to breed. The urge to do this must be very strong because they will leap over waterfalls in their efforts to get upstream. In quiet water near the source of the river the females release their eggs, and the males deposit their sperms over them. The exhausted salmon then die. The eggs develop into young salmon called **fry**. After two or three years they swim down to the sea where they grow into adults. A remarkable fact about salmon is that when they are ready to reproduce they always return to the river where they were born.

Fish as food for man

Vast numbers of fish are caught for food. In Britain alone, the annual catch may be more than a million tonnes. The main fish caught in the fishing industry are mackerel, cod and haddock. Many other fish are also caught though in smaller amounts.

The food value of fish lies mainly in the muscles mentioned earlier, and also in the liver. Fish liver oil is very rich in vitamins A and D, which is why it is given to young children (see page 118). The ovaries and testes are also tasty and nutritious: these are known as hard and soft roes respectively. The eggs of the sturgeon, a large bony fish which lives in the sea but migrates up rivers to breed, are known as caviar which is one of the world's most expensive luxury foods.

For fishing to be worthwhile, the fish population must be dense. Fortunately this is usually the case because fish produce lots of offspring and they tend to swim in large groups or **shoals**.

One of the main ways of catching fish is by **trawling** for which a specially equipped boat called a trawler is used. The trawler pulls a bag-shaped net through the water in which the fish are caught (Figure 11). The bag is kept open by wooden **otter boards**, which work like kites: as the trawler moves forward the water pushes against them and they pull the sides of the bag outwards. Floats and balls round the edge of the bag keep it open like a huge gaping mouth.

The modern trawler is a very efficient vessel. The nets are large and special echo-sounding equipment is used for locating large shoals of fish. In **factory trawlers** there are machines for gutting, skinning, filleting and freezing the fish on board. All this means more fish for more people, but we do have to make sure that we do not take so many fish out of the sea that their numbers begin to dwindle. Avoiding over-fishing is an important aspect of conservation (see page 397). We must also be careful that nothing happens to the plankton on which fishes are dependent for their food.

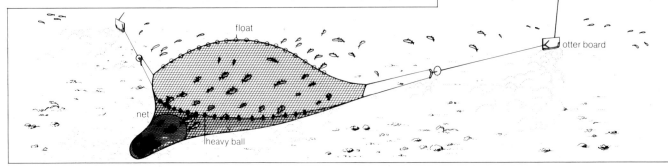

Figure 11 Many of the fish we eat are caught by trawling. The net shown here is one of several different kinds that are commonly used.

Investigation 1

Investigating the structure of a fish

1 Obtain a preserved fish such as a herring.

2 Examine its external structure and list the main features.

3 Force open the mouth and look inside.

What can you see there?

How do the teeth differ from yours?

What are the teeth used for?

4 Pull the operculum towards you.

Can you see the gills underneath?

What are they for?

5 Strip off a piece of the skin from the side of the tail.

Can you see the muscles underneath? What are they for?

6 Draw and label the fish from one side.

Investigation 2

Watching fish swimming

1 Watch a fish, such as a goldfish, swimming in an aquarium tank.

2 As the fish moves along, describe the actions of the tail and the fins.

What makes the fish move forward?

How does it turn left and right?

How does it swim up and down?

How does it stop?

What stops it rolling around in the water?

3 Observe other actions which the fish performs: in particular watch the mouth and operculum.

Explain your observations.

Investigation 3

Looking at the gills

1 Obtain a preserved fish such as a herring.

2 Remove the operculum by cutting round it with scissors.

3 Look at the gills underneath.

How many are there?

4 Observe the structure of one gill in detail.

How is it suited to its job?

How is water made to flow past the gills?

5 Show the route taken by water by threading a piece of cotton through the mouth and out between the gills. The cotton should follow the course of the arrow in Figure 8.

Assignments

1 Each of the words in the left hand column is related to one of the words in the right hand column.

Write them down in the correct pairs.

gill rakers   locomotion
operculum   breathing
scales   sieve
tail fin   floating
swim bladder   skin

2 In plaice the eyes are on the same side of the head.

Why do you think this is?

3 Certain fish live in extremely deep water.

a) What difficulties do such fish have to face?
b) What special features do you think they have?

4 The type of trawl net illustrated in Figure 11 is suitable for catching fish such as cod and haddock which live near the bottom of the sea.

Suggest methods for catching fish such as herring and mackerel which live near the surface.

5 The table below tells us the total mass of fish caught by British fishing vessels in certain years:

Year	mass of fish caught (tonnes)
1938	16 099
1948	45 771
1960	50 977
1970	69 535
1977	224 623

a) Draw a bar chart so these figures can be compared more easily.
b) Why do you think there was a big increase in the amount of fish caught between 1938 and 1948?
c) Why do you think there was a big increase in the amount of fish caught between 1970 and 1977?

6 The picture on the right shows a method of trawling which has become widely used in recent years. What do you think are the advantages and disadvantages of this method compared with the one illustrated in Figure 11?

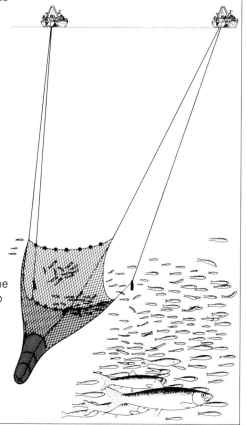

Amphibians

Amphibians include frogs, toads and newts. The name amphibian comes from the Greek word amphibios, which means 'leading a double life'. This is a good name for them, because they live in two different environments: in water and on land.

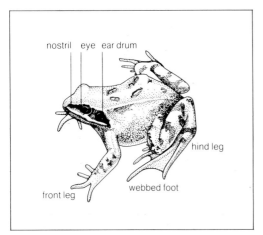

Figure 2 This diagram shows the external structure of the common frog.

Figure 3 The frog jumps by straightening its powerful back legs.

The structure of the frog

You can learn a lot about the way of life of the frog by looking at its external structure (Investigation 1).

Its **skin** is moist. This is because it contains glands which produce slime (mucus) all over the surface. The skin has patches of colour, particularly green and black, which help to camouflage the frog and protect it from being attacked by predators.

The front legs are short and stumpy. The hind legs are much longer and there is skin between the toes; in other words the feet are **webbed** (Figure 2).

Just above the mouth there is a pair of **nostrils**: these open into the mouth cavity and air is drawn in through them when the frog breathes. The **eyes** are high up on the head and bulge out: they give the frog good all-round vision, enabling it to detect movement quickly and escape from danger. The **ear drum** is on the surface of the head and there is no flap of skin behind the ear as there is in man.

Between the back legs there is an opening which leads into a small bag inside the body. This bag is called the **cloaca**. Urine, faeces and sex cells (eggs and sperms) all pass out of the body via this bag.

How do frogs move?

Frogs are difficult to catch: every time you get close they hop away. You can

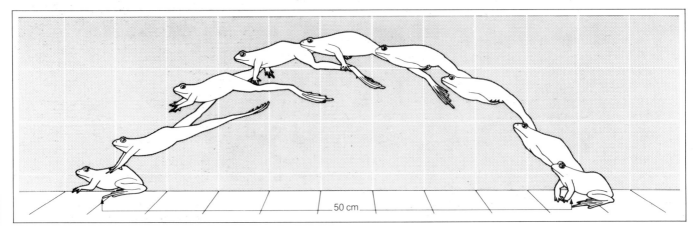

50 cm

discover how frogs move by watching them in the laboratory (Investigation 2).

Frogs are expert hoppers because their back legs are long and have powerful muscles. Figure 3 shows what happens when a frog jumps. The same sort of action enables the frog to swim and the webbed foot gives the frog a good push against the water (Figure 4).

The frog is unique amongst amphibians in being able to hop. Other amphibians such as newts and toads use their legs just for walking.

How do frogs breathe?

Frogs breathe in three different ways: through their **skin**, through the lining of the **mouth cavity**, and by means of **lungs**. Oxygen and carbon dioxide diffuse across these three surfaces (Figure 5).

In order to serve as a breathing surface, the skin must be kept moist: gases will only diffuse through it if this is so. The frog's skin contains many **glands** which secrete a watery slime (mucus) onto the surface. This is why amphibians are always moist and slimy. However, there is a grave disadvantage in having a moist skin: water readily evaporates from it and the animal is liable to dry up. This is the main reason why frogs and other amphibians are normally found only in damp places. The skin has other features which make it suitable as a breathing surface: for example, it is thin and has a good blood supply.

If you watch a frog you will notice that sometimes its throat moves up and down. These movements pump air in and out of the mouth cavity through the nostrils. When this happens, gaseous exchanges take place through the moist lining of the mouth cavity.

When the frog is active and needs a lot of oxygen, it takes a large gulp of air and forces it down into its lungs. The lungs consist of a pair of simple thin-walled sacks situated in the chest region.

How do frogs feed?

If you look inside the mouth of a frog you will see its **tongue**. This differs from the human tongue in that it is attached at the front and points backwards.

The frog uses its tongue for feeding on insects. When an insect flies past, the frog flicks out its tongue and catches it. The tongue is sticky so the insect cannot escape: it is carried to the back of the throat, then quickly swallowed. The whole process is over in about a tenth of a second.

How do frogs reproduce?

Frogs hibernate during the winter amongst rocks and vegetation. In the spring they wake up, and the males go in search of a pond.

When it has found a pond, the male starts croaking. It does this by forcing air through its **voice box** in the throat. The croaking of the male attracts a female who by this time is loaded with eggs. By this stage the male has enlarged thumbs covered with tough black skin, like warts: they are known as **nuptial pads**.

The frogs mate in the water. The male climbs onto the female's back; he places his front legs round her chest and grips her tightly (Figure 6). The skin is slippery, but his large thumbs with their swollen nuptial pads help him to get a firm grip. The pair may remain together like this for two or three days.

Eventually eggs start to pass out of the female's cloacal opening between her back legs. At the same time the male produces a stream of seminal fluid containing numerous sperms. The sperms fertilise the eggs.

Each egg is surrounded by a layer of jelly. This is a protein called **albumen**, which is the same substance that the 'white' of a hen's egg is made of. Soon after coming into contact with the water, the jelly swells up: this protects the eggs from being damaged and from drying out and it makes them clump together in a mass which sticks to weeds and stones. This is the familiar **frog spawn**.

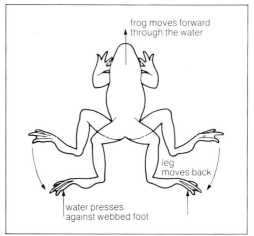

Figure 4 The frog swims by straightening its back legs and pushing its webbed feet against the water.

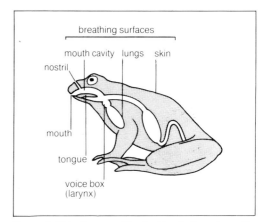

Figure 5 This diagram of the inside of a frog shows its three breathing surfaces.

Figure 6 Amphibians mate and lay their eggs in water. Here two frogs can be seen mating.

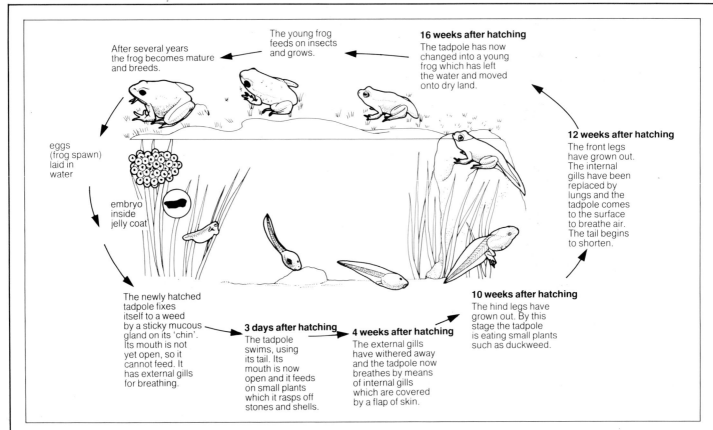

The young frog feeds on insects and grows.

After several years the frog becomes mature and breeds.

16 weeks after hatching
The tadpole has now changed into a young frog which has left the water and moved onto dry land.

eggs (frog spawn) laid in water

embryo inside jelly coat

12 weeks after hatching
The front legs have grown out. The internal gills have been replaced by lungs and the tadpole comes to the surface to breathe air. The tail begins to shorten.

The newly hatched tadpole fixes itself to a weed by a sticky mucous gland on its 'chin'. Its mouth is not yet open, so it cannot feed. It has external gills for breathing.

3 days after hatching
The tadpole swims, using its tail. Its mouth is now open and it feeds on small plants which it rasps off stones and shells.

4 weeks after hatching
The external gills have withered away and the tadpole now breathes by means of internal gills which are covered by a flap of skin.

10 weeks after hatching
The hind legs have grown out. By this stage the tadpole is eating small plants such as duckweed.

Figure 7 The main stages in the development of the frog. Start at the extreme left and work your way round in an anti-clockwise direction.

How do frogs develop?

After about a week the egg hatches into a small fish-like creature called a **tadpole** (Figure 7). You sometimes see tadpoles swimming about in ponds during the spring and they can easily be kept in the laboratory (Investigation 3). The tadpole is an animal in its own right: it occurs in the development of all amphibians. It is an example of a **larva**.

The tadpole swims by flapping its tail from side to side. Its mouth is closed at first, but as soon as it opens it feeds on small plants. To begin with it breathes by means of feathery **external gills** which stick out from the side of the head. Later these are replaced by **internal gills** which are covered by an operculum as in bony fishes (see page 80). Meanwhile it grows.

After about ten weeks the tadpole starts changing into a frog. Legs grow out and the tail shortens. Lungs develop for breathing air: they eventually take over from the gills. You often see tadpoles coming up to the surface to gulp air into their newly formed lungs.

It takes about a month for the tadpole to change completely into a miniature frog. It then leaves the water and begins its life on dry land where it feeds on small insects.

It takes about four years for the frog to become sexually mature. It then returns to the water to breed.

The tadpole and the adult frog look very different and lead quite different lives. The change from one to the other is an example of **metamorphosis**. This is similar to a caterpillar turning into a butterfly (see page 70).

In amphibians, metamorphosis is controlled by a hormone called **thyroxine** (see page 280). This is produced by the thyroid gland in the neck region. For thyroxine to be made there must be iodine in the water. If the water is lacking in iodine, the tadpole cannot change into the adult. On the other hand, if you put extra iodine in the water or inject thyroxine into a tadpole, it will change into the adult more quickly.

Investigation 1

Looking at the frog

1 Obtain a preserved frog.

2 Look at its external structure.

Can you see the structures shown in Figure 2?

What job does each structure do?

Why are its back legs so much longer than its front legs?

3 Open the mouth and look inside.

Can you see the tongue?

How does the tongue differ from yours?

What part does the tongue play in feeding?

Can you see, or feel, the teeth?

What part do they play in feeding?

Investigation 2

Watching frogs moving

In this investigation you will be handling a live vertebrate. When you pick it up, carry it gently but firmly. Try not to frighten it. If you are squeamish or get into difficulties, your teacher will help you.

1 Place a live frog in a cardboard box.

2 Watch it hopping.

What part is played by its hind legs when it hops?

3 Put your frog in a tank of water.

4 Watch it swimming.

How does it differ from a human doing the breast stroke?

In what ways are the frog's hind legs adapted for swimming?

5 Put your frog into a small glass tank (not underwater).

6 Watch its throat region.

What do you observe?

Explain your observations.

Assignments

1 What job is done by each of these structures:

a) the tongue of a frog,
b) the tail of a tadpole,
c) a tadpole's mucous gland,
d) an amphibian's thyroid gland,
e) the nostril of a frog?

2 What is wrong with each of these remarks?

a) 'Frogs feed on frog spawn'
b) 'Tadpoles are useful because they eat harmful insects'
c) 'The frog's feet are webbed so as to enable it to jump higher'
d) 'Frogs swim by doing a breast stroke rather like people do'
e) 'The male frog has nuptial pads on its thumbs in order to attract a female'

3 What are the disadvantages to an amphibian, like the frog, of having a thin moist skin?

4 It has been suggested that the frog is so good at breathing through its skin that it does not need to use its lungs.

Can you think of some evidence that would support this statement?

5 It is said that in a dry atmosphere frogs lose water very quickly.

Suggest an experiment which you could carry out to find out exactly how quickly a frog loses water.

Investigation 3

Watching tadpoles develop

1 Obtain some tadpoles from a pond (see page 353) and bring them into the laboratory.

2 Fill an aquarium tank with pond water, and put in some stones and weeds.

3 Put your tadpoles into the tank.

4 Watch them at intervals over the next few weeks.

How do they change in size and shape?

How does their behaviour change?

Do your observations agree with Figure 7?

5 At the same time as you set up your aquarium tank, set up two further tanks exactly like the first one.

6 Into one of the tanks put a tablet of thyroxine once a week throughout the time the tadpoles are developing.

7 Into the other tank put a few crystals of iodine once a week.

Do either of these treatments affect the rate at which the tadpoles develop?

Explain any observations which you are able to make.

Birds: masters of the air

Apart from insects and bats, birds are the only animals to have developed the power of active flight. It has made them one of the most successful groups of animals.

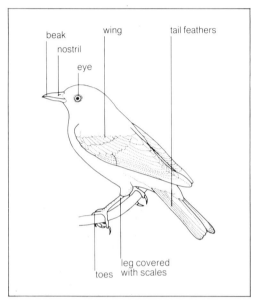

Figure 2 The external structure of a typical bird.

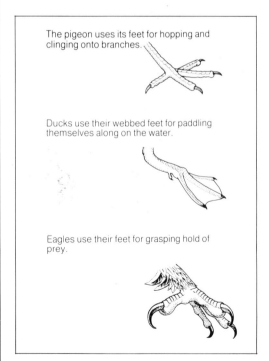

The pigeon uses its feet for hopping and clinging onto branches.

Ducks use their webbed feet for paddling themselves along on the water.

Eagles use their feet for grasping hold of prey.

Figure 3 In some birds, the feet are adapted for jobs other than clinging to branches and hopping on the ground.

Figure 1 A gannet in full flight.

The structure of birds

The external structure of a typical bird is shown in Figure 2. Most of the body is covered with **feathers**. They are made of the protein keratin, the same material that hair is made of.

The front limbs take the form of a pair of **wings** which most birds use for flying. The legs are quite different from the wings. The feet have four **toes**, each ending in a **claw**. Generally, three of the toes point forward and one backwards. This enables the foot to grasp objects such as tree branches when the bird is at rest.

All birds have a **beak**. Like the feather, this is made of keratin. Towards the base of the beak there is a pair of **nostrils** through which the bird breathes. The nostrils open into a cavity rather like the inside of our nose. This is used for smelling. However, birds' sense of smell is rather poor. Their eyesight is much more acute.

The eyes have a **third eyelid** as well as the usual upper and lower lids. This works like a transparent shutter: it slides over the eye from the front to the back and protects the eye from dust without stopping it from seeing.

Some birds have special features which are related to the way they live. Take the feet, for example. Most birds use their feet for standing, hopping and holding onto the branches of trees. However, in some cases they are adapted to do other jobs (Figure 3).

How do birds feed?

Birds have no teeth; they peck at their food, and swallow the pieces whole. The beaks of different birds are adapted for feeding on different kinds of food. In each case, the shape of the beak fits in with the kind of food which the bird eats. This is particularly well shown by the vulture which uses its curved beak for tearing at flesh (Figure 4). Four other examples are shown in Figure 5.

Because the food is unchewed, birds have a special stomach for breaking it up called the **gizzard**. Seed-eating birds such as pigeons keep small stones in their gizzard: these rub against the seeds and grind them up.

Figure 4 These vultures are about to feed on the zebra, tearing at its flesh with their strong curved beaks.

How do birds get rid of waste substances and save water?

You have probably noticed that bird droppings are a mixture of black and white sludge. The black part is the bird's faeces, the remains of food which it hasn't been able to digest and absorb. The white part is a substance called **uric acid**: this is the bird's nitrogenous waste.

The bird's nitrogenous waste is not in liquid form as it is in man; instead it is a semi-solid substance. This is because water has been removed from it before it leaves the body. This is a good way of saving water. It enables birds to live in dry places: certain desert birds can go for weeks on end without water. The same sort of thing occurs in insects (see page 66).

Feathers

Young birds are covered with small fluffy feathers. We call these **down feathers**. As the bird grows, its down feathers fall out and their place is taken by longer and straighter **flight feathers**. An adult bird may keep some of its down feathers, particularly round the tops of the legs.

If you look at a flight feather you will see that there is more to it than appears at first sight (Investigation 1). Running down the centre is the **quill**. The base of the quill is rooted in the skin: muscles are attached to it so the feather can be moved. The flat part of the feather is called the **vane**. This is composed of numerous hair-like structures called **barbs**. The barbs have further branches which interlock with one another as shown in Figure 6.

If you put your finger between the barbs, you will find that it is easy to break the connections between them. They can be connected up again by gently stroking the feather with your finger. If the connections are broken in real life, the bird puts them together again by rubbing its feathers with its beak. This is called **preening**. The bird also uses its beak to spread oil over the feathers and to remove any parasites which might be crawling amongst them. The oil is produced by a gland on the bird's back, close to the tail: the bird rubs its beak in this oil before it preens itself.

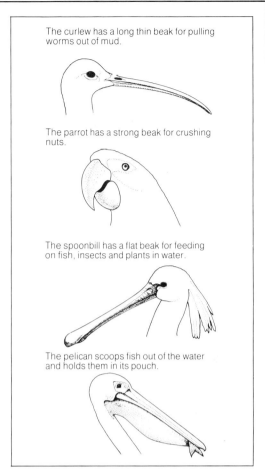

The curlew has a long thin beak for pulling worms out of mud.

The parrot has a strong beak for crushing nuts.

The spoonbill has a flat beak for feeding on fish, insects and plants in water.

The pelican scoops fish out of the water and holds them in its pouch.

Figure 5 The beaks of different birds are adapted for feeding on different kinds of food. Here are four examples.

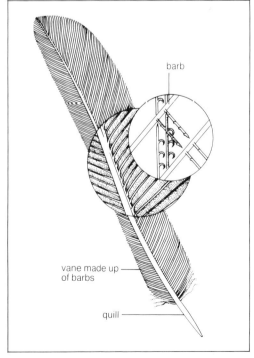

barb

vane made up of barbs

quill

Figure 6 A feather is more complicated than it appears at first sight.

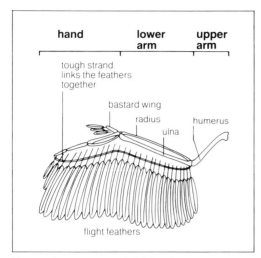

Figure 7 The structure of a bird's wing.

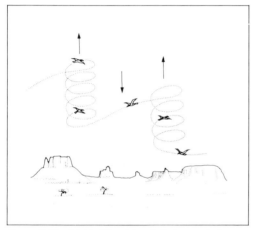

Figure 8 The bird in this picture is being carried up into the sky by warm air currents.

Figure 9 These diagrams show what happens when a bird flies in a straight line.

How do birds fly?

Birds fly by flapping their wings, or by gliding. When fully spread out, the wings have a large surface area. The bird is held up by air pressing against the wings from underneath; the same principle keeps aeroplanes and gliders in the air.

The bird's wing is equivalent to the human arm. However, instead of having five equal fingers, there is just one long one. The others are small or absent (Figure 7). The 'thumb' forms the so-called **bastard wing**: this sticks out in front and smoothes the flow of air over the top of the wing when the bird is flying. When gliding, the bird holds its wings out straight without flapping them. It makes use of rising air currents to hold it up in the air (Figure 8).

When the bird flaps its wings, the feathers behave like the slats of a Venetian blind: they close when the wing goes down and open when it goes up (Figure 9). In this way the bird is given plenty of lift as the wings are lowered, but it is not dragged downwards when they are raised.

The wings are operated by powerful **flight muscles**, which are attached to the **breast bone** (Figure 10). The breast bone has a deep keel to increase the area for the attachment of these large muscles. They make up the 'white meat' which is so good to eat in a chicken or turkey. There are two flight muscles. One pulls the wing down and the other pulls it up. The muscle which raises the wing has a tendon which runs through a hole at the point where the humerus joins the main skeleton. In pulling the wing up, this muscle works like a pulley.

Birds have several other features which help them to fly. They are streamlined and light. To make them light, they have **air sacs** inside their bodies and their bones are hollow. Their tail feathers play an important part in balance, and their muscles help to make them very agile. Can you think of any other ways birds are adapted for flight?

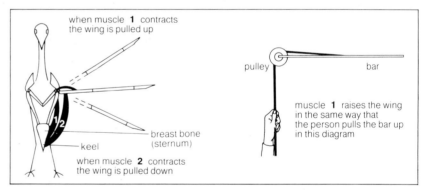

Figure 10 These diagrams show how the muscles work in pulling the bird's wing up and down.

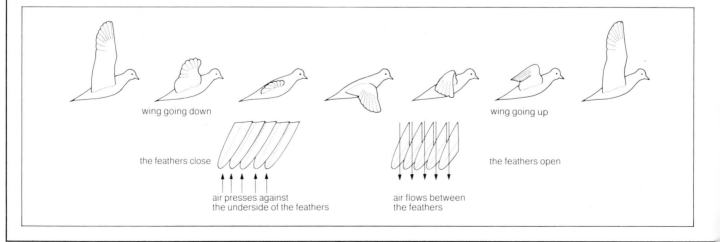

Birds are certainly expert flyers and can achieve high speeds. A racing pigeon can fly at 60 kph for long periods, and swifts can reach speeds of over 150 kph in still air.

What other jobs do feathers do?

The feathers make up the bird's **plumage**. In many ways they are like a person's clothes. Apart from helping the birds to fly, they do three main jobs.

1 Feathers keep the bird warm.
They do this by trapping a layer of air against the skin. Air is a poor conductor of heat, so this layer of air holds heat inside the body.

2 Feathers keep water out.
This is because they are oily. Water tends to run straight off an oily surface without wetting it. This is particularly useful to water birds such as ducks.

3 Feathers enable birds to recognise each other.
This is due to their characteristic colours. Birds are good at seeing colours, particularly red. Plumage is especially important in enabling the males and females to recognise one another. This is taken up in the next Topic.

Investigation 1

Looking at feathers

1 Obtain a flight feather and a down feather.

   How do they differ?

2 Take the flight feather and put your finger between the barbs.

   Do the barbs separate from each other easily?

3 Put the feather between your thumb and forefinger and stroke it gently.

   Do the barbs join up again?

4 With scissors, cut out a small piece of the vane, about 5 mm square, from the flight feather.

5 Put the piece on a slide.

6 With a pipette add a drop of clove oil and cover with a coverslip.

7 Look at it under the microscope, using low power.

   What can you see?

   How do your observations help to explain what happened when you separated and rejoined the barbs?

8 Pipette a few drops of water onto the flight feathers.

   Does the water run off the feather or pass through it?

   Why is this important to the bird?

Investigation 2

Looking at the bird's wing

1 Examine a bird's wing.

   How are the feathers attached to it?

2 Lay the wing on a piece of squared paper, and trace round it with a pencil.

3 Count the number of squares within the outline of the wing, and work out its approximate surface area.

4 Now examine a wing of the same size from which the feathers have been removed.

5 Using the squared paper, work out the surface area of the featherless wing.

   What is the surface of the wing with feathers?

   What is its surface area without feathers?

   By how many times do the feathers increase the surface area of the wing?

   Why is this important to the bird?

6 Look at the featherless wing: this is the bird's 'arm'.

   How does it differ from the human arm?

Assignments

1 Make a list of five ways a bird's body is adapted for flight.

2 You often see gulls gliding on the seaward side of a cliff.

   Why do you think they do this?

3 Where a feather is attached to the bird's skin there are muscles which enable the feather to move.

   Suggest two reasons why it is important for a bird to be able to move its feathers.

4 When eating roast chicken one always finds a lot of meat on the 'breast'.

   What is this meat and what does it do in the life of the bird?

5 The ostrich has a flat breast bone with no keel of the sort shown in Figure 10. On the other hand the pigeon has a very deep keel.

   How would you explain this difference?

6 Look at Figure 10.

   Why do you think muscle 2 is so much larger than muscle 1?

How do birds reproduce?

Birds normally breed in the spring or early summer. They take great care of their eggs and look after their young until they can fend for themselves.

Figure 1 The peacock uses his fan-like tail for displaying to the female.

Figure 2 These two great crested grebes are courting. They have just dived into the water and have fetched some weeds.

Figure 3 The song thrush lays a clutch of about five eggs in a nest made out of pieces of grass woven together.

What happens when birds reproduce?

When birds reproduce, they usually perform a series of actions one after the other.

1 *The male claims a territory.*

This might be a small area of a wood, or perhaps someone's back garden. The male defends his territory by singing and displaying his feathers. For example, the robin sticks out his red breast. If necessary, birds will fight to defend their territory.

2 *The male's song attracts a female.*

When the female approaches, the male displays his feathers to her. One of the most famous displays is put on by the peacock, which has a large, beautiful tail that opens like a fan (Figure 1).

Sometimes the male and female go through various actions together. For example in the great crested grebe, which lives on lakes and estuaries all over Britain, the males and females shake their heads, chase over the water, and occasionally dive down and fetch up weeds (Figure 2).

These displays are called **courtship**. They keep the male and female together and put them in the right mood for mating.

3 *The birds build a nest.*

Often both birds do this, but sometimes the male is lazy and leaves the whole job to the female.

Various materials are used for nest-building, depending on the type of bird: sticks, grass and moss are favourites, and mud is sometimes used. Many birds build their nests in trees or under the eaves of a house. Others make their nests on the ground.

It is remarkable how quickly some birds build their nests. David Lack, an expert on the robin, used to tell this story: a gardener in Basingstoke hung up his coat in a tool shed at nine o'clock in the morning; at one o'clock he took it down to go to lunch and found a robin's nest in one of the pockets.

4 *The two birds mate.*
The male does not have a penis, just an opening. He mounts the female and presses his reproductive opening against hers. His sperms enter her body and fertilise the eggs. Each egg then passes down the female's oviduct and is coated with albumen and a shell.

5 *The female lays her eggs in the nest.*
Usually she lays five or six eggs (Figure 3).

The shell is made of chalky calcium carbonate: it protects the egg and helps to stop water evaporating from it.

Figure 4 shows the structure of a hen's egg. Inside, a tiny **embryo** rests on top of a bag of yolk, the **yolk sac**. The embryo will eventually develop into a chick: the yolk nourishes it while it develops (Investigation 1).

The embryo and yolk sac are surrounded by a thick fluid called **albumen**. This is the 'white' of the egg: its main job is to supply the embryo with water during its development.

One end of the egg is 'blunter' than the other. At the blunt end, just under the shell, there is an **air space**. The shell has tiny holes running through it, which allows oxygen to diffuse into the air space. The oxygen is then carried via blood vessels to the embryo. Carbon dioxide passes out in the reverse direction and in this way the embryo breathes.

6 *The female incubates the eggs.*
She does this by sitting on them: this keeps the eggs warm and is called brooding. Provided it is kept warm the embryo gradually develops into a chick. Figure 5 shows what the inside of the egg looks like after several days. Notice the excretory sac in which the developing embryo deposits its nitrogenous waste.

7 *The eggs hatch.*
Hatching usually takes place a few weeks after the eggs have been laid. Using its beak, the chick breaks through the membranes surrounding it, and cracks open the shell (Figure 6). Then, wet and bedraggled, it clambers out.

8 *The parents look after the chicks.*
The newly-hatched chicks of most birds have no feathers and are helpless. They stay in the nest, and the mother spends some of her time sitting on top of them to keep them warm. The parents feed them on worms and other kinds of food which they collect and bring back to the nest: the chicks open their mouths and the parent bird pushes the food down their throats. Some birds feed their chicks with half-digested food which they bring up from their stomachs.

The chicks soon acquire a covering of downy feathers: they then leave the nest and sit on a nearby branch. Meanwhile, the parents go on feeding them and gradually the young birds grow and develop their flight feathers. Soon they start flying for short distances and eventually they fly away for good.

How long does all this take? It varies from one type of bird to another. In the robin it takes about two months from the time the parents start building the nest to the time when the young fly away. On the other hand ducklings, and the young of many other birds which make their nests on the ground, are born with a covering of feathers and can run or swim straight away.

Migration

In Autumn, birds such as swallows and house martins gather together on roof-tops and telegraph wires. Then suddenly they fly away.

Swallows and house martins are examples of birds which breed in Britain and other parts of the Northern Hemisphere during the summer, and then fly south when winter approaches. These journeys are called **migrations**, and birds which go on them are called **migrants**. In contrast, birds such as robins, sparrows and starlings stay in Britain throughout their lives: we call them residents. Not all British migrants *leave* Britain in the winter – some visit

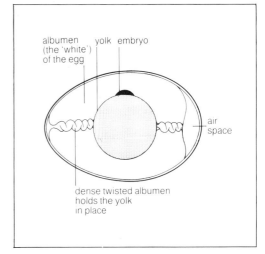

Figure 4 The inside of a fertile hen's egg just after it has been laid.

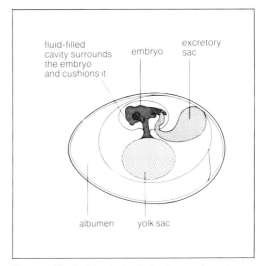

Figure 5 The inside of a fertile hen's egg about three days after it has been laid.

Figure 6 A hen's egg hatching.

Figure 7 This young bird is having its foot ringed so that scientists can find out where it flies when it migrates.

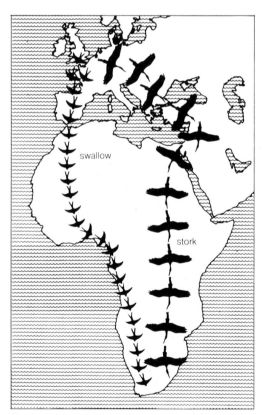

Figure 8 The swallow and the stork both migrate to South Africa from Europe when it is autumn in the Northern Hemisphere but spring in the Southern Hemisphere. This map shows their main routes.

Britain from the Arctic – for example, various kinds of ducks and waders.

To find out where birds migrate scientists have carried out 'ringing' experiments. A numbered band, made out of a metal such as aluminium, is fixed round one of the bird's legs (Figure 7). Anyone finding a ringed bird is asked to report its whereabouts. Migrating birds have also been tracked by radar: this can only be used over short distances, but it has told us how fast and how high they fly.

We know that some birds migrate over very long distances. For example, swallows and storks fly from Europe to South Africa, a distance of more than 8000 km (Figure 8), and the Arctic tern flies from the Arctic to the Antarctic, which is about 18 000 km (Investigation 2).

Why do birds need to migrate? The reason seems to be connected with their food supply. As winter sets in and food runs out the bird moves to a warmer place where it can get plenty of food.

How do birds know when to migrate? No-one knows for certain but scientists believe that a decrease in the amount of daylight creates the necessary urge.

How do migrating birds find their way? This is the most difficult question of all, and many experiments have been carried out to try and find the answer. Here is an example. Scientists have let birds loose in a planetarium, a large dome on which images of the stars are projected. The birds flew in a particular direction. The scientists then altered the position of the stars on the dome. The result was that the birds changed their direction. We conclude that these particular birds were navigating by the stars.

It seems that birds which fly at night navigate by the stars and moon, and those which fly by day use the sun. Some scientists have suggested that birds have a magnetic sense which enables them to use the earth's magnetic field to guide them.

Once they have done a journey, birds remember landmarks, and this makes the return flight easier. It explains why birds such as swallows return to nest in exactly the same place – a garden shed for example – year after year.

Homing

Some birds fly home after they have been taken a long way away. This is called **homing**. For example, adult shearwaters have been taken from their nests in Britain, flown by aeroplane across the Atlantic in dark boxes, and then set free in North America. Without delay they flew straight home on a direct course and were back in their nests within two weeks.

We don't know how birds do this kind of thing. All we can say is that if a man was to make a similar journey using the sun, he would need a clock, a compass and a map. The bird seems to have these inside its brain!

Birds as food for man

For thousands of years man has eaten bird's eggs. Our main source of eggs is the domestic hen. She will lay eggs even if she has not been mated: the eggs are therefore unfertilised and do not contain an embryo. A good breed of hen, well-fed and kept in the right conditions, will lay an egg once every day or two for four or five years.

People sometimes keep hens in their gardens. They roost in small huts and wander about in an enclosure. These are called **free-range hens**.

Nowadays large numbers of egg-laying hens are kept in special poultry farms. They are kept indoors in rows of boxes called batteries, for which reason they are called **battery hens**. A poultry farmer with, say, 600 battery hens may produce as many as 12 000 eggs in a year.

Chickens, turkeys, pheasants and ducks are also used for their meat. Battery chickens, suitable for cooking, are reared and then killed when they are about three months old. These **broiler chickens**, are produced in very large numbers. Because of this, chicken is now one of the cheapest kinds of meat.

Investigation 1

Looking at a chick embryo

1 Obtain a fertilised hen's egg which has been kept in an incubator for 3 to 5 days since it was laid.

2 With plasticine make a 'cradle' for holding the egg.

3 Place the egg in the plasticine cradle as shown in the illustration.

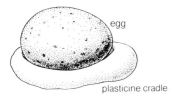

Let it stand for a few minutes.

4 With scissors cut away the shell, piece by piece, from the upper side of the egg.

5 Observe the embryo lying on top of the yolk.

6 Suck away the albumen with a pipette.

7 Examine the embryo under a magnifying glass.

Make a list of all the structures which you can see.

Suggest a function for each structure in your list.

Investigation 2

Why do swallows migrate by a particular route?

1 Study Figure 8.

This shows one of the routes by which swallows migrate from Europe to South Africa.

2 Obtain an atlas and look at the maps of Europe and Africa.

3 Make a list of all the countries the swallow flies over on its route.

4 Now make a list of all the countries the swallow would fly over if it flew right down the middle of Africa instead of down the side.

5 For both routes find out about the climate and vegetation in each country.

Why do you think swallows fly down the side of Africa rather than down the middle?

Assignments

1 If you look at ducks on a pond, you will notice that some of them have a whitish body with a shiny green head and a black posterior, whereas others are dull brown all over.

   a) What kind of ducks are they?
   b) Why do you think they are differently coloured?

2 Why do you think it is an advantage to a pair of breeding birds to have their own piece of territory?

3 Make a list of the advantages and disadvantages of building a nest in each of these places: (a) on the ground, (b) in a tree, and (c) under the eaves of a house.

4 What part do feathers play in helping birds to produce their offspring successfully?

5 Certain birds breed in the Arctic during the summer and then fly south in the winter.

   Why do you think they do this?

6 Before boiling an egg many people

prick the shell at the blunt end with a pin.

Why do you think this is a good idea?

7 A certain bird has a red pouch below its beak as shown below.

Put forward suggestions as to what it might be used for.

How would you test each of your suggestions to find out which one is right?

8 Scientists have taken birds varying distances from their home and have then released them. They have then estimated the percentage of the birds which managed to get back home.

The results are shown below.

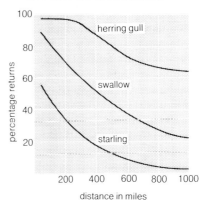

   a) How do you think the birds which returned were recognised as those that had been taken away and released?
   b) What conclusions can be drawn from the graphs?
   c) Suggest three ways by which the birds might have found their way home.
   d) Suggest three things which might have happened to the birds that did not return home.

Mammals

Mammals can be found leading active lives almost everywhere, from tropical forests to the frozen wastelands of the Arctic. This is the group of animals to which man belongs.

The main features of mammals

Let's take the cat as a typical mammal (Figure 1). Other mammals will be mentioned as we go along.

The slim, agile body is covered with **hair**. Hair is made of the protein keratin, the same material that the feathers of birds are made of. The hair keeps the body warm, so mammals are **warm-blooded** and are able to keep their body temperature constant.

Like most mammals, the cat walks on four legs: so it is known as a **quadruped**. The legs are strong and have well-developed muscles which are attached to the skeleton. Cats move swiftly and silently. The feet have five **digits** (toes). Each digit has a claw at the end and a pad underneath. We have nails instead of claws, and some mammals have hooves.

The head contains the main sense organs and feeding structures. The **mouth** is bounded above and below by lips which are muscular and can move in various ways. The mouth is opened and closed by powerful **jaws**.

Just inside the mouth are the **teeth**. These are not all the same, as they are in lower vertebrates, but are of different kinds. Each kind has a particular job to do, either cutting, crushing or tearing the food.

The mouth opens into the **buccal cavity**. Attached to the floor of the buccal cavity is the **tongue** which, like the lips, can be moved by muscles.

Just above the mouth is a pair of **nostrils**; most mammals have a very good sense of smell and the cat is no exception. On either side of the nose are the **whiskers**; these are sensitive to touch and enable the cat to prowl in total darkness without bumping into things. The **eyes** are positioned high up on the head. On each side of the head is an opening leading into the **ear**. Behind this opening is a flap of skin called the **pinna**. The ear is used for hearing, and the pinna directs sound waves into the opening.

At the hind end there is the **tail**. This helps with balance, though many mammals can get along quite well without one. The kangaroo uses its tail to support itself when resting, rather like a 'shooting stick', and certain monkeys use it for swinging from trees. The tail is also used as a means of expression: for example, cats tend to flick their tails when they are cross.

Just below the tail is the **anus**, and below that are the **genital organs**. The male genital organs consist of a **penis** and a pair of **testes** which make sperms. The testes hang from the body in a bag called the **scrotal sac**.

The penis does two jobs: through it urine is passed to the outside, and during mating sperms pass through it into the female. In contrast, the female has two separate openings. Through one of them urine is passed. The other one is the reproductive opening and leads into a tube called the **vagina**: when

Figure 1 This picture shows the main external features of a cat. Which of the structures in bold print on this page can you see in this picture?

Figure 2 Piglets being suckled by a mother pig. Once a baby mammal is able to consume food other than the mother's milk, it is said to be weaned.

mating takes place the male inserts his penis into the vagina and later the young are born through it.

On her underside the female has two rows of **teats**. After the young have been born, they suck the mother's teats and obtain milk from them (Figure 2). This is called **suckling**, and it nourishes the young until they are able to eat solid food. The milk is produced by special **mammary glands** from which mammals get their name. Once the youngsters can eat solid food they are said to have been **weaned**.

Man as a mammal

Man differs from the mammal just described in three main ways.

Firstly, he learns to stand upright on his hind legs: he is **bipedal**. His front legs take the form of arms, with hands and fingers at the ends. The hands are used for grasping and holding things. Monkeys and apes are similar.

Secondly, man has very little hair. In consequence his ability to keep warm in cold weather is very poor compared with other mammals, and this is why we wear clothes.

Thirdly, in the human female there are only two teats: they are called **nipples** and are located on the **breasts**. As in other mammals, their job is to produce milk for feeding the young. The male has nipples too but they are rudimentary and have no function.

Different types of mammals

There are two main kinds of mammals: **pouch mammals** and **placental mammals.**

Pouch mammals, otherwise known as **Marsupials**, live mainly in Australia. An example is the kangaroo (Figure 3). In the belly region of the female there is a fold of skin which forms the **pouch**. The young are born at an early stage when they are helpless, but they manage to crawl into the mother's pouch. The teats are inside the pouch and here the young feeds on her milk and is carried around until it can look after itself.

Most mammals are placental mammals. Dogs, cats, horses, giraffes, elephants, monkeys and man all belong to this group. They get their name from the fact that they possess a **placenta**. This is a special structure which develops in the mother's womb, through which the embryo gets food and oxygen. As a result the young spend a longer time in the womb and are born at a more advanced stage than they are in pouch mammals: indeed in some cases they can walk straight away (Figure 4).

Figure 3 A kangaroo with a baby in its pouch.

Figure 4 This foal was born two hours before this photograph was taken. Though rather unsteady, it can walk straight away.

Name of group	Main features	Examples
Rodents	Small mammals with a pointed nose. Have chisel-like front teeth for gnawing at mainly plant food such as nuts.	Rats, mice, gerbils, hamsters, squirrels. (Rabbits and hares are close relatives but are put in a separate group.)
Carnivores	Meat-eaters. Have sharp claws for bringing down prey, and strong jaws and pointed teeth for eating it. Good hunters.	Cats (including lions and tigers) dogs, wolves, foxes, bears.
Ungulates	Eat plants, particularly grass. Have hooves on their feet. Cheek teeth have a flat ridged surface for grinding food.	Cattle, sheep, pigs, deer, camels, giraffes, hippopotamus, horses, zebras, rhinoceroses.
Proboscideans	Very large, with legs like pillars. The two upper front teeth form a pair of tusks and the nose is lengthened into a trunk. Have huge cheek teeth for grinding up tough plant food.	Elephants. (Some specimens weigh over 10 tonnes.)
Chiropterans	Have wings for flying and small pointed teeth for feeding on insects, fruit etc.	Bats. (The vampire bat sucks blood.)
Cetaceans	Streamlined and fish-like in appearance for swimming. Have sharp pointed teeth for feeding on fish, squid etc. Some whales feed on plankton.	Whales, dolphins, porpoises. (The Blue Whale can be over 30 m long and weigh more than 150 tonnes.)

Figure 5 The main groups of mammals.

Figure 6 The bat's arms take the form of a pair of wings with which it flies through the air.

Figure 7 Besides being superb swimmers, dolphins are intelligent and can be taught tricks.

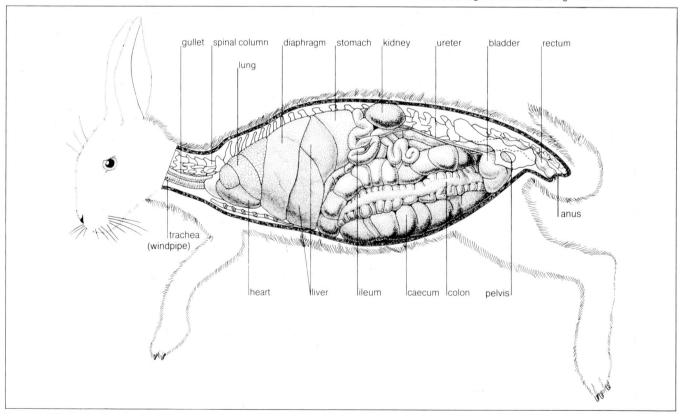

Figure 8 This illustration shows some of the main structures found inside the body of a mammal.

Figure 5 shows the main groups of placental mammals. You will see that there is a great deal of variety, particularly in the way they feed. Most of them walk about on land, but this group also includes some of nature's most expert flyers and swimmers such as the ones shown in figures 6 and 7.

Inside the mammal's body

Figure 8 shows the inside of the body of a mammal. The various organs and other structures shown in this diagram you can see for yourself by cutting open a dead mammal such as a rat (Investigation). This procedure is called dissection, and it is one of the main ways of finding out about the anatomy of animals. Students training to become vets dissect all manner of animals such as dogs, cats, sheep and even horses, and medical students dissect the human body.

In later Topics we will look in detail at some of the organs of the human body and explore how they work.

Investigation

Dissecting a mammal

Your teacher will probably do this as a demonstration.

1 Obtain a mammal such as a rat which has been killed with chloroform.

2 Pin the animal through its legs to a board so its belly side is upwards.

3 With scissors cut through the skin in a line running up the middle of the body.

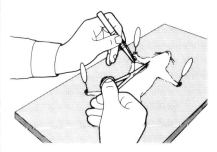

4 Free the skin from the underlying body wall and pin it back on either side.

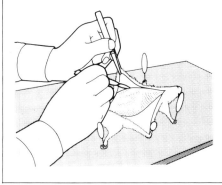

5 Now do the same with the body wall so you can see into the abdominal cavity.

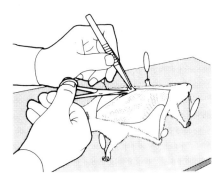

6 With scissors carefully remove the rib cage so you can see into the thorax

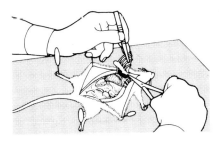

7 Study the internal anatomy of the animal, pushing the organs this way and that to see the structures lying underneath.

Which of the structures shown in Figure 8 can you see?

Assignments

1 Make a list of five characteristics of mammals, not possessed by other animals, which have helped make them so successful.

2 Make a list of all the functions you can think of which are performed by your lips.

In each case, say whether or not the lips perform the same function in other mammals.

Do you think lips have helped to make mammals successful?

3 Choose one animal listed in the right hand column of Figure 5. Find out as much about it as you can, and then write a short essay on it.

4 Give reasons for each of these observations:

a) Mammals are usually born without teeth.
b) Whales have to come up to the surface of the water occasionally.
c) A baby lion is born at a more advanced stage of development than a baby kangaroo.
d) A bat may be stunned by the sound of a gunshot.

5 What job does each of the following structures do:

a) a cat's paws,
b) a bat's toes,
c) a dolphin's front legs,
d) an elephant's trunk,
e) a squirrel's front teeth?

Flowering plants

Of all the kinds of plants which have been described, well over two thirds are flowering plants. The flower is their reproductive device and it produces seeds. In this respect flowering plants differ from all other plant groups

External structure of a flowering plant

Figure 1 shows the structure of a typical flowering plant. The diagram is based on plants like the lupin, foxglove or willowherb. There are many variations on this theme, as you will see if you look at some plants yourself (Investigation 1).

The plant is made up of two main parts: the **shoot** and the **roots**. The main root is called the **taproot** and this gives off lots of short side roots. Near the tip of each root there is a covering of fine **root hairs**, which can only be seen properly with a magnifying glass or microscope.

The roots grow down into the soil. They do two jobs: they anchor the plant, and they absorb water and minerals from the soil.

Some plants, such as grasses, do not have a taproot. Instead they have a bundle of slender roots which spread out from the base of the stem. These are called **fibrous** or **adventitious roots**.

The main part of the shoot is the **stem**. At the top there is an **apical bud**, where growth takes place.

Leaves stick out from the sides of the stem. They are flat and green and their job is to make food by photosynthesis. The green colour of the leaves and the stem is due to the presence of **chlorophyll** inside them.

The leaves have a network of **veins** which stiffen them and help to prevent them drooping. In some plants the leaves and stem are hairy or have spines which help to protect them from attack by insects and other animals which might otherwise damage them.

Each leaf is attached to the stem by a short **leaf stalk**. The leaves are positioned at regular points along the stem: we refer to these points as **nodes**. At each node you can usually see a small bud: this is called an **axillary bud** because it is at an angle, or axil, between the leaf stalk and the stem.

The plant shown in Figure 1 is in flower. The **flowers** are formed towards the upper end of the shoot. You will notice that there is a clear sequence from the top downwards. Right at the top there is a cluster of unopened buds with the apical bud in the middle. Further back you can see flowers. Further back still **fruits** are visible. This sequence reflects how the plant develops: as the stem grows upward new buds are formed, they open into flowers, and the flowers give rise to fruits. The fruits contain the **seeds** which will give rise to new plants.

Eventually, the apical bud develops into a flower. When this happens the main stem stops growing. Axillary buds, on the other hand, are usually dormant and do not develop into anything. In some plants however, they sprout into side branches, giving rise to a bushy kind of plant. In many plants you can make this happen by cutting off the apical bud (see page 293).

Different types of flowering plants

The seed of a flowering plant contains a tiny embryo. Attached to the embryo inside the seed are one or two leaf-like structures, called 'seed leaves' or

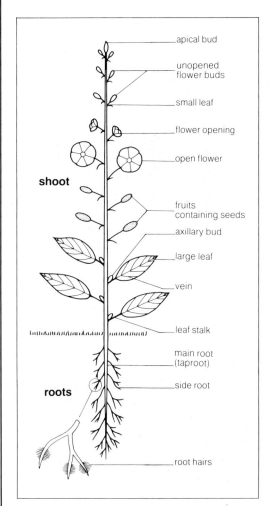

apical bud

unopened flower buds

small leaf

flower opening

open flower

shoot

fruits containing seeds

axillary bud

large leaf

vein

leaf stalk

main root (taproot)

side root

roots

root hairs

Figure 1 Diagram of a generalised flowering plant.

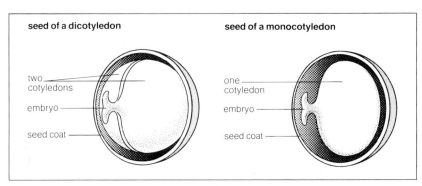

seed of a dicotyledon | seed of a monocotyledon

two cotyledons

embryo

seed coat

one cotyledon

embryo

seed coat

Figure 2 Dicotyledons have two cotyledons inside their seeds whereas monocotyledons have only one.

cotyledons (see page 322). These feed the embryo when it grows into a new plant.

Flowering plants are divided into two great groups depending on how many cotyledons they have inside their seeds (Figure 2). Most flowering plants have two cotyledons in their seeds. They are called **Dicotyledons**, or 'dicots' for short. This group includes buttercups, dandelions, daisies, sweet peas and trees such as oak, beech and elm.

Other flowering plants have only one cotyledon in their seeds. They are called **Monocotyledons**, or 'monocots' for short. This group includes plants like daffodils, tulips, grasses and palm trees. Grasses include important cereal crops such as wheat and barley.

How can we tell whether a plant is a 'dicot' or a 'monocot'? One way is to look inside its seeds and see if they contain one or two cotyledons. However, a much easier way is to look at its leaves (Investigation 2). It so happens that 'dicots' usually have broad leaves with a branching network of veins. 'Monocots' on the other hand, have narrow leaves with parallel veins (Figure 3).

Herbs, shrubs and trees

You have only to look at plants growing in a park or a wood to appreciate that they vary tremendously in size. Close to the ground there are small plants like primroses and forget-me-nots, whilst towering above are massive trees like the oak. We can make a distinction between three kinds of flowering plants: **herbs**, **shrubs** and **trees** (Figure 4).

Herbs do not contain much wood and are generally rather small; they range in height from a few centimetres to about a metre. They include plants such as parsley, sage and thyme whose leaves are used for flavouring food.

Shrubs are larger. They contain a good deal of wood and may reach several metres in height. Usually they have a bushy appearance. Examples are hawthorn which is common in hedgerows, and the familiar privet hedge which many people grow around their gardens.

Trees are larger still. They have an extremely woody main stem or **trunk**, which is very strong and can hold up tremendous weights. The trunks of certain trees provide man with timber.

The size differences between these three types of plants are not all that sharp. A large shrub can be as tall as a small tree, and certain herbs grow surprisingly tall. For example, sunflowers can reach a height of well over 3 metres which is as tall as a hawthorn bush.

Another way herbs differ from shrubs and trees is that the shoot usually dies down at the end of the season. This is discussed in the next section.

How long do flowering plants live?

Some plants get through their life cycle – that is they grow, produce seeds, then die – within one year. Such plants are called **annuals**. Many garden plants are annuals, for example marigolds, nasturtiums and petunias, and so are cereals such as wheat and barley.

Some plants complete their life cycle so quickly that three or four generations are produced within one year. Such is the case with groundsel, a tiresome weed which reproduces and spreads very quickly.

Some plants take two years to complete their life cycle. They are called **biennials**. In the first year they send up a leafy shoot, but they do not produce flowers and seeds until the second year, after which they die. Examples are wallflowers and carrots.

Some plants go on and flower year after year. They are called **perennials**. There are two kinds: in **herbaceous perennials** the shoot produces flowers and seeds, then dies down. However, the underground part of the plant lives through the winter and sends up new shoots the following year (Figure 5). Chrysanthemums and Michaelmas daisies are two well-known examples of this.

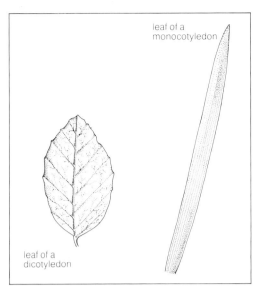

Figure 3 Dicotyledons usually have broad leaves with branching veins, whereas monocotyledons have narrow leaves with parallel veins.

Figure 4 Here you can see herbs, shrubs and trees. As well as being smaller, herbs differ from shrubs and trees in dying down at the end of the season.

Figure 5 A perennial plant like the one shown here produces flowers and seeds every summer. The underground parts of the plant remain dormant in the soil during the winter.

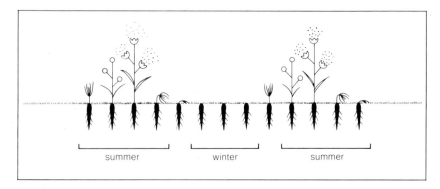

In shrubs and trees the shoot does not die down but remains alive permanently. It produces flowers and fruits every year. We call these **woody perennials**.

Some woody perennials live for a very long time. For example, there are oak trees in England which were already large at the time of Elizabeth I, and some may go back even earlier (Figure 6).

Suppose you have a garden and you are trying to decide whether to grow annuals or perennials. Which would you choose?

Annuals make a brighter splash of colour, particularly if planted neatly and close together along the edge of a lawn or in window boxes. The trouble is that new ones have to be planted every year.

For a lazy gardener, like me, perennials have a great advantage: they may not look as neat as a bed of carefully planted annuals, but they put on a show every year and don't need much attention.

Deciduous and evergreen plants

We all know that before winter comes many plants drop their leaves, and new ones are formed the following spring. Such plants are known as **deciduous**. The best known examples are trees like beech and elm.

Other plants hold onto their leaves throughout the winter. We call them **evergreens**. Holly is an example.

Being evergreen does not mean that the plant never drops its leaves. Eventually the leaves do die and fall off and are replaced by new ones, but not all at the same time. In fact, if you look under a holly tree you will see plenty of dead leaves lying on the ground.

Recognising flowering plants

Scientists have discovered over a quarter of a million different species of flowering plants. How can we tell one from another?

In general, we recognise a plant by looking at three main features: its leaves, its flowers, and, in the case of trees and shrubs, its bark. Perhaps the easiest way of telling one plant from another is by looking at the shapes of their leaves (Investigation 3). The leaves of some common trees are shown below.

Figure 6 This oak tree in Sherwood Forest was probably there at the time of Robin Hood.

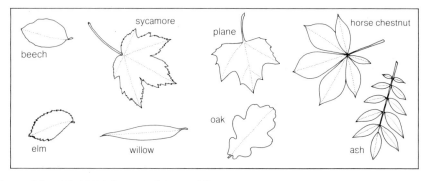

Investigation 1

Looking at a flowering plant

1 Obtain a complete plant (dicotyledon), with, if possible, roots, leaves and flowers.

2 Examine it carefully.

Which of the structures in Figure 1 can you see?

How does your plant differ in appearance from the one shown in Figure 1?

3 Make a diagram to show the position of the leaves, flowers and buds on the stem of your plant.

4 Examine several other plants.

In each case observe how the leaves, flowers and buds are arranged.

Investigation 2

Comparing 'dicots' and 'monocots'

1 Obtain two plants, one a dicotyledon and the other a monocotyledon.

2 Compare their leaves, looking carefully at the arrangement of the veins.

How do they differ?

3 In what other ways do the two plants differ?

Investigation 3

Recognising trees

1 Obtain leaves from various trees.

2 Identify the kind of tree which each leaf comes from.

3 Look at the bark of each kind of tree.

How do the barks differ?

4 Briefly describe each type of bark, so that someone could identify the trees from their bark.

Assignments

1 What jobs are done by each of these structures: leaves, flowers, roots, stem and buds.

2 Each of the activities listed on the left below is related to one of the words on the right.

Write them down in the correct pairs.

sweeping up leaves	annual plants
felling trees	deciduous
pruning roses	evergreen
harvesting wheat	wood
collecting holly	axillary buds

3 Normally, trimming a hedge does not kill it. Why is this?

Are there any circumstances in which cutting a hedge might kill it?

4 What is the difference between annual and perennial plants?

Give one example of each.

What are the advantages and disadvantages of having lots of perennial plants in your garden?

5 How can you tell the difference between:

a) a monocotyledon and a dicotyledon,
b) a herb and a shrub,
c) a shrub and a tree,
d) the leaves of a sycamore and a plane tree,
e) an elm tree and a beech tree in winter?

6 The photograph (right) shows the shoot of a willow herb, the same plant that you can see in the foreground in Figure 4.

a) Why is this plant described as a herb?
b) Give the names of the structures labelled A–E.

Note: this plant is not the same as a willow *tree* which is an entirely different thing.

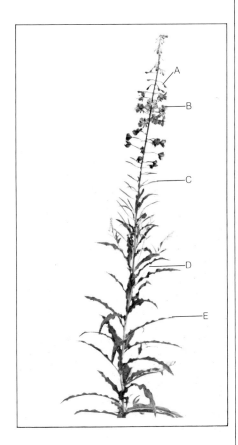

Maintaining life

*This cat has just caught
a sparrow. Feeding is just one
aspect of maintaining life, which
is the subject of the next
series of topics.*

Molecules in motion

In this Topic we shall see how molecules and other tiny particles move about, and why this is important in biology.

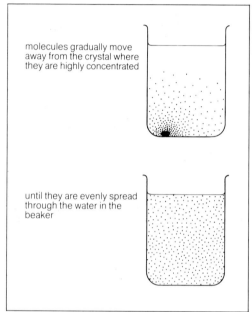

molecules gradually move away from the crystal where they are highly concentrated

until they are evenly spread through the water in the beaker

Figure 1 An example of diffusion in a liquid: this is what happens when you drop a crystal of potassium permanganate in water.

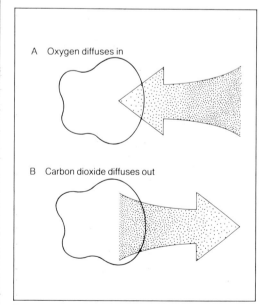

A   Oxygen diffuses in

B   Carbon dioxide diffuses out

Figure 2 Amoeba takes in oxygen and gets rid of carbon dioxide by diffusion. Many other cells and organisms undergo gaseous exchange this way.

Diffusion

Suppose you are sitting at one end of a room. A woman wearing a lot of perfume comes in and sits down at the other end. Before long the smell of her perfume fills the room and reaches your nose. Why does this happen?

Molecules in a gas or liquid are constantly moving about freely, bumping into one another and bouncing this way and that. This takes place randomly, and it results in the molecules being spread out evenly. *The tendency for molecules to become distributed evenly throughout the space they occupy is called diffusion.*

The way a smell spreads through a room is an example of molecules diffusing through air, i.e. through a gas. However, diffusion also takes place in liquids, and this is its main importance in biology.

Diffusion in a liquid

You can watch diffusion taking place in a liquid by dropping a crystal of potassium permanganate into a bowl of water. Gradually the purple colour of the permanganate spreads through the water until eventually all the water is the same shade of purple.

What exactly has happened? In the crystal, the permanganate molecules are packed tightly together, i.e. they are very concentrated. However, to begin with, the surrounding water contains no permanganate molecules at all. As a result the permanganate molecules move away from the crystal until they are evenly distributed throughout the water (Figure 1).

So diffusion results in molecules moving from a region where they are highly concentrated to a region where they are less concentrated. The difference in concentration between the two regions before diffusion occurs is known as the **diffusion gradient**. It's like a hill: molecules at the top (high concentration) roll down the slope into the valley (low concentration). Provided such a gradient exists, molecules will always tend to diffuse in this way.

Diffusion is extremely important in living things, because it is the main way by which substances move from one place to another. Let's take an example.

An example of diffusion in biology

In an animal such as an amoeba, oxygen is continually being used up. The result is that oxygen molecules are less concentrated inside the body than in the surrounding water. As a result, oxygen molecules constantly diffuse in (Figure 2A) and in this way the amoeba gets all the oxygen it needs for respiration.

Meanwhile carbon dioxide is continually being formed. The result is that carbon dioxide molecules are more concentrated inside the body than in the surrounding water. As a result carbon dioxide molecules constantly diffuse out (Figure 2B) and in this way the amoeba gets rid of carbon dioxide.

In order for molecules to diffuse in and out of the amoeba like this, the cell membrane must let them pass through without hindrance – in other words the membrane must be permeable to them.

How fast is diffusion?

Diffusion is a rather slow process. However, it can be speeded up by raising the concentration of the substance which is diffusing – in other words by making the diffusion gradient steeper. Diffusion can also be speeded up by moving the molecules in some way. For example, stirring or heating the contents of the beaker would help to spread out the permanganate molecules, and a gentle breeze would help to spread a lady's perfume around the room. The size of the molecule is also important: small molecules such as oxygen diffuse more rapidly than large molecules such as glucose.

In living things diffusion is sometimes speeded up by pumping the molecules across the cell membrane. This is called **active transport**. No one knows exactly how it takes place, but it requires energy from respiration. In some cases active transport will move molecules from a region of low concentration to a region of higher concentration. For example, this is how plant roots obtain mineral salts from the soil (see page 207).

Diffusion and surface area

Imagine that the box below is an organism:

Now suppose we double the size of the box, like this:

By how much have we increased its volume, and its surface area?

The answer is that we have doubled its volume, but its surface area is *less* than doubled. This is because, in the process of doubling the volume, we have lost part of the original surface (the part shaded in the first diagram), so we can make this statement: *as an object increases in size the amount of surface relative to volume (the surface-volume ratio) gets smaller.*

This is extremely important to organisms which take things in by diffusion. Think of it this way. A small organism, like an amoeba, has a large surface-volume ratio, and so it can take in all the oxygen it needs by diffusion across the body surface. However, a large organism, like a mammal, has a much smaller surface-volume ratio, so it cannot get all the oxygen it needs in this way. Such large organisms need special respiratory organs such as lungs for taking in oxygen. These respiratory organs consist of a sheet of tissue which is folded many times so that it provides a large surface area across which oxygen can be absorbed (Figure 3). The idea of organisms creating a large surface area for absorption is one that we shall meet again and again.

Osmosis

Carry out Investigation 1. This involves making a bag out of a thin membrane, filling it with sugar solution and suspending it in a beaker of water. After a short time, water enters the bag from the beaker, passing through the membrane.

To understand why this happens, look at Figure 4. The sugar molecules are larger than the water molecules. The bag itself has tiny holes in it which are large enough to let the small water molecules through, but too small to let the larger sugar molecules through. Such a membrane is described as a **semi-permeable** membrane.

Now the presence of the sugar molecules in the bag means that there isn't as much room for water molecules there. So the water molecules inside the bag are less concentrated than in the beaker outside. As a result, water molecules *diffuse* into the bag.

This movement of water is called **osmosis**. *Osmosis is the flow of water through a semi-permeable membrane*. It's really a special case of diffusion, in which only the water molecules move from one region to another.

Osmotic pressure

The inward movement of water just described results in a pressure being built up inside the bag. We call this **osmotic pressure**. *Osmotic pressure is the pressure developed by a solution when water moves into it across a semi-permeable membrane.*

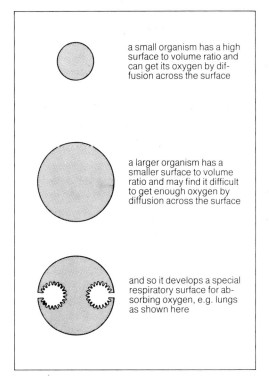

Figure 3 These diagrams show how a large organism can overcome the problem of having a small surface to volume ratio.

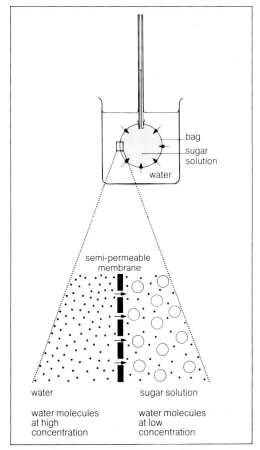

Figure 4 Osmosis, a special case of diffusion.

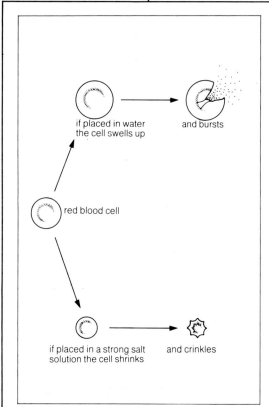

Figure 5 These diagrams show what happens if you put a red blood cell in water or a strong salt solution.

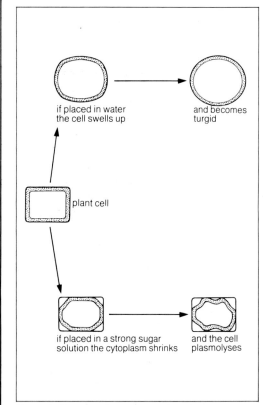

Figure 6 These diagrams show what happens if you put a plant cell in water or a strong sugar solution.

Any solution of sugar, salt or some similar substance is capable of exerting an osmotic pressure. If the solution is in a bottle sitting on a laboratory shelf, it obviously isn't exerting a pressure. However, it *would* do so if it was separated from pure water by a semi-permeable membrane. The pressure which a solution is *potentially* capable of exerting is called its **osmotic potential**.

Strong and weak solutions

Look at Figure 4 again for a moment. For water to move into the bag from outside, the fluid in the beaker doesn't *have* to be water. It could be another sugar solution. All that's necessary is that the solution inside the bag should have a greater concentration of sugar molecules, ie. be stronger, than the solution in the beaker.

What would happen if the situation was reversed, and the solution in the beaker was stronger than the one in the bag? The answer is that osmosis would occur in the other direction, and water would flow *out* of the bag. As a result it would lose mass and volume and go flabby.

Osmosis and animal cells

The effect of osmosis on an animal cell can be investigated by putting red blood cells in solutions of different strengths.

As with other types of animal cells, red blood cells contain a solution of salts and other substances. These are enclosed inside the cell membrane which is semi-permeable.

If a red blood cell is put in water, water enters the cell by osmosis. The cell swells up and eventually bursts, just as a balloon would do if you blew too much air into it.

However, if you put a red blood cell in a salt solution which is stronger than that inside the cell, water leaves the cell by osmosis. As a result the cell shrinks and crinkles (Figure 5).

This is very important in our bodies. It means that the fluid part of the blood (the plasma), in which the cells float around, must have just the right strength to prevent osmosis occurring in either direction. In a later Topic we shall see how this is achieved.

Osmosis and plant cells

Now consider what happens if you put a plant cell in solutions of different strengths.

A plant cell has a cellulose wall as well as a cell membrane. In the centre of the cell there is a vacuole which contains a solution of salts and so on. The thin layer of cytoplasm surrounding the vacuole acts as a semi-permeable membrane. The cellulose wall, however, is fully permeable to salts, as well as water.

If you put a plant cell in water, water enters by osmosis and the cell swells up. However, it doesn't burst. This is because the cellulose wall is tough, like elastic: it stretches but does not break. Eventually the cellulose cannot stretch any more and so the cell stops swelling. It's like trying to blow up a football into which no more air can be pumped. When this point is reached, we say the cell is **turgid**. Turgidity, or turgor as it's called, is very important in plants because it helps to make them firm.

Now what happens if you put a plant cell into a solution which is stronger than that in the vacuole? In this case water is drawn out of the vacuole and the cell shrinks: it loses its turgor and becomes flabby or flaccid. If the external solution is strong enough, the cytoplasm eventually pulls away from the cellulose wall as shown in Figure 6. We call this process **plasmolysis**.

In this Topic we have seen how molecules move as a result of diffusion and osmosis. You will find many examples of these processes in action in other parts of this book.

Investigation 1

Watching osmosis

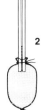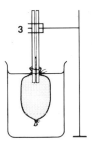

1 Cut a length of visking tubing about 8 cm long.

2 Wet it thoroughly with water.

3 Tie one end with strong thread, so that it forms a bag.

4 Fill the bag with a 20 per cent solution of sucrose (Illustration 1).

5 With a piece of thread, tie the bag to the bottom of a capillary tube (Illustration 2).

6 Clamp the capillary tube to a stand, and lower the bag into a beaker of water (Illustration 3).

7 Mark the level of the sucrose solution in the capillary tube.

8 Five minutes later, with a ruler, measure the distance which the sucrose solution has risen from the original mark. Write down the distance in millimetres.

9 Re-measure the distance every five minutes for about half an hour. In each case write down the distance the sucrose has risen from the original mark.

10 Plot your results on graph paper: put the distance the sucrose has risen on the vertical axis, and time on the horizontal axis.

Why does the sucrose solution rise in the capillary tube?

What property of the visking tubing makes this happen?

Investigation 2

The effect of osmosis on a plant cell

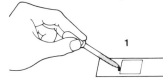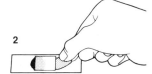

1 Obtain a stem of rhubarb or some other plant with a red epidermis.

2 With forceps, strip off a piece of the red epidermis from the stem.

3 Trim the piece of epidermis with scissors so that it is about one centimetre square.

4 Put the piece of epidermis in a drop of water on a slide, and cover it with a coverslip.

5 Look at your slide under the low power of the microscope.
Can you see the cells clearly?
Notice that each cell is filled with a red substance: this is inside the vacuole and it makes the whole stem look red.

6 With a pipette, place a drop of strong sucrose solution against one side of the coverslip (Illustration 1).
The sucrose solution will flow under the coverslip by capillary action.

7 Put a piece of filter paper against the other side of the coverslip, and draw the sucrose solution across. (Illustration 2).

8 Look at the epidermis cells under the microscope.

How does their appearance change?

What happens to the red substance inside them?

Explain your observations.

Assignments

1 Suggest an explanation for each of the following:

a) If certain kinds of lettuces get floppy, they can be made firm and crisp by putting them in cold water for a while.

b) If you sprinkle sugar on a bowl of strawberries, the juice comes out of them.

2 *Use the index where necessary to answer this question.*
Give one example of each of the following:

a) gaseous diffusion in a leaf;
b) gaseous diffusion in the human body;
c) gaseous diffusion in an insect;
d) diffusion through a liquid in a plant;
e) diffusion through a liquid in a single-celled animal;
f) diffusion through a liquid in the human body;
g) osmosis in a root;
h) osmosis in a single-celled animal;
i) a large surface area for absorption in a plant;
j) a large surface area for absorption in the human.

3 Look up how an amoeba controls its water content (page 25). Write a brief explanation of why water enters the amoeba's body, using these words: semi-permeable membrane, osmosis, osmotic pressure, concentration, salts.

4 A pupil in a school carried out the following experiment. He cut out a rectangular piece of potato 20 mm long, and put it in a dish of strong sucrose solution. Four hours later he found that the piece of potato had shortened so that it was now only 16 mm long.

a) Suggest an explanation for this result.

b) What should the control be in this experiment?

5 When a doctor injects a fluid into a person's bloodstream, he always uses a fluid which has the same osmotic potential as the blood. Why?

Food and diet

*In the course of a lifetime,
a person may eat as much as
100 tonnes of food. The next two
Topics are all about food:
what it is, and what
it does for us.*

Figure 1 A person may eat 100 tonnes of food in his or her lifetime.

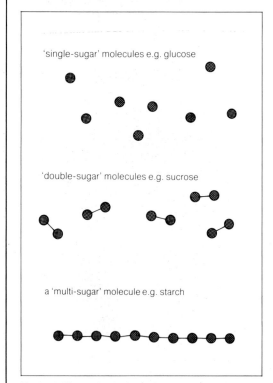

'single-sugar' molecules e.g. glucose

'double-sugar' molecules e.g. sucrose

a 'multi-sugar' molecule e.g. starch

Figure 2 These three kinds of carbohydrate occur in nature.

Why do we need food?

We need food for four main reasons:

1 It serves as fuel, giving us energy and warmth.
2 It enables us to grow and repair and replace our tissues.
3 It provides us with important substances which control our metabolism.
4 It keeps us healthy, helping us to fight disease.

These functions of food apply not just to man, but to other organisms too.

Our diet

The food we consume each day makes up our diet. This includes things we drink as well as those we eat. Whatever we choose to eat, our diet must include the following substances: carbohydrates, fats, proteins, water, minerals and vitamins. These substances give us nourishment, and we call them **nutrients**. Let's look at each in turn, and see what happens if any of them is missing. This is a large subject, so we will deal with the first four substances in this Topic, and the last two in the next one.

Carbohydrates

All carbohydrates contain the elements carbon, hydrogen and oxygen. They include sugar, starch and cellulose: the way they relate to each other is shown in Figure 2.

Sugar

Different kinds of sugar occur in different foods. In fruit the sugar is fructose or glucose, in milk it is lactose. Ordinary table sugar is sucrose, obtained from sugar cane and sugar beet. Normally sugar tastes sweet, which is why it is so popular. Sugar gives us energy, so we call it an energy food.

In its natural state, sugar is normally liquid – think of the sugar in orange juice for example. However, when sugar loses water, it forms crystals. This is what happens when table sugar is made. Juice is extracted from sugar cane or beet and then it is purified (refined). After that, water is evaporated from it, so sugar crystals are formed. The crystals quickly dissolve if placed in water – as you know when you put sugar in a cup of tea.

Brown sugar is less refined than white sugar: it contains various impurities which make it brown and slightly sticky. These impurities do us no harm, indeed they make the sugar better for making cakes, biscuits and toffee.

Starch

Starch is found in bread, potatoes, cereals and many other plant foods. It exists naturally in the form of small granules called **starch grains** (see page 15). We cannot digest starch until it is cooked. Each starch grain is enclosed within a membrane and cooking causes the starch grains to swell up and burst. Like sugar, starch gives us energy, so starchy foods are also energy foods.

Cellulose

Cellulose forms the cell walls of plants, and is very tough. For this reason, plants are often difficult to chew, but cooking softens the cellulose and makes it easier to eat.

Man cannot digest cellulose – we don't have the necessary enzymes in our gut for breaking it down. This means that we cannot get energy from it, but it still performs a useful function: together with other indigestible matter, it forms **roughage** (fibre). Roughage keeps food moving along the gut and prevents constipation (see page 140).

In advanced countries people tend to eat a lot of highly refined foods which

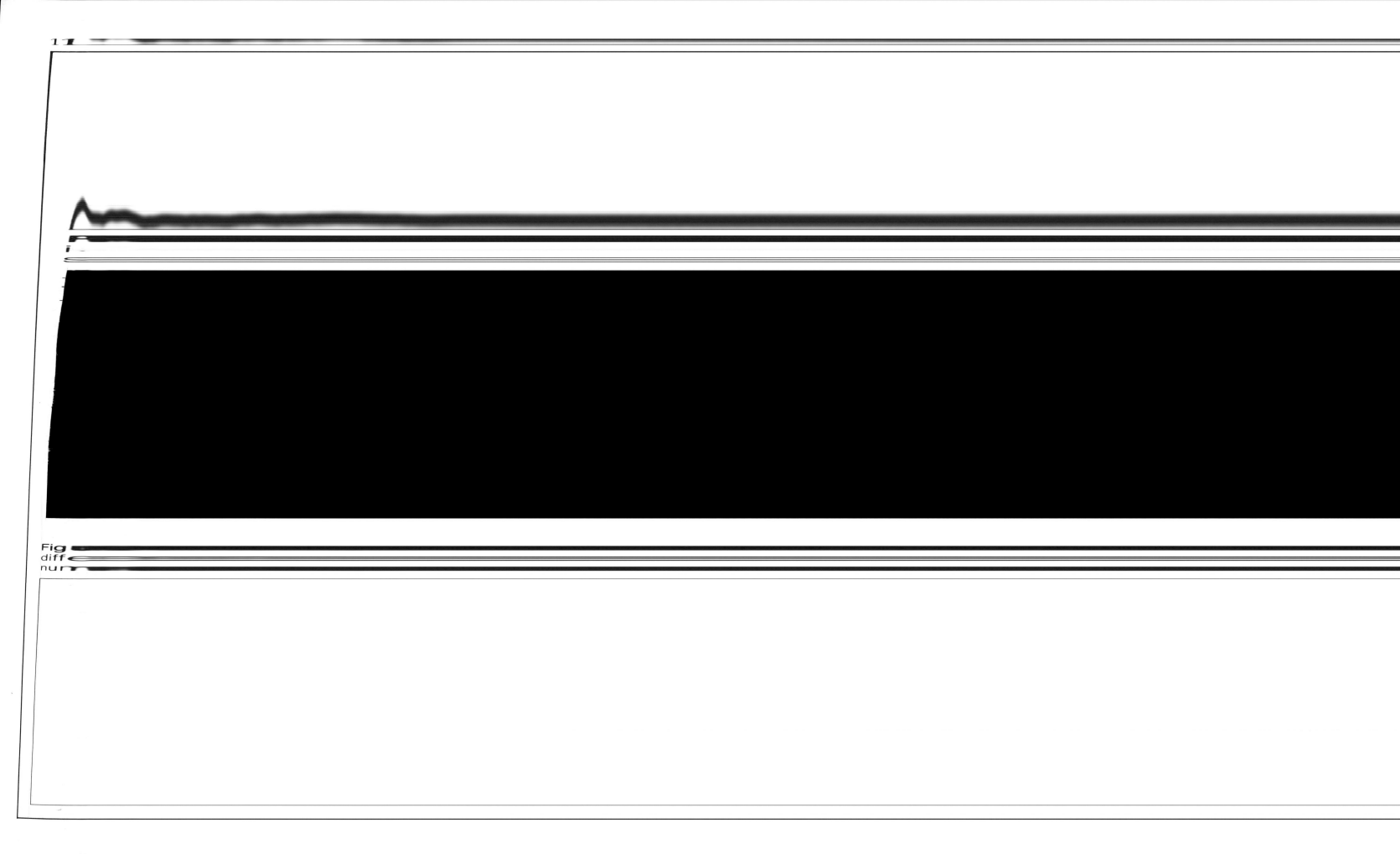

made by the body itself; it is made in the skin provided sunlight is present. In a sunny climate an adult can get all the vitamin D he needs this way.

You will now realise how important vitamins are, and how serious it can be if we don't get enough of them. Unfortunately vitamin deficiency diseases are common in poor countries. Even in advanced countries they occur from time to time. Those at risk include old people living alone, and people who refuse to eat certain kinds of food – vegetarians for example. It's a good idea for such people to buy vitamin tablets from a chemist's shop.

Investigation

Testing food for vitamin C

1 Obtain a lemon, and squeeze some of its juice into a beaker.

2 Pipette one drop of blue DCPIP solution onto a white tile*.

3 With a pipette or syringe add lemon juice to the DCPIP solution, drop by drop, and stir with a needle. Count how many drops of lemon are needed to make the DCPIP solution turn colourless.

The disappearance of the blue colour tells us that vitamin C is present in the lemon juice.

4 Use this test to compare the vitamin C content of different foods. In each case get some juice out of the food. Then find out how many drops of the juice are needed to decolorise one drop of DCPIP solution.

Do you think this is an accurate way of comparing the vitamin C content of different foods?

How could you make the experiment more accurate?

Why can't this test be done with blackcurrant juice?

5 Boil some lemon juice in a test tube and then test it for vitamin C with DCPIP solution.

What effect does boiling have on vitamin C?

*DCPIP is short for 2, 6 dichlorophenol indophenol.

Assignments

1 Each of the diseases in the left hand column is caused by lack of one or more of the substances in the right hand column. Which causes which?

night-blindness	iron
rickets	vitamin A
anaemia	calcium
goitre	vitamin D
xerophthalmia	iodine

2 Explain the reason for each of the following statements.

a) Carrots are good for you.
b) Miners eat salt tablets.
c) Mothers give their children orange juice and cod liver oil.
d) Old people who live alone tend to get scurvy towards the end of the winter.

3 Read how Goldberger discovered the cause of pellagra on page 118, then answer these questions.

a) Does the fact that Goldberger did not get the disease *prove* that it was caused by a poor diet? Explain your answer.
b) Suggest one way in which Goldberger might have confirmed his conclusions.

4 Describe an experiment which could be done to find out if the husk surrounding the rice seed contains a substance which prevents beri-beri.

5 A scientist carried out an experiment to find the effect of heating cabbage on the amount of vitamin C in it. He covered the cabbage with water and brought it to the boil. He then estimated the vitamin C content at ten minute intervals.

Here are his results:

Time after being brought to the boil (minutes)	Vitamin C content (milligrams)
0	50
10	50
20	50
30	50
40	50
50	50
60	50
65	49
70	30
75	5
80	0

a) Plot these results on a piece of graph paper.
b) Suggest reasons why the vitamin C content of the cabbage falls.
c) What advice would you give to a chef about cooking vegetables?

6 The following table shows the daily amount of vitamin D required by different people. The figures are in 'international units' (iu).

Woman during first half of pregnancy	400 iu
Woman during second half of pregnancy	600 iu
Woman breast feeding her baby	800 iu
Child 1–3 years old	400 iu

Explain (a) what each of these people needs vitamin D for, and (b) why a woman's requirement goes up in the second half of pregnancy and when she is breast feeding her baby.

How is food stored?

The carrot shown is one of the largest ever grown in Britain. It weighs nearly 3.5 kilograms! Why should the carrot plant produce a structure like this? In this Topic we will try to answer this question.

Figure 1 In 1979 Mr. A. Howcroft grew a carrot which weighed 3.5 kg (7 lb 11½ oz) – a U.K. record.

Figure 2 An example of a plant storage organ: the onion. This is a closely packed bundle of swollen leaves. The onion in the photograph has been grown in the garden and is ready to be harvested.

Why do organisms store food?

We know that plants and animals use their food for providing energy and doing various other jobs. But any food left over is stored in their bodies. This enables them to survive when food is unavailable or scarce. In fact, by using the food stores in his body, a man can live for several weeks without eating anything. Many animals, particularly hibernating ones, can survive for much longer than this and many of them get through the winter in this way.

Where is food stored?

Organisms store a certain amount of food all over their bodies. However, most of it is packed away in special places. In man one of the main storage places is the liver.

In plants food is often stored in special storage organs which are formed by part of the plant swelling up. Storage organs may be formed from the roots, stems or leaves. The carrot in Figure 1 is a swollen root, and so is the radish. The potato, on the other hand, is a swollen stem, and the onion is a mass of swollen leaves (Figure 2).

A plant's storage organs can survive the winter and give rise to new plants the following year (see page 380). When a new plant sprouts from a carrot, potato or onion, food moves into it from the storage organ, giving it nourishment until it can make its own food by photosynthesis.

Plants also store food in their seeds. When a seed germinates, food passes from it into the new growing plant, giving it nourishment until it can support itself.

The same thing applies to the eggs of animals. Bird's eggs, for instance, contain a rich store of food in the yolk which is used by the growing chick before the egg hatches.

In what form do plants store food?

Green plants make **glucose** by photosynthesis. Some of this is used straight away. The rest is usually turned into **starch**. The starch is converted back into glucose when it's needed:

glucose ⇌ starch

Some plants turn their surplus glucose into other substances, such as oil, and some store it in the form of sugar itself. You can find out what kind of food is present in a particular plant by doing chemical tests on it (see page 115). Although other substances are often present, starch is the main storage substance of plants. Starch gives a blue-black colour with iodine and this is an easy way of showing where it occurs.

What is needed for glucose to be turned into starch?

What do you think happens if you mix a small amount of glucose with some potato juice? (Investigation 1). The answer is that glucose is turned into starch.

Obviously, then, the potato juice contains something which turns glucose into starch. Scientists have discovered that this 'something' is an **enzyme**.

The same enzyme is also present in leaves. You can show its action by putting a piece of de-starched leaf into a glucose solution and keeping it in the dark (Investigation 2). The leaf takes up the glucose and turns it into starch.

How is glucose turned into starch?

The glucose in a plant is in solution. Starch, on the other hand, is in the form of solid grains (Investigation 3). When glucose is turned into starch, the glucose molecules join up to form a long chain like a string of beads, and this chain curls up like a spring (Figure 3). In this way thousands of glucose molecules get packed together into a solid grain of starch (Figure 4).

When starch is converted back into glucose, the chain uncoils and the glucose molecules separate.

How do animals store food?

Animals get glucose from the food they eat. What do they do with surplus glucose which they do not need straight away? They turn it into a substance called **glycogen**. Glycogen is equivalent to starch in plants – in fact it is sometimes called 'animal starch'. Both are carbohydrates. Like starch, glycogen is made by glucose molecules joining together. Glycogen takes the form of tiny granules which are stored in the body's cells, particularly in the liver. Animals also store food as fat which is laid down beneath the skin.

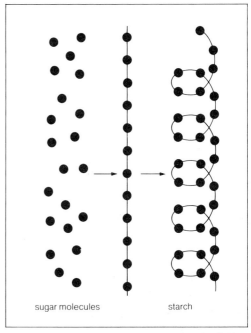

sugar molecules   starch

Figure 3 Starch is formed by sugar molecules linking together into a chain which coils up like a spring.

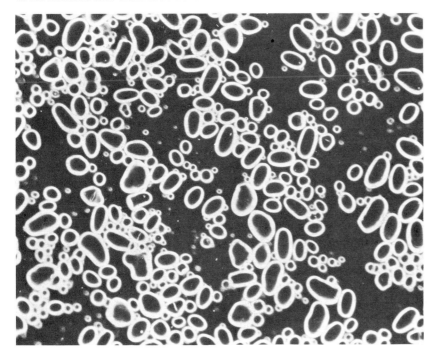

Figure 4 Starch grains inside the cells of a potato seen under the microscope.

Figure 5 This diagram shows what happens to the stored food when a new plant sprouts from a storage organ such as a potato tuber.

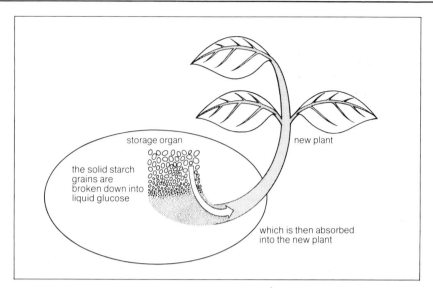

storage organ   *new plant*

the solid starch grains are broken down into liquid glucose

which is then absorbed into the new plant

Mobilising food stores

When food is stored it is usually in a solid form, but when the stored food is needed, it must first be made into a solution. Starch and glycogen are broken down into soluble glucose; fat is broken down into soluble acids. Only then can the stored food be moved to places where it is needed (Figure 5).

When stored food is changed into a form which can be moved, we say the food is being mobilised. It's rather like mobilising an army so it can be moved into action.

Why are food stores important to man?

When an organism stores food, the food substances are usually packed together in one place, often in concentrated form. Any part of an organism where food is stored can therefore be a rich source of food for man.

Take potatoes for example. We have seen that potatoes contain starch, which makes them a useful energy food; they also contain small amounts of other important nutrients. Potatoes are one of our most important food crops. They are easy and cheap to grow and are cultivated commercially as well as in people's gardens. Each plant bears about six to eight potatoes (Figure 6). These can either be eaten or planted so as to grow new potato plants.

Even more important is wheat which is the most widely grown crop in the world. Here starch, protein and a number of other useful nutrients are packed into the ripe seeds (the grains) which are clustered at the tops of the stems.

Wheat is a more concentrated food than potatoes, because wheat grains contain less water – in fact one kilogram of wheat has more food in it than three kilograms of potatoes. It is a general rule that seeds, being drier, are a more concentrated form of food than storage organs.

Another important food plant is sugar cane. This is a giant grass, like bamboo, and may grow to a height of six metres. Sugar, in the form of sucrose, is stored in its thick stem. Much of the world's sugar comes from sugar cane, but it will only grow in hot countries (see page 191). Sugar beet, on the other hand, will grow in cooler climates. This stores sugar in large swollen roots. Sugar beet is becoming more and more important as a source of sugar for man.

What about animal food stores – to what extent do we use them as a source of food? Two of the most valuable animal foods are eggs and liver, which both contain many useful nutrients.

Many other examples could be given, but the important principle is that any localised store of food in an organism can provide us with a valuable source of food.

Figure 6 Potatoes are one of our most important food crops.

Investigation 1

To see if potato juice will turn glucose into starch

1 Put a few pieces of potato pulp (not the skin) in a mortar. Add a pinch of washed sand and a little water.

2 Grind up the potato pulp with a pestle.

3 Filter the contents of the mortar into a test tube. This is your potato juice.

4 Put a drop of the potato juice onto a white tile. Test it for starch by adding a drop of iodine solution. (There should be no starch present in the potato juice.)

5 Put six drops of 0.5 per cent glucose-1-phosphate, side by side, on a white tile.
(Glucose-1-phosphate is an activated form of glucose.)

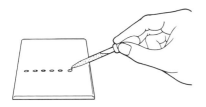

6 To each drop of glucose-1-phosphate add a drop of potato juice and mix.

7 To each drop of the mixture in turn, add a drop of iodine solution after 2, 4, 6, 8, 10 and 12 minutes.

What happens to the colour of each drop?

If it turns blue-black it means there is starch present. In which drops has starch been formed?

What is present in potato juice which turns glucose into starch?

Should this experiment have any further controls?
If so, what should they be?

Investigation 2

To see if a leaf will turn glucose into starch

1 Detach a leaf from a potted plant which has been kept in the dark for at least three days.

2 Test a small piece of the leaf with iodine to be certain there is no starch present (see page 185).

3 Place several small pieces of the leaf in a dish of 5 per cent glucose solution. Label this A.

4 Place several more pieces of leaf in a dish of water. Label this B: it is your control.

5 Put the dishes side by side in a dark place.

6 After several days test the two groups of leaf-pieces for starch.

Has either group of leaf pieces turned black? If they have, starch has been formed.

What is present inside the leaf which turns glucose into starch?

Investigation 3

Looking at starch in a potato

1 Slice open a potato to expose the white pulp.

2 Scrape off a very thin piece of the pulp and place it on a slide.

3 Put a drop of iodine solution onto the tissue.

4 Cover the tissue with a coverslip.

5 Examine it under the microscope.

Can you see starch grains? (They should have stained blue-black with the iodine solution.)

Can you see that they are located inside the cells?

Approximately how many are there inside the cells?

In what respect do they differ from one another?

Assignments

1 Where, and in what form, do the following store food?

a) man, b) a potato plant, c) a radish plant, d) sugar cane, e) sugar beet.

2 Give two reasons why it is useful for organisms to be able to store food.

3 Which is best as a source of food for man: grain or storage organs? Give reasons for your answer.

4 When a potato sprouts into a new plant, the starch has to be turned into glucose before it can be moved into the new plant.

a) Why is this necessary?
b) What is present in the potato which enables the starch to be turned into glucose?

5 Describe an experiment which you would carry out to see whether or not the stem of a particular plant is able to convert glucose into starch.

6 The graph below shows the relative amounts of carbohydrate in leaves and tubers of a potato plant towards the end of the summer.

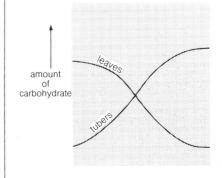

Explain the changes in the potato plant which are illustrated by this graph.

Preserving food

Preserving food is important in modern life. It prevents food going bad, and enables us to store it for long periods In this Topic we shall see how it is done.

Why does food go bad?

Food goes bad because microbes get into it and feed on it. These are the same kinds of microbes that bring about decay (see page 376). They usually break the food down, making it smell and taste unpleasant.

Some of the microbes produce poisonous substances which cause **food poisoning**. When a person gets food poisoning, he usually has an upset stomach, but he soon gets better. On rare occasions the effects are more serious. Such is the case with botulism which is caused by a type of bacteria. The substance produced by the botulism germ is so poisonous that a spoonful of it could kill over a million people.

How can we prevent food going bad?

We can do this either by killing the micro-organisms or by making conditions unsuitable for them. This is the basis of food preservation. There are four main ways of preserving food and they are explored in the Investigation.

1 Cold treatment

For hundreds of years Eskimos have preserved fish by burying it in frozen ground. We do the same when we put food in a food freezer, or when food is frozen in a factory.

Freezing does not kill microbes, but it stops them multiplying and slows their action – and the colder it is the slower they get. However, as soon as the food thaws the microbes start up again, and the food begins to go bad. In the case of meat, decay sets in particularly quickly. This is because during freezing the cells are broken open by ice crystals, and this makes it easier for bacteria to penetrate into the food afterwards. So frozen food should be eaten as soon as possible after it has been taken out of the freezer.

It is important to freeze food quickly. This can be done by subjecting it to a very low temperature (Figure 2). Once it has been frozen, it can be stored at a slightly higher temperature. Most home freezers have two temperature settings: $-24°C$ for freezing food, and $-18°C$ for storing it. At this temperature bacterial action is slowed almost to a standstill.

How long can you leave frozen food in a freezer? It depends on the kind of food, and the temperature at which it is stored. If you look at the front of a

Figure 1 Preserved food is one of the main items in any supermarket.

freezer you will see a row of stars. These indicate how long food can be kept (Figure 3). In a four star freezer some foods can be kept for up to a year.

It is important to distinguish between a food freezer and a refrigerator. The temperature in the main part of a refrigerator is just above freezing. At this temperature bacterial action is slowed down, but nothing like as much as it is when completely frozen. For this reason most foods can be kept in a fridge for only a few days.

Inside most fridges there is a small frozen food storage compartment. The temperature in here is below freezing. You should not use it to freeze food, but it can be used for storing food which is already frozen.

Freezing food is very important nowadays. Frozen foods are transported in refrigerated lorries and ships, and they are a popular item in every super-market.

2 *Drying*

Microbes need moisture. If food is dried they go into a state rather like hibernation. They stop multiplying and their action ceases. If spores land on dried foods they cannot germinate since moisture is needed for this (see page 34). Dried food lasts indefinitely. Samples of dried food, found in Jericho, were preserved over 4000 years ago.

Removing water from an object is called **dehydration**. Nowadays food is usually dehydrated by having hot dry air blown over it. Sometimes other more complicated methods are used. Milk, eggs, potatoes, fish and meat can all be dehydrated. When water is added to them, they quickly take it up and can then be eaten.

Dehydrated foods are compact and light so they can be moved around easily and cheaply. They are particularly useful in wartime and in space travel, and for feeding people in developing countries.

3 *Heat treatment*

Heat kills microbes. It is therefore another good way of preserving food. Getting rid of microbes by killing them is called **sterilisation**.

Let's take an example, bottling fruit. Many people do this in their own homes. A special bottle is used – it has a lid with a rubber rim. You fill the bottle with fruit and syrup and put the lid on. Then you heat the bottle to kill any microbes present. After that you let the bottle cool down. As it cools, the

Figure 2 Peas being frozen in a factory tunnel freezer. A blast of very cold air is blown over the peas.

Figure 3 Temperature scale for refrigerating food. If you want to freeze fresh food you should freeze if at the lowest temperature. It can then be stored in a frozen state at a higher temperature.

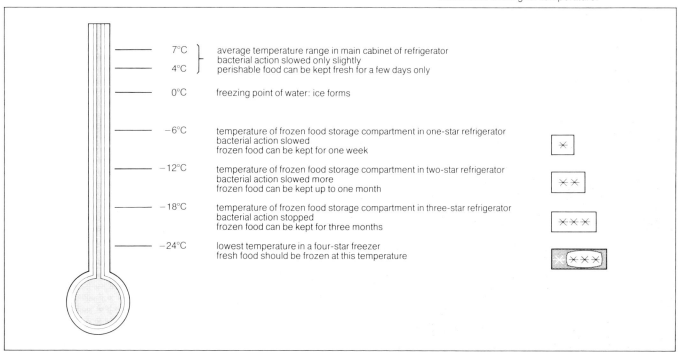

Temperature	Description	Star rating
7°C	average temperature range in main cabinet of refrigerator	
4°C	bacterial action slowed only slightly / perishable food can be kept fresh for a few days only	
0°C	freezing point of water: ice forms	
−6°C	temperature of frozen food storage compartment in one-star refrigerator / bacterial action slowed / frozen food can be kept for one week	✳
−12°C	temperature of frozen food storage compartment in two-star refrigerator / bacterial action slowed more / frozen food can be kept up to one month	✳✳
−18°C	temperature of frozen food storage compartment in three-star refrigerator / bacterial action stopped / frozen food can be kept for three months	✳✳✳
−24°C	lowest temperature in a four-star freezer / fresh food should be frozen at this temperature	✳✳✳

air inside it shrinks and a vacuum is created. This pulls the lid on tightly, so the bottle becomes hermetically sealed.

Canning works in the same kind of way. In a canning factory food is placed in metal cans. Air is then sucked out of the cans, after which they are sealed. They are then heated under pressure for long enough to kill all microbes.

Heat is also used for killing bacteria in milk. Milk can be completely sterilised by boiling it thoroughly and then sealing it. Some housewives buy sterilised milk because it will keep for a long time. However, sterilisation alters the flavour, and most people prefer **pasteurised** milk.

Pasteurisation is named after the famous French scientist Louis Pasteur. He was the first person to realise that heat kills bacteria. When milk is pasteurised it is heated enough to kill any germs, but not so much that it loses its flavour. In one method the milk is heated to about 70°C for 15 seconds, then quickly cooled and put into a sterilised bottle and capped. Some dairies use a quicker method: they heat the milk to 135°C for one second, and then seal it. Some bacteria survive this treatment, but not the ones that cause disease.

4 *Chemical treatment*

The idea here is to add a chemical to the food which kills bacteria but is harmless to man.

Pickling is an example. When food is pickled it is put in a preservative such as vinegar. The acid in the vinegar kills the bacteria and prevents the food from going bad.

Another example is **smoking**. In this process the food is held over a wood fire. The smoke contains substances which kill the bacteria, as well as giving the food a delicious flavour.

Some foods are preserved by **salting**: the food is either soaked in a solution of salt (brine) or salt is rubbed into it. The salt pulls the water out of the bacterial cells by osmosis and kills them.

In hot countries the local people often put fruit or fish out in the sun (Figure 4). Water evaporates from the food, leaving a high concentration of sugar or salt. This raises the osmotic pressure and kills any microbes present. This is why 'dried fruit' such as figs and prunes keep for a long time.

Bacteria and other microbes can also be killed by radio-active rays, a very modern way of preserving food. However, to kill all the microbes very strong rays have to be used, and many scientists feel that this could make the food dangerous to eat. Until we know more about its effects, radiation is unlikely to be a widely used method of preserving food.

Figure 4 A woman places fish to dry in the sun at a nomadic fisherman's village on the Niger river in West Africa.

Investigation

Finding ways of stopping food going bad

1 Obtain six small pieces of fresh meat, about 1 cm square.

2 Obtain six large test tubes. Label them A to F.

3 Set up the six test tubes like this:

A Put a piece of meat in a test tube and plug the tube with cotton wool. Leave it at room temperature. This is your control.

B Put a second piece of meat in a test tube, and plug the tube with cotton wool. Place it in a refrigerator.

C Dry another piece of meat by blowing warm air over it with a hair drier. Then put it in a test tube with a few crystals of silica gel, and plug the tube with cotton wool. Leave it at room temperature.

D Put a piece of meat in a test tube, and plug the tube with cotton wool. Place the tube in an autoclave or pressure cooker, and heat it. Then cool it, and leave it with test tubes A and C.

E Soak a piece of meat in vinegar. Then put it in a test tube and plug the tube with cotton wool. Leave it at room temperature.

F Rub salt into the final piece of meat. Then put It In a test tube and plug the tube with cotton wool. Leave it at room temperature.

4 Can you think of any other ways of preventing meat going bad? If you can, set up extra test tubes and label them G, H . . . etc.

5 Leave the test tubes for several days.

6 After several days look at the pieces of meat and smell them.

Which pieces of meat have gone bad, and which have not?

From this experiment, which methods appear to be good for preserving food?

Assignments

1 Give five examples of foods which go bad within a few days if you don't keep them in a refrigerator.

2 A housewife leaves a loaf of bread, two digestive biscuits and some pickled onions on the sideboard while she goes on holiday. When she gets home she finds that the bread has gone mouldy but the biscuits and pickled onions are unaffected. Can you explain the difference?

3 Explain each of the following:

a) Food left at the South Pole by Captain Scott during his expedition in 1912 was discovered many years later in perfect condition.
b) Sometimes the sides of a can of food bulge out and the can bursts.

4 Here are some simple rules for freezing food.

a) Freeze the food as soon as possible after you have obtained it.
b) Handle the food as little as possible before you freeze it.
c) Set the freezer at its lowest temperature several hours before you put the food in.
d) Don't re-freeze food which has been frozen before.

Give a scientific reason for each of these rules.

5 A fresh fish weighs 120 grams and contains 22 grams of protein. After being dried, the same fish weighs 31.5 grams.

a) What percentage of the fresh fish is protein?
b) What percentage of the dried fish is protein?
c) Why are the two figures different?
d) Why is this important in feeding mankind?

Getting energy from food

We need energy to move, grow, mend our tissues when they are damaged, and just to keep ourselves alive. We get energy from our food.

	kJ/g
margarine	32.2
butter	31.2
peanuts (roasted)	24.5
chocolate (milk)	24.2
cake (plain)	18.0
sugar (white)	16.5
sausages (pork)	15.5
cornflakes	15.3
rice	15.0
bread (white)	10.6
chips	9.9
chicken (roast)	7.7
eggs (fresh)	6.6
potatoes (boiled)	3.3
milk	2.7
apple	1.9
beer (bottled)	1.2
cabbage (boiled)	0.34

Table 1 How much energy is there in various everyday foods? You can find out by looking at this list.

	kJ/day
New-born baby	1885
Child 1 year	3352
Child 2-3	5866
Child 5-7	7542
Girl 12-15	9637
Boy 12-15	11 732
Office worker	11 313
Factory worker	12 570
Coal miner	15 084
Pregnant woman	10 056
Woman breast-feeding	11 313

Table 2 Approximate amounts of energy required daily by different types of people

*The unit of energy used to be the kilocalorie, and in fact this unit is still used in some circles. However, it has now been officially replaced by the kilojoule.

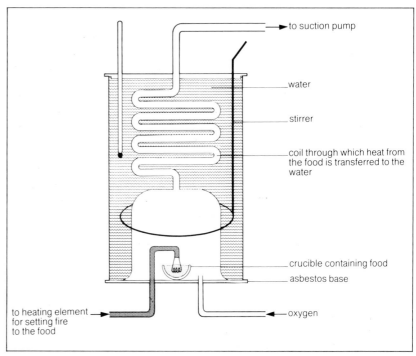

Figure 1 This apparatus can be used to find out how much energy there is in a sample of food. It is called a food calorimeter.

Does food really contain energy?

How can we show that a piece of bread, for example, contains energy? One way is to burn it. When the food is burned the energy contained inside it is set free as heat.

We can use this to find out how much energy a particular piece of food contains. We set fire to it, and estimate how much heat it gives out. This can be done simply, as in the Investigation, or more accurately by the method shown in Figure 1.

A known mass of food is burned. The heat given out heats up a known quantity of water. From the rise in temperature of the water we can work out the amount of energy released by the food. The energy contained in food can be expressed in **kilojoules** (kJ)*. 4.2 kJ of energy are required to raise the temperature of 1 kg of water through 1 °C.

The three main kinds of food are carbohydrate, fat and protein (see page 112). If we estimate the amount of energy in each of these, we can compare their energy values. Here they are:

Carbohydrate	1 g contains 17.1 kJ
Fat	1 g contains 38.9 kJ
Protein	1 g contains 18.2 kJ

You will see that fat contains the greatest amount of energy. Carbohydrate and protein contain about half as much as fat.

How much energy do different foods contain?

Look at Table 1. This tells us how much energy there is in some everyday foods. The amount of energy in a particular food depends on the substances which it contains. Thus margarine and butter contain a lot of energy because they consist almost entirely of fat. At the other extreme, cabbage contains very little energy because it consists of a high percentage of water.

Another thing which determines how much energy a particular food contains is how it is cooked. For example there is three times as much energy in chips as there is in boiled potatoes. Why do you think this is?

How much energy do we need each day?

Imagine someone lying in bed doing nothing. Even in such an inactive state he needs energy to breathe, make his heart beat, and drive all those countless chemical reactions which keep him alive. The rate at which these 'ticking over' processes take place is called the **basal metabolic rate**.

How much energy is needed to maintain the basal metabolic rate? It is difficult to say, because it varies from one individual to another. Very roughly the amount needed is 7000 kJ per day. This is about the same amount of energy that would be needed to boil enough water for one hundred cups of tea.

This figure applies to a person who is completely at rest. It doesn't even include the energy he needs to feed himself. Scientists have tried to work out how much energy an average person needs to get through the day with the minimum effort, i.e. to get up in the morning, eat and drink and do other essential tasks, but no more. The figure is about 9200 kJ per day. A person could get enough energy to satisfy this need by eating one large white loaf a day.

Few of us spend our days like that – most of us do something. Look at Table 2. This tells us roughly how much energy is needed each day by different people. You will see that the amount depends on the person's age, sex and occupation. A person who spends most of his time sitting down needs far less energy than a very active person.

The important thing is that we should eat sufficient food to provide enough energy for our daily activities whatever they may be.

What happens when we eat too much?

Suppose a person eats more food than he needs for producing enough energy. What happens to the food left over? Most of it is turned into fat and stored beneath the skin. The result is that his body weight increases, and he runs the risk of becoming fat. Putting on weight is caused by a person's energy input being greater than his energy output.

The most 'fattening' foods are those which provide the most energy, such as bread and butter, cake and sweets.

How can a person lose weight? The only way is by making his energy input less than his output. This can be done in two different ways:

1 By taking more exercise: this will increase his energy output.
2 By eating less energy-containing food: this will decrease his energy input.

The first method is not very effective – I've tried it! A person has to take a lot of exercise to make much difference to his weight. For example, a man trying to lose weight may play a game of tennis for half an hour. In doing so he loses about 700 kJ of energy. After the game he feels thirsty and has a pint of bottled beer. The result is that he puts back all the energy he has just lost.

The second method is very effective if carried out properly. A person on a well planned weight-reducing diet can lose about 1 kg per week. Such diets contain relatively little high-energy food and a lot of low-energy food, the result of going on such a diet is shown in Figure 2.

The best results can be obtained by combining both methods, i.e. by going on a weight-reducing diet and taking more exercise.

For everyone there is a 'correct' weight. This will depend on his or her age, height and build.

Look at the bar chart in Figure 3. It is based on data obtained in the United States for people between the ages of 15 and 70. It shows that there are more deaths amongst people who are overweight than amongst people of normal weight. In other words, overweight people do not live as long, on average, as people who are the normal weight. An overweight person has a greater chance of having a stroke or a heart attack (see page 175). Other illnesses, too, are connected with overweight. The risk of death is greater for men than for women, and it increases with the amount of overweight.

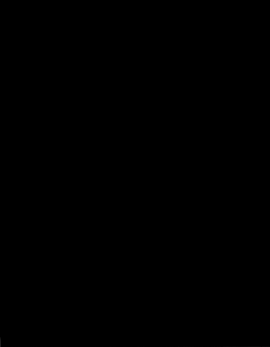

losing weight. This particular person reduced her weight from 183.5 kg (28 st 12 lb) to 64 kg (10 st 1 lb) in under a year, a loss of 119.5 kg (18 st 11 lb). By trick photography she is seen in this picture before and after losing weight.

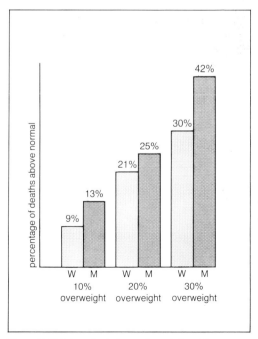

Figure 3 This bar chart shows the relationship between people's body weight and the death rate in the United States.

What happens when someone starves?

What happens if a person eats nothing at all? To begin with he will get energy from his fat stores, As a result he loses weight.

Eventually all his fat gets used up. In order to stay alive the body then starts getting energy from his tissue proteins, particularly the muscles. As a result he 'wastes away', becoming thin and weak (Figure 4). Death usually occurs after about 60 days.

This may happen to victims in concentration camps. Occasionally it happens to a person who goes on a 'hunger strike' in prison, though sometimes an attempt is made to 'force-feed' such an individual to prevent him starving to death.

Some people suffer from a mental condition in which they lose their appetite and eat very little. This is called **anorexia nervosa**. It sometimes happens to people, particularly young women, who are suffering from emotional stress. Such people often become thin and frail.

There are many countries in the world where people do not get enough to eat. Although they may not die from lack of food, they become thin and weak and find it difficult to work.

How many people are starving in the world?

We saw earlier that to get through the day a person needs at least 9200 kJ of energy. Anyone who receives less than this can be said to be starving.

Now look at Figure 5. From this you will see that in many places people get less than this minimum amount of energy. However in other places there are many people who get much more than they require.

Figure 5 This map shows the approximate percentage of the population who are starving in different parts of the world.

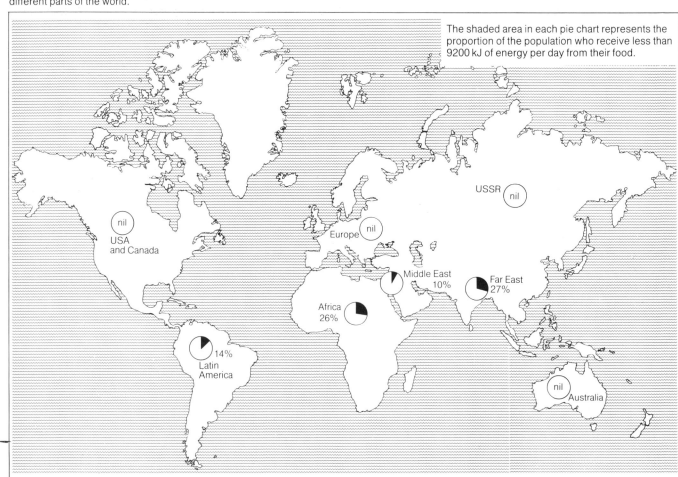

Investigation

A simple way to find out how much energy a piece of food contains

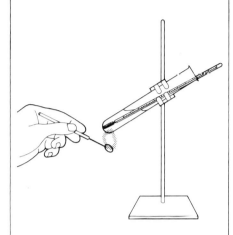

1 Measure out 20 cm³ of water with a measuring cylinder, then transfer it to a large test tube.

2 Clamp the test tube to a stand, and put a thermometer in it, as shown in the illustration.

3 Record the temperature of the water.

4 Weigh a peanut then stick it onto the pointed end of a mounted needle.

5 Hold the nut in a Bunsen flame until it starts to burn, then place it under the test tube as shown in the illustration.

6 When the nut stops burning, record the temperature of the water again.

By how much has the temperature of the water risen?

Energy released from the nut in joules (J) = mass of water in grams x rise in temperature in °C × 4.2

Work out the amount of energy in joules released from the nut.

Convert the joules into kilojoules (kJ) by dividing by 1000.

Knowing the mass of the nut, work out how much energy in kilojoules is contained in one gram of peanut.

Compare your figure with others in your class and find the average.

How does the class average compare with the figure given in Table 1?

Do you think this is an accurate way of finding out how much energy there is in a piece of food?

Assignments

1 What weight of roasted peanuts would the coal miner referred to in Table 2 have to eat in a day to satisfy his energy needs?

2 Work out the total amount of energy you take into your body in a particular day. To do this you will need to weigh each item of food before you eat it, then look up Table 1 on page 128 to find out how much energy it contains. Are you getting more, or less, energy than the amount recommended in Table 2?

3 The table below gives the percentages of carbohydrate, fat and protein in three chocolate products.

	carbo-hydrate %	fat %	protein %
plain chocolate	59	33	4
milk chocolate	54	36	8
cocoa powder	36	26	19

Which one contains the most energy, and which one contains the least?

4 The figures given in Table 1 are the amounts of energy actually present in carbohydrate, fat and protein as measured with a food calorimeter. In practice the amount of energy our bodies get out of each one is slightly less than the figures given. Suggest reasons for this.

5 The data summarised in Figure 3 was compiled by an American life insurance company.

a) Explain in full how you think the data was obtained.

b) Why should a life insurance company want to compile such data?

6 Give examples of the sort of food you would recommend to (a) someone who is going on a hiking holiday, and (b) someone who wishes to lose weight.

7 The following table shows the approximate amounts of energy used up in different activities by a normal man.

sitting	5.88 kJ/min
standing	7.14 kJ/min
washing and dressing	14.7 kJ/min
walking slowly	12.6 kJ/min
walking fairly fast	21.0 kJ/min
walking up and down stairs	37.8 kJ/min
carpentry	15.5 kJ/min
playing tennis	26.0 kJ/min
playing football	36.5 kJ/min
cross-country running	42.0 kJ/min

a) From these figures work out the approximate total amount of energy which you yourself use up in a typical day. Show your working in full.

b) Using Table 1 draw up a menu for breakfast, lunch and supper which would give you just enough energy to satisfy your need. Give the amount of each food item which you would need.

How is energy released?

First we will find out what happens when we burn a piece of food in the laboratory. Then we will see if the same thing happens in our bodies.

What happens when food is burned in the laboratory?

When a piece of food is burned, energy is set free (see page 128). However, for the food to burn certain things are necessary and certain things are produced. Thus food will only burn if oxygen is present. The more oxygen that's present, the better it will burn. One of the best ways of putting out a fire is to stifle it, i.e. to deprive it of oxygen.

In the chemical reaction which takes place when a piece of food burns, a gas is given off. If this gas is bubbled through lime water, the lime water goes milky (Investigation 1). This tells us that the gas is carbon dioxide. So burning food produces carbon dioxide.

We know, too, that burning food produces some water and of course it also produces energy in the form of heat.

So, to sum up, *when a piece of food is burned oxygen is used up, carbon dioxide is given off, water is formed and heat energy is produced*:

food + oxygen → carbon dioxide + water + energy

In other words the food is **oxidised**. In this process it is broken down and energy is released.

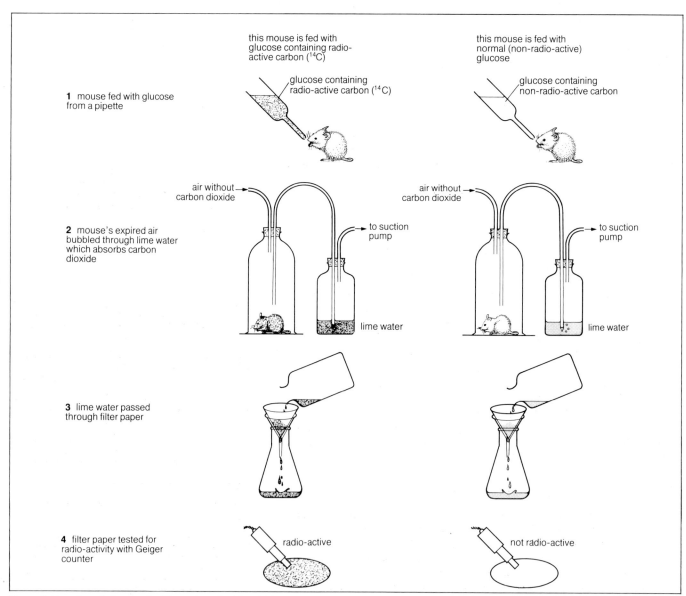

Figure 1 Experiment showing that the carbon dioxide which an animal breathes out comes from its food.

How is energy produced in the body?

We have seen what happens when food is burned in the laboratory. Does the same thing happen in our bodies?

It has been known for a long time that living organisms take in oxygen and give out carbon dioxide. Many different experiments show this (Investigations 2 to 5 for example).

But how can we find out if taking in oxygen and giving out carbon dioxide are connected with the breaking down of food?

One way of doing this is to use radio-active tracers (see page 198). In an experiment scientists made some sugar in which the normal carbon atoms were replaced with the radio-active isotope of carbon. In other words, the carbon atoms in the sugar were 'labelled'.

They then fed this labelled sugar to a mouse and traced what happened to it (Figure 1). The radio-active carbon was detected by means of a Geiger counter.

The scientists found that after a short time the mouse started breathing out radio-active carbon dioxide. They concluded that the carbon dioxide breathed out comes from sugar.

We now know that in our cells food is broken down and oxidised to give carbon dioxide and water. In this process energy is set free, just as it is when a piece of food is burned in the laboratory.

The main substance which is broken down in this way is a form of sugar called glucose. We can summarise what happens like this:

$$C_6H_{12}O_6 \; + \; 6O_2 \; \rightarrow \; 6CO_2 \; + \; 6H_2O \; + \; energy$$
glucose   oxygen   carbon   water
                        dioxide

This process takes place in practically all living cells of animals and plants alike. We call it **cell respiration**. It is vitally important because it gives us energy. Some of the things this energy enables us to do are dealt with on page 129.

The chemistry of respiration

Cell respiration is very complex. Scientists have shown that glucose is not broken down in one jump as the equation given above suggests. It is broken down in a series of small steps, the energy being released bit by bit (Figure 2). Why is this important? Think of it this way. A slice of apple pie contains as much energy as a stick of dynamite. If the energy in the pie was set free in one go, as when dynamite explodes, the person's body temperature would shoot up by at least 10°C and he would die.

Many scientists think that the energy produced by the breakdown of glucose is not used directly. They believe it is linked to activities such as muscle contraction by another chemical substance known as **adenosine triphosphate**, or ATP for short.

However, you need not worry about these details. The important thing is that breaking down food inside our cells gives us the energy we need for our various activities. Our food, therefore, serves as fuel. It drives our bodies, just as the burning of petrol drives a car.

How are the chemical reactions controlled?

Thousands of reactions take place in the body every second. Each reaction is controlled by a substance called an **enzyme**. The enzyme is like a catalyst and it speeds up the reaction. Without enzymes the reactions would be so slow that we would not produce energy fast enough to stay alive.

Enzymes have six important properties which are summarised in Table 1. Enzymes help to keep the body going, they are like oil in a machine, keeping the wheels turning and preventing the whole thing seizing up.

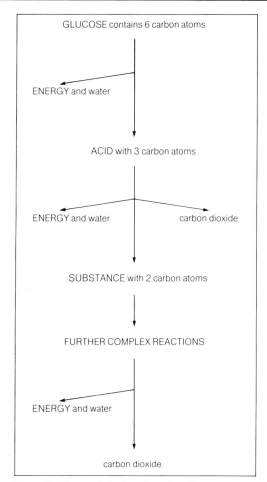

Figure 2 Glucose is broken down in cells to produce energy in a series of small chemical steps.

Property	Notes
1 They are proteins	This is one reason why we need proteins in our food.
2 They are specific	Each enzyme controls one particular reaction, or type of reaction.
3 They can be used over again	This is because they are not destroyed by the reaction.
4 They are destroyed by heat	If the temperature reaches more than about 40°C they stop working.
5 They are sensitive to poisons	This is why poisonous substances such as cyanide and arsenic are lethal.
6 They are helped by vitamins and minerals	This is one reason why these substances are so important in our food.

Table 1 The six important properties of enzymes are summarised in this table.

Investigation 1

To find out if burning food produces carbon dioxide

1 Put one level teaspoonful of sugar (sucrose) into a large test tube.

2 Set up the test tube as shown in the illustration.

3 Place a Bunsen burner under the sugar and heat it with a moderate flame.

What happens to the lime water?

If it turns milky, this means that carbon dioxide gas is being given off by the burning sugar.

Has any liquid condensed on·the side of the test tube?

What does this suggest?

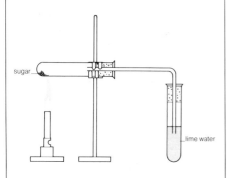

Investigation 2

To find out if a person breathes out carbon dioxide

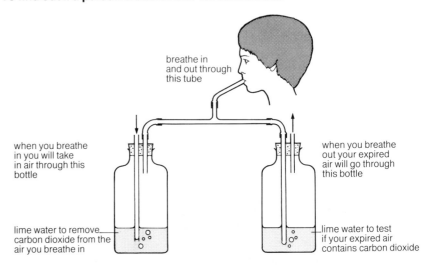

1 Set up the apparatus shown in the illustration.

2 Breathe in and out of the tube.

What happens to the lime water?

If it turns milky, the air you breathe out contains carbon dioxide.

Investigation 3

To find out if a small mammal gives out carbon dioxide

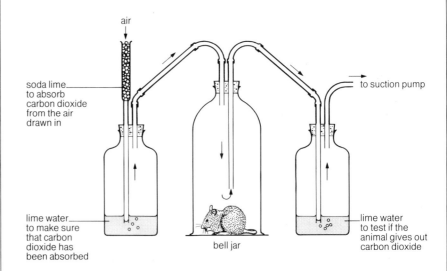

1 Put a small mammal such as a mouse or gerbil on a glass plate under a bell jar.

2 Set up the apparatus shown in the illustration. Use vaseline to make sure the three jars are air-tight.

What happens to the lime water in the right hand jar?

If it turns milky, the animal has been giving out carbon dioxide.

Where is the control in this experiment?

Investigation 4

To find out if small animals and plants give out carbon dioxide

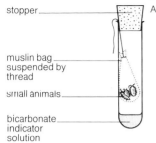

1 Put some small animals such as woodlice, in a muslin bag.

2 Obtain a green leaf.

3 Pour a little bicarbonate indicator solution into three test tubes. Notice that the indicator is reddish-orange.

4 Set up the three test tubes as shown in the illustration.

5 Put test tube B in the dark, e.g. under a cardboard box.

6 Leave the three test tubes for about an hour.

7 After about an hour give each of the test tubes a quick shake.

What has happened to the bicarbonate indicator in each test tube?

If it has turned from reddish-orange to yellow, it means that carbon dioxide has been given off.

What is the purpose of test tube C? Why did you put test tube B in the dark?

Investigation 5

To find out if small animals take up oxygen

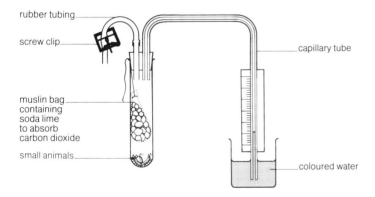

1 Put some small animals, e.g. woodlice, in a test tube.

2 Set up the test tube as shown in the illustration. Make sure the system is air-tight by sealing the stopper with vaseline.

3 Set up another test tube exactly like the first one but without any animals in it. This is your control.

4 Close the screw clip and find out how far the coloured water rises up the capillary tube in 30 minutes.

Has the water risen in the glass tube? If it has, it could be caused by the animals taking up oxygen.

Can you think of any other possible explanation of your results?

How could you use this apparatus to:
a) compare the rate of respiration of different animals, and
b) find the effect of varying the temperature on the rate of respiration?

Assignments

1 What would you conclude from each of these observations?

a) A piece of food will only burn if oxygen is present.

b) When food is burned a gas is given off which turns lime water milky.

2 Describe a simple experiment which could be done to find out if a piece of burning food produces water.

3 In order to show that the air we breathe out contains carbon dioxide, a teacher blows bubbles through a drinking straw into a glass of lime water.

Why is this not as good an experiment as the one given in Investigation 2?

4 Your uncle did not do any science at school, and he does not believe that the air he breathes out contains carbon atoms from the food he eats. Write a short letter to convince him.

5 Give two differences between the way a piece of food releases energy when you burn it in a test tube and when it is broken down inside our cells.

6 A student set up the apparatus shown below to find out if germinating peas give out heat energy.

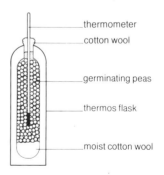

a) Why did he put moist cotton wool in the bottom of the flask?

b) What should the control be in this experiment?

c) Another student did the same experiment but instead of putting the peas in a thermos flask he put them in a milk bottle. Do you think this is as good a method?

Give reasons to support your answer.

How do we digest our food?

Have you ever wondered what happens to the food you eat? In this Topic we will follow what happens to a ham sandwich after it has been put in the mouth.

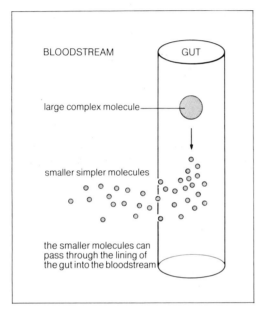

Figure 2 Digestion involves breaking down large molecules into smaller ones which can then be absorbed into the bloodstream.

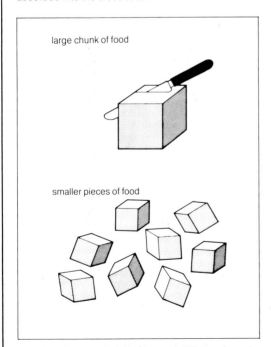

Figure 3 Chopping up food into small pieces increases its surface area.

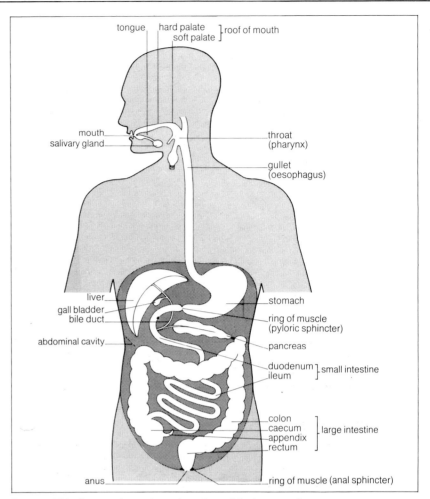

Figure 1 This diagram shows the main regions of the human gut.

The structure of the gut

The mouth leads into the gut or alimentary canal. This is really a tube running from the mouth to the anus. It's between 8 and 9 metres in length – that's about four times an average person's height. Being so long, much of it is coiled up and this enables it to fit into the abdominal cavity.

The gut is divided into a series of regions which are shown in Figure 1. Each region has a particular job to do.

What happens in the gut?

A ham sandwich contains starch, fat and protein: the starch is in the bread, the fat is in the butter and the protein is in the ham.

All three of these substances are solids, but as they pass along the gut they are broken down into liquids. This process is called **digestion**. The liquid substances are then absorbed through the lining of the gut into the bloodstream, and carried round the body to where they are needed (Figure 2).

Digestion is brought about by two distinct processes which occur in the gut:

1 Breaking the food up into small pieces by chewing it and churning it up. This has no effect on the chemistry of the food; it merely breaks it up physically.

2 Mixing the food with substances which dissolve it and break it down into a simpler chemical form. These substances are called **digestive enzymes**, and they are produced by various glands which open into the gut.

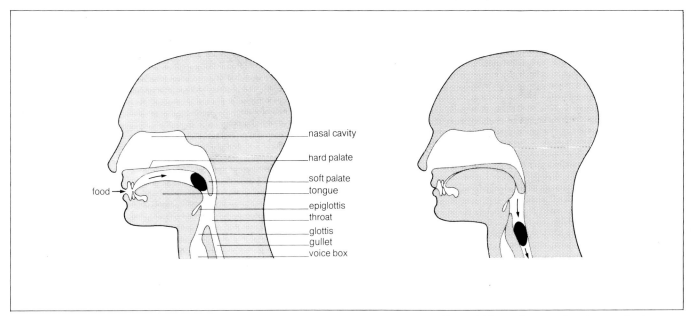

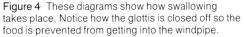

Figure 4 These diagrams show how swallowing takes place. Notice how the glottis is closed off so the food is prevented from getting into the windpipe.

These two processes go on at the same time. Chopping up the food makes it easier for the enzymes to work, because it mixes them with the food and increases the surface area over which they can act (Figure 3).

Now let's look in detail at what happens in each part of the gut.

In the mouth cavity

You bite off pieces of the sandwich with your front teeth, and chew them with your back teeth. At the same time your mouth waters – in other words it becomes filled with **saliva** or 'spit'. This is produced by several salivary glands, each of which is connected to the mouth cavity by a tube, or duct, and it has the effect of moistening the food.

Actually your mouth starts watering *before* you begin to eat the sandwich. This is because the sight, smell and even the thought of food is enough to start saliva flowing. However, the greatest flow occurs when the food is actually in the mouth.

Saliva contains water together with two other main substances:

1 *Mucus*

This makes the food slippery so it slides easily through the throat when it's swallowed. Swallowing a piece of dry food, such as a digestive biscuit, without moistening it with saliva first, can be an uncomfortable experience! A simple chemical test can be carried out to show that saliva contains mucus (Investigation 1).

2 *Amylase*

This is the first enzyme which the sandwich encounters as it travels through the gut. It acts on starch, breaking it down into a type of sugar called maltose (Investigation 2). If you chew a piece of bread for long enough, you can actually taste the sweetness as the maltose is formed.

Through the throat and down the gullet

When you swallow, the food is pushed down your throat into your gullet. Figure 4 explains what happens.

Once swallowed, the food passes down the gullet to the stomach. The gullet has muscles in its wall. A ring of contraction moves slowly downwards, pushing the food in front of it. This process is called **peristalsis** (Figure 5). The mucus from the saliva acts as a lubricant so that the food slips down easily.

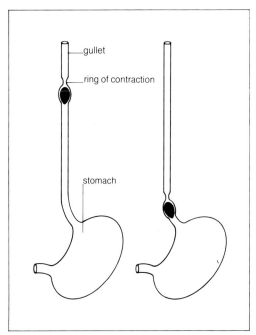

Figure 5 Food is pushed down the gullet by a ring of contraction called a peristaltic wave.

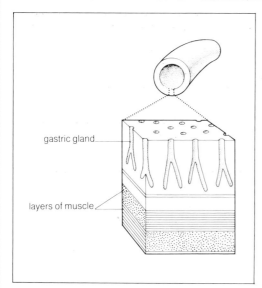

Figure 6 The structure of the wall of the stomach.

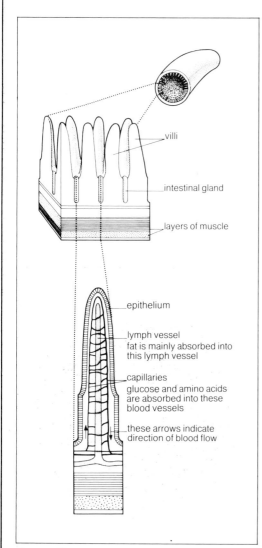

Figure 7 The structure of the wall of the small intestine is well suited to its job of completing the digestion and absorption of food.

In the stomach

The stomach wall is thick and muscular, and its inner surface has numerous holes which lead into narrow cavities called **gastric glands** (Figure 6).

The gastric glands produce a fluid called **gastric juice**. This contains an enzyme called **pepsin** which helps to dissolve the protein in the ham by breaking it down into simpler substances called peptides. Pepsin is quite different from saliva in its action: whereas saliva attacks starch, pepsin only goes for protein (Investigation 3).

Pepsin works best in acid conditions (Investigation 4), so the gastric glands produce large amounts of hydrochloric acid. As well as enabling pepsin to work, the acid helps to kill germs which get into the stomach.

The wall of the stomach also produces large quantities of mucus which protects the stomach lining from being damaged by the acid in the gastric juice.

The food spends three or four hours in the stomach. Every now and again a wave of contraction passes along the stomach and churns the food up. As a result of all these actions, the food is turned into a mushy fluid.

Between the stomach and small intestine there is a ring of muscle. Occasionally this opens and a wave of contraction sweeps some of the food into the first part of the small intestine.

In the small intestine

The small intestine, despite its name, is the longest part of the gut and may be over six metres in length. Here the digestion of the ham sandwich is completed, and the resulting fluids are absorbed into the blood.

The small intestine receives fluids from three different places.

1 The liver

This produces a fluid called **bile**. The bile is stored in the gall bladder and after a meal it is squirted, bit by bit, into the duodenum. Bile contains substances called bile salts. These act on the fat, breaking it up into small droplets. The same kind of thing happens when washing-up liquid comes into contact with fat. We call this process emulsification (Investigation 5).

2 The pancreas

This produces a fluid called **pancreatic juice** which flows down the pancreatic duct into the duodenum. It contains three important enzymes:

Amylase breaks starch down into maltose, and thus continues the process which was begun by saliva in the mouth cavity.

Trypsin breaks down protein into peptides, just as pepsin does in the stomach.

Lipase attacks fat, breaking it down into fatty acids. This completes the digestion of the fat. The action of lipase is made easier by the fact that the fat has already been broken up into droplets by the bile.

3 Intestinal glands

These glands are situated in the wall of the intestine itself. They produce the enzyme **maltase** which breaks maltose down into glucose, thereby finishing off the digestion of starch. They also produce several enzymes called **peptidases** which complete the digestion of protein by breaking up the peptides into amino acids.

The various fluids which are secreted into the small intestine contain a lot of sodium bicarbonate. This neutralises the acid from the stomach and makes the contents of the small intestine alkaline. This is necessary because the enzymes in the small intestine will only work properly in alkaline conditions.

While all this is going on, wave-like contractions of the wall of the small intestine move the food about, and finally sweep it on towards the large intestine.

Where it comes from	Where it works	Name of enzyme	Food acted on	Substances produced	
salivary glands	mouth cavity	amylase	starch	maltose	
stomach wall	stomach	pepsin	protein	peptides	
liver	small intestine	bile salts (not enzymes)	fat	fat droplets	
pancreas	small intestine	amylase trypsin lipase	starch protein fat	maltose peptides fatty acids	} can be absorbed
wall of small intestine	small intestine	maltase peptidases	maltose peptides	glucose amino acids	

Table 1 Summary of the main digestive enzymes found in the human gut. Bile salts are included though they are not really enzymes.

The ham sandwich has now been more or less completely dissolved. The liquid products of digestion are now absorbed through the lining of the small intestine into the blood vessels within its wall.

The structure of the wall of the small intestine is shown in Figure 7. It is well suited to carry out its jobs. It contains numerous pouch-like glands for producing intestinal juice. Thousands of finger-like projections called **villi** (singular: villus) stick into the cavity, greatly increasing the surface area for absorption (Figure 8). Within the villi there are numerous blood capillaries for taking up the absorbed food. Towards the outside of the wall there are muscles for bringing about the contractions mentioned earlier.

Table 1 summarises the various enzymes and other secretions which have helped to digest the ham sandwich.

What happens in the large intestine?

The bread which was used for making the ham sandwich contained a certain amount of cellulose. We refer to this as fibre or roughage (see page 112). Human beings don't have an enzyme to break this down, so it cannot be digested. Along with some fluid, it passes on to the colon.

As material passes along the colon, water is absorbed from it, so it becomes more solid. This solid matter then passes on to the rectum where it is stored as **faeces**. The lining of the rectum produces mucus which eases the passage of the faeces along it. Eventually the faeces are voided through the anus by powerful contractions of the wall of the rectum. We call this **defaecation**.

Although there is much variation, it normally takes between 24 and 48 hours from the time the food is eaten to the time when the faeces derived from it are ready to be voided through the anus.

The caecum and appendix

The caecum and appendix are an offshoot from the first part of the large intestine, a kind of blind alley. They have no function in man, but in grass-eating mammals such as rabbits they contain large numbers of microbes which can digest cellulose and break it down into glucose (see page 146).

Things that can go wrong with the gut

In man the appendix occasionally gets infected with germs. As it is an offshoot from the main part of the gut, the germs do not get flushed out by the normal passage of material along the gut. So they multiply there and may cause severe inflammation leading to **appendicitis**. Normally appendicitis is cured by removing the appendix in an operation.

People often complain of 'indigestion'. This is usually caused by eating food too quickly and not chewing it enough. The gastric glands produce extra large amounts of gastric juice with the result that the stomach contains a lot of

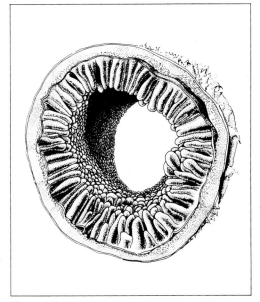

Figure 8 Looking into the small intestine. Notice the finger-like villi projecting into the cavity.

acid. If the person belches, some of the acid comes up the gullet giving a burning sensation which is sometimes called 'heartburn'. Indigestion can be counteracted by taking a tablet or drink which will neutralise the acid.

A person who constantly has too much acid in his stomach may get an **ulcer**. The gastric juice starts to eat into the lining of the stomach which becomes raw and painful. Ulcers tend to develop in middle-aged and elderly people, and they often seem to be brought on by overwork and worry.

A doctor can find out if a person has an ulcer by getting him to drink a thick fluid containing barium. This is called a 'barium meal'. The barium is opaque to X-rays, and if the patient is X-rayed the inside of his gut shows up clearly (Figure 9). An ulcer will appear as a bump in the lining.

Most people suffer from **constipation** at some time or another. The faeces move too slowly along the large intestine, with the result that more water is absorbed from them than usual and they become hard and dry. Constipation is often caused by bad bowel habits. People usually feel the need to defaecate after a meal, particularly breakfast: this is a natural reflex arising from stretching of the gut wall. If you persistently suppress this reflex because you're too busy or can't be bothered to go to the lavatory, the faeces are held in the large intestine and constipation may result.

Doctors believe that constipation is also caused by eating over-refined foods which don't contain much roughage. Roughage adds to the bulk of material in the large intestine, stretching its wall. This stimulates the muscles to contract, pushing the faeces along and keeping them moving.

The opposite of constipation is **diarrhoea**. This results from faeces moving too quickly along the large intestine so there isn't time for the usual amount of water to be absorbed. Diarrhoea is often caused by germs which irritate the lining of the gut, causing it to produce a lot of watery mucus. Diarrhoea is the body's way of sweeping the germs out of the gut as quickly as possible, just as vomiting has the effect of expelling germs from the stomach.

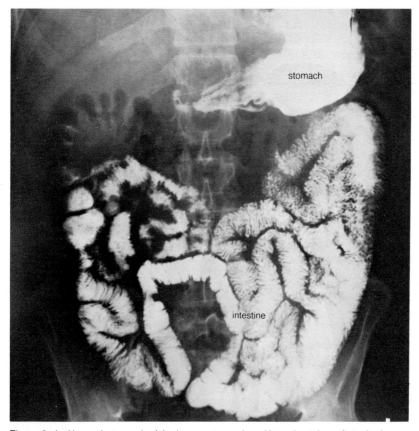

Figure 9 An X-ray photograph of the human stomach and intestine taken after a barium meal was swallowed.

Investigation 1

To find out if saliva contains mucus

1 Collect some saliva in a small beaker.

2 Put a plug of cotton wool into the neck of a funnel. Filter the saliva through the cotton wool into a test tube.

3 Add a few drops of acetic acid to the filtered saliva.

4 Repeat step 3 on a little water in another test tube. This will serve as a control.

What happens in each test tube?

A precipitate means there's mucus present.

How does saliva differ from water in its runniness, and the way it feels?

Why is mucus important in saliva?

Investigation 2

To find out if saliva breaks down starch

1 Collect your saliva in a test tube to a depth of 2 cm.

2 Half fill another test tube with a 4 per cent starch solution.

3 With a pipette place 15 drops of iodine solution, side by side, on a white tile.

4 With a glass rod lift a drop of the starch solution from the test tube and mix it with the first drop of iodine on the white tile. A blue-black colour indicates starch. This will serve as your control.

5 Pour your saliva into the test tube of starch suspension, and shake quickly.

6 With the glass rod place a drop of the starch-saliva mixture with a drop of iodine on the white tile, and mix them together.

7 Repeat step 6 with the other drops of iodine at half-minute intervals. Note the colour given each time.

Explain the colour reactions as fully as you can.

Approximately how long does it take for your saliva to break down the starch?

Investigation 3

To compare the actions of saliva and pepsin

1 Obtain four large test tubes. Label them A to D.

2 Collect some saliva in a test tube.

3 Obtain some acidified pepsin solution.

4 Obtain some hard-boiled egg white (albumen) and some solid starch.

5 Set up the four test tubes like this:

   A starch covered with saliva
   B starch covered with pepsin
   C egg white covered with saliva
   D egg white covered with pepsin

6 Leave the test tubes in a warm place, preferably in an incubator at 37 °C for 48 hours.

7 After 48 hours examine the contents of the test tubes.

In which test tubes has the food material disappeared?

Which of the two food materials is acted upon by:

(a) saliva, and (b) pepsin?

What conclusions do you draw from this experiment?

Investigation 4

To find out if pepsin works best in acid conditions

1 Obtain four large test tubes. Label them A to D.

2 Obtain some hard-boiled egg white (albumen) and cut it up into four pieces.

3 Place one piece of the egg white in each test tube.

4 Cover the egg white with one of the following solutions:

   A pepsin plus acid
   B pepsin plus alkali
   C water plus acid
   D water plus alkali

5 Leave the test tubes in a warm place, preferably in an incubator at 37 °C, for about 48 hours.

6 After 48 hours examine the contents of the test tubes.

What has happened to the egg white in each test tube?

What is the point of setting up test tubes C and D?

Does pepsin work best in acid or alkaline conditions?

Investigation 5

To find the effect of bile salts on oil

1 Obtain three test tubes, and label them A, B and C.

2 Pour some olive oil into each test tube to a depth of about 3 cm.

3 To A add a few drops of water.
To B add a pinch of powdered bile salts.
To C add a few drops of washing-up liquid.

4 Shake the test tubes, then let them stand for a while.

What has happened to the oil in each test tube?

How do bile salts help digestion?

How does washing-up liquid help with washing up?

Assignments

1 Which region of the human gut:

   a) absorbs water from indigestible material,
   b) receives bile from the bile duct,
   c) contains the enzyme pepsin,
   d) is normally acidic?

2 Put forward a reason for each of the following:

   a) A piece of food is dissolved by digestive enzymes more rapidly if it is chewed first,
   b) Eating plenty of roughage (fibre) helps to prevent constipation.

3 What job does mucus do in (a) the throat, (b) the stomach, and (c) the rectum?

4 There is a disease of cattle in which the villi in the small intestine are destroyed and the inner lining of the small intestine becomes smooth. As a result the animal gets weak and wastes away. Why do you think the disease has this effect?

5 It has been suggested that saliva produced *during* a meal digests starch faster than saliva produced *before* the meal. Describe an experiment which could be done to find out if this is true.

6 The diagram, right, shows an experiment which is intended to show what happens in the human gut.

After being set up, glucose but *not* starch passes out of the bag into the surrounding water.

a) How could you show that glucose has leaked out, but starch has not?
b) How would you explain this result?
c) To what extent is this similar to what happens in the human gut?

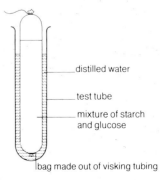

distilled water

test tube

mixture of starch and glucose

bag made out of visking tubing

Teeth

Teeth are one of our most valuable possessions. Here we shall look at their structure, and see what happens if we don't look after them properly.

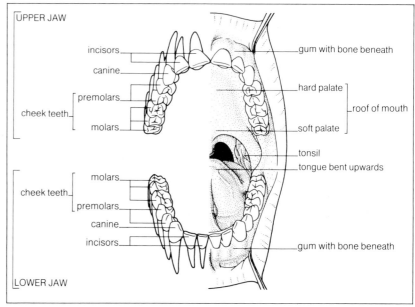

Figure 1 Open wide! Looking inside the mouth of an adult human to see the teeth and other related structures. The roots are shown on the left hand side.

What kind of teeth do we have?

Figure 1 shows the inside of the mouth of an adult man with a full set of teeth.

On both sides of the upper and lower jaws there are, from the middle outwards:

TWO **incisor teeth** which are shaped like chisels and are used for cutting food;
ONE **canine tooth** which is shaped like a dagger and also cuts food;
FIVE **cheek teeth** which have broad tops with bumpy surfaces and are used for grinding food.

The first two cheek teeth are known as **premolars**, and the last three as **molars**. There are 32 teeth altogether.

Now compare the teeth in Figure 1 with your own teeth (Investigation 1).

When the mouth is closed the upper and lower teeth fit together as shown in Figure 2. This biting action is brought about by powerful muscles which run from the lower jaw to the side of the skull.

Looking at the outside of our teeth

If you look at teeth which have been extracted by a dentist, you will see that they can be divided into two parts: the **crown** and the **root** (Figure 3). The crown is the part of the tooth which you can see inside the mouth, that is the part above the gums. The crowns of the cheek teeth have several **cusps** like little mountains on them. The premolars usually have two cusps each, whereas the molars have four.

The root is normally buried in a socket in the jaw bone and is therefore hidden from view. The incisors and canines have roots which consist of a single projection, but the cheek teeth, being larger, generally have two or three projections.

The inside of a tooth

Figure 4 shows the inside of our teeth. The crown is made up of three layers. On the outside is a thin layer of extremely hard **enamel**. Beneath this is a layer of hard ivory-like **dentine**. In the centre is a soft area called the **pulp**

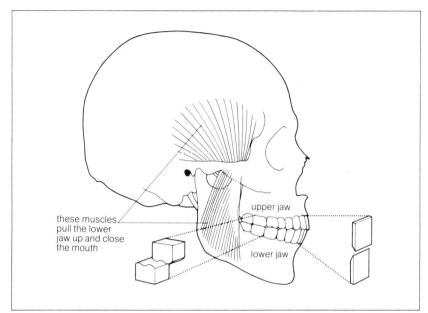

Figure 2 When the mouth closes and you bite something, the teeth fit together as shown in this picture.

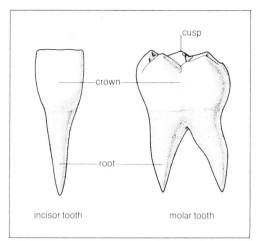

Figure 3 This is what teeth look like after they have been extracted from the jaw.

cavity which contains small blood vessels and a nerve. Tiny channels containing extensions of living cells run out from the pulp cavity into the dentine. This makes the dentine very sensitive. The enamel and dentine both contain calcium phosphate, and it is this that makes them hard.

On the outside of the root is another hard material called **cement**. Attached to the cement are tough fibres which run into the jaw bone. These fibres hold the tooth in its socket, but they permit it to move very slightly and cushion it from excessive jarring when it hits something hard.

When do we get our teeth?

A baby is born without teeth, which is why it must be fed on fluid or soft food which doesn't need to be chewed.

During the next few years the baby develops a set of 20 **milk teeth**. The first tooth breaks through when the baby is about six months old, and usually the set is complete by the age of two or three. There are two incisors and one canine on either side of each jaw, but there are only two cheek teeth as the baby's jaws are too small for any more.

Between the ages of six and twelve the milk teeth fall out, one by one, and are replaced by a set of **permanent teeth**. There are four cheek teeth on either side of each jaw. A fifth cheek tooth may be added after the age of 17. These are known as **wisdom teeth**. The person now has a full set of 32 teeth. This is his final set; if he loses any now, they will not be replaced.

If the person's jaws are small, a wisdom tooth may break through behind the fourth cheek tooth and push it against the third. This is called an impacted wisdom tooth and it can be very painful. Usually the dentist extracts the wisdom tooth together with the tooth immediately in front of it.

Tooth decay

By the age of five, eight out of ten children have **tooth decay**; and by the age of twenty, three people in ten have lost *all* their teeth and wear 'false teeth' or dentures.

Why do our teeth decay? Saliva is normally slightly alkaline, but after a meal, bacteria in the mouth feed on any sugar present and turn it into acids. The acid eats into the teeth. After an hour or so the saliva neutralises the acid and washes it away, but by then the rot has begun.

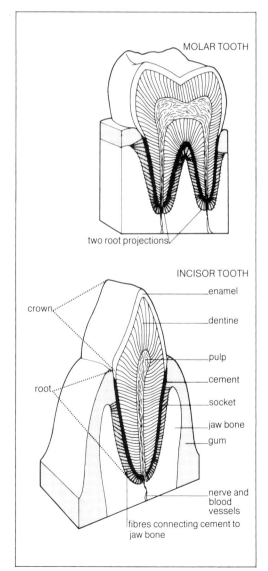

Figure 4 These diagrams show the internal structure of teeth.

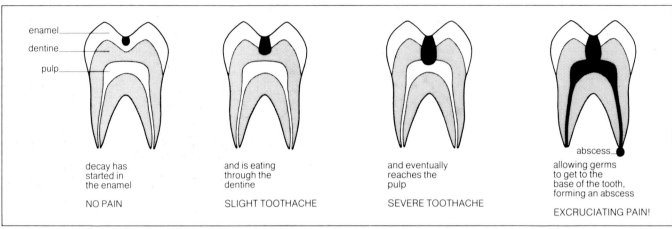

enamel
dentine
pulp

decay has
started in
the enamel

NO PAIN

and is eating
through the
dentine

SLIGHT TOOTHACHE

and eventually
reaches the
pulp

SEVERE TOOTHACHE

abscess

allowing germs
to get to the
base of the tooth,
forming an abscess

EXCRUCIATING PAIN!

Figure 5 The progress of decay in a cheek tooth.

If you look at some teeth which have been pulled out by the dentist, you will see what decay can do to them. Decay usually starts in the crevices between the cusps on the surface of the crowns, and also between the teeth. The acid eats through the enamel into the dentine, thereby enabling the bacteria to infect the pulp cavity (Figure 5). This causes toothache. In severe cases the pulp may be killed and the infection may spread to the base of the tooth, causing an abscess. This can be extremely painful.

Provided the decay hasn't gone too far, the tooth can be repaired by a dentist (Figure 6). The dentist cuts away the decayed part of the tooth with a drill, and fills the hole with a substance which hardens quickly. The hole is always made wider at the bottom than the top: this prevents the filling from falling out.

Normally back teeth are filled with a mixture of metals such as silver and tin, but front teeth are filled with porcelain or a plastic-like material which is the same colour as the teeth.

If the decay has got right into the pulp cavity, the dentist may be unable to save the tooth and he may have to pull it out. Sometimes the dentist will take an X-ray of the person's teeth to find out what state they are in (Figure 7).

Two other common conditions are **gum disease** and **pyorrhoea**. Gum disease, as the name implies, is infection of the gums. In pyorrhoea the fibres which hold the tooth in its socket get infected, with the result that the tooth becomes loose. Pyorrhoea is a major reason why people lose their teeth and have to wear dentures.

Tooth decay and the diet

Many studies have been made which show that tooth decay is caused by eating sugary foods such as cakes, ice cream and sweets, and by drinking sugary drinks. The amount of tooth decay in children in Britain has greatly increased over the last twenty years – so has the amount of sweet-eating. In contrast, the incidence of tooth decay amongst primitive African tribes who don't eat sweets is low. It's also interesting that during the war, when few sweets were available, tooth decay was much less common than it is today.

How can we prevent tooth decay?

The bacteria which cause decay form a thin layer of scum over the surface of the teeth. This scum is called **plaque** (Investigation 2). Tooth decay can be prevented by removing this plaque or stopping it being formed. It takes about 24 hours for plaque to be re-formed after it has been removed, so it's essential to remove it at least once a day.

Here are some tips recommended by dentists:

1 Clean your teeth regularly, particularly after breakfast and before going to bed at night.

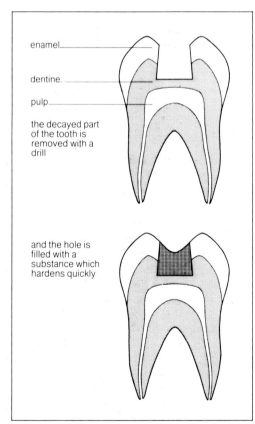

enamel
dentine
pulp

the decayed part
of the tooth is
removed with a
drill

and the hole is
filled with a
substance which
hardens quickly

Figure 6 These diagrams show what happens when a dentist fills a tooth.

2 If possible finish your meal with a rough vegetable or fruit such as a carrot or an apple, then rinse your mouth out with water.

3 Don't eat sweets or drink sugary drinks between meals, and above all don't hold sweets in your mouth and suck them for a long time.

You should visit the dentist every six months, even if you think there's nothing wrong with your teeth. Decay may have started without you realising it. The dentist can then do any necessary fillings *before* the decay gets bad.

There is a growing belief amongst dentists and scientists that fluoride helps to prevent tooth decay by strengthening teeth, particularly when they are forming, and possibly by stopping plaque-formation. Fluoride occurs naturally in the drinking water in some parts of the world, and in these areas the amount of tooth decay is said to be less than elsewhere. In some places very small amounts of fluoride are added to the drinking water, and it's claimed that this has reduced the incidence of tooth decay.

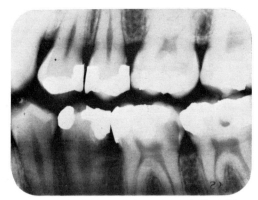

Figure 7 An X-ray picture of human teeth. The white areas are fillings.

Investigation 1

Looking at human teeth

1 Look at the inside of your mouth with a mirror.

Which structures shown in Figure 1 can you see?
How many teeth have you got altogether?
Identify your incisors, canines and cheek teeth.
Which of the teeth shown in Figure 1 have you *not* got?
Which of your teeth are permanent and which ones belong to your milk set?

2 Examine healthy human teeth in detail.

Using Figure 1 to help you, decide whether each tooth is an incisor, canine or cheek tooth.

Draw each type of tooth, showing as many of the structures in Figure 3 as you can see.

3 Examine a human skull and lower jaw.

Whereabouts does the lower jaw move against the upper jaw?

Where would the muscles which close the mouth be attached?

4 Examine decayed teeth which have been extracted by the dentist.

Compare them with healthy teeth. Whereabouts is the decay?

Why do you think the decay is situated where it is?

Investigation 2

To see the plaque on your teeth

This can be done using plaque-staining tablets available from chemist shops.

1 Chew a tablet and spread it over your teeth with your tongue. Then spit it out. (These tablets are not meant to be swallowed, but it will do you no harm if you swallow them.)

2 Rinse your mouth with water.

3 Look at your teeth in a mirror. Any plaque will be stained pink.

Where is the plaque located?

4 Brush your teeth with toothpaste in the usual way, then rinse your mouth out with water.

5 Look at your teeth in the mirror again.

Has all the plaque been removed? If not, where is it still left?

6 Brush your teeth again. Work the toothbrush this way and that, and try hard to remove all traces of plaque. Then rinse your mouth out with water.

7 Look at your teeth in the mirror again.

Has all the stained plaque been removed now?

8 If there is still some plaque between your teeth, try removing it with dental floss. This is a thread-like material which can be pulled backwards and forwards between the teeth.

Does dental floss remove the plaque?

Assignments

1 Each structure in the left hand column below is related to one of the words on the right.
Write them down in the correct pairs.

enamel	crushing
pulp	sharp
tooth fibres	hard
canine	sensitive
molar	pyorrhoea

2 In about a hundred words, give advice to the general public on how to clean their teeth so that all plaque is removed from them.

3 You are employed by a research organisation to test the claim that small amounts of fluoride in drinking water help to prevent tooth decay. How would you tackle this problem?

4 The 'dental formula' of an adult human is

$$i\frac{2}{2} \quad c\frac{1}{1} \quad pm\frac{2}{2} \quad m\frac{3}{3}$$

i stands for incisors, c canines, pm premolars, m molars. The top figures denote the number of teeth on *one* side of the upper jaw, the bottom figures denote the number of teeth on *one* side of the lower jaw.

a) What is your own dental formula at the moment?

b) Which teeth, if any, do you lack and why?

c) What was your dental formula likely to have been when you were aged 5 years old?

d) Explain what has happened to your teeth since you were 5.

Digestion in other mammals

All mammals digest their food in basically the same way as man does. However, the structure of the gut and teeth vary according to the kind of food each animal eats.

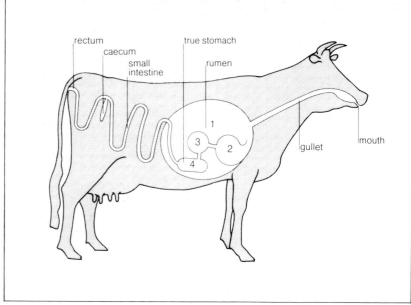

Figure 1 The stomach of a ruminant has four chambers, the first is called the rumen.

Figure 2 Tigers have large canine teeth for tearing the flesh of their prey.

Who eats what?

Animals can be split into carnivores, herbivores and omnivores. **Carnivores** eat other animals: examples include dogs, cats, lions, tigers and wolves. **Herbivores** eat plants: they include cows, sheep and horses. **Omnivores** eat other animals and plants: man and pigs are examples.

Differences in the structure of the gut

Omnivorous and carnivorous mammals have guts which are similar to man's: there is no need for them to be different. Herbivores, however, have a problem because much of their food consists of cellulose which is tough and difficult to digest. Mammals do not possess the necessary enzyme for breaking down cellulose – in fact the only organisms to possess this enzyme are certain microbes. Herbivorous mammals have three special features which enable them to digest cellulose:

1 They have a very long small intestine – as long as 40 metres in the case of the cow. Digestion is slower than in other mammals, and having a long intestine ensures that digestion is complete before the food reaches the end.
2 They have a large caecum and appendix which contain numerous bacteria capable of breaking cellulose down into sugar. Some of the sugar is used by the bacteria, but the rest is absorbed by the herbivore. This is an example of two organisms helping one another: the microbes get shelter and protection, and in return the herbivore gets sugar from cellulose. This kind of partnership is called symbiosis (see page 362).
3 Some herbivorous mammals have a special kind of stomach. Such mammals are called ruminants and they include cows and sheep. The stomach consists of four chambers, the first of which is called the **rumen** and is very large (Figure 1). The animal eats grass and swallows it into the rumen without chewing it first. After a while it stops eating, and regurgitates the grass, a little at a time, into the mouth cavity where it is chewed. This is called chewing the cud. In the rumen the food is churned up by contractions of the muscles, and the cellulose is acted on by microbes similar to those in the caecum and appendix. The food is then passed on to the other chambers where it is further processed, before entering the small intestine.

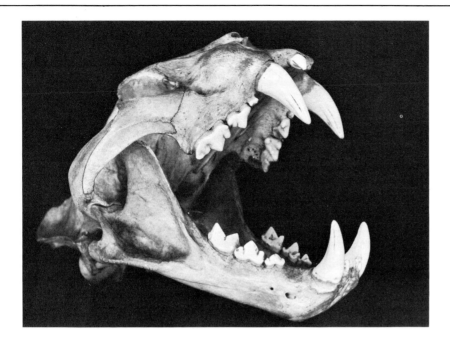

Figure 3 The skull of a lion. Notice the large dagger-like canines.

Differences in the teeth

The structure of the teeth is closely related to the diet (Investigation). Thus lions and tigers have large dagger-like canines for killing their prey and tearing its flesh (Figures 2 and 3). Dogs have an extra-large cheek tooth on either side of each jaw for scraping flesh off bones (Figure 4). Rabbits, mice

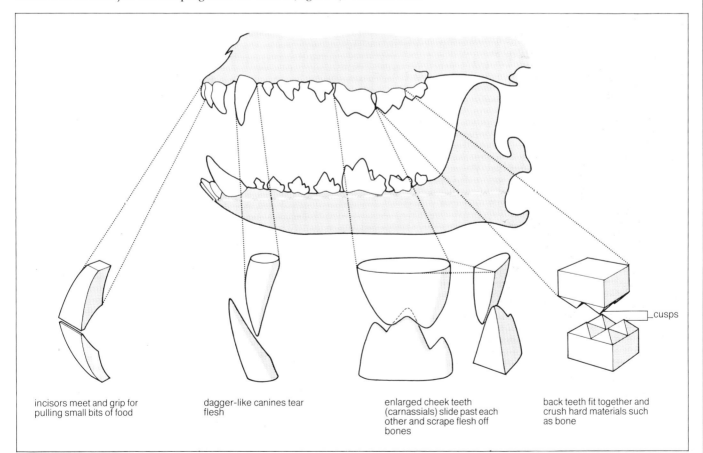

incisors meet and grip for pulling small bits of food

dagger-like canines tear flesh

enlarged cheek teeth (carnassials) slide past each other and scrape flesh off bones

back teeth fit together and crush hard materials such as bone

cusps

Figure 4 These diagrams show the structure and action of the teeth of a carnivore such as the dog.

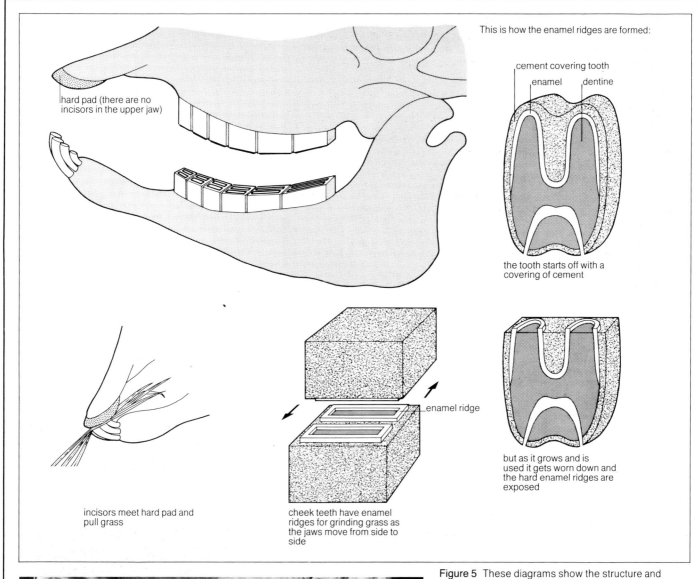

This is how the enamel ridges are formed:

hard pad (there are no incisors in the upper jaw)

cement covering tooth
enamel   dentine

the tooth starts off with a covering of cement

but as it grows and is used it gets worn down and the hard enamel ridges are exposed

incisors meet hard pad and pull grass

enamel ridge

cheek teeth have enamel ridges for grinding grass as the jaws move from side to side

Figure 5 These diagrams show the structure and action of the teeth of a herbivore such as the sheep.

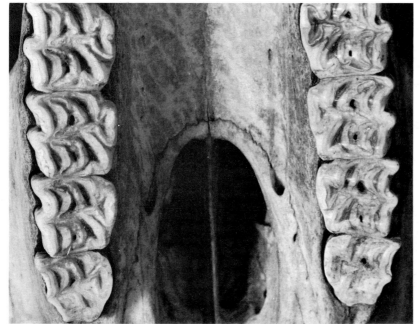

Figure 6 Notice the enamel ridges on these teeth belonging to a horse.

and squirrels have long chisel-like incisors tor cutting or rasping their food; they have no canines: instead there is a gap called the **diastema**. What do you think this is for? And horses, sheep and cows have cheek teeth with ridges on the surface for grinding up grass. The ridges result from the teeth gradually being worn down: the enamel wears down more slowly than the other materials, so it stands up above the rest of the tooth (Figure 5).

If you watch a sheep chewing its food, you will see that its lower jaw moves from side to side. The same is true of many other herbivores such as cows and deer. This fits in with the structure of the cheek teeth whose enamel ridges run longways along the length of the jaw as shown in Figure 5. Obviously the grinding effect will be greatest if the jaw moves from side to side.

Certain herbivores, such as the rhinoceros, have enamel ridges which run transversely across the cheek teeth. For them the best grinding effect will be achieved if they move the lower jaw backwards and forwards – and that's exactly what they do.

One of the most efficient herbivores is the horse. The enamel ridges on its cheek teeth form a complicated pattern and are very good at grinding up grass (Figure 6).

Investigation

Comparing the teeth of different animals

1 Examine the skull of a carnivore such as a dog. Identify the incisors, canines and cheek teeth.

   How do the teeth differ from yours? What job does each type of tooth do?

2 Repeat the above with the skull of a herbivore such as a sheep.

   How do its teeth differ from (a) a human's, and (b) a dog's?

3 Look at the skulls of other animals which your teacher gives you. In each case look carefully at the teeth, and suggest what kind of food the animal feeds on.

4 Write down the dental formula of each animal (see page 145). Are any types of teeth (incisors, canines, etc.) missing?
   How do you think the animal manages without them?

5 When you get an opportunity, watch various mammals eating. Observe the action of the jaws and relate it to the structure of their teeth.

Assignments

1 Why is it that:

   a) man cannot digest cellulose,
   b) herbivores have a particularly long small intestine,
   c) a sheep's jaw moves from side to side when it chews,
   d) the cow has a hard pad at the front of its upper jaw?

2 In what way does (a) chewing the cud, and (b) the grinding up of grass by the teeth, help digestion in a herbivore such as the cow?

3 What is it about the structure of:

   a) a lion's canine teeth that enables it to tear flesh,
   b) a dog's carnassial teeth that enables it to scrape all the flesh off a bone,
   c) a dog's back molar teeth that enables it to break a stick in two,
   d) a horse's cheek teeth that enables it to eat grass?

4 Mammals other than humans rarely suffer from tooth decay. Why do you think this is?

5 Many animals use their teeth for purposes other than feeding. Write down the names of two such animals, and in each case suggest *one* use to which the teeth may be put apart from feeding.

6 The photograph below shows the front part of the skull of a certain mammal.

   a) What kind of mammal do you think it is?
   b) What do you think it feeds on?
   c) What do you think it uses its teeth for?

   Give reasons for your answers.

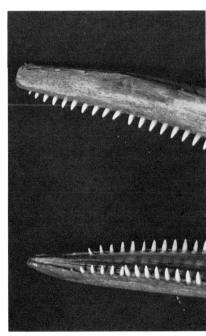

How do we breathe?

Breathing is the process by which we draw air into our bodies. It is an essential part of respiration. If a person is prevented from breathing, he is likely to die within a few minutes.

What happens to the air we breathe in?

When we breathe in, air is sucked into our **lungs**. The lungs are the main organs of our respiratory system. The butcher calls them 'lights'. If you examine them you will see why (Investigation 1). They are light and soft, and are riddled with spaces like a sponge. These spaces contain air. From this air oxygen is taken up into the blood, while carbon dioxide passes in the reverse direction. We refer to this movement of gases as gaseous exchange.

Where are the lungs?

We have two lungs, situated side-by-side in the chest or **thorax** (Figure 1). The sides of the chest are bounded by the **ribs**, which are joined to the backbone (vertebral column) at the back and the breastbone (sternum) at the front. Between the ribs are muscles called **intercostal muscles**. The thorax is separated from the abdomen below by the **diaphragm**. This is a sheet of muscular tissue, shaped like a dome, which is stretched across between the bottom-most ribs.

Each lung is surrounded by two thin sheets of tissue, called the **pleural membranes**. The inner one covers the lungs, and the outer one lines the inside of the thorax. Between them is a narrow space containing a fluid. This fluid serves as a lubricant, allowing the membranes to slide over each other smoothly as we breathe in and out.

The route by which air reaches the lungs

Air is sucked into the lungs through a series of cavities and tubes which together make up the **respiratory system** (Figure 2).

Here are some notes on the main structures which make up the respiratory system.

Figure 1 The lungs are located inside the chest cavity as shown in this diagram.

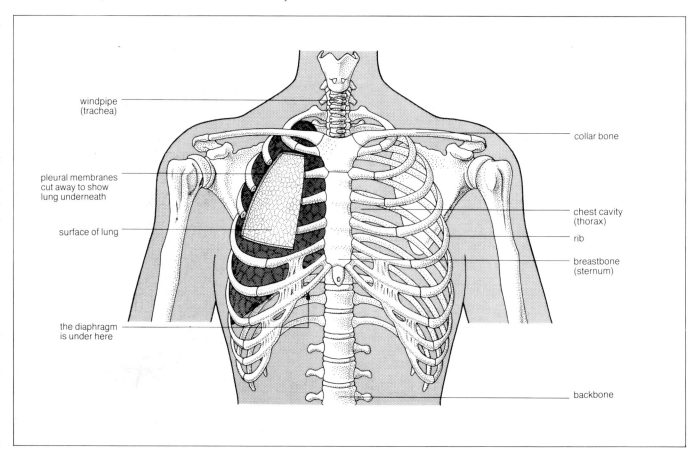

The nose

Air is drawn into the nose through the **nostrils**. The inside of the nose is moist and warm. It is moist because its lining produces slimy mucus. It is warm because there are numerous blood vessels close to the surface. If you've ever had a nose-bleed you will know what a lot of blood there is in the lining of the nose.

At the back of the nose there is a large space called the **nasal cavity**. This is divided up by several bony partitions which give it a large surface area rather like the radiator of a car. The air is warmed and moistened as it passes over these surfaces, and it is cleaned at the same time. Dust and germs get caught in the mucus and are wafted towards the throat by beating cilia. The mucus is then swallowed or coughed up – unless of course you expel it beforehand by blowing your nose. The lining of the nose is also very sensitive to touch, and this may make you sneeze.

In the lining of the nasal cavity there are sensory cells sensitive to smell. Our sense of smell tells us whether or not the air is suitable for breathing. It therefore enables us to test the air before we take it into our lungs.

We can sum up the functions of the nose by saying that it warms, moistens, cleans and tests the air we breathe in. It protects the lungs from germs and harmful substances which might injure them or start an infection.

You can, of course, breathe through your mouth. If you do this the protective functions of the nose are not carried out and the risk of infection is increased. We all breathe through our mouths when we have a cold in the nose, but in general it is a bad habit.

Leading from the nasal cavity at the front of the skull are a number of cavities called sinuses. The sinuses produce mucus which normally drains into the nasal cavity. However, the holes connecting the sinuses with the nasal cavity are small, and if the person has a cold they get swollen and blocked. The sinuses then fill up with fluid, and the pressure may cause a headache.

Figure 2 The respiratory system of man. The lungs are located in the chest cavity or thorax. There are really far more alveoli than are shown in this simplified diagram.

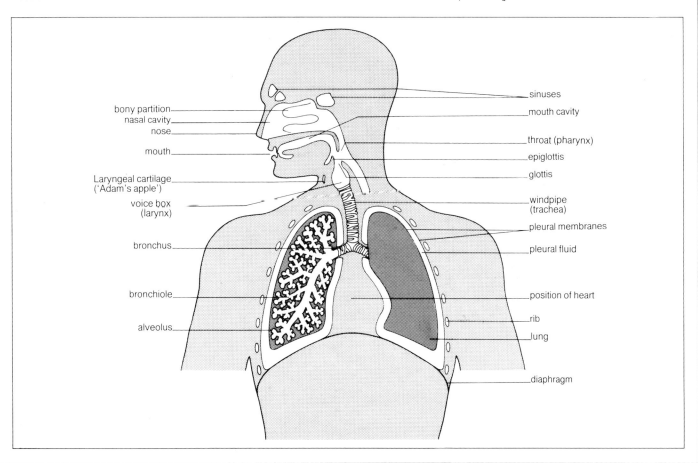

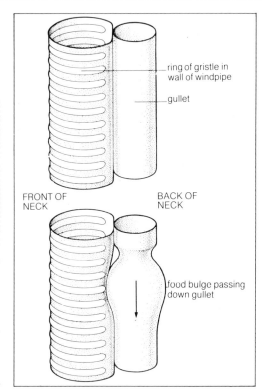

ring of gristle in
wall of windpipe

gullet

FRONT OF
NECK

BACK OF
NECK

food bulge passing
down gullet

Figure 3 The rings of gristle keep the windpipe open
even when food passes down the gullet.

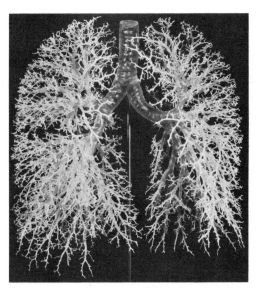

Figure 4 A corrosion preparation of the human lungs.
All the lung tissue except the bronchial tubes has
been dissolved away.

The throat

The throat, or **pharynx**, belongs to both the respiratory and alimentary systems (see page 136). Food passes from the mouth cavity into the gullet and thence to the stomach. Air, on the other hand, passes from the nasal cavity into the windpipe and so to the lungs.

The air enters the windpipe through a small hole called the **glottis**. Obviously it is important that food should not get into this hole. This is prevented by the **epiglottis**, a small flap of tissue stiffened with gristle. When we swallow, the glottis becomes closed off by the epiglottis, and breathing stops. Despite this mechanism, a piece of food may sometimes get stuck in the glottis: we say it has gone down the wrong way. It can usually be dislodged by coughing, helped if necessary by a pat on the back.

The voice box

The glottis opens into the voice box or **larynx**. This shows up at the front of your neck as your 'Adam's apple'. It feels hard because there are pieces of gristle in its wall. The voice box enables us to talk, sing and shout. Thin membranes, formed from its lining, are stretched across the cavity. They are called **vocal cords**. When air is forced through the voice box the cords vibrate, producing sounds in much the same way that a piano produces sounds when the strings vibrate.

The windpipe

If you put your finger below your 'Adam's apple', you can feel your windpipe or **trachea**. It is a straight tube, about 12 cm long, situated just in front of the gullet.

For air to pass freely to and from the lungs, it is important that the windpipe should be open all the time. To keep it open, its wall is stiffened by rings of gristle. These rings are like a pile of C's which face towards the centre of the neck so the open side of the C is next to the gullet: this allows the gullet to expand when food passes down it (Figure 3).

The inner lining of the windpipe produces mucus and is covered with cilia. Some of the dust particles and germs which have escaped being caught in the nasal cavity, get caught in this mucus. The cilia waft the mucus upwards to the glottis so that it can be either swallowed or coughed out. The windpipe thus helps to prevent germs and harmful substances getting to the lungs.

The bronchi

After the windpipe has entered the chest, it splits into two short tubes called **bronchi** (singular: bronchus), one to each lung. The bronchi are similar to the windpipe, except that they are narrower.

The bronchioles

Within each lung the bronchus splits into numerous branches, like a tree: the whole structure is called the **bronchial tree** (Figure 4). The branches are called bronchioles, and they get very narrow towards the ends. Their walls are not surrounded by rings of gristle; instead they contain smooth muscle which allows them to widen or get narrower depending on circumstances.

The alveoli

Each bronchiole leads to a bunch of tiny sacs called **alveoli** (singular: alveolus). The alveoli are surrounded by a network of blood capillaries, rather like a string bag (Figure 5). The capillaries are in close contact with the alveoli, and the membrane separating them is extremely thin. Across this membrane gaseous exchange takes place: oxygen diffuses from the alveoli into the blood, and carbon dioxide diffuses from the blood into the alveoli. The lining of each alveolus is covered by a thin layer of fluid and the oxygen dissolves in this before it passes through into the blood (Figure 6).

There are about 150 million alveoli in each lung, and altogether they cover a very large surface area. Someone has worked out that if you were to open them out and flatten them like a sheet, they would cover an area as large as a tennis court! It is very important that the lungs should have a large surface area, because it means that more oxygen can be taken up by the blood every time we breathe in.

The alveoli always contain air, even when we breathe out as hard as we can. If there was no air inside them, their walls would cave in and stick together. Their surface area would then be reduced and gaseous exchange would be impossible.

How does air get into the lungs?

Breathing takes place by movements of the chest. The chest works rather like a pair of bellows, sucking air in and then forcing it out.

We can divide breathing into two parts:

Inspiration: this is the sucking of air into the lungs, and it is brought about by the chest expanding.

Expiration: this is the forcing of air out of the lungs, and it is brought about by the chest contracting.

This is how inspiration is brought about (Figure 7):

1 The ribs swing outwards and upwards. This is brought about by contraction of the intercostal muscles. This increases the size of the thorax in the side-to-side direction. At the same time the breast bone moves forward slightly, so the size of the thorax is increased in a front-to-back direction as well.
2 The diaphragm moves downwards, so that instead of being dome-shaped, it becomes flattened. This is brought about by contraction of muscles in the diaphragm itself. This increases the size of the thorax in a top-to-bottom direction.

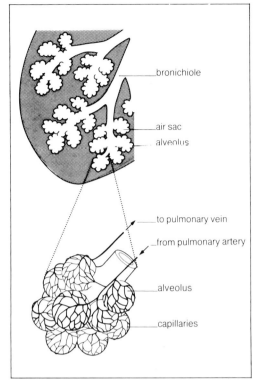

Figure 5 These diagrams show the detailed structure of part of a lung and its blood supply

Figure 6 As the red blood cells go past an alveolus, they give up carbon dioxide and pick up oxygen. These gases move in and out by diffusion.

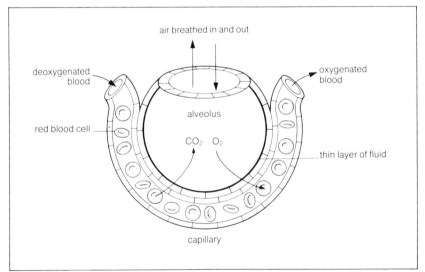

Figure 7 Diagrams showing how the chest expands when we breathe in.

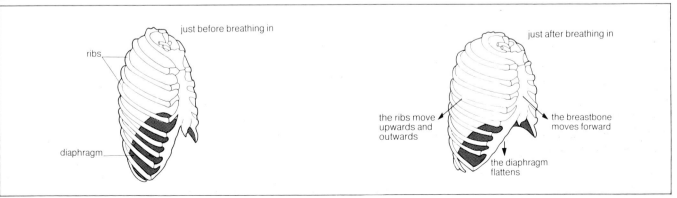

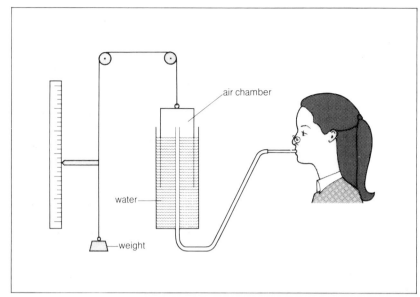

Figure 8 A spirometer measures the amount of air you can take into your lungs. You take the deepest possible breath in, and then exhale as completely as possible into the air chamber. The major divisions on the scale are litres.

All these movements result in a negative pressure – a suction force – being developed inside the thorax. The result is that the walls of the lungs are pulled outwards and air is drawn into them (Investigation 2).

Expiration is brought about by the reverse process: the ribs swing downwards and inwards, the breastbone moves back slightly, and the diaphragm bows upwards. A positive pressure is developed inside the thorax. The result is that air is forced out of the lungs.

How much air do we take into our lungs when we breathe in? About half a litre when we are at rest. However, this is only a fraction of the amount we *can* take in if we want to. You can find out how much air you can take into your lungs by means of a spirometer (Figure 8). The maximum amount for an adult male is usually between 4 and 5 litres, though a trained athlete can often take in more than this. The total amount of air which a person can breathe in is called the **vital capacity**.

How is our breathing controlled?

When we take exercise certain changes take place in our breathing. The most noticeable change is that we breathe more quickly. This is because our muscles are working harder so they need more oxygen. Also the extra carbon dioxide which they produce must be removed quickly, otherwise it might build up and poison our tissues.

You do not have to think about this; it happens automatically. It is brought about by a reflex: the brain senses that there is too much carbon dioxide in the bloodstream, and this automatically causes us to breathe faster.

In fact we do not just breathe faster – we breathe more deeply as well. In this respect there is a difference between fit and unfit people: fit people – trained athletes for example – tend to breathe more deeply when they take exercise; unfit people tend to breathe more quickly, keeping their breathing rather shallow. Are you a deep breather or a shallow breather?

There are many other ways in which our breathing changes: for example, when we cough, sneeze, gasp, yawn or talk. Yawning, for example, occurs when we feel tired: we take a long deep breath in, which has the effect of getting more oxygen into the body. Some of these changes are brought about by involuntary reflexes: we cannot help them happening. Over other activities such as talking, we have voluntary control.

Investigation 1

Looking at the lungs of a mammal

1 Examine the lungs of a pig or sheep obtained from the butcher.

What is the shape of each lung?

2 Press one of the lungs with your finger.

What does it feel like?
How would you explain how it feels?

3 Look at the windpipe.

How is it attached to the lungs?

4 Squeeze the windpipe with your fingers.

What does it feel like?
How would you explain how it feels?

5 Attach a pair of bellows to the cut end of the windpipe. Make sure the joint is airtight.

6 Pump air in and out of the lungs.

What happens?
Explain your observations.

7 Look at a prepared slide of a section of lung under the microscope.

Describe what it looks like.

In what ways does the microscopic appearance of the lung fit in with your earlier observations?

Investigation 2

A working model of the chest and lungs

You will need a bell jar, a rubber stopper with a hole in it, a Y tube, two balloons, a sheet of rubber and some string.

1 Assemble the above items as shown in the illustration. Make sure the bell jar is air-tight.

2 Grasp the rubber diaphragm and move it downwards and upwards.

What happens to the balloons when you move the diaphragm downwards?

What happens to them when you move it upwards?

Explain your observations.

How does the action of this model differ from the action of your own chest?

Mention two ways in which the action of this model differs from your own chest.

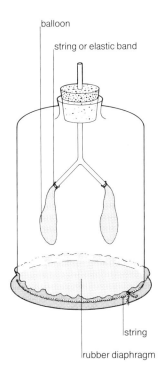

balloon

string or elastic band

string

rubber diaphragm

Assignments

1 Make a list of all the structures that air flows through on its way to the lungs.

2 What function is carried out by each of the following:

   a) the gristle in the wall of the windpipe;
   b) the epiglottis;
   c) the intercostal muscles; .
   d) the diaphragm;
   e) the pleural fluid?

3 Why is it desirable to:

   a) breathe through your nose rather then your mouth;
   b) stop talking when you are about to swallow some food;
   c) breathe as deeply as possible;
   d) blow your nose whenever necessary?

4 The table below shows the percentage volumes of oxygen and carbon dioxide in the air inhaled and exhaled by a man:

	inhaled air	exhaled air
oxygen	20.00	16.00
carbon dioxide	0.03	4.00

Explain how the change in composition of the air is brought about in the lungs.

5 When you yawn you take a deep breath in and then you let the air out quickly. Explain each of the following in terms of breathing in and out: a cough; a gasp; a sneeze; a sigh; a laugh.

6 An experiment was carried out on a young man in which the volume of air taken in at each breath, and the number of breaths per minute, were measured at rest and after running. Here are the results:

	volume of air per breath	breaths per minute
at rest	450 cm³	20
after running	1000 cm³	38

a) What is the total volume of air breathed in per minute at rest and after running?

b) 20 per cent of the air breathed in consisted of oxygen, but only 16 per cent of the air breathed out consisted of oxygen. Assuming that these figures remain constant, work out the volume of oxygen entering the blood per minute at rest and after exercise.

c) Why does the amount of oxygen taken up into the blood need to increase after exercise?

d) How is the increase in the rate of respiration brought about in the body?

7 The diagram below shows the chest (thorax) of a human, and some of the structures which it contains.

Which of the structures labelled:

a) contains air,
b) is filled with fluid,
c) is lined with cilia,
d) is made of bone,
e) moves downwards during inspiration?

In each case identify the structure by its letter.

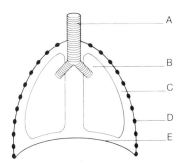

When things go wrong with our breathing

If our lungs become damaged or diseased, or if air is prevented from getting to them, our lives are in danger. In this Topic we will see how this can happen.

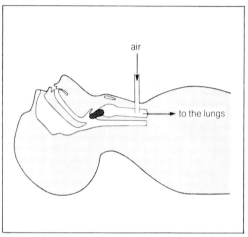

Figure 1 If a person's glottis is completely blocked a tracheotomy may save his life.

What happens if the respiratory tract becomes blocked?

It all depends where the block occurs. If you have a heavy cold, extra mucus is secreted in the nose which may become blocked. However, this isn't serious. You can blow your nose or breathe through your mouth, and if necessary you can clear your nose with nasal drops.

Occasionally a piece of food gets stuck in the glottis so firmly that it cannot be moved. If this happens an emergency operation called a **tracheotomy** may be necessary. A small cut is made into the windpipe below the Adam's apple. A tube is then inserted into it, through which the person can breathe until the obstruction has been cleared (Figure 1).

Artificial respiration

If a person has an accident, he may pass out and stop breathing. Sometimes it is possible to keep the person alive by **artificial respiration**. This must be carried out as soon as possible, otherwise the brain cells may be damaged so badly by lack of oxygen that they never recover. This may happen within a few minutes after the person stops breathing, so speed is essential.

One of the best methods of artificial respiration is the 'kiss of life', known as mouth-to-mouth resuscitation (Figure 2). First you lay the person on his back. You then take a deep breath in, and breathe out into his mouth. As you are forcing your own exhaled air into his lungs, you might think this would do more harm than good. However, there is enough oxygen in your exhaled air to keep him alive. What's more, the carbon dioxide in your exhaled air may stimulate him to start breathing again.

After some accidents the person's brain is so badly damaged that he cannot start breathing for himself. He may then be kept alive by means of a **resuscitator**. He is connected by a tube to a machine which regularly forces air or oxygen into his lungs and then sucks it out. A system of valves ensures that fresh air is sent to the lungs each time. An unconscious person can be kept alive for many weeks or even months on a machine like this. Sometimes the brain recovers sufficiently for the person to start breathing again. On the other hand he may not recover, and the family and doctors have to decide whether to keep him alive on the machine or to switch it off and let him die. Obviously this is an agonising decision to have to make.

Figure 2 Artificial respiration by mouth-to-mouth resuscitation otherwise known as the kiss of life.

1 Pinch the nostrils shut with the fingers of one hand, then tilt the head back and push the lower jaw forward so the chin juts out. This will force the tongue forward and open the air passages.

2 Take a deep breath, then open your mouth and seal your lips against the person's mouth. Breathe out firmly but gently into the person's mouth and so into his lungs.

3 Lift your mouth off, then turn your head so as to look at the person's chest. If you have been successful you will see that it has risen and is now falling as air comes out of the lungs.

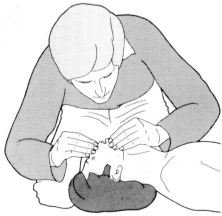
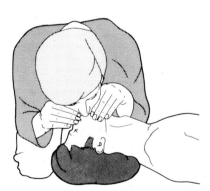
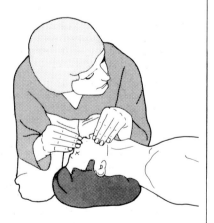

4 Repeat steps 2 and 3 at a steady rate. The person's colour should improve, and eventually he should start breathing for himself.

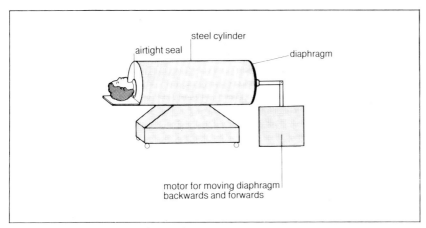

Figure 3 Simplified diagram of an iron lung.

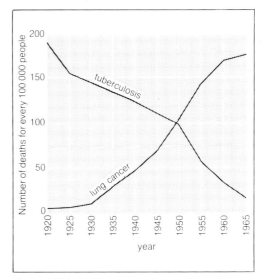

Figure 4 This graph compares the numbers of people dying each year from tuberculosis and lung cancer between 1920 and 1965.

Sometimes people, though fully conscious, cannot breathe because their chest muscles are paralysed. This sometimes happens in poliomyelitis, for example (see page 226). In these circumstances the patient may be put in an **iron lung** (Figure 3). This is a large cylindrical chamber. The patient's head sticks out through an airtight seal at one end. At the other end there is a flexible 'diaphragm' which is moved backwards and forwards by a motor. This has the effect of moving the person's chest wall outwards and inwards, so air is sucked into his lungs and then forced out. Hopefully he will eventually regain the use of his muscles and start breathing for himself.

Respiratory diseases

Despite the mechanisms in the nose for keeping germs out, there are times when most of us get an infection in some part of the respiratory system. The area becomes inflamed and sore, we cough a lot, and a large amount of mucus may be produced which makes it difficult to breathe. Any part of the respiratory system can become infected. Thus **pharyngitis** is inflammation of the pharynx (throat), **tracheitis** is inflammation of the trachea (windpipe), and **bronchitis** is inflammation of the bronchial tubes. **Laryngitis** is inflammation of the larynx (voice box), and this may cause us to become hoarse and lose our voice. Sometimes the pleural membranes surrounding the lungs become inflamed. This is called **pleurisy** and it can make breathing painful.

A severe infection of the lungs may give rise to **pneumonia**. Fluid collects in the alveoli: this cuts down the area over which gaseous exchange can take place, so the patient gets short of breath.

Another serious disease of the lungs is **tuberculosis**, or TB for short. This is caused by bacteria which destroy the lung tissue. Doctors can find out if a person has got TB by doing a chest X-ray which shows up the diseased areas of the lungs. At one time TB, or consumption as it was called, was one of the most common causes of death. Thanks to modern medicine it is now rare.

Today **lung cancer** has taken over from TB as the major killer (Figure 4). In lung cancer a growth develops in the wall of the bronchial tubes. This blocks them, so breathing becomes more and more difficult. Unless the growth is discovered, and destroyed, in time the cancer may spread to other neighbouring organs such as the liver or spine.

Doctors can find out if a person has got cancer of the lungs by doing a chest X-ray (Figure 5). If a growth is visible, the person may have an operation in which the diseased part of the lung is removed, or the growth may be destroyed by radiation treatment. However, these measures are not always successful and unfortunately many patients die eventually.

No-one knows the exact cause of lung cancer, though smoking and pollution of the air seem to be involved.

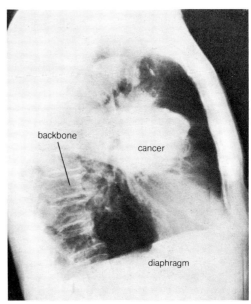

Figure 5 X-ray of human chest taken from the side showing lung cancer. The front of the chest is on the right.

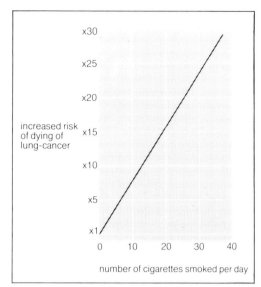

Figure 6 This graph shows the relationship between the number of cigarettes smoked per day and the risk of dying of lung cancer.

Figure 7 These diagrams show how emphesema affects the lungs.

How does smoking affect the lungs?

Most people who smoke inhale the smoke right down into their lungs. What effect does this have on them?

One way of finding the answer is to take two groups of people of the same sex and age. One group are smokers and the other group are non-smokers. You then study each group over a period of many years, and see what happens to their health. If a certain disease is developed by many of the smokers, but not by the non-smokers, this suggests that smoking may cause this particular disease.

In the last thirty years several surveys of this kind have been carried out in Britain by the Royal College of Physicians, and similar surveys have been carried out in other countries. The results all show one thing: amongst people who smoke there are more cases of lung cancer than amongst people who don't smoke. In other words, smoking seems to be associated with lung cancer.

This conclusion is supported by laboratory experiments on animals. These experiments have shown that cigarette smoke contains chemical substances which cause cancer. We call these substances **carcinogens**. The main carcinogens in cigarette smoke are a group of aromatic hydrocarbons which are usually referred to as 'tar'.

Now here is an important point. Being a smoker doesn't mean that you are *bound* to get lung cancer; nor does being a non-smoker ensure that you won't get it. All we can say is that you are more *likely* to get it if you smoke.

Look at Figure 6. This shows that the more cigarettes a person smokes, the greater is the chance of his getting lung cancer. A man who smokes twenty cigarettes a day is about fifteen times more likely to die of lung cancer than a non-smoker; and if he smokes thirty a day, he is about 25 times more likely to get it.

The surveys which have been carried out indicate that a smoker is *less* likely to get lung cancer if he:

1 doesn't inhale,
2 smokes cigars or a pipe rather than cigarettes,
3 takes fewer puffs per cigarette,
4 smokes filter-tips,
5 smokes 'low-tar' cigarettes.

But the best thing is to give it up altogether. If a heavy smoker stops smoking, the risk of his getting lung cancer gradually falls until after a few years it is about the same as for a non-smoker.

Other diseases caused by smoking

When a person smokes, tiny particles in the smoke get caught on the lining of the windpipe and bronchial tubes. Extra mucus is produced and the cilia stop beating: smoking one cigarette is said to stop the cilia beating for about an hour. The mucus collects in the bronchial tubes and this gives rise to a 'smoker's cough'. If the tubes become infected, the person may get **chronic bronchitis**. 'Chronic' means long-lasting: instead of clearing up, the disease persists.

Further unpleasant effects may follow. Repeated coughing may cause the delicate walls of the alveoli to break down into larger air spaces. This cuts down the surface area over which gaseous exchange can take place, so the person gets very short of breath. Doctors call this condition **emphesema** (Figure 7).

Although smoking mainly affects the lungs, it can also cause cancer of other organs such as the mouth, throat, oesophagus and bladder. It is also associated with heart disease and stomach ulcers, and a woman who smokes while she is pregnant is more likely to have a spontaneous abortion or stillbirth or to give birth to an under-sized baby.

Smoking and society

In Britain between 60 and 70 thousand people die of lung cancer, chronic bronchitis and emphesema every year. This is over eight times as many as are killed in road accidents.

The connection between smoking and these diseases is now so firmly established that in Britain cigarette commercials on television have been banned, and every packet of cigarettes has to carry a government health warning. Smoking has been prohibited in many public buildings and in cinemas special areas are often reserved for non-smokers.

Despite this, more and more cigarettes are bought each year. Only amongst doctors has smoking decreased. They understand the risks too well, and they also know how unpleasant it is to die of lung cancer.

Scientists have shown that lung cancer is more common amongst people who live in towns than amongst those who live in the country. There is evidence that it can be brought on by smoke from chimneys, motor car exhaust fumes, asbestos dust and radio-active materials. However, these causes are insignificant compared with smoking. It is claimed that if everyone gave up smoking, deaths from lung cancer would fall to a tenth of what they are at the moment.

Hay fever and asthma

Many people suffer from **hay fever**. This is usually caused by a reaction to pollen, and is therefore particularly common in summer. The lining of the nasal cavity becomes sensitive and inflamed and produces a large amount of mucus, so the nose runs and the person sneezes a lot. The eyes may be affected in the same way, becoming itchy, sore and weepy. When a person reacts adversely to a substance in this kind of way, we say he is **allergic** to it.

Asthma is more serious. The muscles in the walls of the bronchioles contract, so the tubes get narrower. This makes it difficult to breathe and the person wheezes. The attacks are often brought on by pollen or dust, or occasionally by some kind of food to which the person is allergic. With some people the attacks are made worse by nervousness or worry. Asthma can be treated with drugs which make the bronchial muscles relax, so the tubes widen allowing air to be breathed into the lungs more easily.

Assignments

1 Briefly state the cause and symptoms of each of these diseases: bronchitis, pleurisy, tuberculosis, emphesema and asthma.

2 What is meant by mouth to mouth resuscitation? Under what circumstances would it *not* be possible to carry it out?

3 Explain why each of these remarks is unscientific:

   a) 'My dad smokes like a chimney so he's bound to get cancer'.

   b) 'All that stuff about smoking and cancer is nonsense: my uncle died of lung cancer and he never touched a cigarette all his life'.

4 In 1952 two British scientists carried out a survey in a large hospital. They selected two groups of patients, both the same sex and roughly the same age. The patients in group A all had lung cancer, whereas those in group B had other diseases. The scientists then found out how many patients in each group smoked. Here are the results:

	Percentage of patients who smoked more than 15 cigarettes per day
Group A	25
Group B	13

	Percentage of patients who were non-smokers
Group A	0.5
Group B	4.5

   a) What do you think the scientists were trying to find out?

   b) Group B is called the control group. Why was it necessary to investigate this group of patients as well as Group A?

   c) What conclusion would you draw from the results?

   d) Suggest one other way the scientists might have carried out their survey.

5 It has been suggested that smoking and lung cancer appear to be connected, not because smoking *causes* cancer, but because people who need to smoke are the kind of people who get cancer. What sort of investigations would have to be carried out to show that this is *not* the correct explanation?

Living without oxygen

What makes dough rise when bread is made? It is because of an organism which can respire without oxygen.

Figure 1 In making bread yeast causes the dough to rise.

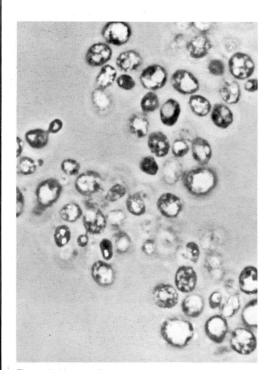
Figure 2 Yeast cells seen under the microscope.

Respiration without oxygen

Organisms obtain energy by respiration. Normally sugar (glucose) is broken down in the presence of oxygen into carbon dioxide and water (see page 133):

glucose + oxygen → carbon dioxide + water + energy
$C_6H_{12}O_6$ $6O_2$ $6CO_2$ $6H_2O$

For sugar to be broken down like this oxygen is essential. For this reason we call it **aerobic respiration**.

Now what happens if no oxygen is available? Usually the organism will suffocate. However in some cases sugar may still be broken down and energy released. Instead of being broken down into carbon dioxide and water, the sugar is turned into other products. Oxygen is not needed for this, so we call it **anaerobic respiration** – respiration without oxygen.

Anaerobic respiration occurs in both plants and animals, but the end products are different in each case.

How do plants respire without oxygen?

Plants which can respire without oxygen break sugar down into **ethanol (alcohol)** and carbon dioxide gas.

sugar → ethanol + carbon dioxide + energy
$C_6H_{12}O_6$ $2C_2H_5OH$ $2CO_2$

We call this process **alcoholic fermentation** (Investigation 1).

Anaerobic respiration does not produce as much energy as aerobic. In aerobic respiration the sugar is broken down completely; in anaerobic respiration it is only partly broken down – a lot of energy is still locked up in the alcohol. This can be shown by burning some alcohol, the energy in it is then released as heat. Though inefficient, alcoholic fermentation is a useful way of getting energy when oxygen is scarce.

Another plant which does this is yeast. Fresh yeast bought from a shop looks like putty. But it is really a fungus consisting of millions of tiny living cells (Figure 2). Wild yeast grows on the surface of fruit, feeding on sugar. In the right conditions it multiplies rapidly by budding: each cell pinches off new ones. A large number of cells can be formed in a short time.

For centuries man has used yeast for making alcohol and bread.

Making alcohol

Basically all you need for making alcohol is sugar, yeast and water. But to make a pleasant alcoholic drink it is not quite so simple, as any wine-maker will tell you.

Wine is usually made from grapes. The grapes are crushed and the juice extracted. The juice contains sugar and wild yeast. The yeast ferments the sugar and gradually turns it into alcohol.

Wine-making is an art which has been practised for over 4000 years. Although the alcohol is always the same, every wine has its own flavour. This depends on the type of grapes used and the conditions in which fermentation is carried out.

People sometimes make home-made wine from other plants such as elderberries, turnips or dandelions. Fermentation is usually carried out in a large glass jar which is put in a warm place because fermentation occurs more quickly in warm conditions. The jar is fitted with a valve which allows carbon dioxide to escape but prevents bacteria from getting in (Figure 3). If bacteria do get in they may turn the alcohol into vinegar.

Beer is made from barley. The process is known as brewing. The grain, which contains malt sugar, is mashed with water, and the resulting liquid is boiled with hops to give it the right flavour. Then yeast is added and fermentation commences (Figure 4). The sugar is gradually converted into alcohol.

One problem is that alcohol is poisonous in large amounts. If the concentration of alcohol in fermenting wine or beer gets more than about 14 per cent it kills the yeast and fermentation stops. This is why beer and wine never contain more than this amount of alcohol. The only way to produce a stronger alcoholic drink is to distil it after fermentation is complete. The fermented liquid is heated to a certain temperature in a special flask: the alcohol vaporises and condenses on the cool sides of the flask. The drops of alcohol are then collected. This is how spirits such as whisky, gin and vodka are made. A spirit may contain as much as 60 per cent alcohol.

Wine and beer making used to be carried out mainly by monks in monasteries, but now it is a major industry. In Great Britain over 6000 million litres of beer are drunk each year. Yeast too is manufactured on a large scale. For the pleasure (and the problems!) which we gain from a night in the pub we have to thank this tiny organism.

Making bread

When yeast ferments, carbon dioxide gas is given off. Man makes use of this in baking bread.

Imagine you are a baker. You make some dough by mixing flour and water. To the dough you add a small amount of sugar and yeast (Investigation 2). You then leave the dough for an hour or so in a warm place. During this period the living yeast cells multiply and give off carbon dioxide gas. This should make the dough rise, more or less doubling its size (Figure 1). Then you bake the dough in a hot oven: the heat kills the yeast and evaporates the alcohol. Result? A crisp golden loaf if you're lucky – or a brick if you're not.

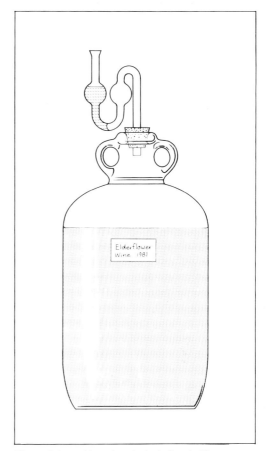

Figure 3 In making wine the jar is fitted with a valve which allows carbon dioxide to escape but prevents bacteria from getting in.

Figure 4 Beer fermenting in fermentation tanks in a brewery.

Figure 5 A sprinter can hold his breath while running a hundred metres.

How do animals respire without oxygen?

If there is no oxygen available, animals break sugar down into **lactic acid**. In this case carbon dioxide is *not* given off.

$$\text{sugar} \rightarrow \text{lactic acid} + \text{energy}$$
$$C_6H_{12}O_6 \quad\quad 2C_3H_6O_3$$

Animals which live in places where there is little or no oxygen respire anaerobically. Here are some examples: worms living in mud at the bottom of stagnant lakes; diving mammals such as whales and seals which can stay under water for long periods; and parasites like the tapeworm which live in the gut. And we can do it too.

If we take strenuous exercise our muscles need extra oxygen. Unfortunately we cannot breathe fast enough, nor pump our blood sufficiently quickly to get oxygen to our muscles. So the muscles produce energy by making lactic acid.

Think of running a 100 metre sprint (Figure 5). During the race lactic acid builds up in your body. Lactic acid is a mild poison and it causes our muscles to ache (Investigation 3). When the race is over we have to get rid of it. This is done by breaking it down into carbon dioxide and water. Oxygen is needed for this, and it is why we pant immediately after the race. The oxygen required to get rid of the lactic acid is called the oxygen debt. If we incur an oxygen debt during a race, we must pay it off immediately afterwards. Because the muscles can work for a short time without oxygen, a sprinter can hold his breath while running a hundred metres.

In a long distance race lactic acid builds up to begin with, but later it is removed while the athlete is actually running. When this happens we say that the person has got his 'second wind'.

Anaerobic respiration produces far less energy than aerobic and it cannot go on indefinitely. However, it can make the difference between life and death. An antelope fleeing from a cheetah may owe its life to the fact that for a short time its muscles can produce energy without oxygen.

How long can organisms respire without oxygen?

For man the time is very short – a matter of seconds. Some lower animals and plants can respire anaerobically for much longer periods, but eventually they must return to aerobic respiration.

Certain bacteria and parasites can respire anaerobically all their lives. They can live permanently in places where there is no oxygen. In fact some of them are actually poisoned by oxygen.

Investigation 1

Finding out about alcoholic fermentation

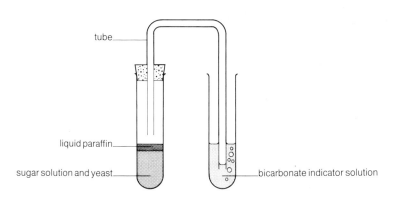

tube

liquid paraffin

sugar solution and yeast

bicarbonate indicator solution

1 Put some 10 per cent glucose into a large test tube to a depth of 2 cm.

2 Boil the glucose to expel any oxygen present in it.

3 Cool it, then add a little yeast.

4 Pour a thin layer of liquid paraffin on top of the glucose to stop oxygen getting to the yeast.

5 Set up the test tube as shown in the illustration.

6 Set up a second test tube exactly like the first one but do not add any yeast to the glucose. This is your control.

7 Leave the test tubes in a warm place for at least an hour. Then examine them.

Has the lime water gone cloudy? If it has, carbon dioxide has been given off.

Sniff the contents of the test tubes. Does either smell of alcohol?

Feel the two test tubes. Is one warmer than the other?

What conclusions do you draw from this experiment?

How might the experiment be improved?

Investigation 2

To find out if yeast makes dough rise

1 Make a small amount of dough as follows. Add about 10 g of sugar to 50 g of flour. I hen add water a little at a time and mix with a knife. Don't add too much water.

2 Put some yeast in a test tube and shake it up with some water.

3 Divide the dough into two portions. To one portion add the yeast suspension and mix it in well with your hands. Do not put any yeast into the second portion.

4 Grease the inside of two beakers. Into one beaker put the dough which contains yeast.

5 Into the second beaker put the dough which does not contain yeast. This is your control.

6 Leave both beakers in a warm place for about an hour.

7 After an hour compare the appearance of the dough in the two beakers.

Has either risen? If so which one – and why?

Design, and carry out, an experiment to find out the effect of temperature on the rising of dough.

Investigation 3

To show the effect of lactic acid in our muscles

1 Raise one arm above your head.

2 Clench and unclench your fist twice a second for as long as you can.

3 Notice the feeling in your arm as you exercise your muscles.

4 When you can continue no longer, rest your arm on your lap and follow what happens to the feeling.

How would you describe the feeling in your arm during and after the exercise?

Does your experience fit in with the idea that lactic acid builds up in the muscles during exercise and is removed afterwards?

Assignments

1 A winemaker always makes sure that his equipment is absolutely clean before he starts. Why do you think this is important?

2 Mr. Smith and Mr. Brown both make their own wine. Mr. Smith leaves his to ferment in the airing cupboard whereas Mr. Brown puts his in the cellar. Who's would you expect to ferment first and why?

3 A housewife makes some marmalade and stores it in jars in a cupboard in the kitchen. When she opens one of the jars several months later the marmalade looks frothy and smells of alcohol. What do you think has happened, and why? How might she prevent this occurring in the future?

4 Why does anaerobic respiration produce less energy then aerobic respiration?

5 The following animals are all able to respire anaerobically:

a) whales, b) the beef tapeworm, c) threadworms, d) the coelacanth, e) the worm *Tubifex*.

Find out where each of these animals lives, and then explain why it is useful to it to be able to respire anaerobically.

Blood, the living fluid

An average sized man contains about five litres of blood – that's nearly a bucket full. In this Topic we will deal with the main components of blood and what they do.

Figure 1 Blood cells greatly magnified.

What does blood consist of?

To the naked eye blood looks like a simple liquid. However, if you look at some blood under the microscope, you can see that there is more to it than that (Investigation 1). In fact it is a very special kind of tissue in which numerous cells float about in a fluid (Figure 1).

The cells are of two types: **red blood cells** (erythrocytes) and **white blood cells** (leucocytes). The fluid part of the blood is called **plasma**.

Red blood cells

The red blood cells are extremely numerous: a single drop of blood contains millions of them. Their job is to carry oxygen and carbon dioxide around the body.

The red blood cell has a very distinctive shape: it is a disc which looks as if it has been pressed in on either side, like the wheel of a car (Figure 2). This gives it a large surface area so it can take up more oxygen.

Another peculiar feature of red blood cells is that they do not have a nucleus. The inside is filled with the red pigment **haemoglobin** – this is what makes blood look red. Haemoglobin is a remarkable substance, and is responsible for carrying the oxygen.

How does the blood carry oxygen?

When a red blood cell passes through the lungs, the haemoglobin readily takes up oxygen. However, when it reaches the tissues it gives it up equally readily (Figure 3). The movement of oxygen in and out of the red blood cells takes place by diffusion.

How does haemoglobin work? When it combines with oxygen it is turned into a compound called **oxyhaemoglobin**. In this form the oxygen is carried by the blood from the lungs to the tissues. Once it has reached the tissues, the oxyhaemoglobin releases the oxygen and is turned back into haemoglobin. Haemoglobin contains iron, and this plays an important part in the way the oxygen is carried. This is why we need iron in our food.

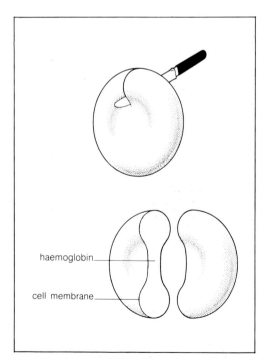

haemoglobin

cell membrane

Figure 2 A human red blood cell sectioned to show its characteristic shape.

How does the blood carry carbon dioxide?

The red blood cells also carry carbon dioxide from the tissues back to the lungs. Most of the carbon dioxide combines, not with the haemoglobin, but with water to form carbonic acid. There is an enzyme inside the red cell which causes the carbon dioxide to combine with the water extremely quickly. Once the carbonic acid has been formed it undergoes further reactions which we need not go into here.

Red blood cells and carbon monoxide posioning

Carbon monoxide is a gas which is given off by burning fuel. It is poisonous because it affects our blood. It combines with haemoglobin about three hundred times more readily than oxygen does. The result is that if we breathe it in, it displaces the oxygen from the red blood cells, so oxygen can't be carried to the tissues. Breathing concentrated carbon monoxide gas for more than a few minutes can be fatal.

Carbon monoxide is present in coal gas, but not in natural gas such as is now used in Britain. It is also present in motor car exhaust fumes, which is why it is dangerous to run a car engine in a closed garage. Small amounts of it are also present in cigarette smoke, which is why a person who is not used to smoking feels faint when he smokes a cigarette.

How are red blood cells produced?

Red blood cells live for only about four months, after which they are destroyed. To keep up the full number in our bloodstream, new ones are constantly being produced. They are made in the bone marrow, the soft tissue in the centre of certain bones (see page 270). About two million are manufactured every second!

In certain circumstances red blood cells are produced at an even faster rate, so the number in the blood increases. This happens, for example, when people live at high altitudes where there is not so much oxygen in the air. Their extra red blood cells help them to get enough oxygen to their tissues.

If a person's blood does not contain enough red blood cells, or enough haemoglobin, he suffers from **anaemia**. An anaemic person is tired and pale. Anaemia can be caused by not getting enough iron in our food, or by losing a lot of blood.

White blood cells

There are fewer white blood cells than red: for every 700 red blood cells there would be about one white blood cell. They do not contain haemoglobin and they have a nucleus just like most other kinds of cell. Their job is to kill germs which get into the body, so they help to defend us against disease.

There are several kinds of white blood cell (Figure 4). They fight disease in different ways: some of them are like little amoebas and they kill germs by eating them. We call these cells **phagocytes**. Others produce chemical substances called **antibodies** which stick to the surface of the germs and kill them (see page 232).

Phagocytes are produced in the bone marrow, whereas antibody-producing cells are produced in the lymph glands (see page 179). In both cases extra cells are produced when we are ill and need a good supply.

Some unfortunate people develop a disease known as **leukaemia**, a sort of cancer of the blood. The number of white blood cells increases greatly, and they start destroying the red blood cells. There are different kinds of leukaemia, some more serious than others. Some kinds can be treated quite successfully with certain drugs.

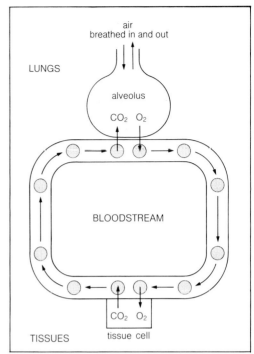

Figure 3 Red blood cells carry oxygen from the lungs to the tissues, and carbon dioxide from the tissues to the lungs.

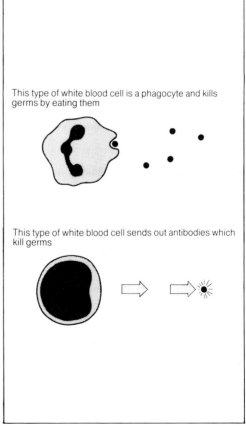

This type of white blood cell is a phagocyte and kills germs by eating them

This type of white blood cell sends out antibodies which kill germs

Figure 4 The two main types of white blood cell found in the human bloodstream. The dark object in each of the cells is the nucleus.

Plasma

This is the fluid part of the blood in which the cells float around (Investigation 2). It consists mainly of water, but many important substances are dissolved in it. They include salts, food substances such as glucose, and an important group of substances called plasma proteins. There are three kinds of plasma proteins and each has a particular job to do:

1 Albumen
This is the same protein as is found in the 'white' of an egg. It makes the blood thick and viscous.

2 Globulin
This kind of protein is produced by the white blood cells for destroying germs. It constitutes the antibodies.

3 Fibrinogen
This protein plays an important part in the clotting of blood which is discussed in the next Topic.

Fibrinogen can be taken out of the plasma by allowing it to clot and then removing the clot. What's left is a colourless fluid called **serum**.

It's essential that the plasma should contain just the right amount of water and salt and other substances. If these are allowed to vary, the blood cells may gain or lose water as a result of osmosis, and this could damage them (Investigation 3).

Summary of the functions of blood

Blood does three different kinds of job: transport, protection and regulation.

Transport
1 It carries oxygen from the lungs to the tissues and carbon dioxide from the tissues to the lungs.
2 It carries dissolved food substances from the gut to the various parts of the body.
3 It carries unwanted substances to the kidneys which then get rid of them.
4 It carries hormones and antibodies from one part of the body to another.

Protection
1 By clotting it prevents fluid being lost from cuts and wounds.
2 It protects us against disease by killing germs.

Regulation
1 It helps to control the amount of water in the tissues.
2 It helps to regulate the amounts of various chemical substances in the tissues.
3 It helps to keep our body temperature constant by spreading warmth evenly around the body.

Many of these functions are discussed in more detail in later Topics.

Investigation 1

Looking at blood

This experiment involves drawing your own blood. *This must be done under proper supervision in strictly hygienic conditions to avoid any possibility of transmitting infection.*

1 Put an elastic band round one of your fingers

2 Clean the skin of your finger by rubbing it with cotton wool soaked in ethanol.

3 With a sterile lancet prick the tip of your finger with a firm jab, so a drop of blood comes out.

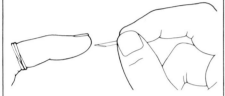

4 Place the drop of blood at one end of a microscope slide.

5 With another slide spread the blood over the surface of the slide so it forms a smear.

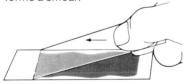

6 Let the blood smear dry, then examine it under the microscope: low power first, then high power.

   Can you see red blood cells?

   White blood cells will only show up if they are stained.

7 Cover the smear with Leishman's stain and leave it for five minutes. Then gently wash the stain off with tap water.

8 Let the slide dry, then examine it under the microscope again.

Can you see any white blood cells like those in Figure 4?

Investigation 2

Separating the components of blood

This investigation involves using a centrifuge, a machine in which a liquid can be spun round and round very rapidly.

1 Take some blood which has been obtained from a blood bank.

2 Fill two centrifuge tubes with the blood.

3 Spin the tubes for five minutes in the centrifuge. This will throw any solid objects to the bottom of the tubes.

4 Stop the centrifuge and take out the tubes. Notice the clear fluid towards the top of the tube, and the red sediment at the bottom.

   What is the clear fluid?
   What does the sediment consist of?

   Approximately what percentage of the blood is made up of red blood cells?

Investigation 3

The effect of osmosis on red blood cells

This experiment involves drawing your own blood. *This must be done under proper supervision in strictly hygienic conditions to avoid any possibility of transmitting infection.*

1 Label three microscope slides A, B and C.

2 Obtain a drop of blood from your finger or thumb with a sterile lancet.

3 Place a small drop of your blood on each slide.

4 To A add a drop of distilled water.
   To B add a drop of 0.75 per cent salt solution.
   To C add a drop of 3.0 per cent salt solution.

5 Put a coverslip on each slide.

6 Examine each slide under the microscope at regular intervals.

   What happens to the red blood cells?

Explain your observations.

Assignments

1 Write down three ways red and white blood cells differ in their structure. What job does each do?

2 Why is it dangerous to breathe in motor car exhaust? Explain the reason for your answer.

3 There are approximately 5 million red blood cells in a cubic millimetre of human blood, and the total volume of blood in the whole body is about 5 litres. Each red blood cell has a surface area of about 20 square micrometres.

   a) How many red blood cells are there in the entire bloodstream?
   b) What will be the total surface area of all the red blood cells together in square metres?

4 In the lungs there is a steep diffusion gradient favouring the passage of oxygen from the alveoli into the blood.

   a) What do you understand by the term 'diffusion gradient'?
   b) How is this steep diffusion gradient maintained?

5 In what form is carbon dioxide carried in the blood?
   Laboratory experiments have shown that the greater the amount of carbon dioxide present in the blood, the less firmly haemoglobin holds onto oxygen. Why do you think this is important in the body?

6 A scientist investigated the number of red blood cells possessed by people living at sea level and in a mountainous region at a height of 5860 metres. Here are his results:

   Sea level   5.0 million per mm³
   5860 metres   7.4 million per mm³

   How would you explain the difference?

7 A human red blood cell has a diameter of 8.0 micrometres. What is the approximate magnification of the blood cells in Figure 1 on page 164?

More about blood

Occasionally we cut ourselves, or perhaps a blood vessel bursts, and we bleed. This Topic is about what happens when we lose blood.

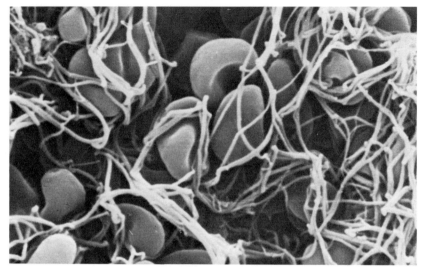

Figure 1 When blood clots a meshwork of fibres is formed as shown in this photograph which was obtained by means of an electron microscope. The objects which look like deflated footballs are red blood cells.

Blood clotting

Normally when you bleed from a cut, the blood soon hardens and the bleeding stops. The hardening of the blood is called **clotting**. Clotting is important because it stops too much blood being lost through cuts and wounds, and is the first step towards healing.

How does clotting take place?

The damaging of a blood vessel, or the exposure of blood to air, triggers off a chain reaction. In the last step of this process, the plasma protein fibrinogen is turned into a meshwork of solid fibres called **fibrin** (Figure 1). For this to happen a substance called thrombin has to be formed in the blood first.

Many different chemical substances are needed in our blood to make it clot. We are born with some of them; others we get from our food – calcium and vitamin K for example. Vitamin K is found in cabbage and spinach. Also important are the small bodies found in the blood called platelets.

Occasionally a person is born without one of the substances needed for blood-clotting. This may result in bleeder's disease or **haemophilia** in which the blood takes an exceptionally long time to clot, so that the person may lose a great deal of blood from even a small cut. This is an inherited disease and runs in families. At one time the royal families of Europe suffered from it.

It is obviously desirable that blood should clot when we are cut or wounded. However, it would be fatal if this happened while the blood is flowing through the blood vessels. To prevent this, our blood vessels contain substances which prevent clotting. These are **anti-coagulants**. Anti-coagulants are also added to blood which is kept in hospitals (Investigation 1).

What happens if we lose a lot of blood?

Despite the clotting process, a person may lose a lot of blood after an accident, or if one of his blood vessels bursts. Losing blood is called a **haemorrhage**. A person can lose a litre or two of blood without ill effects, but if more than this is lost he is in danger for two reasons:

1 His blood pressure is reduced, and this slows down the flow of blood round the body.
2 The number of red blood cells is reduced, so the oxygen-carrying power of his blood is lowered.

All sorts of consequences follow, but the main one is that not enough oxygen

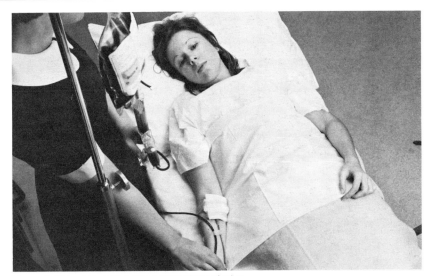

Figure 2 This girl is being given a blood transfusion.

gets to the brain, so the person may pass out and eventually die. However, his life may be saved by giving him a **blood transfusion**.

Blood transfusions

During a blood transfusion the person is given blood which has been taken from other people. The blood is put into a vein in his arm through a narrow tube (Figure 2). Usually 'whole blood' is given, but sometimes plasma alone is used. This restores the blood pressure, so the blood will move round at its normal speed. Over the next few weeks the patient makes new red blood cells to replace the ones that he has lost.

Do blood transfusions always work?

Blood transfusions were first carried out during the First World War. In some cases they worked, but in many cases the results were disastrous: the red blood cells in the transfused blood stuck together, blocking the patient's blood vessels and causing death. This sticking together of the red blood cells is called **agglutination**.

We now know that for a blood transfusion to be successful, the two lots of blood must be compatible – that is they must be able to mix together without sticking together. In practice this means that they should belong to the same **blood group**.

What are blood groups?

Everyone's blood belongs to one of four different groups which we call A, B, AB and O. The letters A and B refer to certain substances which are present in the red blood cells: AB means that both are present, and O means that neither is present. If bloods of different groups are mixed together, agglutination of the red blood cells may occur. The reason for this is explained in Figure 3.

In addition to the A and B substances, there is another substance in red blood cells called the **Rhesus factor**, so-called because it was first discovered in a type of monkey known as the Rhesus monkey. People who have this substance in their blood are described as Rhesus positive; people who don't are Rhesus negative. If two transfusions of Rhesus positive blood are given to a Rhesus negative person, one after the other, agglutination will occur after the second transfusion.

Before a blood transfusion is carried out, doctors always make sure that the

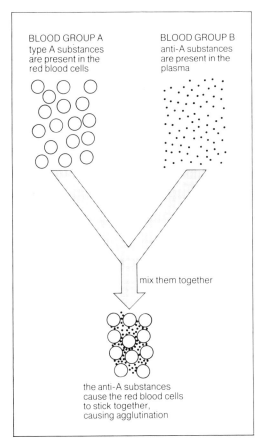

Figure 3 Diagram showing why agglutination occurs when blood of different blood groups are mixed together.

patient's blood is compatible with the transfused blood, with respect to both the ABO and Rhesus systems. The blood groups are determined by means of a simple compatability test (Investigation 2).

We inherit our blood groups from our parents. In Britain the percentages of white people with different blood groups are as follows:

ABO system		*Rhesus system*	
Group	percentage	Rh +	85%
O	47%	Rh −	15%
A	41%		
B	9%		
AB	3%		

These figures are important to doctors because they tell them which particular blood groups are likely to be needed most for blood transfusions.

Giving blood

Hospitals always need a supply of blood for use when needed. Many people give blood at Blood Donation Centres. They are known as **blood donors**. A blood donor must be healthy and aged between 18 and 65. First a drop of his blood is tested to find out what group he belongs to. Then about half a litre of blood is taken from a vein in the arm and drained into a bottle (Figure 4). He is then given tea and biscuits and rests for about half an hour. After that he can resume his normal activities. The blood which he has lost will soon be replaced by his own body.

Meanwhile an anti-coagulant is added to the blood which the donor has given, to stop it clotting. The blood is then stored at a temperature just above freezing in a blood bank. The blood is normally kept for about a month. It cannot be used for transfusions after that because too many red blood cells will have died by then.

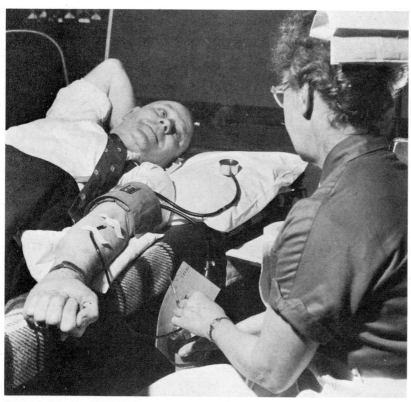

Figure 4 This man is giving some of his blood in a blood donation centre.

Investigation 1

To find out how long blood takes to clot

This investigation involves drawing your own blood. *This must be done under proper supervision in strictly hygienic conditions to avoid any possibility of transmitting infection.*

1 Prick the tip of a finger with a sterile lancet and squeeze out a little blood (see Investigation 1, page 167).

2 Place two drops of blood side by side on a white tile.

3 To one drop of blood add a very small drop of sodium citrate solution.

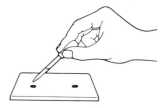

4 To the other drop of blood add a very small drop of water.

5 Stir each drop with a needle, and keep stirring until the blood begins to clot.

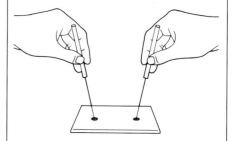

How long does it take for the blood to clot?

What effect does the sodium citrate have on clotting?

When a person gives blood at a blood-donation centre, a small amount of citrate is added to the blood: why is this done?

Investigation 2

Finding your blood group

Use a blood group test card: examine it carefully first. You will need to get a drop of blood from your finger or thumb. *This must be done under proper supervision in strictly hygienic conditions to avoid any possibility of transmitting infection.*

1 Pipette one drop of water onto each of the test panels.

2 Mix the water and reagent in each panel with the flat end of a plastic stick. *Clean the stick thoroughly between finishing one panel and moving to the next.*

3 Obtain a drop of blood from your finger or thumb with a sterile lancet (see Investigation 1, page 167).

4 Place the blood on the flat end of the plastic stick as shown.

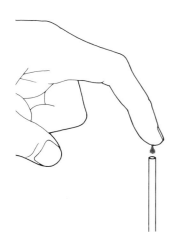

5 Mix the blood with the reagent in the left hand test panel, spreading it evenly over the whole panel.

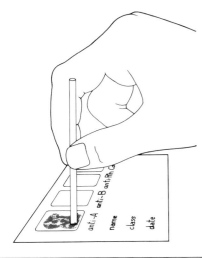

6 Wipe the stick, then mix another drop of blood with the reagent in each of the other panels. Wipe the stick between each one.

7 Tilt the card backwards and forwards so as to mix the blood thoroughly with the reagents in the test panels.

In which test panel or panels has the blood agglutinated?

8 Find your blood group from this table:

Agglutination in anti-A panel means you are group A

Agglutination in anti-B panel means you are group B

Agglutination in both panels means you are group AB

Agglutination in neither panel means you are group O

Agglutination in anti-Rh panel means you are Rh positive

No agglutination in anti-Rh panel means you are Rh negative.

9 Write down the blood groups of everyone else in your class and work out the percentage of students belonging to each group.

Compare the percentages with those given on page 170.
In each case indicate whether your class figure is higher or lower than the national average.

Assignments

1 Give two reasons why it is dangerous to lose more than two litres of blood.

2 Explain the reason for each of the following:

a) Not more than half a litre of blood is normally taken from a blood donor.

b) After a person has given blood he or she is advised to sit down quietly for about half an hour.

c) A little sodium citrate is usually added to blood which has been given at the blood donation centre.

d) Complete blood is only kept for about a month after it has been obtained from a blood donor, but plasma may be kept much longer.

3 The poison of certain snakes causes the blood of their victims to clot inside the blood vessels, thereby blocking the vessels. Suggest two different ways by which this effect might be brought about.

4 In trying to find their own blood groups, four pupils in a school mixed drops of their blood with different kinds of serum:
John got agglutination with anti-A serum but not with anti-B;
David got agglutination with anti-B serum but not with anti-A;
Susan got no agglutination with either sera; and Tim got agglutination with both sera.

a) Which blood group does each pupil belong to?

b) Whose blood group is needed most in blood donation centres, and why?

c) What causes agglutination?

5 A scientist fed chickens on a patent diet of pellets, and he found that the chickens died of bleeding as a result of their blood clotting extremely slowly. However, when he fed his chickens on a diet which included cabbage, the chickens did not suffer from this condition.

How would you explain these observations?

How does blood move round the body?

Blood constantly flows round the body, and this is called the circulation. The various structures through which the blood flows all belong to the circulatory system.

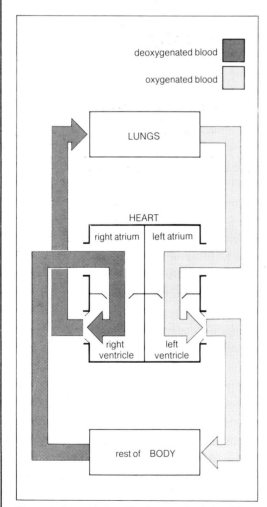

Figure 1 General plan of the human circulation. it is usual to show pictures of the human anatomy from the ventral (belly) side, so in this diagram the right hand side of the heart is on the left, and the left side of the heart is on the right.

The general plan of the human circulation

The main organ in our circulatory system is the **heart**, which is situated in the chest between the lungs. The heart is hollow and its wall contains muscles: its job is to pump the blood round the body.

The blood flows round the body in tube-like **blood vessels** which eventually lead back to the heart. The blood vessels which carry blood away from the heart are called **arteries**. Those that bring blood back to it are called **veins**. The arteries and veins are connected by narrow, thin-walled vessels called **capillaries**.

As blood flows through the capillaries, oxygen and other useful substances pass out of them to the surrounding cells, and unwanted substances pass in the reverse direction. In this way the capillaries bring life to the cells and maintain them in a state of health and repair.

The capillaries are extremely numerous and every organ contains thousands of them: no cell is more than a twentieth of a millimetre from the nearest one. If a person's capillaries were laid end to end, they would stretch round the world 2½ times!

A closer look at the circulation

If you look at Figure 1, you will see that there are really two circulations: one serves the lungs and the other serves the rest of the body. The heart is divided by a partition into left and right halves. Blood is pumped from the right side of the heart to the lungs where it takes up oxygen. The oxygenated blood is taken back to the left side of the heart and is then pumped to the rest of the body. The oxygen is taken up by the various organs as the blood passes through them. The deoxygenated blood then returns to the right side of the heart, and the cycle is repeated.

Each side of the heart consists of two chambers: an **atrium** (plural: atria) and a **ventricle** (Investigation 1). Both have muscles in their walls, but the walls of the ventricles are much thicker and more muscular than those of the atria. The ventricles play the most important part in pumping blood round the body. A detailed diagram of the human heart and circulation is shown in Figure 2.

The heart as a pump

The heart contracts approximately 70 times a minute throughout our life: that's over 100 000 times a day. This is made possible by the muscles in its wall. Heart muscle differs from other kinds of muscle in that it does not get tired. Try clenching your fist at the rate of 70 times per minute and your hand muscles will soon give up. Heart muscle, however, has no difficulty in working at this rate.

Each contraction of the heart is followed by relaxation during which the heart wall returns to its original position. When it relaxes, blood is sucked into it from the veins. When it contracts, the blood is pumped out of the heart into the arteries. So blood flows through the heart in only one direction. This is made possible by **valves** which prevent the blood flowing backwards (Figure 3).

Every time the heart beats it sets up a wave of pressure which travels along the main arteries. This is called a **pulse wave** and if you put your finger on your skin just above the artery in your wrist you can feel it, as a slight throb (Investigation 2). Doctors and nurses often feel a patient's pulse to see if the heart is beating at its normal rate. It is also possible to *hear* the heart by putting your ear, or better still a stethoscope, against a person's chest (Investigation 3).

In order to contract repeatedly and powerfully throughout life, the heart muscles must have a good supply of oxygen. They get this through a system of arteries which spread over the heart wall. These are called the **coronary vessels**.

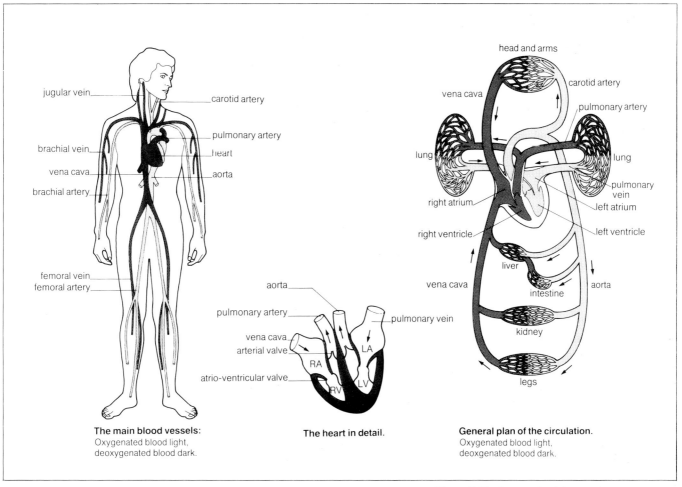

The main blood vessels:
Oxygenated blood light,
deoxygenated blood dark.

The heart in detail.

General plan of the circulation.
Oxygenated blood light,
deoxgenated blood dark.

Figure 2 The human circulatory system.

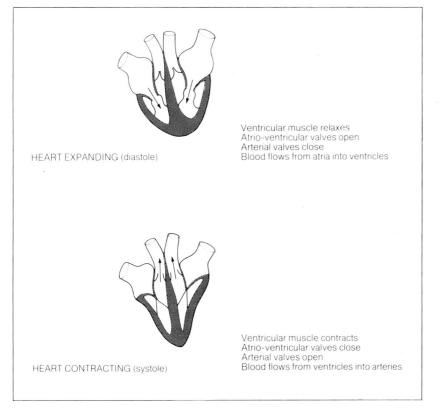

HEART EXPANDING (diastole)

Ventricular muscle relaxes
Atrio-ventricular valves open
Arterial valves close
Blood flows from atria into ventricles

HEART CONTRACTING (systole)

Ventricular muscle contracts
Atrio-ventricular valves close
Arterial valves open
Blood flows from ventricles into arteries

Figure 3 These diagrams show how blood flows
through the heart. The valves prevent the blood
flowing backwards.

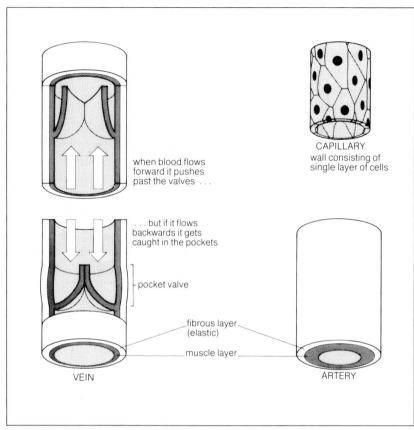

Figure 4 The three types of blood vessel found in the circulatory system.

The blood vessels

Figure 4 shows the structure of an artery, vein and capillary. Their structure suits them to their particular jobs.

The arteries have tough elastic walls containing smooth muscle (see page 18). This makes the walls press back against the blood as it flows through them. This helps to force the blood along quickly, much as water is forced along a narrow hosepipe.

The capillaries are only just wide enough to allow the red blood cells to pass along in single file. Their walls are very thin, consisting of a single layer of flattened cells. This enables oxygen and other substances to pass through easily.

By the time the blood gets through the capillaries into the veins, the pressure pushing it along is greatly reduced, and it is now mainly moving against gravity. This makes it difficult for the blood to get back to the heart. However, the veins have thinner walls than the arteries, so they have more 'give' and let the blood along more easily. Also they contain valves which prevent the blood slipping back. Moving around also helps to keep the blood going because the contraction of the leg muscles squeezes the blood along the veins.

Blood pressure

The pumping action of the heart, combined with the narrowness of the smaller vessels, results in a considerable pressure being built up in the arteries. This is what we mean when we talk about 'blood pressure'. It's important that our blood pressure should be reasonably high because it keeps the blood on the move.

A person's blood pressure varies according to what he is doing. In general

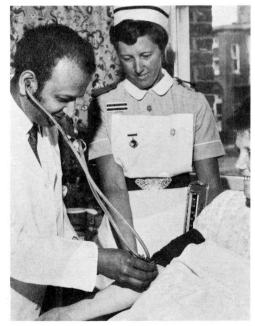

Figure 5 A doctor taking a patient's blood pressure. The doctor is listening to the pulse with a stethoscope, and the blood pressure is registered by the mercury manometer on the right.

anything which makes the heart beat faster, or the arteries get narrower, will increase the blood pressure. For example, anger, excitement and exercise all have this effect.

If a person goes to the doctor complaining of feeling tired and run down, one of the first things the doctor does is to measure his blood pressure (Figure 5). He wraps an inflatable band round the upper part of the patient's arm; then he pumps up the arm band until it is so tight that it stops blood passing down the main artery of the arm. When this point is reached the pulse disappears. He now lets the air very slowly out of the arm band. The pressure which *just* allows the pulse to reappear is the same as the patient's blood pressure.

The circulation during exercise

Suppose you run a race. During the exercise your heart beats faster (Investigation 4). What makes this happen? The answer is that extra carbon dioxide is produced by your muscles and this starts to build up in your bloodstream. The brain senses this is happening, and it sends nerve impulses to the heart making it beat faster. This is an automatic reflex: it happens without you having to think about it.

The result of the heart beating faster is that more blood can be sent to the muscles. The arteries serving the muscles widen, whereas those serving less needful organs get narrower. The result is that extra blood is diverted to the muscles.

What can go wrong with the circulation?

One of the most common defects of the circulation is high blood pressure. We all develop a high blood pressure at one time or another, when we take exercise for example, but some people have high blood pressure all the time. This puts an extra strain on the heart, and may lead to **heart failure**. It also pushes out the walls of the arteries, and may burst them – just as a balloon will burst if you blow it up too much. The risk of this happening is greatest in old people whose arteries have become fragile. Sometimes an artery bursts inside the brain, and the spillage of blood kills the cells in that part of the brain. This results in a **stroke**, and it may leave the person partly paralysed and unable to speak properly. A severe stroke can be fatal.

What causes high blood pressure? We don't know, but it is frequently associated with the stress and tensions of modern life and with eating and drinking too much.

Another defect of the circulation is **hardening of the arteries**. This is caused by fatty substances being laid down in the walls of the vessels, making them narrower and slowing the flow of blood through them. Where this happens, a blood clot may occur inside the artery, completely blocking it. The structures served by that particular artery will no longer receive any oxygen. If this happens in one of the coronary vessels, the part of the heart deprived of oxygen stops contracting and the result is a **heart attack**. If only a small area of the heart is affected, the person may recover, but if a large part is involved, the attack may be fatal. The person's life may be saved by massaging the heart (Figure 6).

What causes hardening of the arteries? There is some debate about this, but eating large amounts of animal fat appears to increase the chances of it happening (see page 113). Other possible causes include smoking, drinking and stress.

The defects mentioned so far are all serious, but other less serious things may go wrong with our circulation. For example, the flow of blood through some of the veins may become sluggish and the valves may not work properly. The back-pressure of blood stretches the walls of these veins which become flabby, like thin bags. These are called **varicose veins**, and they are particularly liable to develop just under the skin at the back of the legs. Sometimes the same thing happens in the wall of the rectum where it gives rise to piles or **haemorrhoids**.

Press on chest: blood expelled from heart

Release pressure: blood enters heart

Figure 6 A person who has had a heart attack can sometimes be saved by cardiac massage.

Investigation 1

Looking at the heart

1 Look at the heart of a mammal such as a pig or sheep obtained from the butcher. The heart has been cut open, so you can look inside.

2 Decide which side of the heart is dorsal, and which side is ventral. The more rounded (convex) side is the ventral.

3 Identify the two atria, and the ventricles.

   How do they differ in size and shape?

4 Feel the atria and ventricles with your fingers.

   How do they differ in the way they feel?
   Explain the reason for the difference.

5 Look at the large blood vessels attached to the heart.

   Can you recognise the vessels shown in Figure 2?
   Which ones are arteries and which ones are veins?
   How do the arteries and veins differ from each other?

6 Observe the narrow blood vessels ramifying over the surface of the ventricles, and notice where they come from. These are the coronary vessels.
   What is their function?
   What would happen if one of them became blocked?

7 Look at the cut which has been made in the wall of the ventricles.

   What is the wall made of?
   Has one of the ventricles got a thicker wall than the other?
   Why do you think they differ in this way?

8 Look inside one of the ventricles.

   Which structures shown in Figure 2 can you see?
   In particular notice the valves.
   What are their functions?

In what ways is the heart suited to its job of pumping blood round the body?

Investigation 2

Finding how fast your heart is beating

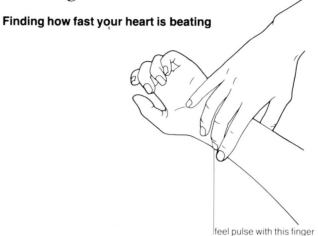

feel pulse with this finger

You can find how fast your heart is beating, that is your heart rate, by feeling your pulse.

1 Sit down comfortably in a chair with the palm of your hand facing upwards.

2 Gently place the middle finger of your other hand on your wrist as shown in the illustration. Can you feel your pulse as a repeated throb?
   If necessary change the position of your finger, until you can feel your pulse really well.

3 Count the number of heart beats in one minute.

4 Repeat step 3 four times.

Write down the number of beats per minute each time.
Work out your average heart rate.
As this is your heart rate when sitting down, it is called the resting heart rate.

5 Stand up for one minute.

6 Still standing, take your pulse another five times.

   Work out your average heart rate in beats per minute. This is called your standing heart rate.

How do your resting and standing heart rates differ?

Why do you think they are different?

Investigation 3

Listening to the heart

Work in pairs, one person acting as the subject. The subject should sit down comfortably.

1 Put the bell of a stethoscope against the chest wall and listen.

   Can you hear regular thud-like sounds?

2 Listen with the bell of the stethoscope in different positions.

   Where is the best place to put the stethoscope for the sounds to be loudest?
   What kind of information do you think the doctor can get about a patient's heart by listening to it with a stethoscope?
   What do you think causes the heart sounds?

3 Learn how to feel the pulse (see Investigation 2)

4 Now feel the pulse, and listen to the heart sounds at the same time.

   Notice that there is a time lag between the heart sounds and the pulse.

   What is the time lag caused by?

   In what circumstances would you expect the time lag to be shorter?

Investigation 4

To find the effect of exercise on the heart rate

1 Measure your standing heart rate by feeling your pulse (see Investigation 2). Write down your heart rate in beats per minute.

2 Do steady walking on the spot for 3 minutes.

3 Immediately after walking, measure your heart rate again. Write down your new heart rate in beats per minute.

How does it differ from the standing heart rate?
How would you explain the difference?

4 Stand still and wait until your heart rate returns to its normal standing rate.

5 Do some hard exercise for 3 minutes. Your teacher will tell you what kind of exercise to take.

6 Immediately after the exercise, measure your heart rate every minute until it returns to the normal standing rate. Write down your heart rate in beats per minute for each minute.

How does your heart rate immediately after the hard exercise differ from the standing rate?

How would you explain the difference?

How long did it take for your heart rate to return to its normal standing rate?

Assignments

1 A person's blood pressure can be recorded continuously by means of an electronic pressure gauge placed inside one of the arteries.

Here is a recording obtained in this way:

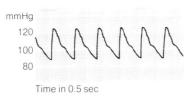

a) Why do you think the pressure goes up and down all the time?
b) Give two circumstances in which you would expect the frequency of the waves to increase.

2 Suggest a reason for each of the following:

a) the left ventricle has a thicker, more muscular wall than the right ventricle;
b) arteries have more muscle in their walls than veins;
c) capillaries have very thin walls;
d) veins contain valves;
e) doctors advise people with varicose veins to walk as much as possible.

3 Devise an experiment which you could do to test the suggestion that veins have more stretchable walls than arteries.

4 The average speed of the blood in the arteries is 45 cm/s, but the average speed in the capillaries is only 0.5 mm/s.

a) Give the speed in the capillaries as a percentage of the speed in the arteries.
b) What do you think causes the difference?
c) Why is it desirable for blood to flow through the capillaries comparatively slowly?

5 The chart below shows the pulse rate of a hospital patient measured at four hourly intervals every day.

a) Can you detect a regular pattern in the way the pulse rate changes? If so, describe the pattern.
b) Do you have any criticisms of the way the pulse rate is graphed in the chart?
c) What were the highest and lowest values of the pulse rate during the period in question and when were they recorded?
d) Give possible reasons why the pulse rate reached these particular values.

6 The blood system has been likened to the London Underground railway. In this comparison, each of the items listed on the left below is equivalent to one of those on the right. Write them down in the correct pairs.

Circulation	*Underground*
heart	tunnels
blood cells	electrified third line
vessels	trains
oxygen	stations
capillaries	passengers

Look at a map of the London Underground: which line is most like the human blood system?

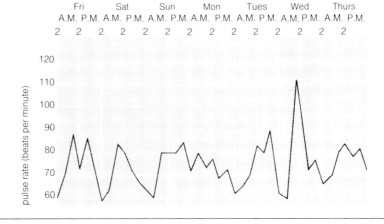

Tissue fluid and lymph

Our bodies contain other important fluids besides blood. These include tissue fluid and lymph.

Tissue fluid

In between our cells there are narrow spaces filled with a watery fluid. This is **tissue fluid**.

Tissue fluid is extremely important. It bathes the cells and keeps them in the right condition. The cells get all the substances they need from the bloodstream, *via* the tissue fluid. The tissue fluid is therefore an essential link between the bloodstream and the cells, and it comprises the immediate surroundings of the cells.

How is tissue fluid formed?

Tissue fluid is formed from the blood (Figure 1). As blood flows along the capillaries, a certain amount of fluid leaks through the capillary walls into the spaces between the cells. Once it has left the capillaries, it becomes the tissue fluid.

The process by which tissue fluid is formed involves a kind of filtration: the blood cells and plasma proteins are too large to go through the capillary walls, so they stay in the bloodstream. What passes through is therefore a colourless fluid consisting of blood plasma minus the proteins.

Lymph

Once formed, the tissue fluid seeps around amongst the cells. If there is too much of it, it either returns to the capillaries, or is drained into a system of narrow channels called **lymph vessels**. The fluid in these vessels is called **lymph**.

The body is permeated by lymph vessels: some of them can be seen in Figure 2. They eventually lead to the veins, so sooner or later lymph gets back into the bloodstream. The lymph vessels contain valves, which help to keep the lymph flowing in the right direction.

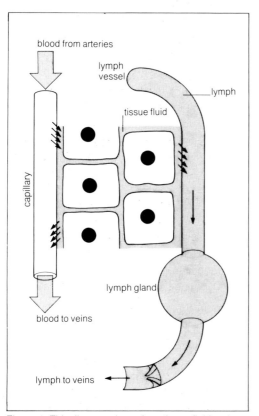

Figure 1 This diagram shows how tissue fluid and lymph are formed.

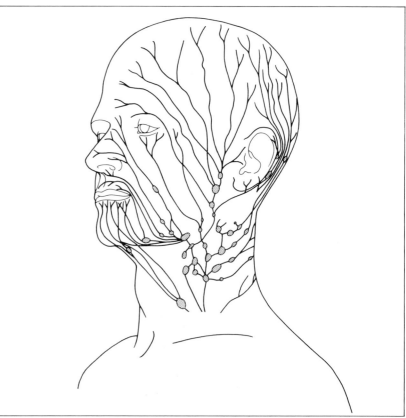

Figure 2 Some of the main lymph vessels and glands in the head and neck.

Occasionally tissue fluid is formed faster than it can be drained away in the lymph vessels. The result is that fluid builds up in the tissues, causing them to swell up. This is called **oedema**. It sometimes happens in old people, particularly in their legs (Figure 3).

Lymph glands

If you look at Figure 2 you will see that there are little swellings at intervals along the length of the lymph vessels. These are called **lymph glands**.

Each lymph gland is full of tiny spaces like a sponge, and the lymph has to filter through these spaces before it can continue on its journey back to the bloodstream.

The lymph glands help us to fight disease. They contain cells which attack and destroy germs in the lymph as it filters through. These cells are the same as the white blood cells mentioned on page 165: some of them are phagocytes and eat up the germs; others kill the germs by producing antibodies against them.

The positions of our main lymph glands are shown in Figure 4. The largest ones are located in the neck, armpits and groin.

Suppose you have a severe throat infection. The germs get trapped in the nearby lymph glands in your neck where your phagocytes and lymph cells do their best to kill them and prevent them getting into the rest of your body. This causes the glands to swell up and become tender and painful. Lymph tissue is also found in the **tonsils** and **adenoids** in the throat. Sometimes these organs get repeatedly infected and swollen, making it difficult to breathe. This may make it necessary for them to be removed in an operation. At one time children had their tonsils and adenoids out almost as a matter of course but nowadays it is only done if it is really necessary. After all, these are useful organs which help to defend us against disease, and it's best to keep them if we can.

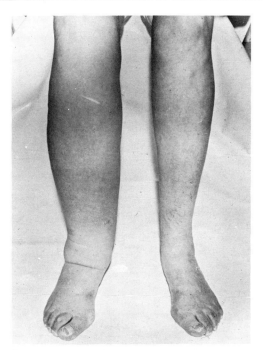

Figure 3 The legs of this person are swollen as a result of tissue fluid accumulating in them.

Assignments

1 In what respect do blood, tissue fluid and lymph differ in what they contain? Explain the reason for the differences.

2 State two functions which are performed by the lymph system.

3 A person cuts his foot and the cut goes septic. Within a short time his groin hurts whenever he touches it. Explain the reason for the pain.

4 At the arterial end of a capillary, the blood pressure is high, but the osmotic pressure of the blood is low. However, at the venous end, the blood pressure is low, but the osmotic pressure is high.

   a) Why do you think the blood pressure is higher at the arterial end than at the venous end?

   b) Why do you think the osmotic pressure is higher at the venous end than at the arterial end?

   c) What part do you think these differences play in the formation and movement of tissue fluid?

5 Sometimes elderly people get tissue fluid accumulating in their feet and legs which consequently become swollen. Suggest two possible reasons why this may happen.

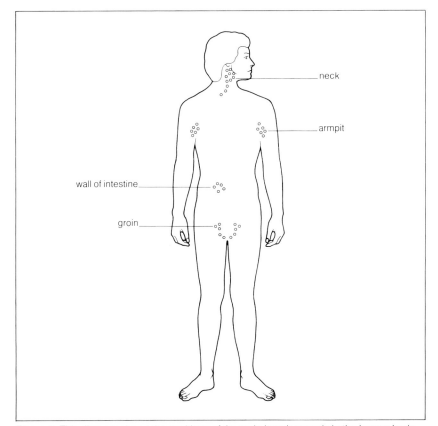

Figure 4 This diagram shows the positions of the main lymph vessels in the human body.

What do plants need to grow?

When a gardener or a farmer puts manure or fertiliser in the soil, he is giving his plants essential materials which they need for healthy growth.

Plants need certain major elements

It has long been known that plants need the elements carbon, hydrogen, oxygen, nitrogen, phosphorus, sulphur, magnesium, potassium, calcium and iron. These ten elements are needed in quite large amounts by all green plants. For this reason we call them 'major elements' or **essential elements**.

How do plants obtain the major elements?

Think of a plant growing in your garden or in a park. It obtains carbon and oxygen from the air. It obtains hydrogen from water in the soil, which it absorbs through its roots. The other elements are also absorbed by the roots. They are present in the soil in the form of mineral salts, such as calcium nitrate and potassium phosphate.

One of the first people to realise that plants need these mineral elements was a German scientist called Willhelm Knop. In 1965 he made up a solution which appeared to be ideal for plant growth. Knop's solution contained salts of all the mineral elements listed above. All he had to do was to dangle the roots of a young seedling in his solution, and it grew. However, if he left out one of the major elements, the seedling would not grow properly (Figure 1). You can repeat Knop's experiment for yourself (Investigation).

Why do plants need the major elements?

Some of the elements help to form important structures in the plant. Nitrogen, for example, occurs in proteins. Proteins make up much of the structure of the plant's body. If a plant does not get nitrogen it cannot make proteins, and so it will not grow. Sulphur and phosphorus also occur in proteins, so lack of these elements has a similar effect. Sometimes you see plants with curly brown edges to the leaves: this is a sign of lack of phosphorus.

Magnesium occurs in chlorophyll, the green pigment in leaves. If a plant lacks magnesium, its leaves cannot make chlorophyll and so they go yellow. This is known as chlorosis.

Other elements help the action of enzymes, and therefore help to control metabolism (see page 2). If any of them is missing the plant shows poor growth, and its leaves, buds or roots may wither and die.

Some minor elements are needed too

We know that most plants need, in addition to the major elements, certain other elements as well. These are required in only tiny amounts, so we call them **trace elements** or minor elements. These elements include boron, zinc, copper, aluminium, molybdenum, sodium, chlorine, silicon, manganese and cobalt. As with the major elements, they are obtained in the form of mineral salts. Plants absorb them through their roots.

If any of these minor elements is absent from the soil, plants may show poor growth. Look at Figure 2: this shows the effect of depriving tomato plants of the element molybdenum. In certain parts of Australia crops grew very badly until it was discovered that there was no molybdenum in the soil. The soil was then sprayed with a very dilute solution of molybdenum, and this made all the difference – the plants grew splendidly. Very little molybdenum was needed: one teaspoonful was enough for an area the size of a tennis court. Putting on too much of a trace element may have a damaging effect on plants.

What makes soil short of minerals?

Think of a natural plant community – a forest if you like, or a field. When plants die they rot: the various chemicals in their bodies are set free and put back into the soil. They can then be absorbed and used again by new plants.

Now think what happens in a field with a crop in it, such as wheat. The

Figure 1 The effect of depriving barley seedlings of certain major elements. The plants in Jar A were grown in Knop's solution; those in B were deprived of potassium, in C of nitrogen, and in D of phosphorus.

wheat is harvested and the plants are taken away. The chemicals are not returned to the soil, and so the soil becomes poor. The soil is made even worse if heavy rain washes useful chemicals out of it.

How can we overcome this problem? One way is to grow crops on one piece of land for several years and then move somewhere else. This is what nomadic tribes do in certain parts of the world. But it can't be done in a developed country. A better solution is to make sure that the soil does not become poor in the first place. This can be achieved in two ways: by rotating the crops or putting on a fertiliser.

Rotation of crops

Have you noticed that a farmer does not usually grow the same crop in a field year after year? For a year or two he may grow wheat, then turnips perhaps, then barley–and so on. This is known as **rotation of crops,** and it has been carried out since Roman times.

This is a good idea because some plants take more of certain chemicals out of the soil than others. If the same crop is grown in a field year after year, a particular element – nitrogen say – may eventually be removed altogether. Rotating crops prevents this.

Every now and again a farmer may grow a crop of clover, or a similar plant, in his field. These plants make the soil richer in nitrates (see page 375). They therefore have a good effect on the soil.

It isn't only farmers who rotate crops–a gardener may do it too. He plants his various vegetables in different places each year. Peas and beans have the same effect on the soil as clover, so it's wise to grow these in different parts of your vegetable garden (Figure 3).

Fertilisers

The best way of preventing the soil from becoming poor is to put back into it what the plants take out. This can be achieved by putting fertilisers into the soil.

A fertiliser is any substance containing chemical elements needed for plant growth. We can divide them into two groups: **organic fertilisers** and **inorganic fertilisers**. Let's take each in turn.

One of the most natural organic fertilisers is farmyard manure. This consists of the dung and urine produced by farm animals, mixed with straw. It is spread on the ground where it decays. As it rots, nitrates and other inorganic nutrients are released from it into the soil. These can then be used by plants.

Another natural organic fertiliser is compost. This consists of the rotting remains of vegetable matter: old cabbage stalks, grass cuttings, and so on. Many people make compost heaps in their gardens (see page 377). As with farmyard manure, the decay process releases inorganic nutrients into the soil which can then be used by plants.

Because of their colour, farmyard manure and compost are referred to as **brown manure**. Sometimes, however, a farmer will grow a crop of green plants and then plough them into the soil. This is called **green manure**. Once ploughed in, it rots and the nutrients are set free. Plants such as peas, beans and clover make particularly good green manure because they enrich the soil with nitrates.

Organic fertilisers have one disadvantage: they have to decompose first before the inorganic nutrients can be released. This makes them slow to act.

Quicker results can be achieved by using inorganic fertilisers. These contain mineral nutrients which can be absorbed by plants straight away. They are manufactured in fertiliser factories. They are either made from natural materials such as bone and horns, or by special chemical processes. Thanks to fertilisers a rigid rotation of crops is no longer necessary and the same crop can be grown in a field year after year. This is known as **monoculture.**

Figure 2 The tomato plants on the left were given everything they need. The ones on the right were deprived of the trace element molybdenum.

Figure 3 Bean plants enrich the soil in nitrates.

Figure 4 The Broadbalk Field at Rothamsted Experimental Research Station.

Do fertilisers work?

In the Broadbalk Field at Rothamsted Experimental Research Station there are strips of soil where wheat is grown each year (Figure 4). In one strip wheat has been planted and harvested every year for over a hundred years. During this time no fertiliser has ever been added to the soil. Since the first crop was harvested way back in 1843, the annual yield of grain has fallen to less than half what it was originally.

In other strips, however, different kinds of fertiliser have been added to the soil. In some of these strips the yield has more than doubled (Figure 5). So fertilisers certainly help.

Which is better, natural manure or artificial fertiliser? The Rothamsted results suggest that it does not matter much: equally high yields have been obtained with both. However, what applies to wheat at Rothamsted may not apply to Mr. Smith's onions in Barnsley. Some gardeners swear by farmyard manure. Others feel that artificial fertilisers are better and a lot easier to use. An undoubted advantage of manure is that it improves the texture of the soil and this helps the soil particles to stick together.

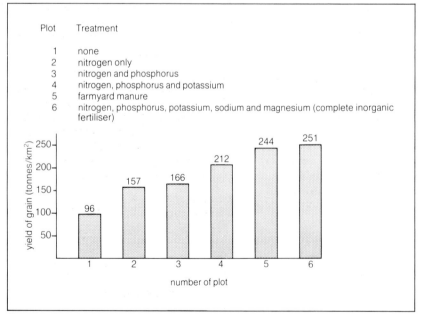

Plot	Treatment
1	none
2	nitrogen only
3	nitrogen and phosphorus
4	nitrogen, phosphorus and potassium
5	farmyard manure
6	nitrogen, phosphorus, potassium, sodium and magnesium (complete inorganic fertiliser)

Figure 5 This bar chart shows the average yearly yield of grain given by six of the plots of wheat in the Broadbalk Field at Rothamsted Experimental Station in Hertfordshire between 1852 and 1967.

Investigation

To find out which elements are needed for plant growth

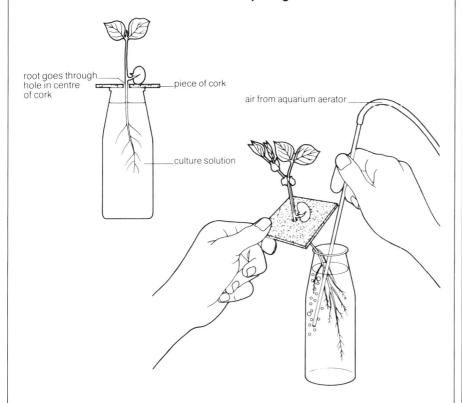

root goes through hole in centre of cork

piece of cork

air from aquarium aerator

culture solution

In this experiment we will grow plants in a series of solutions. One of the solutions contains all the chemical elements believed to be needed for plant growth. This serves as a control. The other solutions each lack one particular element.

1 Obtain 8 bottles, and label them A to H.

2 Fill each bottle with the following solutions:

A Knop's solution: this contains nitrogen, phosphorus, sulphur, magnesium, potassium, calcium and iron.
B Knop's solution without nitrogen.
C Knop's solution without phosphorus.
D Knop's solution without sulphur.
E Knop's solution without magnesium.
F Knop's solution without potassium.
G Knop's solution without calcium
H Knop's solution without iron.

3 Obtain 8 seedlings of e.g. wheat, barley or broad bean.

4 Put one seedling in each bottle as shown in the top illustration.

5 Wrap a sheet of black paper round each bottle to shield it from light. This will prevent algae growing inside.

6 Put the bottles in a warm, light place, e.g. close to a window or in a greenhouse.

7 Observe the seedlings at intervals over the next 2 – 3 weeks.

Note You may need to bubble air into the solutions from time to time to make sure the roots get enough oxygen. Do this by first connecting a narrow glass tube to an aquarium aerator via a length of rubber tubing. Then insert the glass tube into the solution as shown in the bottom illustration.

After 2 – 3 weeks, how do the eight seedlings differ from each other in appearance?
Can you explain the differences?

Assignments

1 Gardening shops sell a special liquid medium which can be used for growing pot plants. Make a list of the major chemical elements which you think it should contain.

2 Give two reasons why soil may become poor in mineral nutrients.

3 Why is it a good idea:

a) to plant your vegetables in a different part of the garden each year,
b) to dig compost into the soil,
c) to give a potted plant some 'plant food' every few days?

4 Give two advantages which farmyard manure has over artificial fertilisers, and two advantages which artificial fertilisers have over farmyard manure.

5 The barley in a certain area of East Anglia is giving a lower yield of grain than would be expected. You have been called in to find out the cause. What would you do?

6 Look at Figure 5, then answer these questions:

a) Express the yield of grain given by plots 2 to 6 as a percentage increase over that given by plot 1.
b) From the data it might be concluded that artificial fertiliser is better than farmyard manure. Do you think this conclusion is justified? Give reasons to support your answer.

7 At Rothamsted Experimental Station scientists have investigated the effect on the annual yield of wheat grain of leaving a field bare (fallow) every fifth year. Here is a sample of their results: the yield is expressed as a percentage of what it is when wheat is grown continuously.

Years after fallow	1	2	3	4
Percentage increase	101	65	48	50

a) Explain in words what is meant by an increase of 101 per cent.
b) Suggest reasons why the yield increases and then gradually decreases after the fallow year.

How do plants feed?

To answer this question we must ask two questions: what sort of organic substances do plants contain and where do they get them from?

What sort of organic substances do plants contain?

You can answer this by testing an iris leaf for sugar. This can be done with Benedict's solution (Investigation 1). The plant must be given all the things it needs beforehand, such as plenty of light and well watered soil.

What about starch? You can find out if a leaf contains starch by testing it with iodine (Investigation 2). Try it on a piece of geranium leaf. As with the iris, it is important that the geranium should be given all the things it needs beforehand, such as plenty of light and well watered soil.

These two experiments, and many others besides, tell us that plants contain organic substances such as sugar and starch. Normally plants convert sugar into starch for storage. The iris is exceptional in that it stores sugar in its leaves and does not convert it into starch.

Where do plants get their organic substances from?

It is possible that your iris plant might obtain sugar from the soil. How could you find out if this is so? One way would be to test a small sample of the soil with Benedict's solution to see if there is any sugar there (Investigation 3).

You will probably find that no sugar can be detected in the soil. In fact neither the soil enveloping its roots, nor the air surrounding its leaves, contains sugar.

So our iris plant contains sugar, but it does not take it in. How, then, has the sugar got there?

Van Helmont's experiment

In 1692 a Dutchman called Van Helmont did an interesting experiment which helps us to answer this question. He weighed a young willow tree and planted it in a pot containing a known mass of soil. He then left the tree to grow, giving it nothing but water. After five years he weighed the tree, and the soil, again. He found that the tree had gained 74 kg in mass, but the soil had only lost 56 g.

Although Van Helmont did not realise it at the time, the willow tree had absorbed simple substances from the air and soil and had built them up into food. We now know that all green plants can do this provided they are kept in the light. It is their method of feeding, and we call the process **photosynthesis**. It is a remarkable process and for over 100 years man has tried to repeat the process in the laboratory, but with very little success. Yet it happens naturally in the leaves of a green plant.

Why is photosynthesis important?

Think of it in this way. Animals cannot make complex food substances for themselves. The only way an animal can get these substances is by eating plants – or animals which have eaten the plants – or animals which have eaten the animals which have eaten the plants.

So animals are dependent on plants for their food. When you eat a chunk of beef steak, you are able to do so only because the cow ate grass. We can sum this up by saying that plants manufacture food which can then be consumed by animals.

To give you an idea of the importance of this, here are some facts and figures: a hectare (nearly 2½ acres) of corn can make more than 20 000 kg of sugar in a year. If it was in the form of ordinary table sugar, this would be enough to sweeten well over a million cups of tea.

Or looking at it another way: if the food made by all the world's plants was amassed in the form of sugar for three years, it would form a heap the size of Mount Everest.

Investigation 1

Testing a leaf for sugar

1 Put a few pieces of leaf into a mortar. Add a pinch of sand and cover with water.

2 Grind up the pieces of leaf with a pestle.

3 Filter the contents of the pestle into a test tube to a depth of about one centimetre.

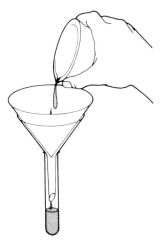

4 Pour the same amount of Benedict's solution into the test tube. Stand the test tube in a beaker of boiling water until its contents boil.

5 Repeat step 4 on some water in a test tube to serve as a control.

What happens to the solution in the test tubes?

A green, brown or red colour means there's sugar present.

Is there any sugar in the leaf?

Investigation 2

Testing a leaf for starch

1 Dip your leaf into a beaker of boiling water for about ten seconds. This will kill it and make it soft.

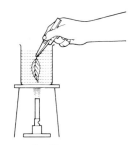

2 **Turn out the Bunsen.**
Put the leaf into a test tube of ethanol. Stand the test tube in the beaker of hot water for about ten minutes. The ethanol will boil and this will decolourise the leaf.

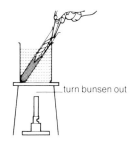

turn bunsen out

3 Wash the leaf by waving it to and fro in the beaker of water.

4 Put the leaf in a petri dish and cover it with dilute iodine solution.

A black colour shows that starch is present.
Is there any starch in the leaf?

Investigation 3

Testing soil for sugar

1 Put a small sample of soil in a mortar. Add a little water.

2 Grind up the soil with a pestle.

3 Filter the contents of the mortar into a test tube.

4 Pour a little Benedict's solution into the test tube.

5 Heat to boiling in a beaker of water over a Bunsen flame.

6 Repeat steps 4 and 5 on some tap water in a test tube: this will serve as a control.

A green, brown or red colour means there is sugar present.

Is there any sugar in the soil?

Assignments

1 How does an oak tree differ from a human being in the way it feeds?

2 Plants are generally rooted to the ground and do not move about. How does this fit in with their method of feeding?

3 'When you eat a chunk of beef steak, you are able to do so only because the cow ate grass.' Explain the reason for this.

4 Using the index, find out what substances are present (a) in the air, and (b) in the soil. List them, and put a tick against those which you think the plant uses for making food.

5 Someone has worked out that the total amount of organic matter made by all the world's plants in a year is 125 thousand million tonnes. But the total amount of food consumed by the earth's human population is only 1/200th of this. If plants make more food than man needs, why are people starving?

How do plants photosynthesise?

Plants make their own organic food such as starch, but how do they do it? In this Topic we will look at this question.

Finding out what plants need in order to produce starch

One way of discovering how plants make food is to find out what they need in order to produce starch. Just from a general knowledge of plants we can say that the following might be necessary: light, carbon dioxide, chlorophyll, and water.

We can do experiments to find out if these four factors are required for starch-formation.

The principle behind the experiments is quite simple. We take a potted plant and remove all the starch from its leaves. This can be done by putting it in the dark for a few days. To make sure it has been completely destarched, we do an iodine test on one of its leaves.

We then give the plant everything it needs except the one factor we want to investigate. After a time we again do an iodine test on one of the leaves to find out if it has been making starch. If it has not made any starch, we can conclude that this particular factor is needed for starch-formation.

As with other biological experiments we must have a control with which to compare the result. The control plant is given everything it needs, including the factor which we are investigating.

Now let's look at the individual experiments in detail.

Do plants need carbon dioxide to make starch?

We can investigate this by removing carbon dioxide from the air surrounding one plant, while another plant, the control, is given air containing plenty of carbon dioxide (Investigation 1). Later a leaf from each plant is tested with iodine to see if it has made any starch.

Do plants need chlorophyll to make starch?

The ideal way of investigating this would be to remove the chlorophyll from a leaf and see if this stops it making starch. However, it is impossible to remove the chlorophyll without killing the leaf!

So what can we do? Luckily nature comes to our aid. It so happens that the leaves of certain plants are green in some places but yellow in others: chlorophyll is present in the green areas, but absent from the yellow areas. Such leaves are described as **variegated**. Good examples are certain types of ivy and geraniums (Figure 1).

To find out if chlorophyll is needed for starch-formation, all we have to do is to carry out a starch test on a variegated leaf (Investigation 2).

Do plants need light for making starch?

We can investigate this by putting one plant in the dark and another plant, the control, in the light (Investigation 3, method a). After a few days each one is tested for starch.

Another way is to take a plant and cover part of one leaf with a piece of black paper. We then leave the plant in the light (Investigation 3, method b). Later we test the leaf with iodine to see if the covered area has been prevented from making starch.

Do plants need water to make starch?

There is no simple experiment which can be done to answer this question. You certainly cannot do it by depriving the plant of water because it is impossible to take all the water out of a plant and in any case this would certainly kill it. The importance of water has been investigated by more complicated methods involving the use of isotopes. This is described on page 198.

Figure 1 Variegated leaves of a house plant.

What do these experiments tell us?

The results of these and many other experiments tell us that plants need carbon dioxide, water, light and chlorophyll in order to make starch. If the plant is deprived of any of these essential factors, it cannot make starch. Even if a single leaf, or just part of the leaf, is deprived, starch fails to be made in that region. This is seen most strikingly in the experiment where part of a leaf is covered with black paper to prevent light getting to it (Investigation 3, method b). On testing the leaf with iodine you get the characteristic black colour only where the leaf was uncovered. This is called a **starch print** and an interesting example is shown in Figure 2.

What does photosynthesis produce?

We have seen that photosynthesis produces food substances such as starch. But is anything else formed in the process?

Figure 3 illustrates an experiment which helps us to answer this question. A lighted candle is placed in a sealed chamber. After a while it goes out.

A sprig of mint is then introduced into the chamber without any air being let in. It is then left in the light. After about ten days the candle, on being lit, burns again.

This experiment was first carried out by Joseph Priestley in 1771. He did not understand why the mint should enable the candle to be re-lit. However, we now know that the burning candle had used up all the oxygen. Putting the mint into the chamber had the effect of putting oxygen back into the air, so that the candle could burn again. This was the first demonstration that plants give out oxygen.

A more direct way of finding out if plants give out oxygen is to use a water plant such as Canadian pondweed (Figure 4). These plants obligingly produce bubbles when put in the light. The bubbles can be collected and tested for oxygen (Investigation 4).

There is now a lot of evidence that *all* green plants give off oxygen in the light. The observation that they will do this only in the light strongly suggests that it has something to do with photsynthesis.

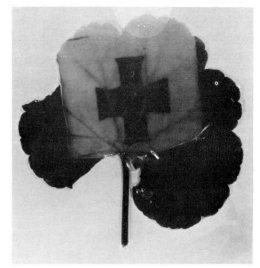

Figure 2 A starch print made on a geranium leaf.

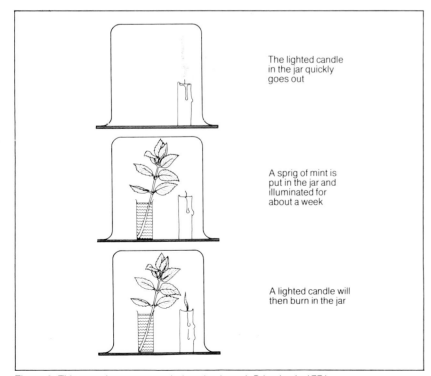

The lighted candle in the jar quickly goes out

A sprig of mint is put in the jar and illuminated for about a week

A lighted candle will then burn in the jar

Figure 3 This experiment was carried out by Joseph Priestley in 1771.

Figure 4 Canadian pondweed is found in ponds and in slow-flowing streams.

What happens during photosynthesis?

The experiments in this Topic tell us that plants need carbon dioxide, water, light and chlorophyll in order to make food; and starch and oxygen are produced.

Carbon dioxide and water are the **raw materials** of photosynthesis. They react in some way to produce starch and oxygen, the **products**. We now know that this is not a simple reaction, but takes place in a series of steps. The reactions need energy, and this comes from the light. The chlorophyll enables the plant to use light energy in this way. Light and chlorophyll are therefore essential 'helpers' in the process.

Although starch is made in the end, it is not the first substance to be formed. Glucose is formed first and this is then turned into starch.

Photosynthesis is therefore a complicated process. However, it is usually summed up by this simple equation:

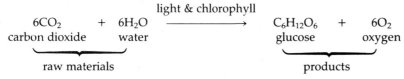

$$\underbrace{6CO_2 \quad + \quad 6H_2O}_{\text{raw materials}} \quad \xrightarrow{\text{light \& chlorophyll}} \quad \underbrace{C_6H_{12}O_6 \quad + \quad 6O_2}_{\text{products}}$$

carbon dioxide   water   glucose   oxygen

In later Topics we will explore some of the details of this reaction.

Investigation 1

To find out if a plant needs carbon dioxide in order to make starch

1 You will need two potted geranium plants which have been de-starched.

2 Put a dish of dampened soda lime on the soil beside one of the plants.
Cover the plant with a polythene bag as shown in the upper illustration.
The soda lime will absorb carbon dioxide from the air inside the bag, so this plant will be deprived of carbon dioxide.

3 Put a dish of saturated sodium bicarbonate solution on the soil beside the other plant.
Cover the plant with a polythene bag as shown in the lower illustration.
The sodium bicarbonate will slowly give out carbon dioxide into the bag, so this plant will have plenty of carbon dioxide.

4 Place both plants side by side in a well lit place for about 48 hours.

5 After about 48 hours take a leaf, or part of a leaf, from each plant. Test them for starch (see page 185).

Which plant contains starch?

Is carbon dioxide needed for starch-formation?

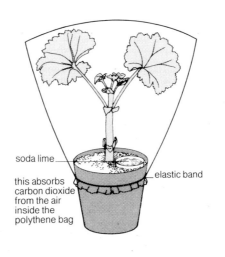

soda lime
this absorbs carbon dioxide from the air inside the polythene bag
elastic band

saturated sodium bicarbonate this slowly gives out carbon dioxide into the air in the polythene bag

---

Investigation 2

To find out if a plant needs chlorophyll to make starch

1 You will need a potted geranium plant with variegated leaves. The plant should have been put in the light for several days.

2 Detach one of the leaves and draw its upper side, making a clear distinction between the green and yellow areas.

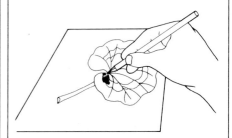

3 Now carry out a starch test on the whole leaf (see page 185).

Which parts of the leaf turn black when treated with iodine?
Indicate your answer in your drawing by writing B in the black areas.
Where is the control in this experiment?
Is chlorophyll needed for starch-formation?

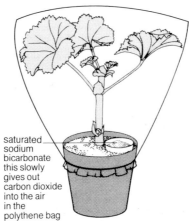

Investigation 3

To find out if a plant needs light in order to make starch.

Method a

1 You will need two potted geranium plants which have been de-starched.

2 Place one of them in the dark, and the other in the light. The plant in the light is your control plant.

Method b

1 You will need a potted geranium plant which has been de-starched.

2 Attach a strip of black paper or metal foil to the upper and lower sides of a leaf, as shown in the illustration.

3 Put the plant in a well-lit place.

4 After several days detach the leaf and test it for starch (see page 185). Make a drawing of the leaf to show your result.

This is called a 'starch print'. What conclusion do you draw?

3 After several days take a leaf (or part of a leaf) from each plant. Test them for starch (see page 185). Don't forget which leaf is which!

Has either plant formed starch?
Is light needed for starch-formation?

Investigation 4

To find out if a water plant gives off oxygen

1 Put some Canadian pondweed or some other suitable water plant into two separate beakers of water.

2 Cover the weed with an upturned funnel and test tube, as shown in the illustration.

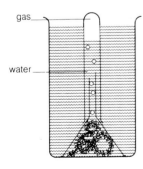

gas

water

3 Place one of the beakers in the light and the other one in the dark.

4 After a few days compare what has happened in the two cases.

Do you find that the illuminated pondweed has produced some gas,

whereas the darkened one has not? How can you find out if the gas is oxygen?
The simplest way is to test it with a glowing splint.

5 Remove the test tube from the beaker. Quickly insert a glowing splint to the far end as shown in the illustration. If it flares up, there is a high proportion of oxygen there.

Has the illuminated pondweed produced oxygen?

Assignments

1 Explain how the starch print in Figure 2 was made.

2 Some people feel that from a scientific point of view a starch print is not a good way of finding out if a plant needs light for making starch. What do you think?

3 In Investigation 1 the plants should be de-starched first.

a) How are they de-starched?
b) Why is this necessary?
c) How could you make sure they have been completely de-starched before you begin the experiment?

4 Elizabeth wants to find out if a potted plant needs carbon dioxide in order to make starch. She is not very satisfied with the method given in Investigation 1, so she tries a different way. She selects two leaves on the plant and, *without cutting them off*, she encloses each one in a small polythene bag. In one bag she puts some soda lime, and in the other bag she puts some saturated sodium bicarbonate solution.
Make a diagram of the set-up. Do you think Elizabeth's method is as good as the one given in Investigation 1? Give reasons for your answer.

5 One way of showing that carbon dioxide is necessary for starch-formation is illustrated below. Study the picture, then answer the questions underneath it.

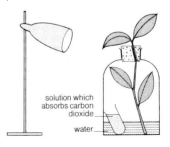

solution which absorbs carbon dioxide

water

a) What should be done to the plant beforehand, and why?
b) Where is the control in this experiment?
c) At the end of the experiment, how would you find out if the plant has made starch in its leaves?
d) Is this a satisfactory experiment? Give reasons for your opinion.

What controls the rate of photosynthesis?

Photosynthesis occurs rapidly or slowly, depending on circumstances, and this will determine how much food is made in a certain period of time. This is important for man because we depend on plants for our food.

Figure 1 Primroses are able to photosynthesise efficiently in shady places.

Figure 2 The lettuces on the left were grown in an atmosphere containing more carbon dioxide than those on the right.

What factors affect the rate of photosynthesis?

Four factors have a particular effect on the rate of photosynthesis: light, carbon dioxide, temperature and water. We will look at them in turn.

Light

We can find out the effect of light on photosynthesis by using Canadian pondweed, the plant which produces bubbles in the light. If the brightness of the light affects the speed of photosynthesis, we would expect the weed to produce more bubbles the brighter the light. We can do an experiment to test this (Investigation).

This and other experiments tell us that, up to a point, the brighter the light the faster the rate of photosynthesis.

How does light affect plants in their natural surroundings? Light varies from day to day and from place to place. On a bright sunny day plants photosynthesise faster than on a dull overcast day. And a plant growing in an open meadow photosynthesises more quickly than a plant growing in the shade.

This is important to gardeners. If a gardener wants his vegetables to do well, he plants them in a place which gets the sun. Sometimes bright lights are shone on indoor plants to increase their rate of photosynthesis. However some plants, such as primroses, thrive in shady places such as a wood (Figure 1). They have special adaptations which enable them to photosynthesise even in dim light.

As with many other things in life, it is possible to have too much of a good thing: in *very* bright sunshine photosynthesis actually slows down. Very bright sunshine contains a lot of ultra-violet light which can damage plants.

Carbon dioxide

Experiments have shown that the more carbon dioxide there is in the air surrounding a plant the faster the plant photosynthesises. How does this affect plants in their natural surroundings? The amount of carbon dioxide in the atmosphere is about 0.03 per cent and it does not vary very much. Even so, there are slight differences from place to place which may affect the rate of photosynthesis. For example, the concentration of carbon dioxide close to the ground in a dense forest is higher than in an open field. Why do you think this is?

Extra carbon dioxide is sometimes pumped into greenhouses, or produced by a 'burner' so as to increase the rate of photosynthesis. This is useful to gardeners, as well as to scientists, who want to increase the speed at which plants make food (Figure 2).

Temperature

Up to a certain point, the higher the temperature, the faster a plant will photosynthesise. Normally a rise in temperature of 10 °C doubles the rate. Temperature can therefore have a marked effect on the rate at which plants make food.

In the natural world there are tremendous variations in temperature, both from place to place and at different times of the day and year. One of the main reasons why plants do so well in a greenhouse or a sheltered garden is because of the warmth there (Figure 3).

Raising the temperature up to about 40 °C increases the rate of photosynthesis. However if the temperature gets above this, photosynthesis slows down and soon stops altogether. This is because the heat destroys the enzymes which are responsible for the chemical reactions.

Water

Plants need water for photosynthesis and if they do not get enough of it they will not photosynthesise so quickly. A plant which is beginning to droop through lack of water may photosynthesise at only half the normal rate. However, water is needed for many other purposes besides photosynthesis, and the effect of water-shortage on photosynthesis may be indirect.

Which places provide the best conditions for photosynthesis?

The answer is the tropical rain forests of South America, Central Africa and South East Asia. Lots of sunshine, warmth and a high rainfall ensure maximum photosynthesis and prolific growth of plants. So lush is the vegetation in these areas that the tropical rain forest has been described as a 'vegetative frenzy' (Figure 4).

Crop plants grown in places where light, temperature and moisture are at their most suitable for photosynthesis, make particularly large amounts of food. This is true of sugar cane, for example, which has the highest yield of all crop plants. It is one of the largest members of the grass family, reaching heights of six metres or more. Most of the world's sugar comes from sugar cane (Figure 5).

Sugar cane needs a hot, moist climate with temperatures averaging around 21 °C (70 °F) and an annual rainfall of about 150 cm (60 inches). This it gets in places such as the West Indies, parts of South America, India and South East Asia. When grown in drier places like North America and Southern Africa, water must be supplied by irrigation.

In some parts of the world plants are grown in special air conditioned greenhouses in which all the factors which affect photosynthesis and plant

Figure 3 The warm conditions inside the greenhouse enabled this excellent crop of melons to be produced.

Figure 4 The tropical rain forest has been described as a vegetative frenzy. Notice the dense vegetation in this photograph of the Everglades in Florida, USA.

Figure 5 Sugar cane, the world's foremost photosynthesiser is seen here being harvested by a farmer in the West Indies.

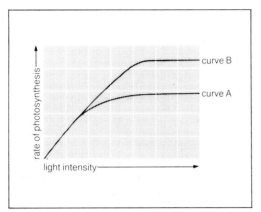

Figure 6 This graph shows how a plant's rate of photosynthesis is affected by the light intensity. Curve A was obtained with the plant in a low concentration of carbon dioxide. Curve B was obtained at a higher concentration of carbon dioxide.

growth are carefully controlled. In this way scientists can make sure that the plants are given exactly what they need.

Which places provide the worst conditons for photosynthesis?

The answer is dark or dimly lit places, particularly if they are cold as well. For example, grass may be lush in an open meadow, but in a corner which is always in the shade it may be sparse. Have you ever noticed how bare the soil is under beech trees? This is because the leaves of the beech fit closely together and don't let much light through (see page 200). The soil is even barer under overhanging rocks and in caves. However, you have only to provide an artificial light and a few green plants will soon pop up if there's enough moisture for them.

We have seen that a plant's rate of photosynthesis is influenced by light intensity, carbon dioxide concentration, temperature and water supply. But they do not act separately; in practice they interact and influence each other.

How do the factors influence each other?

Look at the graph in Figure 6. This shows the results of an experiment which was designed to find the effect of raising the light intensity on the rate of photosynthesis – like the experiment in Investigation 1.

First look at curve A. Notice that as the light intensity is gradually raised, the curve rises, i.e. the rate of photosynthesis increases.

However, there comes a point when the curve flattens out – in other words the rate of photosynthesis does not increase any more, however much the light intensity is raised. Why do you think the rate of photosynthesis stops increasing? The answer is that some factor other than light is preventing photosynthesis from going any faster. We say that this factor is now **limiting** the rate of photosynthesis.

What might this factor be? Well, it could be carbon dioxide. How could we find out if it is carbon dioxide? One way would be to raise the amount of carbon dioxide in the atmosphere surrounding the plant and repeat the experiment.

The result of doing this is shown in curve B. This time a much higher rate of photosynthesis is achieved. What does this tell us? It tells us that carbon dioxide must have been limiting the rate of photosynthesis when the curve flattened out in the first experiment.

From experiments of this kind we can draw this general conclusion. The rate of photosynthesis is controlled by several different factors. At a particular moment the rate is determined by whichever factor is closest to its minimum value. This is called the **law of limiting factors**, and is of great importance to plants.

In a particular place, a wheat field for instance, different factors limit photosynthesis at different times of the day. At the beginning and end of the day, when the light is dim, light limits the rate of photosynthesis. In the middle of the day, when the light is good, carbon dioxide limits photosynthesis.

The same kind of thing applies to the seasons. Take the British summer and winter for example. In summer, when the light is good, carbon dioxide limits photosynthesis most of the time. In winter, on the other hand, light is the limiting factor most of the time.

What about a small plant growing in a forest? Here the light is dim but the concentration of carbon dioxide is high. Result? Light limits photosynthesis all the time.

We have already seen that in the closed atmosphere of a greenhouse extra carbon dioxide can increase the rate of photosynthesis. However, the law of limiting factors must always be born in mind. It is no use pumping extra carbon dioxide into a greenhouse if the light is poor. It simply will not make any difference.

Investigation

To see if raising the light intensity increases the rate of photosynthesis

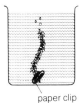
paper clip

heat shield

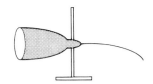

1 Darken the room so the light from the windows does not vary.

2 Cut off a sprig of Canadian pondweed about 5 cm long.

3 Attach a paper clip to the top end to weigh it down.

4 Put it in a beaker or jam jar of water as shown in the illustration.

5 Add a pinch of potassium bicarbonate to the water: this will ensure that the pondweed has a good supply of carbon dioxide.

6 Place a lamp to one side of the jar.

7 Fill a narrow aquarium tank with water and place it between the jar and the lamp. This will serve as a heat shield and will prevent the pondweed from heating up.

8 Illuminate the weed with a lamp placed a long way away (say 50 cm).

9 Wait a few minutes, then count the number of bubbles given off during a one minute period. Do this three times and work out the average.

10 Now bring the lamp closer, wait a few minutes, then count the number of bubbles again. Do this three times and work out the average.

How many bubbles are given off per minute (a) with the lamp a long way away, and (b) with the lamp close?

Do you find that the closer the lamp, the greater is the rate at which bubbles are given off?

Does raising the light intensity increase the rate of photosynthesis?

Assignments

1 Mr. Smith plants his onions in a shady place whereas Mr. Jones plants his in the sun. Whose onions would you expect to do best, and why?

2 Mr. Jones left a bucket on his lawn for several weeks. When he lifted it up he found that the grass underneath was dead. What might have killed the grass?

3 Someone observed that wheat grows taller, and gives a higher yield of grain, close to a certain coal-burning factory in the Midlands than further away. Suggest a reason for this. What investigations would you carry out to find if your suggestion is right?

4 The following figures give the total annual amounts of organic matter produced per hectare by plants in different parts of the world:

sugar cane (Java)	87 tonnes
tropical rain forest	59 tonnes
pine forest, England	16 tonnes
birch forest, England	8.5 tonnes

Can you account for the differences?

5 An experiment was carried out to investigate the effect on a plant's rate of photosynthesis of increasing the amount of carbon dioxide in the air. The light intensity and temperature were kept constant throughout the experiment. The results are shown in this graph:

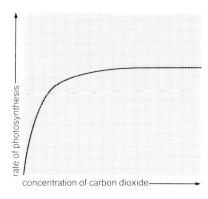

rate of photosynthesis

concentration of carbon dioxide

a) Say in your own words what the graph shows.
b) Why does the curve rise to begin with?
c) Suggest reasons why the curve eventually flattens out.
d) What experiments would you do to find out which of your reasons in (c) is correct?

6 A scientist grew some cereal plants in a field. During the course of one day he took several plants every four hours and measured the amount of sugar in the leaves. The sugar concentrations, expressed as a percentage of the dry mass of the leaves, are given below:

Time of day	Sugar concentration
4 am	0.45
8 am	0.60
12 noon	1.75
4 pm	2.00
8 pm	1.4
12 midnight	0.5
4 am	0.45

a) Plot the data on graph paper, putting sugar concentration on the vertical axis.
b) What is the probable concentration of sugar in the leaves at (a) 10 am and (b) 2 am?
c) At what time of the day is sugar probably at a maximum in the leaf?
d) Explain the changes which occur in the sugar concentration over the 24 hour period.

Chlorophyll the miracle molecule

The green colour of plants is caused by the chemical substance chlorophyll which occurs inside them. Chlorophyll does a remarkable job, as we shall see.

What is chlorophyll?

To find out what chlorophyll is, we can extract it from leaves and make a solution of it (Investigation 1). By doing this we can see that it's a green substance. Coloured substances of this sort are called **pigments**.

Scientists have analysed this pigment. It turns out to be a complex substance containing magnesium (see page 180).

Chlorophyll plays a vital part in photosynthesis. To understand what it does, we must first examine its effect on light.

What does chlorophyll do to light?

We can investigate this by observing what happens to light as it passes through a solution of chlorophyll (Investigation 2).

Ordinary white light, such as sunlight, is made up of different colours or wave-lengths. We don't normally see these colours except, for example, when there's a rainbow. However, in the laboratory, light can be split up into its colours by means of a prism. The colours form a series which we call a **spectrum**. The colours of the visible spectrum are:

Red Orange Yellow Green Blue Violet

Now when light passes through chlorophyll, certain colours disappear. Which ones? The experiment in Investigation 2 tells us that they are **blue** and, to a lesser extent, **red**.

These two colours disappear because they are absorbed by the chlorophyll. Other colours, particularly green, pass straight through it or are reflected. The reason why leaves look green is that chlorophyll reflects the green part of the spectrum.

What colours are used in photosynthesis?

Chlorophyll absorbs blue and red light. It would therefore seem likely that these two colours are used in photosynthesis.

Can you think of an experiment which could be done to test this suggestion? One way would be to shine different coloured lights onto plants. We could then find out which colours are most suitable for photosynthesis. As an indication of how much photosynthesis had been going on with each colour, we could measure either the volume of oxygen given off or the amount of starch formed.

Experiments of this kind show that the two colours which are best for photosynthesis are blue and red – the very same colours that are absorbed. A plant which is deprived of these two colours cannot photosynthesise properly and doesn't make much starch. Sunlight provides these two colours in the right proportions.

Other pigments

Leaves contain several pigments besides chlorophyll. They can be separated from each other by a process called **chromatography** (Investigation 3). In addition to chlorophyll there are yellow and grey pigments.

Separating the pigments like this is useful because after they have been isolated each one can be investigated on its own. In this way scientists can find out what each one does.

Careful experiments of this sort have shown that they all play a part in photosynthesis, but the most important is the green chlorophyll.

Certain plants which are known to photosynthesise are not green. How can we explain this?

Take seaweed, for example. Most seaweeds are brown. This is because they possess a brown pigment called **fucoxanthin**. Chlorophyll is present too, but the brown pigment is so abundant that it completely masks the green colour of the chlorophyll. Both are used in photosynthesis.

Another example is copper beech. The leaves of this beautiful tree contain chlorophyll, but they also contain large quantities of a purple substance. This belongs to a group of pigments called **anthocyanins**. There are several different anthocyanins, and they help to give flowers and fruits their characteristic colours, but they play no part in photosynthesis.

Where does chlorophyll occur?

If you look at a simple leaf under the microscope (Investigation 4), you will see that its cells contain lots of small green bodies (Figure 1). These bodies are called **chloroplasts**. They are packed together in the cytoplasm round the edges of the cells. Each one is filled with chlorophyll. It's here that photosynthesis takes place and starch is formed.

Inside the chloroplast

Chloroplasts are extremely small: about ten thousand of them would fit onto the full stop at the end of this sentence. They are therefore invisible to the naked eye, but under the light microscope each one can be detected individually. However, the light microscope does not magnify them enough for us to see any detail inside them. With an electron microscope, however, we can see much more.

Figure 2 is a picture of a chloroplast based on its appearance in the electron microscope. It is magnified about 30 000 times. If a whole moss leaf was magnified to this extent, it would be the size of an oak tree.

The most noticeable features are the piles of disc shaped objects. These are very thin membranes. By careful analysis scientists have shown that millions and millions of chlorophyll molecules are attached to these membranes.

The chlorophyll molecules are laid out on the chloroplast membranes rather like library books are stacked on shelves. In this way a great many chlorophyll molecules are packed together inside a small space.

Think of an oak tree. It has a large number of leaves. Inside each leaf are numerous chloroplasts; within each chloroplast are numerous membranes; and covering each membrane are numerous chlorophyll molecules. So there is a lot of chlorophyll in a single plant like an oak tree and it covers a huge surface area. This is very important in view of the job it has to do.

What does chlorophyll do?

Chlorophyll absorbs light energy and enables it to be used by the plant for building up sugar. The overall effect is that energy is transferred from sunlight to sugar molecules.

The energy contained inside molecules is called chemical energy. So chlorophyll's job is really to convert light energy into chemical energy.

There is nothing particularly mysterious about this in itself: it is a well known law of physics that one form of energy can be changed into another.

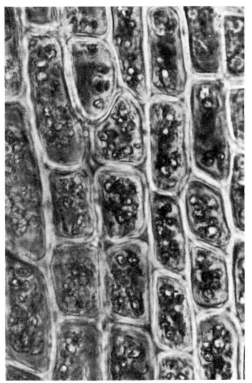

Figure 1 Part of a moss leaf as seen under a light microscope. Notice the numerous chloroplasts in the cells.

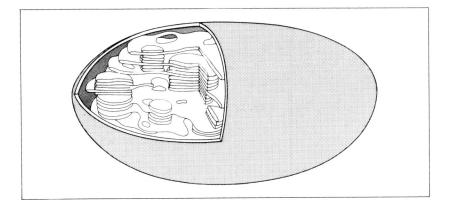

Figure 2 This drawing shows the structure of a chloroplast based on its appearance in the electron microscope.

Investigation 1

How to make a solution of chlorophyll

There are several ways of doing this.
Here is one of the simplest methods:

1 Cut up a few green leaves into small pieces.

2 Put them in a mortar with a pinch of washed sand.

3 Cover them with ethanol and grind them up with a pestle. This will break open the cells, and the chlorophyll will dissolve in the ethanol.

4 Filter the fluid into a beaker (see illustration). The green chlorophyll solution will pass through the filter paper, leaving any bits of leaf behind.

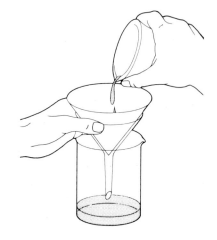

5 If necessary add some water to the chlorophyll solution to make it less concentrated.

What colour is the chlorophyll?

Does its colour differ from that of the leaf from which it was obtained?

If so, in what way does it differ, and why?

Investigation 2

To find the effect of chlorophyll on light

1 Prepare a solution of chlorophyll as instructed in Investigation 1.

2 Pour the solution into a narrow transparent container as shown in the upper part of the illustration.

3 Set up a projector, prism and screen as shown in the lower part of the illustration. The prism splits the light into its different colours.

What colours can you see?

4 Now place your chlorophyll solution between the projector lens and the prism (see arrow in the illustration).

Do you find that certain colours disappear?

If so, which ones?

What effect does chlorophyll have on the light from the projector?

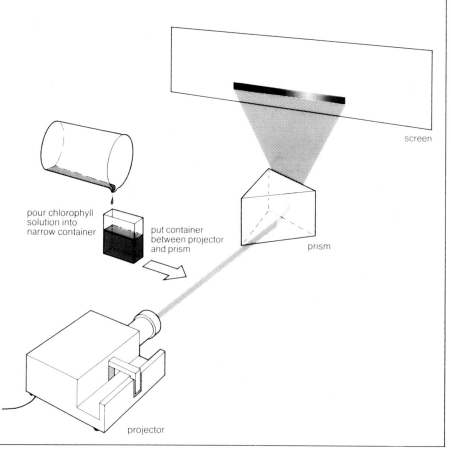

pour chlorophyll solution into narrow container

put container between projector and prism

prism

screen

projector

Investigation 3

To separate the pigments present in a leaf

1 Prepare a solution of leaf pigment as instructed in Investigation 1. The solution should be as strong as possible.

2 Dip a stick of white blackboard chalk into the solution so the end of the chalk goes thoroughly green.

3 Let the chalk dry. Meanwhile, pour a little ethanol into the bottom of a small beaker.

4 Stand your piece of chalk, green end downwards, in the ethanol.

What happens?

Can you see that the solution is made up of at least two different coloured pigments?

What colours are these pigments?

Investigation 4

Looking at chloroplasts in a moss leaf

1 With a pair of tweezers, carefully detach one small leaf from a sprig of moss.

2 Put the leaf in a drop of water on a slide, and cover it with a coverslip.

3 Examine the leaf under a microscope.

Where is the green pigment?
Look at Figure 1. Can you see the various structures shown in this picture in your own moss leaf?

4 Lift up the coverslip and put a drop of iodine on the leaf.

5 Put the coverslip back.

6 Examine the leaf under the microscope again.

A black colour indicates starch. Is there any starch in the cell? If there is, where is it?

What conclusions would you draw from this experiment about the function of chlorophyll?

Assignments

1 Why do leaves generally look green?

2 Describe *in detail* an experiment which you would do to find out which colours of the spectrum a potted geranium plant uses in photosynthesis.

3 A man works in a windowless office lit by a single light bulb. To cheer himself up he puts an aspidistra in the room. Instead of growing tall, as he hoped it would, it remains short and stunted. Suggest explanations.

4 Observe the leaves of various indoor and outdoor plants. Are they always green? If they are not, can you suggest why they are some other colour?

5 If you put a green plant in the dark for a week or so, the leaves turn yellow. Several days after returning the plant to the light the leaves turn green again.

　a) Suggest an explanation for this.
　b) How would you find out if yellowing of the leaves prevented them from photosynthesising?
　c) What do you think would happen to the plant if you left it in the dark for ever, and why?

6 The following diagram shows light being shone onto a screen from a slide projector.

　a) What colour would you expect the light to be on the screen?
　b) What would happen if you put a prism in the beam of light at A?
　c) What effect would be produced if you then put a glass jar containing a solution of chlorophyll at B?
　d) What conclusions can be drawn from this experiment?

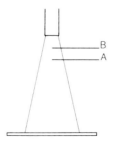

More about photosynthesis

In recent years scientists have discovered a lot about the chemical reactions which take place during photosynthesis thanks to the use of isotopes

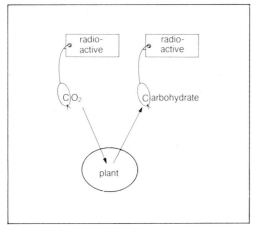

Figure 2 A scientist, wearing protective gloves, transferring some radio-active bicarbonate solution into a bottle containing *Chlorella*. The bicarbonate solution contains radio-active carbon atoms. This provides the plant with a source of labelled carbon dioxide.

Figure 3 If *Chlorella* is supplied with carbon dioxide whose carbon is radio-active, the radio-active carbon gets into the carbohydrate which the plant makes.

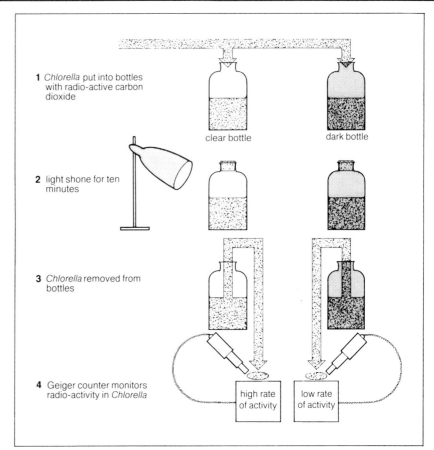

1 *Chlorella* put into bottles with radio-active carbon dioxide

clear bottle   dark bottle

2 light shone for ten minutes

3 *Chlorella* removed from bottles

4 Geiger counter monitors radio-activity in *Chlorella*

high rate of activity   low rate of activity

Figure 1 This diagram summarises an experiment which scientists have carried out with *Chlorella*.

What are isotopes?

Consider the element carbon. The normal form of carbon has an atomic mass of 12 (^{12}C). However, there is another kind of carbon which has an atomic mass of 14 (^{14}C), and is **radio-active**. These two forms of carbon are known as **isotopes**. The radio-active form can be detected by means of a **Geiger counter**. This has a probe: when the probe is brought close to a radio-active substance it clicks repeatedly.

One of the first places to use isotopes for studying photosynthesis was Berkeley, California. There in the Lawrence Radiation Laboratory, scientists have used radio-active carbon to study photosynthesis in *Chlorella*. *Chlorella* is a single-celled plant which is easily grown in the laboratory.

Tracing what happens to carbon dioxide in photosynthesis

Here is a simplified account of what the Berkeley scientists did. First they made some carbon dioxide whose normal carbon had been replaced by its radio-active isotope. By substituting the radio-active isotope for the normal carbon, they were able to 'label' the carbon dioxide and follow what happened to it.

The labelled carbon dioxide was supplied to *Chlorella*, and a Geiger counter was used to detect it. A simplified version of the experiment is summarised in Figure 1, and one of the steps is illustrated in Figure 2.

The Berkeley scientists found that in the light the radio-active carbon was taken up by the plant. However, they went further than this. They carefully extracted the chemical compounds from the plant and tested them for radio-activity. They found that the radio-active carbon had got into the carbohydrate (sugar) which the plant had made (Figure 3). From this experiment it was concluded that the carbon in the carbohydrate made by plants comes from carbon dioxide:

Carbon dioxide + H₂O ⟶ Carbohydrate + Oxygen

Tracing what happens to oxygen

Carbon dioxide contains oxygen as well as carbon. What happens to the oxygen? Labelling experiments have given us the answer to this too. There is a rare isotope of oxygen whose atoms are slightly heavier than those of normal oxygen. They can be detected by a machine called a **mass spectrometer**.

Scientists have given plants carbon dioxide whose normal oxygen has been replaced by this heavy isotope. What happens? The heavy oxygen gets into the carbohydrate which the plants make. This tells us that the oxygen in the carbohydrate made by plants comes from carbon dioxide:

Carbon diOxide + H₂O ⟶ CarbOhydrate + Oxygen

The conclusion from these experiments is that in photosynthesis carbon dioxide is somehow converted into carbohydrate.

How is carbon dioxide converted into carbohydrate?

If you compare the formulae of carbon dioxide and a carbohydrate, you will find that the carbohydrate contains hydrogen whereas carbon dioxide does not. Where does the hydrogen in the carbohydrate come from? There is really only one possible answer: water.

The formula of water is H₂O. It is possible to label the oxygen in water by replacing it with its heavy isotope. If such water is given to an illuminated plant, the heavy oxygen is given off as a gas. This tells us that the oxygen which plants give off during photosynthesis comes from water:

Carbon dioxide + H₂O ⟶ Carbohydrate + Oxygen

The water must therefore be split into its constituent hydrogen and oxygen atoms. Careful experiments have confirmed that this really does happen in photosynthesis, though the details are very complicated.

Photosynthesis occurs in two stages

We now know that photosynthesis occurs in two stages (Figure 4). In the first stage water is split into oxygen and hydrogen. In the second stage hydrogen combines with carbon dioxide to form carbohydrate.

Scientists have discovered an interesting thing about these two stages: only the first requires light; the second can occur in the dark. So we call these the **light** and **dark stages** respectively.

The job of the light stage is to split water and provide hydrogen atoms for the dark stage. Energy for this comes from sunlight and is trapped by the chlorophyll. Neither light energy nor chlorophyll are needed for the dark stage.

How do plants make proteins?

The scientists who did the experiment with *Chlorella* (Figure 1) found that radio-active carbon quickly got into carbohydrates, but later on it got into other more complex compounds such as protein. The plant makes carbohydrate first and then converts some of this into other things.

To make proteins plants need the extra elements nitrogen, sulphur and phosphorus, in addition to carbon, hydrogen and oxygen. It obtains these extra elements from its surroundings in the form of mineral salts.

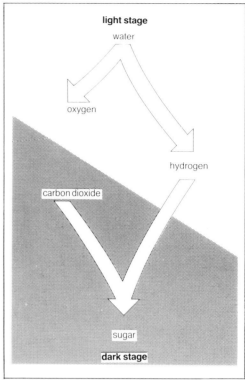

Figure 4 Photosynthesis occurs in two stages. First water is split into oxygen and hydrogen. Then the hydrogen combines with carbon dioxide to form sugar (carbohydrate).

Assignments

1 How has heavy oxygen helped scientists to understand photosynthesis?

2 If you transfer a green plant from the light to total darkness, do you think it stops making sugar straight away? Explain your answer.

3 A farmer gives his crops a nitrogen fertiliser. Why is this desirable?

4 What criticisms, if any, do you have of the experiment illustrated in Figure 1?

5 Experiments with isotope tracers enable us to say the following about photosynthesis:

a) The carbon and the oxygen in the sugar come from carbon dioxide.
b) The hydrogen in the sugar comes from water.
c) The oxygen gas that's given off comes from water.
d) Some water is formed.

Give a balanced chemical equation for photosynthesis that takes all these observations into account.

The leaf: organ of photosynthesis

Plants make food wherever chlorophyll is present, which is mainly in the leaves. This Topic is about the structure of leaves, and how it fits their job of feeding the plant.

The external structure of leaves

You can learn much about leaves simply by looking at them from the outside (Investigation 1). The leaf is attached to the stem or branch by a leaf stalk or **petiole** (Figure 1). The leaf stalk is continuous with the veins (Figure 2).

Here is a summary of the leaf's main adaptations for photosynthesis which you can see from the outside.

Leaves have a large surface area

Leaves come in all manner of shapes and sizes, but they are generally flat, sometimes large, and usually numerous. The result is that they cover a large surface area (Investigation 2). This makes them good at absorbing carbon dioxide from the air, and light energy from the sun.

Leaves are arranged in the best way

Leaves are usually positioned in such a way that they get the maximum amount of light. Moreover, they may fit together snugly so very little light passes through to the ground below. This is why it is so dark under many trees and shrubs.

In a large tree with a lot of leaves there is always a risk that the leaves at the top may shade those lower down. This is avoided by having leaves which are divided into leaflets, or have jagged edges, so light can get between them. Another way is by the leaves at the top arranging themselves so their edges are directed towards the sun. This allows light to pass between them to the leaves lower down (Figure 3).

Leaves have pores

A leaf's 'skin' (the **epidermis**) is pierced by tiny air pores known as **stomata** (singular: stoma). They occur mainly on the lower side of the leaf. They allow carbon dioxide to enter the leaf. Some leaves have a very large number of air pores (Figure 4). Air pores are discussed in more detail on page 204.

Leaves are thin

Leaves are usually less than a millimetre thick. This cuts down the distance through which carbon dioxide has to diffuse after it has entered the leaf.

But there's a problem. Being so thin, leaves would be liable to droop, but this is prevented by the **veins** which serve as a kind of skeleton holding the leaf out flat (Figure 5).

Figure 1 Leaves are the main place where plants make food.

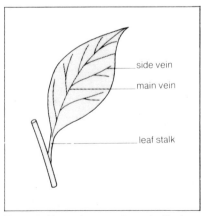

Figure 2 The main parts of a typical leaf.

Figure 3 Some plants have their leaves arranged like this.

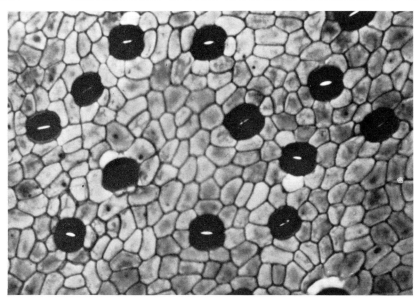

Figure 4 Air pores on the lower side of a leaf of a box plant.

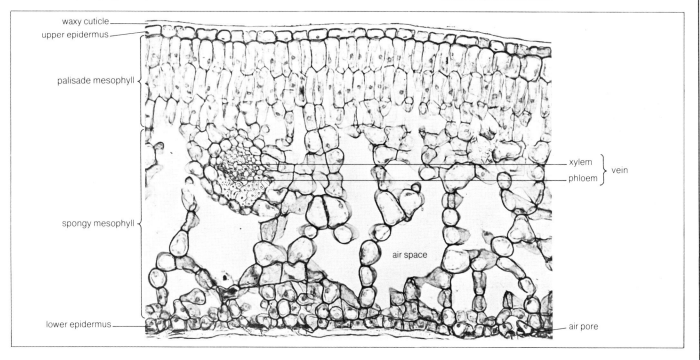

Figure 5 Underside of a sycamore leaf showing the branching veins.

The internal structure of leaves

You can study the inside of a leaf by cutting thin sections of it and examining them under a microscope (Investigation 3).

Figure 6 shows the inside of a holly leaf as it appears under the microscope. The leaf is contained above and below by the **epidermis**. In between are lots of cells which together make up the **mesophyll**. The mesophyll is divided into the **palisade mesophyll** on the upper side and the **spongy mesophyll** below. The cells of the palisade mesophyll are shaped like bricks and are arranged neatly side by side. The mesophyll cells are rounded and appear haphazard in their arrangement.

In the middle of the leaf in Figure 6, there is a small vein. This consists of two main tissues: **xylem** towards the top and **phloem** below. The xylem consists of pipe-like **vessels**, and the phloem consists of elongated cells called **sieve tubes** (see page 205).

Figure 6 Cross-section of part of a holly leaf as seen under a light microscope.

waxy cuticle
upper epidermus
palisade mesophyll
spongy mesophyll
lower epidermus
xylem ⎫ vein
phloem ⎭
air space
air pore

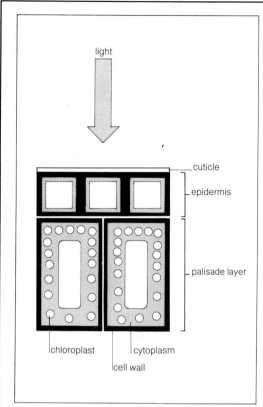

Figure 7 The surface cells of a leaf showing the palisade layer. Notice how the chloroplasts are bunched up towards the top of the cells where the greatest amount of light is.

Figure 8 The food-making mesophyll cells in the leaf take up carbon dioxide which diffuses in through the air pores, and water and minerals which come to them from the roots via the vessels. They give out oxygen which escapes through the air pores, and they make sugar some of which is carried away in the sieve tubes.

Here are the main adaptations for photosynthesis which you can see inside the leaf.

The mesophyll cells contain chloroplasts

The palisade and spongy mesophyll cells all contain chloroplasts and can photosynthesise. However, most of the chloroplasts are located in the palisade layer, so it is here that photosynthesis mainly takes place (Figure 7).

The photosynthetic cells are mainly on the upper side of the leaf

The palisade cells, where most of the photosynthesis takes place, are near the surface of the leaf on the side which gets most light. Inside these cells, the chloroplasts – as if greedy for light – are often clustered towards the upper side.

There are air spaces between the mesophyll cells

The cells making up the spongy mesophyll are loosely packed, with large air spaces in between. Carbon dioxide diffuses readily through the air pores into these spaces. It then circulates freely inside the leaf, passing through the moist cell walls into the cells.

The leaf contains transport tissues

As well as giving strength, the veins serve as the leaf's transport system. The vessels carry water and mineral salts from the roots to the leaves. The sieve tubes carry sugar and other food substances which have been made by photosynthesis, from the leaves to other parts of the plant.

The dense network of veins, typical of most leaves, ensures that none of the leaf cells are far away from the transport system.

What happens to the sugar which a leaf makes?

Some of the sugar is broken down straight away to provide energy for the leaf's own needs. Some of it is converted into starch and stored. The rest is sent to other parts of the plant, either to supply energy there or to be stored.

As the plant's food-manufacturing device, the leaves must be as extensive as possible and in full communication with the rest of the plant (Figure 8).

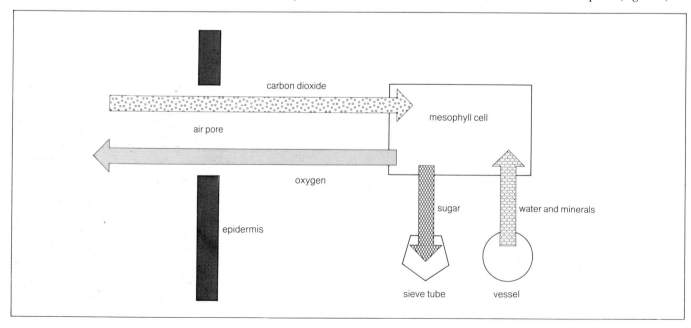

Investigation 1

Looking at leaves

1 Examine a leaf of e.g. privet. Which of the structures shown in Figure 2 can you see?

2 How does the colour of the upper side of a leaf differ from the lower side? Why the difference?

3 Look at leaves from different trees. Can you explain their different shapes?

4 Examine the veins of different leaves. What sort of pattern do they form?

5 Tear leaves in two. Does this tell you anything about the function of the veins?

A method for looking at the air pores of a leaf is described in Investigation 5 on page 209.

Investigation 3

Looking inside the leaf

You will need either a prepared slide, or a thin section of a leaf which you can mount on a slide yourself. If you are mounting a section, proceed as follows:

1 Put a drop of water on a microscope slide.

2 Carefully transfer the leaf section to the drop of water on the slide.

3 Cover the section with a coverslip.

4 Examine it under the microscope.

Can you see the structures shown in Figure 6? (You may be looking at a different kind of leaf from the one shown in the figure, so watch out for differences).

In what respects is the inside of the leaf adapted for photosynthesis?

Assignments

1 Make a list of all those features of green plants in general which help the leaves to get as much light as possible. (It may help you to do this if you observe plants living around your home or school.)

2 Each word in column A, below, is related to one or more words in Column B.

a) Against each word in column A write down the appropriate word, or words, from column B.

A	B
air pores	carbon dioxide
vessels	light
chloroplasts	water
air spaces	chlorophyll

b) What do the four words in column A have in common?

c) What do the four words in column B have in common?

3 Why is the lower side of a leaf often a paler green than the upper side?

4 Why do the palisade mesophyll cells contain more chloroplasts than the spongy mesophyll cells?

5 What part is played by each of these structures in photosynthesis:

a) the xylem, b) the air pores, and c) the air spaces in the spongy mesophyll?

6 The photograph below shows part of a tree with the leaves in their natural position. In what way might the positioning of the leaves help the tree to survive?

Investigation 2

Finding the leaf area of a plant

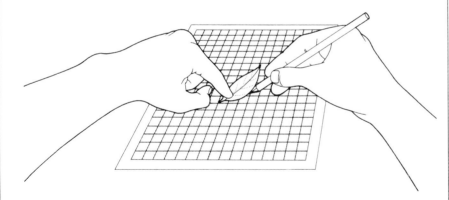

1 Select a large plant (tree or shrub) whose leaves are approximately equal in size.

2 Detach a leaf. Lay it on squared paper and trace round it with a pencil.

3 From the number of squares which the leaf covers, work out the surface area of the leaf.

4 Count the number of leaves on the plant. (If your plant is a tree, you will have to make a rough estimate).

5 Multiply the area of one leaf by the number of leaves.

What is the total surface area of the leaves of your plant?

How does the area compare with the floor of the room where you are working?

Why is it useful to the plant to have a large leaf area?

(A quick way of estimating the area of a leaf is to measure its length and maximum width in millimetres. Its area in square millimetres is the length × width × 0.75)

How are things moved about in plants?

The inside of a plant is the scene of much activity. Substances are constantly being moved from one place to another. To understand how this occurs, we must first look at the structures inside the plant.

Inside the plant

A flowering plant consists of three main parts: the roots, stem and leaves. To find out about the internal structure of the plant, you need to cut thin sections of it and look at them under the microscope (Investigation 1). The internal structure of a flowering plant is shown in Figures 1 and 2. You will see that it is composed of different kinds of tissues. These are the main ones:

The **epidermis**: this is the outermost layer of cells and forms a kind of skin. The epidermis of the leaves and stem is covered with a waxy **cuticle** which prevents water evaporating through it, and it is pierced by a variable number of **air pores** or stomata (singular: stoma). The root does not have a cuticle or air pores, but the cells towards the tip have extensions called **root hairs**.

Packing tissue: This is composed mainly of rounded cells packed close together. This makes up the bulk of the inside of the plant.

Xylem: this contains long tubular vessels. These are dead structures, and their walls are made of wood (lignin). They are narrow, like capillary tubes (Figure 3A).

Phloem: this contains elongated living cells with cellulose walls. They are called **sieve tubes** because the end walls between one cell and the next are perforated by tiny holes, like a sieve (Figure 3B).

The xylem and phloem together make up the plant's **vascular tissue**, which plays a very important part in transporting substances within the plant.

Transpiration

Carry out Investigations 2 and 3. From these kinds of experiments we conclude that water evaporates from the leaves, and is replaced by the roots which absorb it from the soil. The evaporation of water from the above-ground parts of the plant is called **transpiration,** and the flow of water–or sap as it's called–through the plant is called the **transpiration stream.** Scientists have done experiments which show that this movement of water occurs in the xylem vessels (Investigation 4).

In keeping the water moving, the air pores have an important part to play, for they provide the main route by which water can escape from the plant into the atmosphere.

The air pores

You can find out about the air pores by looking at the surface of a leaf under the microscope (Investigation 5). In most plants, the air pores are mainly on the undersides of the leaves: there may be as many as four hundred of them in a square millimetre. Because of this, water generally evaporates more quickly from the lower side of the leaf than from the upper side of the leaf (Investigation 6).

The structure of an air pore is shown in Figure 4. It is bounded by a pair of sausage-shaped **guard cells**. How does the air pore open? The guard cells take in water by osmosis from the neighbouring epidermal cells. As a result, the guard cells swell up and bend, so a gap develops between them. The bending is accentuated by the fact that the inner wall of the guard cell is thicker and less elastic than the outer wall. The air pore closes by the reverse process: water is drawn *out* of the guard cells by osmosis, so they straighten.

The air pores of most plants open during the day and close at night. When they are open, water vapour can escape from inside the leaf, and oxygen and carbon dioxide can diffuse in and out.

How does water move through the plant?

Figure 5 summarises how water passes through a plant. Water is drawn into the root from the surrounding soil. The root hairs help by increasing the surface area. The concentration of salts in the root hairs is greater than that in

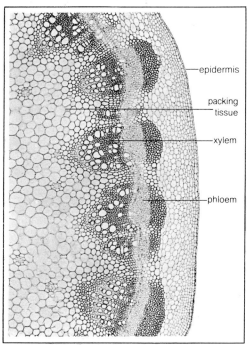

Figure 1 Transverse section of a sunflower stem

epidermis

packing tissue

xylem

phloem

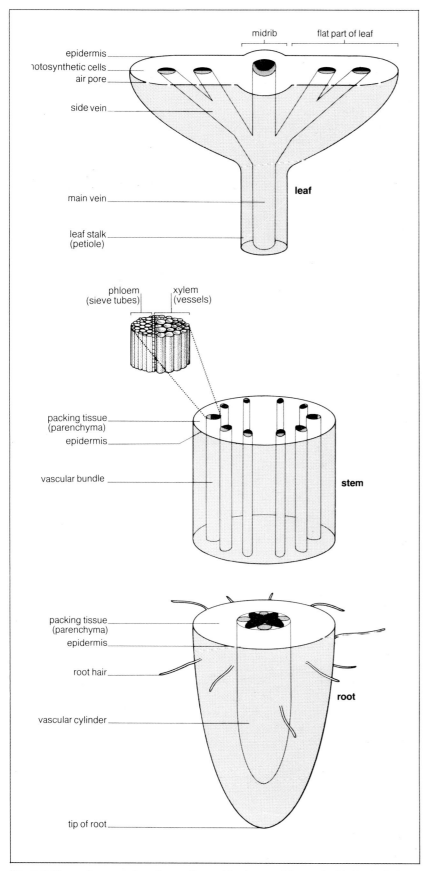

midrib flat part of leaf

epidermis
photosynthetic cells
air pore
side vein
main vein
leaf
leaf stalk
(petiole)

phloem
(sieve tubes) xylem
(vessels)

packing tissue
(parenchyma)
epidermis
vascular bundle **stem**

packing tissue
(parenchyma)
epidermis
root hair
vascular cylinder **root**
tip of root

Figure 2 These diagrams show the positions of the transport tissues inside the leaf, stem and root of a flowering plant, xylem black, phloem grey.

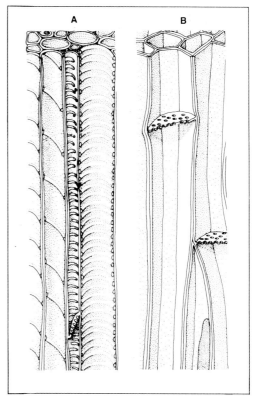

A B

Figure 3 Vessels (left) and sieve tubes (right) inside a plant stem as they appear in a thin section cut longways and viewed under a microscope.

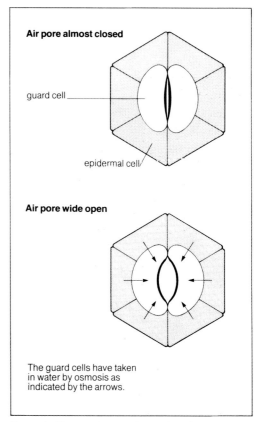

Air pore almost closed

guard cell

epidermal cell

Air pore wide open

The guard cells have taken in water by osmosis as indicated by the arrows.

Figure 4 Air pores (stomata), greatly enlarged, as they appear in a surface view of a leaf.

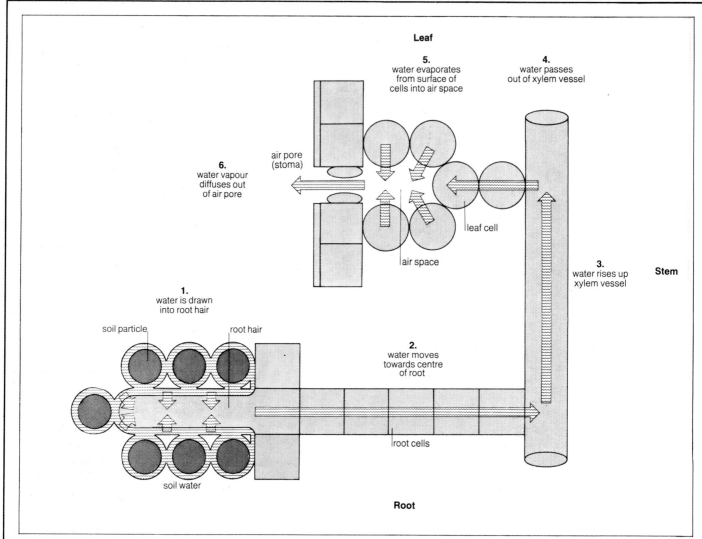

Leaf

5.
water evaporates
from surface of
cells into air space

4.
water passes
out of xylem vessel

6.
water vapour
diffuses out
of air pore

air pore
(stoma)

leaf cell

air space

Stem

3.
water rises up
xylem vessel

1.
water is drawn
into root hair

soil particle

root hair

2.
water moves
towards centre
of root

root cells

soil water

Root

Figure 5 This diagram summarises how water passes through a flowering plant. There are three different pathways through which water may be transported in the root and leaf. Most of it flows along the cellulose cell walls; some travels in the cytoplasm of the cells; and the rest passes from vacuole to vacuole.

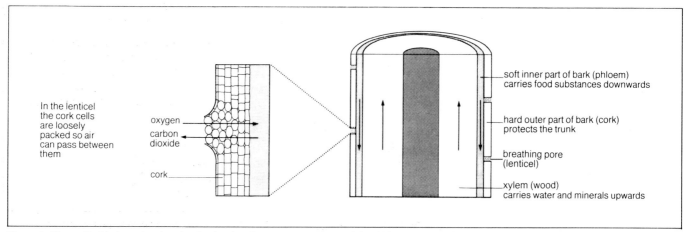

In the lenticel
the cork cells
are loosely
packed so air
can pass between
them

oxygen

carbon
dioxide

cork

soft inner part of bark (phloem)
carries food substances downwards

hard outer part of bark (cork)
protects the trunk

breathing pore
(lenticel)

xylem (wood)
carries water and minerals upwards

Figure 6 The main structures in a tree trunk, showing the movement of materials within it.

the soil water, so water is drawn into them by osmosis. The water then moves towards the centre of the root.

The water rises up the stem partly by being pushed from below, and partly by being pulled from above. The pushing force can be shown by cutting a stem near its base: provided there is plenty of water in the soil, water will ooze out of the stump for a long time. This is known as **root pressure**.

The pull from above is created by the evaporation of water from the leaf. If you stop this pulling force by, for example, cutting off the leaves, the passage of water up the stem is slowed down. This is why little water is taken up by deciduous trees in winter when they drop their leaves.

Uptake of mineral salts

Mineral salts are drawn into the roots along with the water. They are taken up partly by passive diffusion.

However, they can be absorbed by the roots even when they are more dilute in the soil than they are inside the root cells. In these circumstances they are taken up by **active transport** which requires energy from respiration (see page 109).

Transport of food substances

If you turn to page 198, you can read about an experiment in which scientists gave a plant carbon dioxide containing radio-active carbon. Eventually the radio-active carbon got into the sugars and other food substances which the plant made during its photosynthesis. Using the same technique, scientists have traced what happens to the food substances later on. They have shown that some of them move out of the leaves to other parts of the plant such as the roots. What's more, they travel in the sieve tubes which, as we saw earlier, belong to the phloem tissue.

The importance of the phloem in transporting food substances can be seen in trees (Figure 6). In a tree trunk, the phloem tissue is located in the soft inner part of the **bark**. If a ring of bark is cut out from right round a tree trunk, food substances cannot get down the trunk (Figure 7), so the roots are starved and eventually the tree dies. The reason why grey squirrels and other small mammals kill trees is that they gnaw the bark and destroy the phloem underneath.

Many other experiments and observations indicate that the phloem is the pathway by which food substances are transported inside the plant. However, no one knows for certain *how* it takes place except that it definitely requires energy from respiration and if the sieve tubes are killed it stops immediately.

The phloem must therefore have an adequate supply of oxygen. The corky part of bark is impervious to gases, but scattered around are breathing pores called **lenticels** which allow oxygen to diffuse in to the phloem and carbon dioxide to diffuse out. A lenticel is shown on the left hand side of Figure 6.

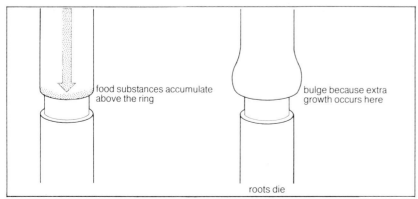

food substances accumulate above the ring

bulge because extra growth occurs here

roots die

Figure 7 The effect of removing a ring of bark from a tree trunk.

Investigation 1

Looking at the transport tissues in a stem

1 Obtain a short length of stem, about 6 cm long.

2 With a sharp razor blade, cut thin slices of the stem as shown in the Illustration. These slices are transverse sections.

3 Float the sections in a dish of water.

4 Pour some phloroglucin stain into a watch glass and add three drops of concentrated hydrochloric acid.

5 With a paint brush, transfer one of your thinnest sections into the stain, and leave it there for three minutes.

6 Put a drop of water in the centre of a slide.

7 Lift the section out of the stain, and place it in the water on the slide.

8 Cover the section with a coverslip, and examine it under the low power of the microscope.

   Which structures shown in Figure 2 can you see?
   (The phloroglucin should have stained all woody structures red, including the xylem vessels).

9 Try cutting sections longways as shown in the illustration below. Stain them with phloroglucin as before. This will enable you to see the transport tissues in side view.

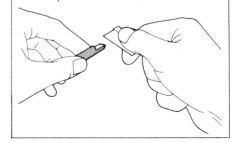

Investigation 2

Observing transpiration

1 Obtain two potted plants such as geraniums.

2 Cut one of them at the base of the stem: this is your control. Leave the other one intact.

3 Put each pot in a polythene bag and tie the bag round the base of the stem.

4 Cover each plant with a bell jar and leave overnight.

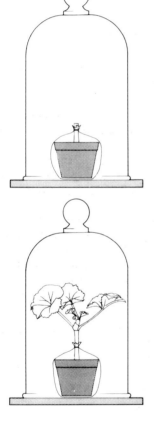

5 Next day look at the inner surface of the two bell jars.

   Which one has moisture on it?
   How could you find out if the moisture is water?
   Explain your observations.

How could you measure the amount of water vapour given out by a potted plant in a given time?

Investigation 3

To find out how much water a plant loses and gains

1 Obtain a leafy shoot of a non-woody plant.

2 Stand it in a 20 cm³ measuring cylinder.

3 Pour water into the measuring cylinder up to the top mark.

4 Carefully run a little oil into the measuring cylinder so that it forms a thin layer over the surface of the water. This will prevent water evaporating from the measuring cylinder.

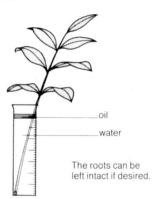

oil

water

The roots can be left intact if desired.

5 Weigh the plant and measuring cylinder together.

6 Leave the plant for about 24 hours.

7 After about 24 hours, weigh the plant and measuring cylinder again. Write down their mass in grams.

8 Read off the new level of the water in the measuring cylinder.

How much mass has been lost?

What volume of water has been taken up?

One cm³ of water weighs one gram: from this work out the mass of water which has been taken up by the plant.

Does this figure equal the loss in mass?

What do you think happens to the water which the plant takes up?

Are there any other reasons why the plant might lose mass besides losing water?

Investigation 4

Showing the passage of water through the xylem

1 Obtain a plant with its leaves and roots intact.

2 Wash the soil off the roots.

3 Stand the plant in a jar of water containing a coloured dye such as eosin, for about 24 hours.

eosin

4 After 24 hours, cut the stem in two with a sharp knife.

Whereabouts is the dye?

Compare the appearance of the stem with that of a plant which has been standing in water.

Explain your observations.

5 Cut the stem longways so as to find out more about where the dye is within the stem.

6 Stand a 'Busy Lizzie' in a jar of dye, and leave it for 24 hours. This plant has a tranparent stem, and you will be able to see where the dye is inside it.

7 Obtain a plant with white flowers, such as the deadnettle, and stand it in a jar of dye for several days.

   Does the dye eventually reach the flowers?

The use of a potometer for measuring the uptake of water by a leafy shoot is described on page 359.

Investigation 5

Looking at the air pores in a leaf

1 Cut a green leaf off a plant.

2 With a paintbrush apply a thin layer of clear nail varnish to a small area on the lower surface of the leaf.

3 When the nail varnish is dry, peel it off with a pair of forceps. The nail varnish will have made an exact replica of the leaf surface.

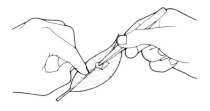

4 Put the nail varnish in a drop of water on a slide, and cover it with a coverslip.

5 Examine it under the low power of the microscope.

Can you see the air pores? Approximately how many air pores are visible in the field of view?

6 Now look at a single air pore under the high power.

Can you see the guard cells?

7 Repeat the experiment on the upper surface of the leaf.

How many air pores are visible in the field of view this time?

Which side of the leaf has the greater number of air pores, the upper side or lower side?

Why do you think the two sides of the leaf differ in this respect?

Investigation 6

To see how quickly the two sides of a leaf lose water

1 Obtain two small pieces of dry cobalt thiocyanate paper. This is blue when dry, but turns pink when moist.

2 Obtain a leafy shoot of a tree such as sycamore and stand it in water.

3 With Sellotape stick one piece of the cobalt thiocyanate paper to the upper side of a leaf. Completely cover the piece of paper with the sellotape.

4 Stick the other piece of cobalt thiocyanate paper to the lower side of a different leaf in the same way.

5 Note the time, and observe the two pieces of cobalt thiocyanate paper at intervals.

How long does it take for the first trace of pink to appear on each piece of paper?

How long does it take for each piece of paper to go completely pink?

Which side of the leaf loses water faster, the upper side or the lower side?

Why do you think one side loses water faster than the other?

Assignments

1 What part do the leaves and roots play in helping water to pass through a plant?

2 Before going to bed, a lady placed a potted plant on the windowsill and pulled the curtains. It was a chilly night. The following morning she opened the curtains and found that the glass behind the plant was covered with drops of moisture.

　a) What do you suppose the moisture was?
　b) Where precisely had it come from?
　c) What caused it to appear on the window glass?

3 Suggest a reason for each of the following:

　a) In summer it is better to water plants in the evening than earlier in the day.
　b) After transplanting a plant it is a good idea to remove some of the leaves.
　c) Water moves up a stem more quickly on a hot dry day than on a cool wet day.
　d) When a greenfly feeds on a plant it sticks its proboscis into the phloem.

4 How could a five-year old child kill a tree with a penknife?
Explain your answer fully.

5 In order to show how the air pores of a plant open, a student makes an artificial air pore as follows. He fills two short lengths of visking tubing with a 20 per cent solution of sucrose (sugar) and ties the ends together as shown in the left hand diagram below. He then places the artificial air pore in a dish of distilled water and leaves it there for three hours. At the end of the three hour period it looks like the right hand diagram.

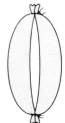

　a) What structures do the two lengths of visking tubing represent?
　b) Explain how the change in appearance has been brought about.
　c) Mention two ways in which the working of this artificial air pore differs from a real one.
　d) What part do the air pores play in the movement of water through a plant?

How do plants support themselves?

The tree below is over 80 metres tall and has a mass of over 600 tonnes. How can such a huge structure stand up? In this Topic we shall see how plants support themselves.

Figure 1 The General Grant Tree, a giant conifer, in California, is estimated to be 3500 years old and at 60 m above the ground the trunk is 3 m thick.

Why do plants need to stand upright?

There are two main reasons:

1 It puts the leaves in the best position to get plenty of light for photosynthesis. This is particularly important in a forest where plants are competing with one another for light.
2 It lifts the flowers into a high position from which the fruits and seeds can be scattered over a wide area. This helps the species to spread to new places.

The main way plants stand upright is by having a strong stem (Investigation 1).

What makes the stem strong?

In general, strength is achieved by the stem containing three different structures, namely packing cells, cellulose strands, and wood. Let's look at each of these in turn.

Packing cells

If you look at the inside of the stem of a plant like a sunflower or lupin under the microscope you will see that it is full of large rounded cells (Figure 2). These are packing cells, they are full of a watery fluid and are blown up like balloons. The epidermis or 'skin' of the stem holds the packing cells in place, and causes them to press against one another, making the whole stem firm yet flexible. Similar cells inside the leaf help to keep that firm too.

What keeps the packing cells full of fluid? The packing cells draw in water by osmosis which makes them **turgid** (see page 110). However, this will only happen if the plant has a good supply of water from the soil.

If the packing cells don't get enough water they become flabby or flaccid, just as a balloon does if you let air out of it. When this occurs the whole plant droops (Figure 3). We call this **wilting**. It happens on hot dry days when water evaporates from the leaves more rapidly than it can be replaced by the roots.

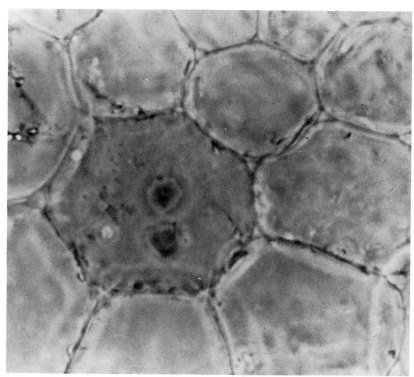

Figure 2 Packing cells seen in a transverse section of a plant stem.

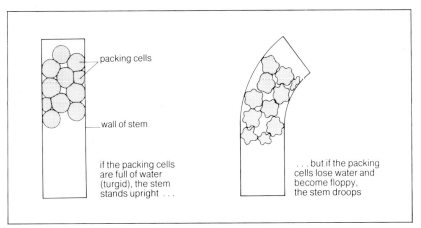

Figure 3 These diagrams show how the packing cells in a stem help it to stand upright.

Cellulose strands

A plant like a sunflower or lupin contains strands of cellulose just beneath the epidermis in the stem. These strands are formed by the thick cellulose walls of living cells which lengthen as the stem grows. The cellulose strands are tough and rubbery, and they help to make the stem strong and flexible.

In the stem of the deadnettle, which looks square in cross section, the cellulose strands are concentrated at the corners. The corners thus serve as buttresses, strengthening the stem and helping it to stand erect (Figure 4).

Wood

When a plant grows, certain cells in the stem lengthen, and a substance called **lignin** is added to the cellulose in their walls. Lignified cells are wood. As lignin won't let water through the cells die, so all that's left in the mature stem are long strands of wood. These strands are of two types. Some of them are narrow tube-like **vessels** and their job is to carry water and minerals through the plant as well as to provide support (see page 204). Others, known as **fibres**, do not transport anything and their job is only to support the plant.

Plants vary in the amount of wood their stems contain (Investigation 2). In herbaceous plants that only last one year, there is not much wood and it is confined to special regions called the vascular bundles (Figure 5). On the other hand, in shrubs and trees, which go on year after year, the wood more or less fills the entire stem.

Wood makes stems strong and rigid. Think of a tree, for example. The branches and leaves, which together make up the canopy, are held up by a single trunk. This may be very tall: some of the giant conifers in California are over a hundred metres high (Figure 1).

As a tree gets taller, its trunk gradually broadens, helping it to support the increasing weight of the canopy. Some of the giant conifers have trunks up to eleven metres wide, and there is a cypress tree in Mexico whose trunk is over 34 metres wide: twelve double-decker buses could be lined up side by side behind this tree-trunk without being seen.

Inside a tree trunk

If you look at the cut end of a felled tree trunk you may see that there is a dark region towards the centre and a lighter region further out. The dark central region is called the **heartwood** and the lighter region the **sapwood** (Figure 6).

The heartwood is extremely dense and hard and its only job is to support the tree. The sapwood is less dense and therefore softer than the heartwood. It provides support too, but it also carries water and mineral salts (sap) up the trunk. It is therefore much wetter than the heartwood.

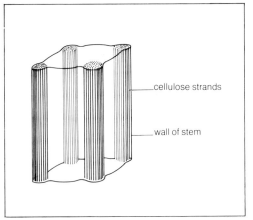

Figure 4 In a deadnettle the cellulose strands are concentrated at the corners of the stem, making it strong and helping it to stand upright.

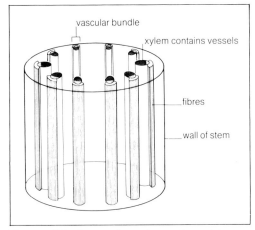

Figure 5 This diagram shows the positions of the woody tissue in the stem of a herbaceous plant such as a sunflower.

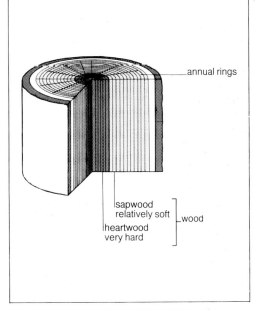

Figure 6 The inside of a tree trunk showing the two kinds of wood.

Wood and the timber industry

Wood is extremely strong for its weight, and so it has been used for building and similar purposes for centuries. Its main disadvantage is that it's liable to be attacked by fungi or insects. For example 'dry rot' is caused by a fungus.

For building, heartwood is better than sapwood because it is stronger, drier and more resistant to decay. Also being drier, it is less likely to shrink and warp.

Every type of tree has its own particular kind of wood which varies in appearance and strength. The arrangement of the cells gives the wood its characteristic grain. Generally the wood of coniferous trees, such as pine, is softer than the wood of flowering trees such as oak.

Different trees have different uses. For example, the wood of the ash tree is strong and springy, which makes it ideal for the handles of tennis racquets and hockey sticks. On the other hand the oak has very hard and durable wood, which makes it more suitable for building ships and houses.

Growing trees for timber

In Britain and many other countries, the trees most often used for their wood are conifers such as pines, spruces and firs. They grow comparatively quickly and their wood is hard enough for most purposes. The seeds are sown in nurseries, and when the seedlings are large enough they are transplanted to plantations in the country. They are usually grown in areas which are unsuitable for other kinds of crops, such as steep mountain slopes. The seedlings are planted quite close together, but later they are thinned out and the healthiest ones are left until they are large enough to be cut down for their timber. All this takes about thirty years in a warm country, but more like fifty years in a cool country like Britain.

After felling, the trunks are sawn up (Figure 7). Before it can be used the wood must be allowed to dry out, that is seasoned. In this process its weight may be halved. If wood isn't seasoned properly it is likely to shrink and warp later.

Figure 7 A sawmill in Malaysia.

Investigation 1

To find how strong a stem is

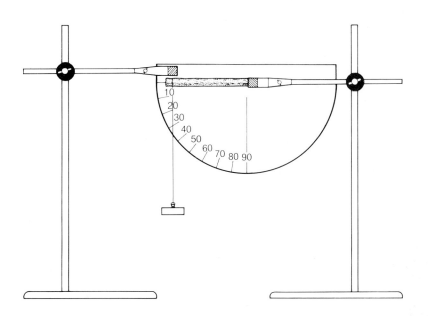

1 Cut off a 5 cm length of a stem which is approximately 5 mm wide.

2 Clamp a protractor to a stand as shown in the illustration.

3 Clamp the stem horizontally to another stand in front of the protractor.

4 Hang a 10 g weight on the end of the stem.

5 From the protractor, note the angle through which the stem bends.

6 Repeat this experiment on the stems (or branches) of different plants. They should all be the same length and thickness, and the same weight should be used for all of them.

What are the advantages and disadvantages to a stem of being able to bend?

Investigation 2

To find out how much wood there is in a stem

1 Obtain a short length of the stem of a plant such as a sunflower.

2 Cut one end cleanly with a knife.

3 Pour some phloroglucin stain into a watch glass and add three drops of concentrated hydrochloric acid. Acidified phloroglucin stains wood red.

4 Dip the cut end of the stem into the stain and leave it for five minutes.

5 Take the stem out of the stain and look at the cut end. Draw the cut end to show where wood is situated.

Approximately what proportion of the cut end of the stem is taken up with wood?

6 Repeat this investigation on the stems or branches of different plants. In each case try breaking the stem with your hands before you stain it.

Is it true that the more wood a stem contains the more difficult it is to break it?

Assignments

1 Mention three structures which help stems to stand erect.

2 It was a hot dry day in the middle of summer. By midday all the plants in the flowerbed were drooping. That evening the gardener sprayed the soil with water, and within an hour the plants were standing up straight. Explain in detail what caused the plants to droop, and then stand up straight again.

3 Why do plants tend to droop on cold frosty days?

4 How do heartwood and sapwood differ in their properties and in their functions in a mature tree.

5 Give one reason each why:

   a) heartwood is preferred to sapwood for building purposes;

   b) ash is used for making tennis raquets;

   c) oak is used for ship-building;

   d) nowadays conifers are most commonly grown for timber.

6 If you cut the end of a piece of celery longways, and put it in water, the ends bend outwards as shown in the illustration below.

 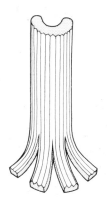

before   after

   a) Why do you think this happens?

   b) What sort of solution could you put the celery in to make the ends bend inwards again.

Protecting the body

*The hedgehog
in this picture shows
one of the most obvious ways
by which an animal's body can be
protected. In the next group of
topics we shall explore other
aspects of protection
as well.*

The skin and temperature control

We usually think of the skin as just a covering which holds the body together and keeps the things inside from falling out. However, there's much more to the skin than that, as we shall see.

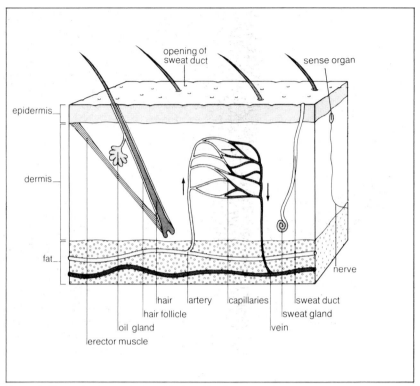

Figure 1 The main structures found in the skin of a mammal.

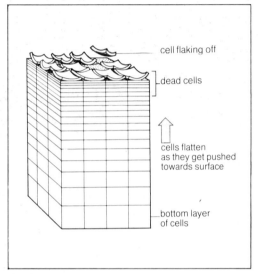

Figure 2 Diagrammatic view of the epidermis of human skin.

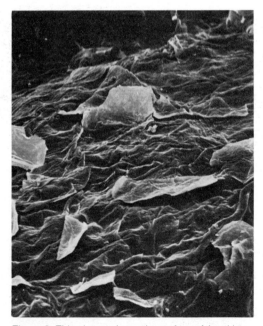

Figure 3 This picture shows the surface of the skin greatly magnified in the scanning electron microscope. The structures which look like dead leaves are epithelial cells flaking off.

The skin

The detailed structure of the skin is shown in Figure 1. It is divided into two main layers: a thin **epidermis** at the surface and a thicker **dermis** beneath.

The epidermis is made up of layers of cells like a brick wall (Figure 2). New cells are constantly being formed by the bottom-most layer and they push the older ones towards the surface. As the cells get pushed upwards, they become flat and hard and eventually die, forming a dead layer at the surface. The cells right at the top are like overlapping tiles and are constantly flaking off (Figure 3). These cells contain the protein keratin, the same substance that the scales of reptiles and feathers of birds are made of. It makes the skin waterproof and protective.

The epidermis contains a dark pigment called **melanin**. White people don't have much melanin, though the amount can be increased by the action of ultraviolet light on the skin, which is why people go brown when they sunbathe. Black people have much more melanin, and oriental people have an additional pigment called carotene which gives their skin a yellowish colour.

The dermis is composed of a network of tough connective tissue fibres. In amongst the fibres are blood capillaries and sense organs. Towards the bottom of the dermis there are **sweat glands** from which narrow sweat ducts, run to the surface of the skin.

Sticking out of the skin are **hairs**. Each hair projects from a deep pit called the hair follicle, and its root is situated deep in the dermis. Hairs are made of keratin, as are other skin structures such as nails and claws.

Opening into the hair follicles are glands which produce oil. This keeps the hair supple and helps to make the skin waterproof.

A slender muscle runs from the side of each hair to the base of the epidermis. When this **erector muscle** contracts, the hair stands upright. When it relaxes, the hair lies down flat. This is important in temperature control, as we shall see presently.

Below the dermis is a layer of cells containing fat which varies in thickness from one part of the body to another.

Figure 4 The spines of this West African crested porcupine protect it from attack.

Figure 5 The tiger's stripes camouflage it by breaking up its surface.

What does the skin do?

We will answer this mainly in relation to man, with occasional reference to other animals.

1 It protects the body

The keratinous layer of the skin, and the hairs, play the most important part in this. In some mammals the hairs are thickened up into sharp spines or flat plates which protect the animal from attack (Figure 4), and in animals like the rhinoceros the horns are made of lots of hairs all fused together. The skin also protects the body from germs, and the melanin pigment stops harmful ultraviolet rays from penetrating into the body.

2 It camouflages the animal

In many mammals the hairs vary in colour and pattern, camouflaging the animal in its natural surroundings (Figure 5).

3 It keeps water in

The keratinous layer of the skin is waterproof, and this prevents the body drying out. It also stops water getting in by osmosis when, for example, we go swimming.

4 It is sensitive to stimuli

The skin is sensitive to touch, pain, temperature and pressure. Different sense organs in the skin are responsible for detecting each of these stimuli (see page 248).

5 It keeps the body warm

The hair plays a very important part in this. Of course, humans have very little hair and that's why we wear clothes. But other mammals have plenty of hair.

The fat under the dermis also helps to keep the body warm. Animals such as whales and seals which live in cold water have a specially thick layer of fat called **blubber** (Figure 6).

Although the skin certainly keeps the body warm, it does more than that: it helps to *control* the body temperature, keeping it constant.

Figure 6 This sea lion has a thick layer of fat, or blubber, in its skin to help keep it warm.

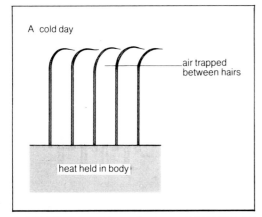

A   cold day

air trapped
between hairs

heat held in body

B   hot day

less air
between hairs

heat escapes from body

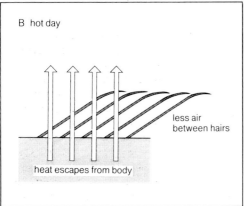

Figure 7   These diagrams show the part played by the hairs on the skin in controlling the body temperature of a mammal.

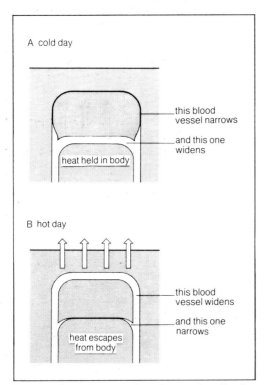

A   cold day

this blood
vessel narrows

and this one
widens

heat held in body

B   hot day

this blood
vessel widens

and this one
narrows

heat escapes
from body

Figure 8   These diagrams show the part played by the blood vessels in the skin in controlling the body temperature of a mammal.

Temperature control

Except when we're ill, the body temperature stays at just under 37 °C all the time. This is the temperature at which our body functions best; indeed, if the body temperature gets much above 40 °C, death is likely to occur.

If it's cold, the following things happen which keep the body warm:

1   *The hairs are raised*
You may have noticed that on cold days a cat's hairs are ruffed up. This is brought about by contracting the erector muscles. As a result, the hairs stand up and a layer of air is trapped between them (Figure 7A). Air is a poor conductor of heat, so this helps to insulate the body and prevents heat being lost from it (Investigation). Of course, this response is not much use to man with his sparse covering of hair; nevertheless our erector muscles still contract in cold weather, giving rise to goose pimples.

2   *Blood is held back from the surface of the skin*
Instead of flowing through the capillaries just under the epidermis, the blood is diverted through blood vessels deeper down. This prevents heat being lost from the blood as it flows through the skin. This response is brought about by the surface blood vessels getting narrow so blood can't flow through them so easily (Figure 8A). This is why people tend to go pale in cold weather.

3   *More heat is made by the body*
In cold weather our metabolic rate increases and we produce extra heat. The liver plays an important part in this. Also we shiver and may run about so as to keep warm. Shivering is caused by an involuntary contraction of our muscles.

If it's hot, various things happen which keep the body cool:

1   *The hairs are lowered*
This is brought about by relaxing the erector muscles. As a result, the hairs lie down flat: a layer of air is no longer held between them, so heat is lost more easily from the body (Figure 7B).

2   *Blood flows close to the surface of the skin*
In warm weather, the surface blood vessels widen so that more blood flows through them, and heat is lost as it flows close to the surface (Figure 8B). This is why people go pink in warm weather.

3   *Sweating or panting occur*
In hot weather, our skin gets covered with sweat, which is secreted by the sweat glands. When the sweat evaporates, it cools the skin and the blood flowing through it.

Evaporation occurs more quickly in dry air than in wet, humid air. That's why we feel hot and sticky on a humid day. Movement of the air helps to evaporate the sweat, so a gentle breeze has a cooling effect.

Not all mammals sweat. Dogs, for example, have sweat glands only on their pads and they cool themselves mainly by panting. When a dog pants, water evaporates from its mouth and tongue.

The various heating and cooling mechanisms just mentioned happen without our having to think about them. They are controlled by a special centre in the brain. In addition to these automatic responses, we can of course take deliberate steps to control our temperature by, for example, putting on more clothes if it's cold, or bathing in cool water if it's hot.

The warmth of our clothing depends on the fact that it traps a layer of air against the skin. As we have seen, air is a poor conductor of heat, so this helps to prevent heat being lost from the body. Woolly clothes are particularly warm because air gets trapped in the meshes.

Animals which can keep their body temperature constant, irrespective of the temperature of their surroundings, are described as **warm blooded** (or homoiothermic). They include all mammals and birds.

All other animals are described as **cold blooded** (or poikilothermic): their body temperature is the same as that of their surroundings. The only way that such an animal can control its body temperature is by making sure that it is always in a place where the temperature is suitable.

Adjusting the body temperature so that it is kept constant is an example of what biologists call **homeostasis**. This word is applied to any control process which keeps conditions the same all the time. We shall meet it again in the next two Topics.

Investigation

To see the effect of insulation on heat loss

1 Obtain two identical water baths with lids.

2 Put a wad of cotton wool on top of the lid, and wrap another wad round the sides. Leave the second water bath uncovered.

3 Pour hot water into the two water baths, the same amount into each.

4 Put a thermometer into each water bath, and take the temperature of the water at one minute intervals for at least fifteen minutes.

5 Plot the temperatures on graph paper, putting temperature on the vertical axis and time in minutes on the horizontal axis. Use the same sheet of graph paper for both water baths, so you can compare them easily.

Which water bath loses heat fastest?

How does the cotton wool help to prevent heat being lost?

What structures in a mammal are equivalent to the cotton wool?

Assignments

1 A small girl remarked that her cat looked larger on cold days than on warm days. How would you explain this?

2 What is the dark colour of a black person's skin caused by, and why is it useful?

3 Explain each of the following:

a) You feel cooler on a hot dry day than on a hot humid day.

b) Dogs pant when they're hot.

c) The metabolic rate of a naked human increases if the surrounding air temperature is lowered.

d) On a cool day on the beach you feel warmer after a swim in the sea even though the temperature of the sea is lower than that of the air.

e) While you are asleep at night your body temperature falls by about 2 °C.

4 A scientist carried out an experiment which showed that in cold conditions the amount of heat lost per unit mass from a small mammal was greater than that from a large mammal, although their insulation mechanisms were equally efficient and their body temperatures stayed the same.

a) Why do you think the small mammal lost more heat than the large mammal?

b) How do you think the small mammal managed to keep its body temperature as high as the large mammal?

c) The small mammal ate more than its own weight in food each day, whereas the large mammal ate only a small fraction of its weight in food. How would you account for this difference?

5 The graph below shows how the air temperatures, and the body temperature of a human and a lizard, varied in the course of a hot sunny day in the desert.

a) Explain what the lizard was probably doing at 8 am, 2 pm and 6 pm.

b) How was the human's body temperature controlled between 12 noon and 6 pm?

c) What do these results tell us about the way warm blooded and cold blooded animals control their body temperature?

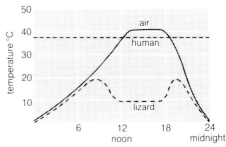

The liver

'Is life worth living? It all depends on the liver'. This is what a teacher I once knew used to say, because the liver performs many functions which affect our day to day health.

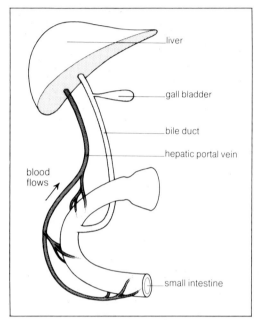

Figure 1 The liver and its connections with the gut.

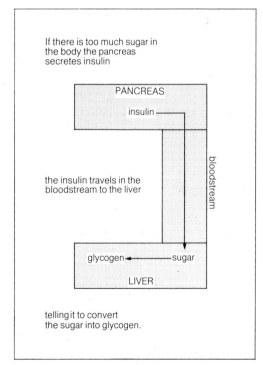

If there is too much sugar in the body the pancreas secretes insulin

the insulin travels in the bloodstream to the liver

telling it to convert the sugar into glycogen.

Figure 2 A simple scheme to show how the hormone insulin controls the amount of sugar in the body.

The structure of the liver

The liver is the body's largest organ weighing well over a kilogram. It is situated at the top of the abdominal cavity just under the diaphragm; you can feel it as a hardish area just below your bottom rib. Leading from the liver to the small intestine is the **bile duct**, attached to which is the **gall bladder** (see page 136).

The liver has a very good blood supply: about a litre of blood flows through it every minute – this is more than is sent to any other organ. Much of its blood comes from the **hepatic portal vein** which brings blood to it from the gut (Figure 1). All the food which is absorbed into the bloodstream from the gut is taken straight to the liver in this vein. This is because one of the liver's main jobs is to 'process' the food before it goes on to the rest of the body.

The liver itself is composed of masses of small cells, and the whole organ is riddled with narrow passages and blood vessels which give it a soft spongy texture.

What does the liver do?

The liver does many jobs. These are the main ones:

1 It helps with digestion
It does this by producing **bile** which is held in the gall bladder before being sent to the small intestine. Bile contains bile salts which emulsify fat, breaking it up into tiny droplets (see page 138).

2 It produces heat
Many chemical reactions take place in the liver, and this makes it produce a lot of heat. As blood flows through the liver, it is warmed up and this keeps the inside of the body warm.

3 It gets rid of poisons
Suppose you eat some food which happens to contain a mild poison: when the poison gets to the liver, the liver turns it into a harmless substance. We call this process **detoxication**. Poisonous substances are constantly being formed in our bodies as by-products of the many chemical reactions which take place inside our cells. The liver detoxifies these too.

4 It makes urea
Most people eat more protein than they need. The liver cannot store the unwanted protein. Instead, it breaks it down, getting some energy from it in the process. The nitrogen part of the protein is turned into ammonia. This is very poisonous and would kill you if it was allowed to build up. So the liver quickly turns it into a less poisonous substance called **urea**. This is carried by the blood to the kidneys, and is then expelled in the urine.

5 It destroys old red blood cells
After about four months, red blood cells wear out and stop working properly. The liver then breaks them up and any unwanted haemoglobin is converted into coloured substances which pass out with the bile. In the intestine these bile pigments are turned into a brown substance which gives the faeces their characteristic colour.

6 It stores food substances
After a meal, glucose is carried to the liver by the hepatic portal vein, and the liver turns any unwanted glucose into **glycogen**. This is stored inside the liver cells in the form of tiny granules. The glycogen can be turned back into glucose when the body needs it. The liver also stores many other food substances, including various vitamins and minerals.

7 It controls the amount of sugar in the blood

We have just seen that the liver stores surplus glucose. By storing surplus glucose, the liver ensures that there is never too much sugar in the blood. This is most important because, if sugar was to build up in our bloodstream our cells would be unable to work properly. What makes the liver turn sugar into glycogen? The answer is a hormone called **insulin** which is secreted into the bloodstream by special cells in the pancreas: insulin is a protein, and the cells which produce it occur in little groups called the Islets of Langerhans. Figure 2 summarises how it works. It is an example of homeostasis.

Diabetes

Some people have too much sugar in their blood. They are suffering from **diabetes** and are known as diabetics. The extra sugar in their blood makes them tired and thirsty. If nothing is done about it, the person loses weight and may eventually die. The kidneys try to get rid of the extra sugar, so one of the signs of diabetes is that sugar is present in the urine. In the old days, doctors used to tell whether or not a patient had diabetes by tasting his urine to see if it was sweet. Nowadays, a simple chemical test is used.

Diabetes is caused by the pancreas not producing enough insulin. The result is that the liver does not turn as much sugar into glycogen as it normally would. A person may inherit this condition or he may develop it as he gets older. It cannot be cured, but it can be controlled by:

1 following a restricted diet: the aim is to eat foods which do not contain much carbohydrate, so you don't get too much sugar in your blood.
2 taking tablets: certain tablets have the effect of lowering the amount of sugar in the blood.
3 insulin treatment: the diabetic takes a certain amount of insulin every day. This makes the liver turn his blood sugar into glycogen.

Insulin Treatment

Unfortunately insulin cannot be taken by mouth, because it is broken down by digestive enzymes in the gut. So it must be injected through the skin with a hypodermic needle. Diabetics are taught to do this for themselves (Figure 3).

The trouble is that it's sometimes difficult to get the dose exactly right. What sometimes happens is that the diabetic gives himself too much insulin with the result that his blood sugar falls too low. This can produce all sorts of effects such as trembling, sweating and weakness. The diabetic learns to recognise these signs and, if they come on, he eats a few lumps of sugar or glucose tablets to bring his blood sugar up to the right level.

There are over half a million known diabetics in the United Kingdom. The British Diabetic Association, founded in 1934, provides information and advice for diabetics. With proper medical help, the diabetic can learn to control his affliction and to work, play games and lead a full and active life. Some leading sportsmen are diabetics.

Gallstones and Jaundice

Normally the various substances present in the bile are in solution, but sometimes they solidify in the gall bladder or bile duct forming gall stones. These may block the bile duct and stop the bile getting into the intestine. One effect of this is that the skin goes yellow. This is a type of jaundice and is caused by the bile pigments getting into the bloodstream.

Jaundice can also be caused by the liver not working properly as happens in certain diseases such as viral hepatitis.

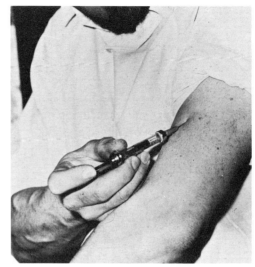

Figure 3 A diabetic injecting himself with insulin.

Assignments

1 Which functions of the liver listed in this topic:
 a) help in temperature regulation,
 b) assist the action of a digestive enzyme,
 c) make bile coloured,
 d) help the body to get rid of nitrogenous waste,
 e) make liver a useful food?

2 Look at Figure 1. Which of the various tubes in this illustration would you expect to:
 a) contain a lot of glucose after a meal.
 b) carry the hormone insulin,
 c) contain an emulsifying agent,
 d) contain digestive enzymes?

3 A person who is suspected of having diabetes is asked to produce a sample of urine. The urine is then tested for sugar.
 a) Describe a suitable test which could be carried out.
 b) What would be the cause of sugar being present in the urine?

4 Insulin cannot be taken by mouth because it would be broken down by digestive enzymes in the gut.
 a) Give the names of two digestive enzymes which would attack the insulin.
 b) What would these enzymes break the insulin down into?
 c) How is insulin taken by a diabetic?
 d) Mention one danger of taking insulin this way.

How do we get rid of waste substances?

Like a chemical factory, the body produces many waste products, some of which are poisonous. The body must get rid of these unwanted substances. This is known as excretion.

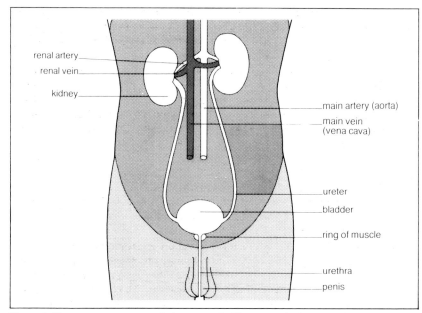

Figure 1 The excretory system of a human male. The arrows indicate the direction of flow of blood.

The excretory system

The main organs in the excretory system are the **kidneys**. We have two; they are reddish bean-shaped organs situated towards the back of the abdominal cavity just above the waist. If you put your hands on your hips, your kidneys are just about where your thumbs are.

Figure 1 shows how the kidneys are connected with the rest of the excretory system. A narrow tube called the **ureter** runs from each kidney to the **bladder**, a muscular bag situated towards the bottom of the abdominal cavity. Leading from the bladder is a tube called the **urethra** which runs down the middle of the penis in the male, and opens close to the vaginal opening in the female (see page 303).

The kidneys have a good blood supply: blood is carried to each one by the **renal artery** and away from it by the **renal vein**.

The kidneys produce a watery fluid called **urine** which contains substances which the body does not want. The urine trickles down the ureters to the bladder which gradually expands like a balloon as more and more urine collects inside it.

How is the bladder emptied? If you look at Figure 1 you will see that the top of the urethra is surrounded by a ring of muscle. Normally this muscle is tightly contracted, so urine cannot get out of the bladder. When the bladder is emptied this ring of muscle relaxes, and at the same time the muscles in the wall of the bladder contract, so urine is forced out of the body. This process is called **urination**.

How is urine formed?

The main waste substance in urine is **urea** (see page 220). Scientists have done tests which show that there is about sixty times more urea in our urine than there is in the blood. The explanation is that as the blood passes through the kidneys, urea is taken out of it and passed into the urine.

However, the kidneys do more than simply cleanse the blood of urea. They also regulate the amount of water and salt in the blood (Investigation 1).

Suppose you drink a lot of water quickly. The water is absorbed from your gut into the bloodstream, and it has the effect of *diluting* the blood. The diluted blood reaches the kidneys which take the water out of it and pass it into the urine.

Salt is dealt with in the same kind of way. Suppose you have a very salty meal. The salt is absorbed into your blood, which thus becomes very concentrated. The salty blood reaches the kidneys which remove the salt from it and pass it into the urine.

The relative amounts of water and salts in the blood give the blood a particular osmotic pressure. By regulating the water and salt, the kidneys make sure that the osmotic pressure of the blood stays more or less the same all the time. The name for this process is **osmo-regulation** and it is an example of homeostasis. If the osmotic pressure of the blood was allowed to fluctuate wildly our cells would not work properly.

Inside the kidney

The kidney is divided into two areas: a light outer area called the **cortex**, and a darker inner area called the **medulla** (Investigation 2). The medulla is connected to the ureter as shown in Figure 2.

Inside the kidney there are about a million microscopic devices called **nephrons**. The structure of an individual nephron, together with its blood supply, is shown in Figure 3. It consists of a little cup-like **capsule** which is connected to a narrow **tubule**. The tubule twists and turns, doubles back on itself and eventually leads to a **collecting duct**. About twelve nephrons share the same collecting duct, and all the collecting ducts open into the ureter.

The nephron's blood supply comes from a branch of the renal artery. This enters the capsule and splits up into a little bunch of capillaries called the **glomerulus**. These then join up again to form a vessel which leaves the capsule and splits up into further capillaries which are wrapped round the tubule. These then join up to form a vessel which leads to the renal vein.

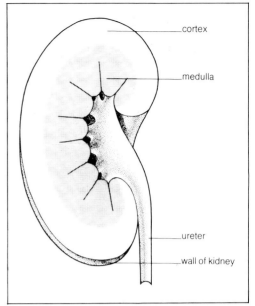

Figure 2 A kidney sliced horizontally to show the inside.

Figure 3 The structure of a nephron. The black arrows indicate the direction of blood flow; the open arrows indicate the direction in which the urine flows.

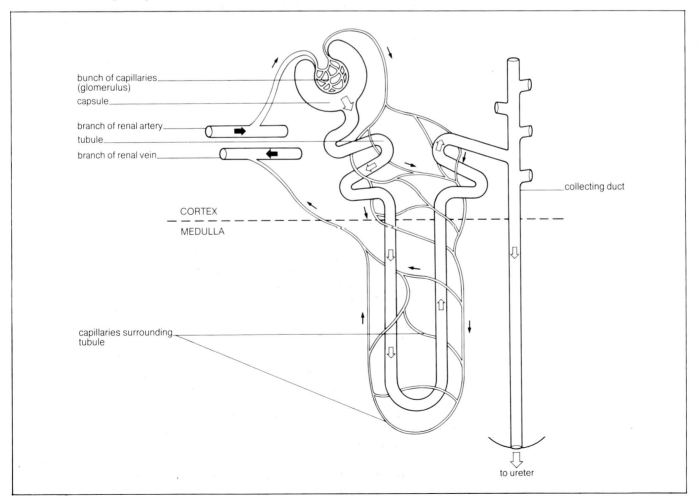

Figure 4 A corrosion preparation of the kidney. All the tissues except the main blood vessels have been dissolved away. The two kidneys contain about 16 km of tubules and 160 km of blood vessels.

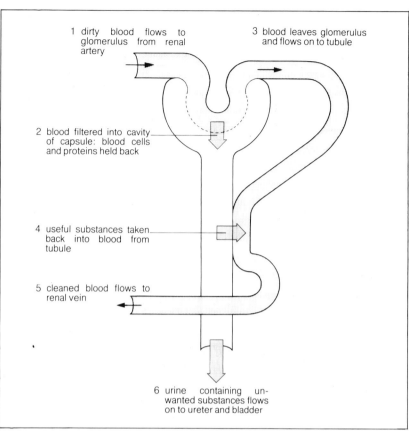

1 dirty blood flows to glomerulus from renal artery

3 blood leaves glomerulus and flows on to tubule

2 blood filtered into cavity of capsule: blood cells and proteins held back

4 useful substances taken back into blood from tubule

5 cleaned blood flows to renal vein

6 urine containing unwanted substances flows on to ureter and bladder

Figure 5 This diagram shows very simply how the nephron cleans the blood and makes urine.

How do the nephrons work?

Figure 5 will help you to understand how the nephron works.

The blood which reaches the glomerulus is under high pressure. This is because the vessel which leaves the capsule is narrower than the one which enters it. As a result, the fluid part of the blood is forced through the walls of the capillaries into the space inside the capsule. The fluid which goes through contains urea, glucose, water and salt. However, the blood cells and proteins are too large to go through, so they stay in the capillaries. In this way the blood is *filtered* as it passes through the glomerulus.

The filtered fluid, or filtrate, then trickles along the tubule. The urea remains in the tubule and eventually passes via the collecting duct into the ureter. However, all the glucose, most of the water and some of the salts are taken back into the capillaries wrapped round the tubule. Just enough water and salts are taken back to give the blood its correct composition.

So the kidneys work by first filtering the blood, and then reabsorbing back into it those substances which the body needs. Reabsorption is an extremely important aspect of how the kidney works. If the kidneys stopped reabsorbing water, the body would become completely dehydrated in less than three minutes.

What happens if the kidneys fail?

Occasionally one or both kidneys stop working properly. This may happen if they become infected, or sometimes after a severe shock such as a car accident.

A person can manage with only one kidney, but if both fail the blood soon becomes full of urea and other waste substances, and if nothing is done about it the person will die.

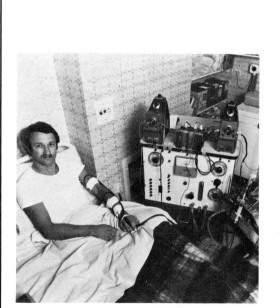

Figure 6 A person connected to an artificial kidney (kidney machine).

One way of saving the person's life is to attach him to an **artificial kidney** (Figure 6). This is a machine which filters and cleans the blood. A tube is connected to an artery in the patient's arm. His blood is then drawn off and made to flow over the surface of a thin sheet of cellophane on the other side of which is a watery solution. Urea and other unwanted substances in the blood pass through the cellophane into the solution on the other side, while larger components of the blood, including the blood cells, are held back. The blood is then returned to the patient by a tube inserted into one of his veins. As well as cleansing the blood of urea, the kidney machine adjusts the amount of salt in the blood before it is returned to the patient.

A person with complete kidney failure needs to spend about twelve hours on a kidney machine twice a week, either in hospital or at home. He can then lead a more or less normal life.

Investigation 1

To find the effect of drinking on urine production

1 Urinate as completely as possible, preferably after going for a long time without drinking.

2 Fifteen minutes later urinate again into a measuring cylinder, and estimate the volume of urine produced.

3 Fill a test tube with a sample of the urine, then throw the rest away.

4 Now drink a known volume of water. Your teacher will tell you how much to drink.

5 Fifteen minutes later urinate and measure the volume of urine, as before.

6 Fill a second test tube with a sample of this new lot of urine, then throw the rest away.

7 Repeat steps 5 and 6 at fifteen minute intervals for as long as possible. Compare the volume of urine produced in each case, and also its colour.

8 Plot your results on graph paper. Put volume of urine on the vertical axis, and time on the horizontal axis.

How do the urine samples differ in colour?

Why do they differ in this way?

How would you explain the volume differences?

What organ is responsible for controlling the water content of the body?

Investigation 2

Looking at the kidney

1 Obtain a kidney of a mammal such as the pig, obtained from a butcher.

2 With a sharp knife slice the kidney across the middle as shown in the illustration.

3 Which of the parts shown in Figure 2 can you see?

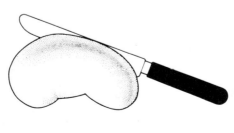

Whereabouts does urine leave the kidney?

Where does the urine go after it has left the kidney?

Assignments

1 What effect, if any, would you expect each of the following to have on the quantity and composition of the urine?

a) Eating a large quantity of salty food.
b) Having a bath.
c) Drinking a lot of beer.
d) Playing a hard game of squash.
e) Eating four cream cakes.

2 Explain the meaning of the terms excretion and osmo-regulation? What job does the kidney do (a) as an excretory organ, and (b) as an organ of osmo-regulation?

3 In the course of one hour, 72 litres of blood enters both kidneys in a man, but only 0.072 litres of urine leaves the kidneys and passes down into the bladder.

a) What fraction of the blood which enters the kidneys, gets into the bladder as urine?
b) What happens to the rest?

4 Which of the substances listed in column A are found in each of the fluids listed in column B:

Column A	Column B
protein	blood entering kidney
glucose	blood leaving kidney
urea	fluid filtered into capsules
water	urine leaving kidney

5 It has been suggested that in summer a person passes less urine than in winter.

a) Describe an experiment which could be done to find out if this is true.
b) How would you explain it?

Germs and Disease

What causes diseases, and why do they sometimes spread so quickly? In this Topic we will try to answer these questions.

What is disease?

Disease is the condition which arises when something goes wrong with the normal working of the body. As a result we become ill. The signs of the disease may include a headache, rashes on the skin, sore throat and fever. We call these the **symptoms** of the disease. When a doctor examines a sick patient he looks for these symptoms as they will probably tell him what's wrong with the person. He can then suggest a remedy: in medical language, he prescribes some kind of treatment. Sometimes the doctor gives the patient a prescription which he then takes to a chemist to obtain some medicine.

The symptoms of disease are usually physical. Sometimes, however, they may be mental: the person may feel confused or depressed, or he may behave in a peculiar way. In this Topic we will deal only with physical disease.

What causes disease?

Most diseases which afflict us are caused by parasitic **microbes**. Doctors describe them as **pathogenic**. This word comes from Greek, and literally means 'gives rise to suffering'. Here we will refer to them by their everyday name – germs.

Germs get into the body mainly through the mouth and nose, or sometimes through cuts and wounds. Once inside, they may multiply very quickly. This is called the **incubation** period and several days or even weeks may go by before the person actually starts feeling ill.

Germs harm us in one of two ways. Some of them attack and destroy our cells, others release poisonous substances into the bloodstream. Their intense

Name of disease	Caused by	Main symptoms	Spread by	Notes
Typhoid	bacteria	headache, cough, fever, coma	contamination of food or water with faeces	very severe disease in countries with poor hygiene
Whooping cough	bacteria	severe coughing ('whoops')	breathing onto other people, or intimate contact	one of the most serious diseases of children
Syphilis	bacteria	sores on genitals, more serious symptoms later	sexual intercourse	one of several venereal diseases which can be serious unless treated promptly
Chicken pox	virus	mild fever, red pimples on skin	breathing or slight contact	mild, but very infectious, disease, especially of children under 10
Poliomyelitis	virus	paralysis of muscles, particularly legs	breathing, or contamination of food	highly infectious disease which can kill or cripple people
Yellow fever	virus	sickness, fever, yellow skin (jaundice)	certain type of mosquito	serious disease of the tropics
Typhus	rickettsias	severe fever, muscle pains, rash	lice, mites, ticks, fleas	serious disease transmitted to man from rats and other rodents
Dysentery (Amoebic)	protozoan	diarrhoea, blood in faeces	contamination of food or water with faeces	unpleasant disease common in countries with poor hygiene
Athlete's foot	fungus	itching between toes	contact with floor	mild but irritating skin disease
Malaria	protozoan	severe recurring fever	certain type of mosquito	serious disease of the tropics which still kills millions of people every year
Bilharzia	flatworm	skin rash, cough, sickness, diarrhoea	water snail	another serious disease of the tropics

Table 1 Summary of the diseases mentioned in this and the next Topic.

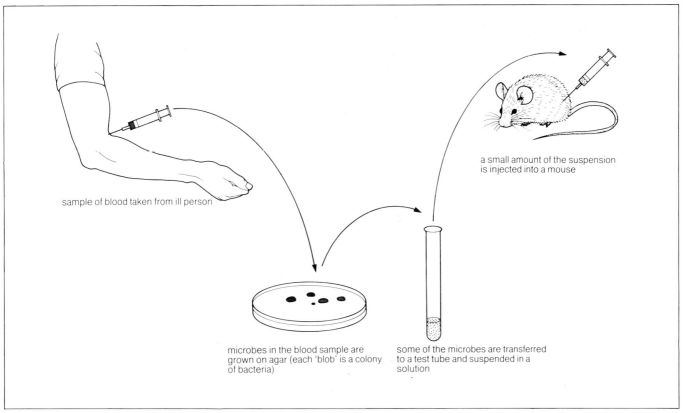

sample of blood taken from ill person

a small amount of the suspension is injected into a mouse

microbes in the blood sample are grown on agar (each 'blob' is a colony of bacteria)

some of the microbes are transferred to a test tube and suspended in a solution

Figure 1 This diagram shows how a scientist can find out the cause of a particular disease.

activity produces a lot of heat – this is one reason why our temperature usually goes up when we are ill. Can you think of any others?

There are two main kinds of germs: bacteria and viruses (see pages 32 and 36). They are responsible for most of the diseases which attack man and his domestic animals. Various other diseases are caused by parasitic protozoans, worms and fungi (Table 1).

Not all diseases are brought about by organisms. Some are caused by not eating the right kind of food, or by breathing in chemicals which injure the body. Others are inherited – people are born with them. Finally there are many diseases whose cause is not yet known.

How do we know which germs cause particular diseases?

Anthrax is a serious disease of farm animals and man. Suppose you are a scientist, and you want to find out what causes this particular disease. How could you do it?

Here is one way. You take a small amount of blood from an individual who has got anthrax, and you examine it under a microscope. You find the blood is teeming with a certain kind of microbe. You then examine the blood of lots of individuals who have got anthrax. You discover that they all have this particular microbe in their blood. This leads you to suspect that these microbes are responsible for the disease.

But you're not absolutely sure, so you do a further experiment (Figure 1). You take a sample of blood from an individual with anthrax, and grow the microbes from it. You then inject these microbes into a mouse – and wait anxiously. If the mouse gets anthrax, you can be pretty sure that these particular microbes cause the disease.

One of the first people to do experiments of this kind was a German doctor, Robert Koch (Figure 2). He discovered the bacteria responsible for diseases such as tuberculosis and cholera. Other scientists including Louis Pasteur carried out similar experiments, and by the end of the 19th century the causes of many diseases were known.

Figure 2 Robert Koch (1843–1910) discovered the cause of many diseases.

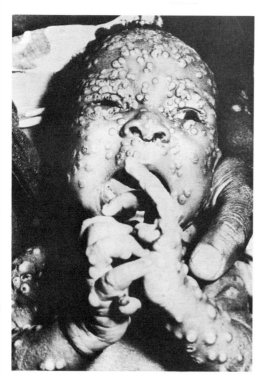

Figure 3 In crowded places people readily infect one another with their germs.

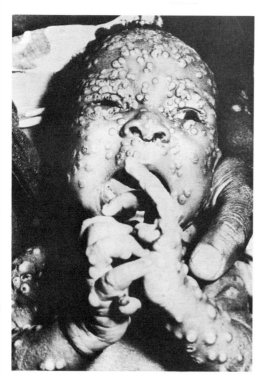

Figure 4 Every disease has its symptoms. Smallpox is caused by a virus, and it is characterised by spots on the skin. Fortunately this terrible disease has now been successfully wiped out.

How are germs spread?

In 1918 there was an outbreak of 'flu' in Spain. Within a few months it had spread all over the world. Between April and November over 21 million people died of it – twice as many as were killed in the whole of the First World War. When a large number of people go down with a disease, we say there is an **epidemic**. If it's worldwide it's called a **pandemic**.

Diseases spread because germs get passed from one individual to another: a healthy person 'catches' the disease from someone else – or maybe from an animal. Diseases which are spread like this are called **infectious** diseases. Sometimes a person may have germs in his body without showing the symptoms of the disease. Such a person is called a **carrier**.

Here are the main ways that germs are spread from one individual to another:

1 *By droplets in the air*

When you cough or sneeze, thousands of tiny drops of moisture shoot out of your mouth and nose, and if you have a disease these droplets may be swarming with germs. If they are breathed in by other people, the disease is likely to be spread to them. This can happen even if an infected person breathes into someone else's face, or talks to him. Colds and flu spread rapidly this way, particularly in crowded places such as buses and underground trains (Figure 3).

2 *By dust*

Some diseases can be spread by dust, for example smallpox. Germs stick to the dust particles and float through the air. Eventually they settle on surfaces which may be a long way from where they arose. People can catch the disease by breathing in the dust, or getting it in their mouths. Smallpox can be caught from the clothes or bedding of an infected person (Figure 4).

3 *By touch*

Impetigo is a skin disease which occasionally breaks out in schools. It is caused by a bacterium. You can catch it by touching an infected person, or even by brushing against his clothes or sharing his hairbrush or towel. Another skin disease, athletes' foot (caused by a fungus), can be picked up from the floor of changing rooms and showers. In both these cases infection is by contact: such diseases are said to be **contagious.**

4 *By faeces*

The faeces of an individual with a disease may be teeming with germs. If the faeces get into food or drinking water, the disease will quickly spread to other people. Epidemics of typhoid and cholera have been caused this way. Food can become contaminated with faeces if it is handled by a person with dirty hands. This is why you should always wash your hands after going to the toilet, particularly if you are about to prepare food for other people. Drinking water may become contaminated if sewage gets into it. This happens in places where sewage is not disposed of properly. We tend to associate this with backward countries, but it can happen anywhere if there is a disaster such as an earthquake or flood.

5 *By animals*

Germs are brought onto food by animals such as rats and mice, cockroaches and flies. Take flies for example: these little animals are equally happy feeding on dung or sugar lumps (Figure 5). Their legs may be covered with germs. Moreover, they have the delightful habit of spitting onto their food before they feed on it. In this way germs may be transferred from faeces to food.

Many diseases are spread by animals which suck blood. An example is the mosquito which transmits malaria and yellow fever. In 1897, a doctor in the Indian Medical Service, Sir Ronald Ross, examined the stomach of a certain

Figure 5 A fly feeding on a lump of sugar.

kind of mosquito. He found malarial parasites there. In this way he showed that the mosquito spreads the disease.

It was more difficult to prove that mosquitoes transmit yellow fever. This is because the yellow fever germ is too small to be seen under the microscope. During the building of the Panama Canal, yellow fever killed so many people that the project had to be abandoned for several years. Obviously it was important to find the cause. In 1900 an experiment was carried out by an American army doctor, Jesse Lazear. He suspected that yellow fever was transmitted by mosquitoes, but he wasn't sure. To settle the matter, he allowed himself to be bitten by a mosquito. A few days later he went down with the disease, and three weeks later he died.

A number of diseases are spread by pets such as dogs and cats. By far the most serious is rabies. This can be caught by being bitten or even licked, by an infected dog. Fortunately rabies has not yet occurred in Britain, but pets can carry other less serious diseases. The family dog may look innocent, but its tongue is covered with germs. It is unwise to let it lick your face.

6 By cuts and scratches

Suppose Jean scratches herself with a needle, and then the needle scratches Ann. Certain diseases may be passed from Jean to Ann in this way. Sometimes people prick their fingers to obtain blood for observing under the microscope. If you ever do this you should use a sterilised lancet which has not been used by anyone else. This is not a common way of spreading disease, but it can happen. For example, viral hepatitis, a disease of the liver, can be spread this way.

Disease and modern travel

The incubation period for smallpox is 8–12 days. In the old days a person travelling from one country to another by ship would probably show signs of the disease before docking, so he could be isolated before he had a chance to mix with other people. With modern travel, particularly by air, this is no longer the case. An infected person may arrive in a country and move about freely for over a week before he starts feeling ill and goes to the doctor. At least one outbreak of smallpox in Britain has been caused this way.

Assignments

1 Give five examples of places where diseases are likely to spread by people coughing and sneezing.

2 What part is played by each of the following in spreading disease:
   a) flies,
   b) rats,
   c) mosquitoes,
   d) needles,
   e) aeroplanes?

3 Write down the missing words (a to e) in the following passage. It is what a doctor says to one of his patients when he visits him.
   'David, you have chicken pox. The symptoms are quite clear, specially the ____(a)____ on your chest. I'm going to give your mother a ____(b)____ so she can get you some lotion from the chemist: it will stop your spots itching so much. The disease is caused by a ____(c)____ . It's not too serious, so don't worry. But I'm afraid it's very ____(d)____ so you won't be allowed to mix with your friends for the time being, otherwise they may ____(e)____ it.'

4 Mr. X makes sausage rolls in a small town. Though he does not know it, he is a carrier of typhoid. Mr. X is usually very clean, but one morning he is late for work so he does not bother to wash his hands after going to the toilet. That day he puts the meat into 600 sausage rolls, all of which are sold in his shop. Two weeks later several hundred people in the town go down with typhoid.

   a) There were germs on Mr. X's hands. Where did they come from?
   b) What sort of germs were they?
   c) Name two symptoms which you would expect to be shown by the people who got the disease.
   d) Name two other ways this disease could be spread round a town.
   e) Suppose you were the Health Officer for the area in which this town is situated. What steps would you take to prevent the disease spreading further?

Defence against disease

In this Topic we will look at the various ways we are defended against disease. Some of the methods are man-made, others are natural. They are all important in helping us to survive.

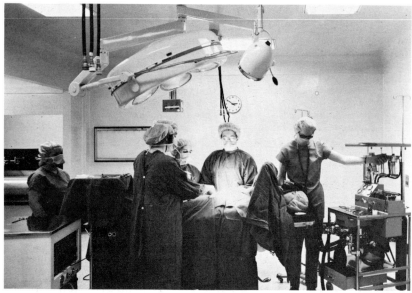

Figure 1 A modern operating theatre — out of bounds to all germs!

Germs are destroyed by sterilisation

When an object is freed of germs, it is said to be **sterilised**. One of the best ways of sterilising things is to heat them under pressure.

Why under pressure? If you heat water at ordinary atmospheric pressure, the water boils at 100 °C and that's the highest temperature it will reach. Unfortunately this is not hot enough to kill all germs. However, if you raise the pressure, the water boils at a higher temperature, and the germs are killed.

The process is carried out in a metal chamber which works in the same way as a pressure cooker in the kitchen. It is called an autoclave. Heating at 120 °C for 15 minutes is sufficient to kill most germs.

Another way of sterilising things is to clean them in chemicals such as dettol. We call these chemicals **disinfectants**, and many different ones are available. The trouble is that they don't kill *all* germs. However, they are suitable for general use, particularly in the home.

The most germ-free place is the hospital operating theatre (Figure 1). Before entering the theatre, the air passes through a special filter. The surgeons and nurses wear sterilised gowns, head covers and face masks, and all the instruments are sterilised beforehand. There is therefore no risk of the patient becoming infected.

Animals which spread diseases are exterminated

Great efforts have been made to get rid of disease-spreading animals such as rats, fleas, lice and mosquitoes. The battle against insects has been helped enormously by insecticides. These are chemical substances which kill insects. One of the most useful insecticides has been DDT which was first used during the 1939–45 war. DDT has now been banned in many countries because it may be dangerous to man. However, there is no doubt that it has been enormously useful in the fight against diseases such as malaria and yellow fever.

Infectious individuals are isolated

A person who has a serious infectious disease, or is a carrier of it, must be kept away from other people. So he is isolated until he is no longer infectious. This is called being put in **quarantine**. Occasionally a person entering our country is placed in quarantine because it's thought he may be carrying a

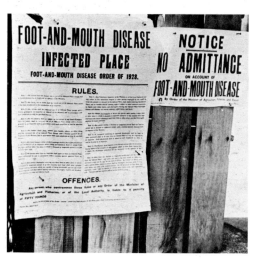

Figure 2 When there is an outbreak of foot and mouth disease, the whole area is put in quarantine.

serious disease. This is only done if the person comes from an infected area and has not been vaccinated against the disease. With animals it is *always* done. Cats and dogs brought into Britain have to be put in quarantine for six months. This is to make sure they don't bring rabies, a very serious disease of animals and humans, into this country. Outbreaks have occurred in Europe, but so far it has been kept out of Britain. Anyone who breaks the quarantine law and smuggles a pet into the country could start a rabies epidemic.

An animal which gets rabies is destroyed immediately, to prevent the disease spreading further. This is also done with foot and mouth disease which occasionally strikes at cattle and other livestock. When there is an outbreak of foot and mouth disease on a farm, the whole area is put in quarantine (Figure 2). All infected animals are killed and then buried in quicklime. It is one of the most distressing things that can happen to a farmer.

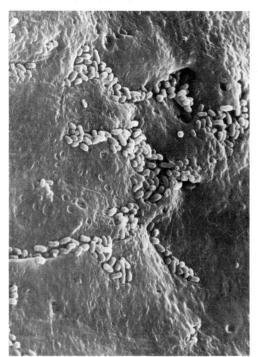

Figure 3 This enormously magnified picture of the surface of human skin shows groups of bacteria living in the crevices.

Germs are kept out by the skin

The skin serves as a barrier to most germs. However, some germs can get through the pores in the skin, and certain parasitic worms can bore through it (see page 52 for example).

The surface of the skin is very uneven, and thousands of tiny organisms make their homes in its nooks and crannies (Figure 3). Some of these organisms kill harmful germs, so they help to protect us against disease. However, others are harmful and can cause unpleasant skin diseases. For good health it is important to wash the skin regularly with soapy water. When you cut your skin, you open a door to germs. Fortunately nature comes to our aid: the blood hardens (clots) and a protective scab is formed. Later the two surfaces of the skin knit together, and the cut heals. In this way germs are kept out.

The trouble is that germs may get in before the blood has had time to clot. The cut is then likely to go septic. However this can be prevented by quickly applying a substance which kills germs. Such substances are called **antiseptics** and iodine is an example.

Antiseptics were discovered in the 1860s by the English surgeon Joseph Lister (Figure 4). In Lister's day more than half the people who had operations died: many of them got a bacterial infection of their wounds, called gangrene. Lister discovered that if he sprayed the patient's wound with carbolic acid during the operation, it did not go septic. Thanks to Lister, the number of people who died after operations was enormously reduced.

If you cut yourself, the wound should be covered with a piece of elastoplast or a bandage. These are called dressings. They prevent germs getting in, and they also bring the cut surfaces of the skin close together which speeds up the healing process. Sometimes a cut may be so large that it has to be stitched.

Germs are killed or expelled by the mouth and nose

Germs which enter your mouth get caught up in saliva. This contains a substance which kills germs. Some germs escape this lethal substance and are swallowed. Many of these will be killed by hydrochloric acid in the stomach. Of course, if the food is bad, you will probably be sick. Vomiting is the body's way of getting rid of germs and harmful substances from the gut.

Germs which get into your nose get trapped in the mucus lining the nasal passages. Many germs are killed by it, for it contains the same lethal substance as saliva.

The windpipe is also lined with mucus. The mucus is kept moving away from the lungs towards the throat by thousands of tiny beating hairs called **cilia**. When you breathe in, germs and dust particles get caught in the mucus. The beating cilia carry them to the throat where they are either swallowed or spat out.

Occasionally there may be so much mucus in your nasal passages that you blow your nose, or sneeze. If a lot of it collects in your throat or windpipe, you cough. These are all ways of preventing germs getting into your lungs.

Figure 4 Joseph Lister, the first person to use antiseptics.

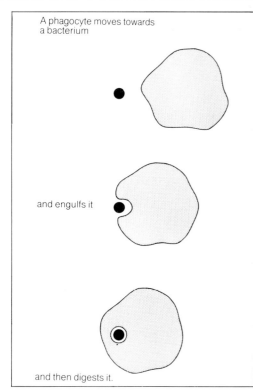

A phagocyte moves towards a bacterium

and engulfs it

and then digests it.

Figure 5 In the diagram a phagocyte is seen ingesting a germ.

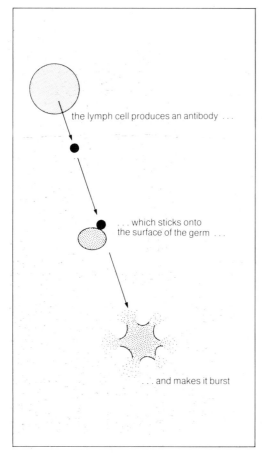

the lymph cell produces an antibody . . .

. . . which sticks onto the surface of the germ . . .

. . . and makes it burst

Figure 6 This diagram shows how one kind of antibody kills germs.

Germs are killed by phagocytes

In our bodies there are millions of cells called **phagocytes**. A phagocyte is a remarkable cell. It moves and feeds like an amoeba (see page 165). When it comes into contact with a germ, it engulfs it and takes it into its body. The germ is then killed and digested (Figure 5).

Phagocytes are the policemen of the body. They patrol the bloodstream and tissues, waiting for germs to arrive. If the body becomes infected they wriggle to the site of the infection. Then they attack and destroy the germs. It's like a battle, and the place where it happens – the battlefield – becomes red, swollen and painful. We call this **inflammation**. As in a real battle, there are casualties on both sides: the remains of dead bacteria and phagocytes accumulate to form pus. Sometimes the inflamed area swells up into a boil. Pressure builds up inside the boil, so that eventually it bursts, releasing the pus.

Germs are killed by antibodies

Suppose you are attacked by the virus which causes measles. When the virus gets into your bloodstream it is detected by certain cells called **lymph cells**.

When a lymph cell detects the viruses, it produces a chemical substance which kills them. Not all the viruses are killed instantly. Some of them multiply and damage your tissues, and this makes you feel ill. But gradually your lymph cells overcome them and you begin to feel better.

What causes the lymph cells to behave in this way? Germs contain chemical substances which we call **antigens**. When an antigen comes into contact with a lymph cell, it causes the lymph cell to produce another chemical substance called an **antibody**. The antibody then combines with the antigen, and this kills the germ. Exactly how the germ is killed varies from one kind of antibody to another.

Some antibodies make the germs burst (Figure 6); others make them stick together in clumps, after which they may be eaten up by phagocytes. In the case of germs which release poisonous substances, the antibodies combine with the poison making it harmless. Such antibodies are called **antitoxins**.

Becoming immune to diseases

Once you have had a disease like measles you are protected against getting it again. A doctor would say that you have now developed **immunity** against the disease.

We can explain immunity like this. The first attack taught your lymph cells how to make antibodies against measles. Once your lymph cells have learned how to do this, they will make antibodies more quickly in future. When a second attack comes, the lymph cells leap into action so quickly that the germs are destroyed before they have a chance to do any damage.

Having a particular disease will protect you against that disease in the future. However, it won't protect you against other diseases. This is because the antibodies you produce against, say, measles will act only against the measles germs – they will not act against any other kinds of germs.

Sometimes people get a mild attack of a disease when they are young – so mild they don't even notice it. However, it causes them to make antibodies, so they are protected from this particular disease when they are grown up. This is how many people gain immunity to diseases such as tuberculosis.

Some diseases – mumps for example – are mild in children but severe in adults. It's a good thing to get such diseases when we are young. This enables us to build up immunity, so we don't get the disease again later.

There are a few diseases which we seem never to become immune to – the common cold for instance. This is because colds are caused by many different types of viruses, and one type is constantly changing into another. When you get a cold, it may give you immunity against that particular virus in the future. However, your next cold will probably be caused by a different type of virus, against which you have no protection. Much the same applies to 'flu'.

Sometimes people make antibodies against things which are harmless, such as pollen. The antibodies attack the pollen grains in the tissues lining the nose and eyes. This makes the person sneeze and his eyes run. We call it hay fever. When a person reacts in this kind of way to a substance, we say that he is **allergic** to it. Some people are allergic to some kinds of food or to substances on the skin such as cosmetics.

Being immunised

We have seen that when a particular germ gets into your bloodstream, it causes you to produce antibodies which kill it. Now suppose a small amount of fluid obtained from some dead germs is injected into your blood before you've had the disease. What effect will this have? The fluid contains antigens, so it causes you to make antibodies: you will then be protected against the disease. This is what doctors do to make people immune to various diseases. The process is called **immunisation**.

The first person to immunise someone against a disease was the English physician Edward Jenner. In 1796 he immunised a young boy against the dreaded disease smallpox. He did this by giving him serum from a girl who had a related disease called cowpox or *Vaccinia*. For this reason the process of being immunised is called **vaccination.** The material which is injected into the bloodstream is called the **vaccine.**

Since Jenner's day immunisation has been extended to many other diseases. When a doctor immunises you, he puts a small quantity of vaccine into your bloodstream. Normally this is done with a hypodermic needle (Figure 7). In some cases he does it by scratching the skin, and occasionally the vaccine can be taken by mouth. The vaccine itself is generally made from *dead* germs. The germs must be dead, otherwise they might give you the disease the doctor is trying to protect you from.

When you were a baby you were immunised against various serious diseases. Most of these immunisations will protect you from the disease for the rest of your life. However, in a few cases protection only lasts for a limited time, and you need to be given further doses of vaccine from time to time if you are to keep up your protection. These are called **boosters**. Table 1 summarises the main diseases people may be immunised against in the course of their lives.

Immunisation programme in Great Britain:

All babies are immunised against:
   diphtheria
   tetanus (regular boosters required)
   whooping cough
   polio (by mouth)
   measles

All girls at age 11 are immunised against:
   German measles
   This is a mild disease but if a woman gets it in the early stages of pregnancy, it can
   damage her baby.

All teenagers are immunised against:
   tuberculosis
   unless skin test shows that they are already immune to it

Everyone travelling abroad must be immunised against:
   Yellow fever (certain tropical countries)

People travelling abroad are advised to be immunised against:
   typhoid and paratyphoid
   cholera (only lasts for 6 months)
   tetanus
   Smallpox (very few countries still require it)

Table 1 Diseases we are immunised against.

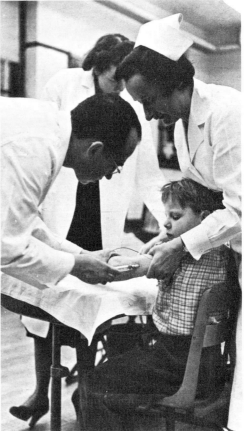

Figure 7 Dr. Jonas Salk, discoverer of the vaccine against poliomyelitis, inoculates a boy against this disease.

Figure 8 Alexander Fleming in his laboratory.

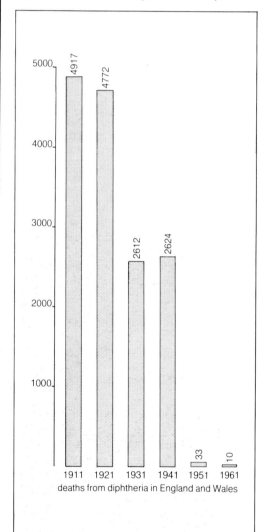

Figure 9 Between 1911 and 1961 the number of deaths each year from diphtheria fell from nearly 5 000 to only 10.

Receiving ready-made antibodies

Tetanus is a serious disease which kills over a hundred people each year in Britain. The muscles, particularly those working the jaws, go into painful spasms – the disease is sometimes called lockjaw. Tetanus germs can be picked up if you cut yourself with a dirty instrument such as a penknife.

Suppose you cut your finger with a dirty knife; the doctor wants to make sure that you don't get tetanus. It's too late to give you an injection of vaccine; by the time your body had made the necessary antibodies you would probably be dead! What can the doctor do to give you quick and immediate protection?

The answer is to give you some ready-made antibodies, that is antibodies which have already been made by someone else or by an animal. Fortunately the doctor has such a preparation in his cupboard. It is called **anti-tetanus serum**. He injects some of this into your arm and sends you home, confident that your cut won't cause you to get tetanus.

Giving a person ready-made antibodies is useful in an emergency. However, the protection does not last long. This is because the antibodies are gradually broken down and got rid of from the body. For long-term protection the person must make his own antibodies.

Germs are killed by antibiotics

In 1928 a Scottish bacteriologist called Alexander Fleming was working at St. Mary's Hospital, London (Figure 8). He was growing bacteria on plates of jelly (see page 22). The bacteria multiplied and spread over the jelly, forming colonies.

Normally Fleming covered his bacteria with a lid to prevent them becoming contaminated. But one night he forgot to do this, and left one of his dishes uncovered. When he returned next morning he had a surprise. His bacterial colonies had been killed.

What had killed them? Fleming had no idea, but he was determined to find out.

After a great deal of searching, Fleming discovered that his bacteria had been killed by a mould. It seemed that some spores of this mould had got into the laboratory and had landed on his bacteria, and then destroyed them.

The mould was identified as *Penicillium*, a fungus related to pin mould (see page 46). Fleming realised that the mould must have produced a chemical substance which killed the bacteria. If this substance could be extracted from the mould, it might be used to cure people of bacterial diseases. It took scientists twelve years to obtain it in a usable form. In 1940 it was tried out on patients in hospital. The results were dramatic: people who were dying of bacterial diseases recovered almost immediately. This 'miracle substance' was christened **penicillin**.

Today hundreds of substances are used by doctors to treat bacterial diseases. Some, like penicillin, are obtained from moulds and other microbes: we call them **antibiotics**. Others are drugs which are made in chemical laboratories. These substances have saved countless millions of lives.

Is the battle won?

Look at Figure 9. This shows the number of people in Britain who died of diphtheria each year between 1911 and 1961. You will see that there has been a tremendous fall in the number of deaths from this disease. The same is true of smallpox and many other infectious diseases. In some parts of the world these diseases are virtually extinct.

This happy state of affairs has been brought about by immunisation and antibiotics, and by improvements in personal and community hygiene. To an extent it is also due to people being better fed. If you are well fed, clean and healthy, you are less likely to succumb to disease.

Unfortunately the situation is not so good in many developing countries where infectious diseases still kill a lot of people. This is due partly to poor food, overcrowding and dirty habits, but also to shortage of nurses and doctors.

Investigation

Preventing the growth of bacteria

1 Obtain three petri dishes containing sterile agar. Label them A, B and C.

2 Your teacher will give you a tube or bottle containing a culture of bacteria.

3 Transfer some of the bacteria to the agar in each petri dish, using the method shown in the illustration.

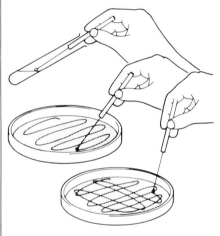

4 Cut out two pieces of filter paper, about 1 cm square.

5 Onto the agar in petri dish A, lay a piece of filter paper which has been soaked in a disinfectant, e.g. Dettol.

6 Onto the agar in petri dish B, lay a piece of filter paper which has been soaked in an antiseptic, e.g. iodine.

7 Onto the agar in petri dish C, lay a piece of filter paper which has been soaked in an antibiotic, e.g. penicillin.

8 Cover each petri dish with a lid and fix it firmly with a piece of Sellotape.

9 Put the petri dishes in an incubator at 37 °C for at least 24 hours.

10 After this time, examine each petri dish for the presence of bacteria.

Which substances prevent the growth of bacteria?
Have you any criticisms of this experiment?

Assignments

1 Why is it particularly important that the following places should be as free of germs as possible:

a) operating theatres,
b) public lavatories,
c) hotel kitchens,
d) swimming pools,
e) doctors' surgeries?

2 Explain the reason for each of the following:

a) A pet which is brought into Britain from overseas is put into quarantine for six months.
b) If you graze your knee it's sensible to wash it immediately and put iodine on it.
c) If you cut yourself badly the wound should be covered with elastoplast or a bandage.
d) Many of the food items in a supermarket are wrapped in cellophane.
e) Chlorine is sometimes added to drinking water.

3 What part is played by each of the following in getting rid of germs from the body:

a) saliva, b) cilia, c) phagocytes, d) antibodies, e) antibiotics?

4 Explain each of these statements:

a) 'Coughs and sneezes spread diseases'
b) 'It's a good thing to have mumps when you're young so as to get it out of the way'

5 Until the late 1940s there were special 'isolation hospitals' for patients with infectious diseases. Such hospitals no longer exist in Britain because they are not needed any more.

a) Why were isolation hospitals necessary in the old days?
b) Why are they no longer needed?

c) Write down three particular difficulties which you think there might have been in running an isolation hospital.

6 During the influenza epidemic of 1918 people were given face masks like the one in the picture below to protect them from breathing in the influenza germs. These masks proved to be useless. Why do you think they were of no use?

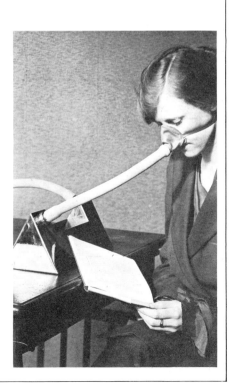

Responding to stimuli

In reacting
to one another
these sparring hares
are showing an important
characteristic of life, namely
response to stimuli. This
is the subject of the next
group of topics.

The nervous system and reflex action

If you put your hand on a hotplate, you pull it away quickly. This response is brought about by messages which are sent at high speed through the nervous system.

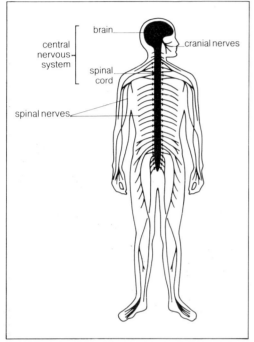

Figure 1 The main parts of the human nervous system.

Figure 2 This doctor is testing a patient's knee jerk.

General plan of the nervous system

The nervous system consists of two parts: the **Central Nervous System (CNS)**, and a series of **nerves** which link the CNS with the various organs (Figure 1).

The CNS is divided into two parts: the **brain** and the **spinal cord**. The brain is enclosed within the **cranium** or brain case which is part of the skull. The spinal cord runs down the centre of the backbone. The whole of the CNS is therefore protected by a covering of bone.

The nerves are of two kinds. Some of them come out of the brain, and go mainly to structures in the head such as the eyes and jaws: these are called **cranial nerves**. Others come out of the spinal cord, and go to the arms, legs and various structures in the trunk. They are known as **spinal nerves**.

Nerve messages

The main job of the nervous system is to carry messages from one part of the body to another. Scientists have carried out experiments to find out about these messages, and they have discovered that they consist of tiny pulses of electricity which travel rapidly through the CNS and along the nerves. We call these messages **nerve impulses**.

Reflex action

Let's consider what happens when you pull your hand away from a hot object. First of all, sensory endings in your fingers are stimulated by the heat. This causes impulses to pass up the nerve in your arm to the spinal cord and brain: the actual feeling of pain occurs when the impulses reach the brain. Further impulses then pass back down the arm to the muscles, causing them to contract. The contraction of the muscles has the effect of pulling your arm away from the unpleasant stimulus. The whole response only takes a fraction of a second and this shows how quickly the impulses travel through the nervous system (Investigation 1).

Pulling your hand away from a hot object is an example of a **reflex action**. *A reflex action is an immediate response of the body to a stimulus.* Many other reflexes are shown by humans and other animals (Investigation 2). For example, if you tap your knee in a certain place, your leg gives a little kick. This is called the **knee jerk**, and it is often used by doctors to find out if the patient's spinal cord is working properly (Figure 2). Another well known reflex is the scratching movement of the hind leg of a dog when you tickle its tummy.

The reflex arc

Scientists have worked out the route by which impulses travel through the nervous system in bringing about a reflex action. This comprises what we call a **reflex arc**.

The structure of the reflex arc involved in pulling your hand away from a hot object is shown in Figure 3. Notice that it is made up of three distinct nerve fibres:

1 A **sensory nerve fibre** which carries impulses from the sensory endings to the spinal cord.

2 An **intermediate nerve fibre** which carries the impulses from the upper to the lower side of the spinal cord.

3 A **motor nerve fibre** which carries the impulses from the spinal cord to the muscle.

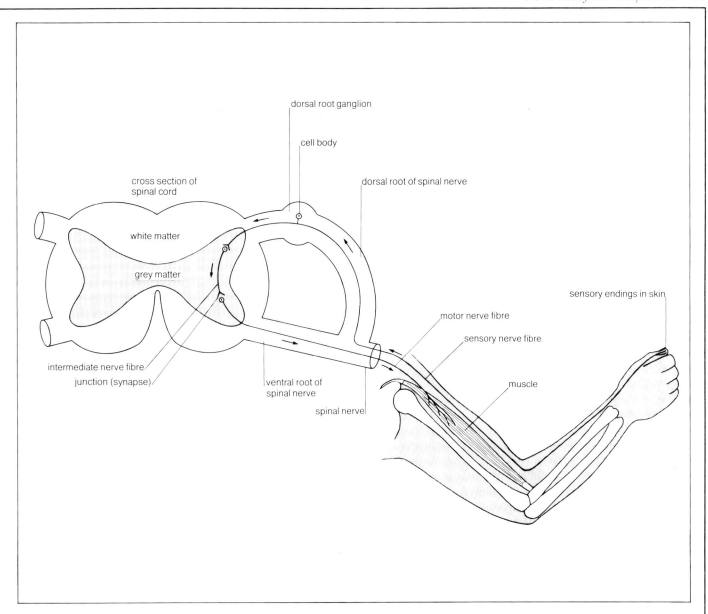

Labels on diagram:
dorsal root ganglion
cell body
cross section of spinal cord
dorsal root of spinal nerve
white matter
grey matter
sensory endings in skin
motor nerve fibre
sensory nerve fibre
intermediate nerve fibre
junction (synapse)
ventral root of spinal nerve
spinal nerve
muscle

These three nerve fibres are connected to each other by junctions inside the grey matter of the spinal cord (Investigation 3). These junctions are called **synapses**.

When an impulse travels through a reflex arc it has to cross these synapses. They will only let the impulses pass in one direction, and this ensures that the impulses always go the right way, i.e. *from* the sensory endings *to* the spinal cord and on to the muscles.

All three fibres must be working properly if the impulses are to get through to the muscle. If one of the fibres dies the reflex cannot occur. In the disease poliomyelitis ('polio') a certain kind of germ (a virus) attacks some of the motor nerve cells in the spinal cord. As a result, impulses cannot reach the muscles and the person becomes paralysed. Which particular muscles are affected depends on what part of the spinal cord is attacked by the virus. Polio victims often lose the use of their legs. Fortunately people can now be immunised against this disease (see page 233).

Most reflexes involve the brain as well as the spinal cord. The messages travel into the spinal cord, and then up to the brain. They then travel back down again, and out to the muscles. A few reflexes, for example the knee jerk, do not involve the brain and can occur in an animal whose brain has been completely destroyed.

Figure 3 This diagram shows the pathway through which nerve messages pass when you pull your hand away from a hot object. It is called a reflex arc.

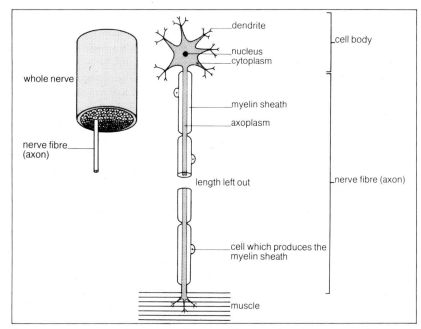

Figure 4 The structure of a nerve cell. This kind of nerve cell carries messages from the spinal cord to the muscles.

Nerve cells

The nervous system is composed of millions of **nerve cells** and it's these that transmit the impulses.

A typical nerve cell is shown in Figure 4. It is made up of two parts: a **cell body** and a long thread-like **nerve fibre** or **axon**. The cell body is situated in the CNS, and the nerve fibre extends out into a nerve.

The nerve cell differs from a typical cell in that it has a number of branches protruding from it. These are called **dendrites**. They link up with other nerve cells to form a complex network. This enables impulses to be sent in many different directions within the nervous system.

The nerve fibre is enveloped by a layer of fat called the **myelin sheath** which speeds up the impulses and prevents them leaking out. If the myelin sheaths don't work impulses cannot be transmitted properly, so the person gradually loses the use of his muscles. This happens in the disease multiple sclerosis.

Voluntary and involuntary responses

Scratching your head, crossing your legs or walking across the room are all voluntary actions which you can do or not do as you wish. They are brought about by muscles attached to the skeleton, and the part of the nervous system which controls them is known as the **voluntary system**. In contrast, various processes are constantly occurring in our bodies over which we have no voluntary control – for example, the beating of the heart and the movements of the gut. We cannot *make* these things happen, nor can we stop them of our own free will. These activities are brought about by muscles which are controlled by the **involuntary nervous system**.

As an example, let's take the heart. The heart receives two nerves: when messages reach it through one of these nerves, the heart beats faster; when messages reach it through the other nerve, the heart slows down. Many other involuntary organs have a double nerve supply of this kind and in general the two nerves produce opposite effects.

There are some actions which we cannot control early in life, but we gradually learn to control them as we get older. The emptying of the bladder and bowels are two examples.

Finally some activities, such as breathing, are partly voluntary and partly involuntary: we breathe automatically without thinking about it, and yet we can alter the rate of our breathing if we want to.

Investigation 1

Measuring your reaction time

Various instruments can be used for measuring reaction time. If such a gadget is available, use it and follow the instructions given by your teacher; alternatively hold a metre rule vertically and find out how far it falls before being caught by your partner. Here is another thing you can do:

1 Divide into two teams, with the same number of people in each.

2 Stand in a row, one person behind the other, with your arms at your sides.

3 When your teacher says 'go' the person at the back should touch the person in front, and then he touches the person in front of him, and so on. Your teacher will time how long it takes for the chain reaction to reach the front.

4 Trace the nervous pathway through which the impulses pass in bringing about each person's response.

Investigation 3

Looking at nerve cells in the spinal cord

1 Obtain a prepared transverse section of the spinal cord, which has been specially stained to show up the nerve cells.

2 Examine the slide under the low power of the microscope. Identify the grey and white matter.

   Can you see nerve cells in the grey matter?

3 Go over to high power, and focus on a single nerve cell. Draw the nerve cell to show its shape.

   What part of the nerve cell shown in Figure 4 does your drawing correspond to?

Investigation 2

Some human reflexes

Work in pairs, one person acting as the subject.

1 The knee jerk.

Sit on a table with your legs hanging loosely. With a heavy instrument such as a metal rod, your partner should gently tap your knee just below the knee cap.

What happens?

2 The ankle jerk.

Kneel on a chair and let your feet hang loosely. Your partner should tap the back of your foot just above the heel.

What happens?

3 Repeat the above reflexes but this time make a conscious effort to *prevent* them taking place.

Do you succeed?

What conclusions do you draw?

4 The blink reflex.

Open your eyes and look straight ahead. Your partner should suddenly wave his hand in front of your eyes.

What happens?

5 The swallowing reflex.

Swallow the saliva in your mouth, then immediately try swallowing again.

Is it difficult to swallow the second time?

Suggest an explanation.

6 Repeat the above experiment, but this time swallow your saliva and then swallow a mouthful of water.

What difference does this make?

What conclusions do you draw?

Assignments

1 In Figure 4 what job is done by each of these structures: the dendrites, the axon, the myelin sheath, the cell body?

2 Explain the reason for each of the following:

a) If you tickle a dog's tummy it 'scratches' with its hind leg.
b) Messages travel through a reflex arc in only one direction.
c) In multiple sclerosis the person gradually becomes extremely weak.
d) In 'polio' the patient may lose the use of his legs.

3 The diagram below shows the pathway through which messages (impulses) travel in bringing about the knee jerk in man. When the tendon is tapped, receptors in the muscle are stretched and this causes the messages to be sent off.

a) Which structure in the diagram is the receptor? (Identify it by its letter.)

b) Which structure carries impulses away from the spinal cord?
c) Which structure shortens as a result of the reflex?
d) The diagram is not drawn to scale. What would be the approximate length of the structure labelled E in the actual reflex arc?
e) Assuming that the impulses travel at 100 metres per second, how long would it take for the impulses to travel right through this reflex arc?

4 A person is walking across a room in bare feet and he treads on a drawing pin. He lets out a cry. Explain as fully as you can what happens in his nervous system in bringing about this response.

5 Devise a method for measuring how long it takes for a knee jerk to begin from the moment the knee is tapped. (Be as inventive as you like!)

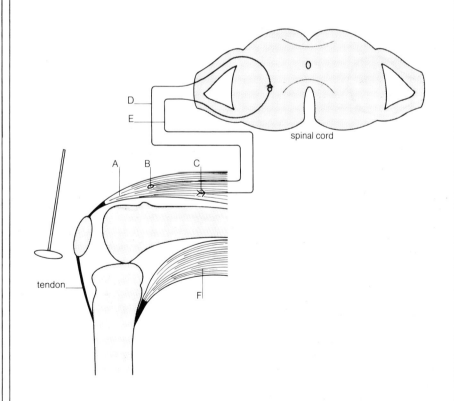

Assignment 3 The nervous system and reflex action.

The brain and behaviour

We all have to think and make decisions every day: this comes into almost every aspect of our lives. The organ which enables us to do these things is the brain.

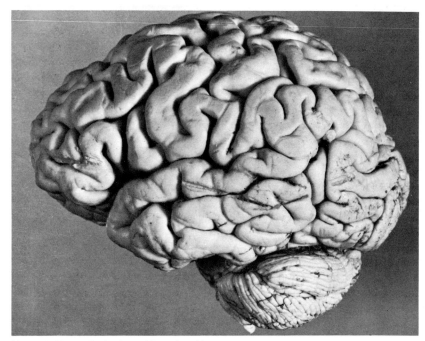

Figure 1 Human brain viewed from the side.

What does the brain consist of?

Figure 1 shows a human brain. If you were to slice it open you would not see much, just a soft whitish material. However, if you looked at a thin section of it under the microscope you would see that it's made up of countless millions of **nerve cells** (Figure 2). Scientists have worked out that there are over one thousand million nerve cells in the brain. Each cell may be connected with 25 000 others and the total number of connections in the entire brain is around ten to the power of three million. This number is so enormous that if it was written out fully as a figure it would fill a book as large as the one you are reading at the moment! With so many connections, the brain is like an extremely complex computer with thousands of electrical messages travelling from place to place. But it is much more than a computer because it makes us conscious beings with feelings and emotions.

The structure of the brain

The brain is shaped like a large mushroom (Figure 3). The cap of the mushroom is called the **cerebrum**, and the stalk is called the **brain stem**. Sticking out of the top side of the brain stem just below the cerebrum is a protuberance rather like a little cauliflower: this is called the **cerebellum**.

The entire brain is enclosed within the bony brain case or cranium which is part of the skull. Surrounding the brain inside the cranium are two membranes with fluid in between. This is called **cerebro-spinal fluid**, and it is formed from two masses of fine blood capillaries called **plexuses** in the roof of the brain. The cerebro-spinal fluid nourishes the brain and also serves as a shock absorber, so the brain is cushioned from damage when the person jumps around or bangs his head.

If you look at Figure 3 you will see that there is a cavity in the centre of the brain: cerebro-spinal fluid is found in here too. It is also found inside and around the spinal cord which is continuous with the brain stem.

What does the brain do?

We have seen that the brain is divided into three main parts: the cerebrum, cerebellum and brain stem. Scientists have discovered what each of these

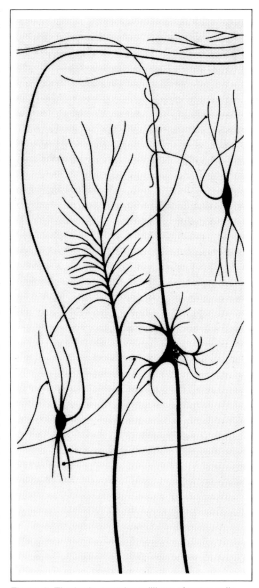

Figure 2 The brain contains millions of nerve cells which are connected with one another in a most complex way.

parts does by observing the behaviour of people whose brains have been damaged in accidents. Here is an example:

A motor cyclist has a crash and his cerebellum is badly damaged but the rest of his brain is unaffected. He gradually recovers, but he keeps toppling over when he stands up, and he finds it difficult to make accurate movements with his hands.

From this kind of observation we can say that the cerebellum controls our sense of balance and allows us to make precise and accurate movements. In doing this, it works in conjunction with various sense organs (see page 265).

The cerebrum registers various sensations, such as seeing and hearing, and it makes our legs and arms move. It also enables us to think, speak and remember things. This is such an important part of the brain that we must study it in more detail.

The brain stem controls various processes which go on without our thinking about them, such as breathing and the beating of the heart.

Figure 3 The human brain seen in its natural position inside the head.

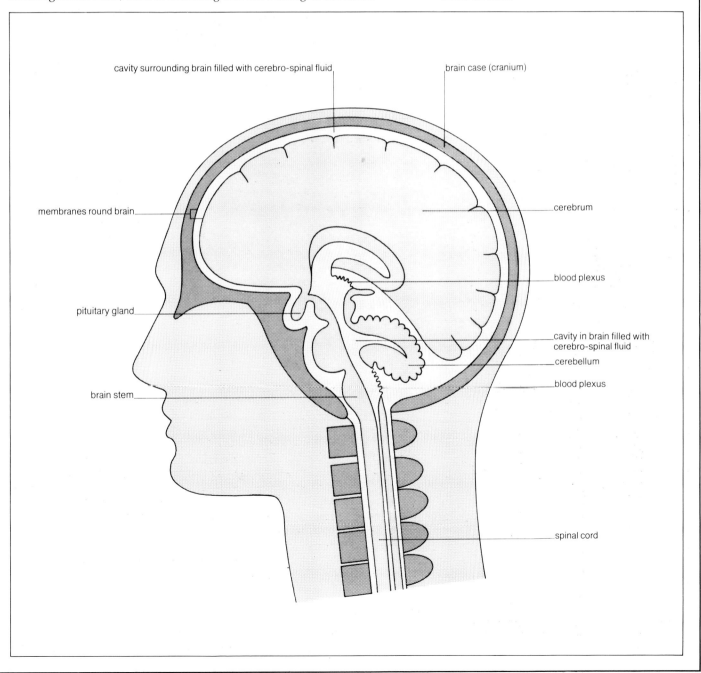

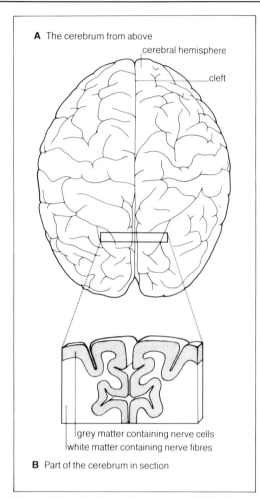

A The cerebrum from above

cerebral hemisphere

cleft

grey matter containing nerve cells
white matter containing nerve fibres

B Part of the cerebrum in section

Figure 4 The structure of the cerebrum. Notice the folding of the grey matter.

Figure 5 Who is this?

A closer look at the cerebrum

If you look at Figure 4A, you will see that the cerebrum is divided into two halves by a cleft which runs down the middle. The two halves are called **cerebral hemispheres**. A person's actions tend to be dominated mainly by one side of the brain. However, the two hemispheres are connected by nerve fibres and are very much in communication with one another.

Most of the nerve cells in the brain are situated in the cerebrum, where they are concentrated in a thick layer towards the surface. This is the grey matter or cerebral cortex. If you look at a slice of the cerebrum you will notice that the grey matter is folded like a piece of crumpled paper (Figure 4B). This has the effect of increasing the surface area, so a greater number of brain cells can be packed into the brain.

What does the cerebrum do?

The function of the cerebrum can be illustrated by considering a certain dog called Oliver. Oliver is sitting outside his kennel when suddenly he sees a cat at the bottom of the garden. He is just about to chase it when his owner calls him in for his supper. So he delivers a quick menacing bark at the cat and runs indoors.

What has been happening in Oliver's brain? On seeing the cat, impulses were sent from the eyes to the cerebrum. When a moment later he heard his owner calling, impulses were sent from his ears to the cerebrum. Oliver's cerebrum understood that his owner's call meant food, so it sent impulses to the leg muscles and made him run indoors.

We can sum it up like this: the cerebrum receives impulses from various sense organs, it sorts them out and then sends impulses to the particular muscles which are needed.

Oliver's behaviour illustrates two activities for which the cerebrum is responsible: **learning** and **instinct**.

Learning

How did Oliver *know* that when his owner called it was supper time? Ever since he was a puppy, his owner had clanked his dish on the floor when calling him for supper. In the course of time Oliver had learned to associate this clanking noise with food.

In the same way he had learned not to urinate in the house: he had come to associate doing *that* with a smack on the backside.

We call this kind of behaviour **conditioning**. It was first described in the early 1900s by a Russian scientist called Pavlov who did some interesting experiments on dogs. Normally a dog's mouth waters when it is given food: this is a straightforward reflex of the kind described on page 238. In one experiment Pavlov rang a bell just before giving the dog its food. After repeating this several times, the dog learned to associate the bell with the food, and it would salivate as soon as it heard the bell, i.e. before the food appeared. Pavlov called this response a **conditioned reflex**. Many examples of this kind of thing are shown by humans and other animals. In general, conditioning can be speeded up if the animal is rewarded for doing the right thing, and punished for doing the wrong thing. This principle is used in training dogs and other animals.

Conditioning is a form of **learning.** We can define learning as a change in behaviour resulting from past experience. Learning is made possible because the cerebrum can store information, or, to put it simply, it can remember things (Investigation 1).

Learning enables us to get to know our surroundings and to respond in the right way to new situations. Learning also enables us to recognise objects, even when they are not very clear (Figure 5). In fact the brain is remarkably good at filling in gaps, provided that it has some idea of what to expect (Investigation 2).

Instinct

Returning to Oliver, if it hadn't been for his supper arriving, he would have chased the cat. Oliver hadn't learned to chase cats. He'd done it all his life, and was born with the knowledge of how to do it.

This kind of behaviour is called **instinct**. Instinctive behaviour is inherited and does not need to be learned. All that's necessary is that it is set off by the right kind of stimulus. For Oliver the appearance of the cat was the stimulus, unleashing the chasing behaviour. In fact, dogs will instinctively chase any swiftly moving object they happen to see. This is made use of in greyhound racing in which the dogs run after a dummy hare propelled along a track (Figure 6).

Instinctive behaviour is common in many animals: we see it in the courtship behaviour of birds, for example (see page 94). However it is less obvious in human beings, though in babies sucking the mother's nipples for milk and pushing things into the mouth are probably instinctive.

Many scientists feel that it is wrong to make a sharp distinction between learning and instinct. The behaviour of higher animals, and humans in particular, results from a combination of these and other types of activity. Greyhound racing, though based on instinct, depends to some extent on learning in that the dogs are trained to chase the dummy hare as fast as they can.

Figure 6 A greyhound chasing the 'hare'.

Investigation 1

Short-term and long-term memory

First experiment

1 Your teacher will give you ten objects to look at for a minute.

2 After the minute is up, write down the names of as many of the objects as you can remember.

3 About half an hour later, try writing them down again.

How many did you get right the first time?

And the second time?

Second experiment

1 Make a simple drawing of the front of any building which you know well and see regularly.

2 Compare your drawing with the actual building.

In what respects is your drawing right?

In what respects is it wrong?

Which of these two experiments demonstrates your short-term memory, and which one demonstrates your long-term memory?

Investigation 2

Recognising things

1 Your teacher will give you a box containing ten objects, some well known to you, others less well known.

2 Put your hands in the box, *but do not look inside*.

3 Feel each object in turn and write down what you think it is.

4 When you have finished making your list, look at the objects in the box.

Which ones have you got right?

To what extent has past experience helped you to recognise these objects?

5 Your teacher will give you various incomplete or fuzzy objects or pictures.

6 In each case write down what you think the object is.

Which ones did you find difficult to recognise?

In each case say why it was difficult.

Which ones are made easier to recognise by looking at them from further away?

What does this experiment tell us about the brain?

Assignments

1 Give one function which is performed by each of the following:

a) the cerebro-spinal fluid,
b) the cerebellum,
c) the cranium,
d) the brain stem,
e) the cerebrum.

2 The ability to learn is associated with certain parts of the cerebrum. How do you think we know this?

3 How do you think scientists know that there are over one thousand million nerve cells in the human brain?

4 A chimpanzee is put in a cage in which there is a lever. Every time the chimpanzee presses the lever, he is given a banana. After a time the chimpanzee realises that if he wants a banana all he has to do is to press the lever.

In what respect is the chimpanzee's behaviour (a) similar to, and (b) different from the behaviour of Pavlov's dogs?

5 What is meant by instinctive behaviour? Describe an experiment which could be done to test the suggestion that a dog chasing a cat is an example of instinct.

Drugs and mental illness

The brain is easily affected by outside influences, particularly drugs. In this Topic we will look briefly at how drugs affect the brain, and what happens when the brain does not work properly.

Drugs

A drug is any substance which alters the way the body works. Here we will concentrate on drugs that affect the brain. Such drugs fall into four groups:

1 Sedatives
These drugs slow down the brain and make you feel sleepy. They include tranquillisers and sleeping pills. Tranquillisers have a calming effect and are often given to people suffering from anxiety. Sedatives include a group of chemical substances called **barbiturates** which are so powerful that they are used as anaesthetics. Alcohol is also a sedative.

2 Stimulants
These drugs speed up the action of the brain and make you more alert. They include 'pep pills' which are sometimes given to people who are suffering from severe depression. Coffee and tea contain a mild stimulant called caffeine, but it doesn't do you any harm. Another mild stimulant is nicotine, the drug found in tobacco.

3 Hallucinogens
These drugs cause **hallucinations**. An hallucination is something which a person senses but which does not actually exist. Drugs that cause hallucinations include cannabis (nicknamed 'pot') and LSD (lysergic acid diethylamide).

4 Pain-killers
These drugs suppress the part of the brain responsible for the sense of pain. They include two powerful drugs called **morphine** and **heroin** which are obtained from opium, a substance found in a certain type of poppy. Morphine is often given to people suffering from severe pain.

Why are drugs dangerous?

If taken under doctor's orders certain drugs can be very helpful to sick people, but if taken in the wrong circumstances they may be extremely harmful. There are three main reasons for this:

1 They may impair the person's judgements and make him clumsy.
Often they lengthen the reaction time, so the person takes longer to respond to a stimulus. Such is the case with alcohol. In an experiment, a group of bus drivers drove their buses between two rows of posts, and then did the same again after drinking some whisky. It was found that after drinking they knocked over more posts than they had done before.

2 The person may become dependent on the drug and crave for it.
Such is the case with cigarette smoking and cannabis. Drugs such as heroin get such a grip on the body that if the person has to go without regular doses, he may develop **withdrawal symptoms** such as fever, sickness and severe cramp. People who reach this state are said to be **addicted** to the drug.

It's particularly easy to become dependent on alcohol. Such a person is called an **alcoholic**. There are over 300 000 known alcoholics in Great Britain. Once a person becomes addicted to a drug it is very difficult to give it up, as Figure 1 makes only too clear.

3 It may injure the body by damaging the cells.
For example, in a heavy drinker the alcohol goes to the liver and gradually kills the cells which then become replaced by fibrous connective tissue. This is called **cirrhosis of the liver**, and a heavy drinker may eventually die of it. Cannabis is dangerous too: there is evidence that it damages the brain cells.

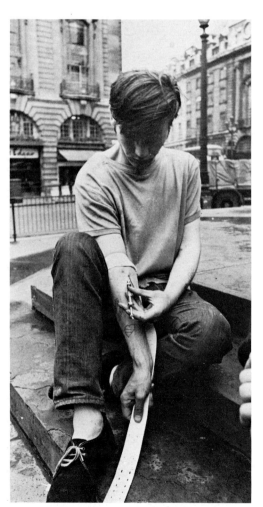

Figure 1 An addict injecting himself with a drug.

Mental illness

Mental illnesses develop when the brain does not work properly. Some are the result of accidents in which the brain is damaged, others are by-products of diseases such as syphilis; and sometimes a person is born with a mental disorder.

Types of mental illness

Doctors recognise two main kinds of mental illness:

1 *Neurosis*

Neurotic illnesses are fairly mild and it's not normally necessary for the person to go into hospital. They often involve the person being obsessive about something. For example, he may have an obsession about being clean, or an obsessive fear of being shut into a small space such as a crowded underground train. Such fears are called **phobias**. Usually the person realises he has a problem and wants to overcome it.

2 *Psychosis*

Psychotic illnesses are more serious and usually necessitate going to hospital. Often the patient does not realise there's anything wrong with him. He may think that everyone is getting at him, or that he is a famous character such as Napoleon or Hitler, and he will act accordingly. Violent crimes are sometimes committed by psychotic individuals.

There is much variation in the symptoms of mental illnesses and how long they last. For example, in **epilepsy** the patient suffers from occasional 'fits' which may be mild or severe depending on what kind of epilepsy it is.

In epilepsy and many other mental disorders the person may be perfectly normal at times when he is not having an attack of the illness. For example, there is a condition called **schizophrenia**. This word comes from Greek and literally means 'split mind'. A typical schizophrenic has two different personalities: a normal one and an abnormal one. Much of the time he behaves just like anyone else, but at times he becomes psychotic and his behaviour changes accordingly.

Treating people with mental illnesses

Doctors who specialise in mental illness are called **psychiatrists**, and they can do much to help people overcome even the most severe illnesses.

One of the difficulties is knowing exactly what's wrong with the patient. This can sometimes be found out by recording the electrical waves from the patient's brain with a machine called an **electroencephalograph** or **EEG**. The waves are picked up by electrodes which are placed on the patient's head, and they are recorded on a roll of moving paper by a series of pens (Figure 2).

Everyone has the same kind of brain waves, but if the brain is not working properly the waves are abnormal and this shows up on the paper.

What is a nervous breakdown?

This is a rather unscientific term, but it usually refers to a type of mental illness which is brought on by stress, worry or overwork. Usually the person is overwhelmed by a feeling of utter despair, a condition which is known as **depression**. Of course we all feel depressed at times, but the kind of depression which occurs in a nervous breakdown is particularly intense and may be accompanied by various obsessions and hallucinations. Sometimes the periods of depression alternate with periods of extreme elation.

It's said that in the United Kingdom approximately one person in six has a nervous breakdown at some stage of his or her life.

Much can be done to help people through such an illness. With proper treatment the patient may be able to return to a normal life within a few months.

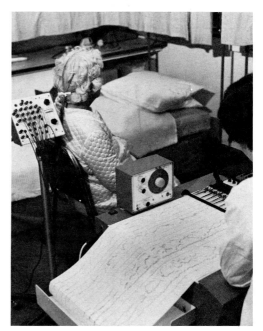

Figure 2 This person is having her brain waves recorded by an electroencephalograph.

Assignments

1 Each of the words in the left hand column is related to one of the words in the right hand column. Write them down in the correct pairs.

cannabis	stimulant
morphine	tobacco
caffeine	hallucination
barbiturate	anaesthetic
nicotine	pain-killer

2 Nowadays doctors try to avoid giving people barbiturates except in special circumstances. Why do you think this is?

3 Briefly explain each of the following terms: sedative, hallucinogen, opium, withdrawal symptoms.

4 The number of people attending psychiatric clinics in Great Britain is much greater now than it was fifty years ago. Suggest reasons for this.

Receiving stimuli

We are constantly subjected to all sorts of stimuli. It is vital that we should be able to detect these stimuli and respond to them in the right way.

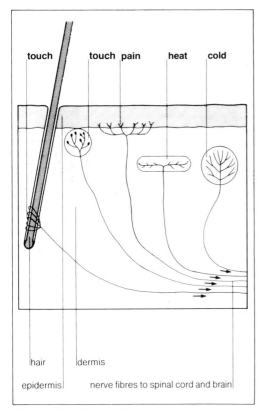

Figure 1 This diagram shows the main receptors in human skin and the sensations they respond to.

Labels in figure: touch, touch, pain, heat, cold, hair, dermis, epidermis, nerve fibres to spinal cord and brain

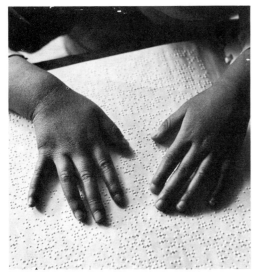

Figure 2 A blind person reading braille.

How are stimuli detected?

We have a number of different **receptors** or **sense organs**, and each is sensitive to a particular kind of stimulus: some respond to touch, others to light, and so on.

A receptor is usually composed of a group of **sensory cells**. These link up with nerve fibres which lead to the brain or spinal cord. When a receptor is stimulated, impulses are sent to the brain where they give rise to a corresponding **sensation**. For example, when you put your hand on something warm, impulses pass from certain receptors in your skin to the brain, giving rise to a feeling of warmth.

Now let's look at some everyday sensations and see how they are produced. We will start with the skin, whose receptors are shown in Figure 1.

Touch

If you place the tip of a needle in contact with your skin, you will feel the sensation of touch. You get the same kind of feeling if you wiggle one of the hairs on the back of your hand. Our sense of touch is explored in Investigations 1–4.

The sense of touch is spread all over the skin, though some areas are more sensitive than others.

If you put on a hairy shirt, it tickles at first but gradually the sensation wears off until eventually you stop noticing it. This is because after a time the touch receptors stop sending impulses to the brain. This is called **sensory adaptation**. Why do you think it's useful?

Most of us don't use our sense of touch as much as we might. This is because we depend more on our eyes for finding out what's round about us. However, blind people develop their sense of touch to a much greater degree, particularly in their finger tips which they use for identifying objects and for reading braille (Figure 2 and Investigation 4).

Pain

There are special receptors in the body which, when stimulated, give rise to the sensation of pain. In the skin these receptors take the form of free nerve endings. However, pain is also caused by excessive stimulation of other kinds of receptor. Pain also results from muscle spasms, as in cramp, and when an organ is short of oxygen. For example, the pain which is felt by people with heart trouble is caused by the heart muscle not getting enough oxygen.

People with certain kinds of heart trouble get pain in the left arm a long way from the heart itself. This is because impulses from the heart and the left arm go to the same part of the spinal cord. Pain which is felt some distance from its true origin is called **referred pain**.

People who have had a leg amputated often say that they can feel pain in the missing leg. This is called **phantom pain**. It is caused by the severed nerve healing and then sending impulses to the brain again.

Temperature

In our skin there are receptors for telling us whether it's hot or cold. In fact there are two different receptors, one for detecting heat and the other for detecting cold.

Our temperature receptors are not very good at telling us what the actual temperature is. What they really do is to tell us when the temperature changes, and what we actually feel depends on how quickly the skin gains or loses heat (Investigation 5).

If we *feel* uncomfortably hot or cold we do something to remedy the situation; for instance if it is cold we put on more clothes. This helps us to maintain a constant body temperature (see page 218).

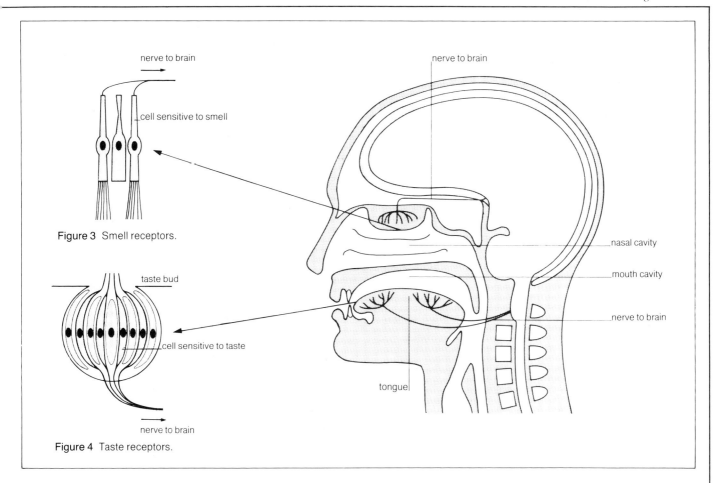

Figure 3 Smell receptors.

Figure 4 Taste receptors.

Smell

The receptors responsible for our sense of smell are shown in Figure 3. They consist of groups of sensory cells in the roof of the nasal cavity. The cells have minute bristles like little brushes, and they are stimulated by molecules which float through the air and land on them. For something to be smelled, it must go into solution first, and so the lining of the nasal cavity is covered with a thin layer of fluid. Our nose is sensitive to many different smells, but our sense of smell is very poor compared with that of other mammals such as dogs. Smells which are far too faint to be detected by a man will be picked up by his dog.

Taste

If you stick your tongue out and look at it in the mirror, you will see that it is covered with hundreds of short hair-like structures towards the front and wart-like bumps towards the back. These are your **taste buds**, and they contain receptors sensitive to certain chemicals (Investigation 6). Each receptor is a tiny flask containing about half a dozen sensory cells (Figure 4). As with the sense of smell, substances must be in solution before they can be tasted.

Experiments show that the tongue is sensitive to only four kinds of stimuli: sweet, sour, bitter and salt. Each of these stimuli is detected by a different part of the tongue (Investigation 7).

How can we explain the wide variety of taste sensations which we experience when we eat and drink? The answer is that our sense of smell also plays an important part: when you think you're tasting something, you're also smelling it. Have you noticed that if you have a heavy cold and your nose is blocked, your sense of taste is impaired as well as your sense of smell?

Investigation 1

Which parts of the skin are sensitive to touch?

Work in pairs, one person acting as the subject.

1 With a fine ball-point pen, rule a grid of 25 squares on the back of your partner's hand; the sides of the squares should be 2 mm long so each one will have an area of 4 mm^2.

2 Obtain a bristle which has been mounted on a wooden holder.

3 Press the tip of the bristle against the skin in one of the squares until it just bends.

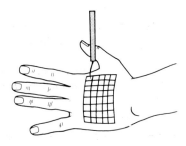

Does your partner feel the bristle?

4 Now touch the skin in the other squares, one by one.

In each case note whether or not your partner feels the bristle.

5 If some of the squares were insensitive, try touching them more strongly with a blunt needle.

Does your partner feel anything now?

6 Repeat this experiment on different parts of the skin, for example on the arm and leg.

Are some parts of the skin more sensitive than others?

How could you find out which parts of the skin are sensitive to other stimuli, such as pain and temperature?

Investigation 2

Getting used to a stimulus

1 With a mounted needle, wiggle a hair on your arm or hand.

What does it feel like?

2 Keep wiggling the hair until you can no longer feel it.

Suggest a possible reason why the sensation disappears. Why is it an advantage to get used to this kind of stimulus?

3 Wiggle another hair and note the sensation.

4 Now rub the skin with your finger for about 15 seconds.

5 Wiggle the hair again.

Can you still feel it?

Suggest an explanation for what has happened.

Investigation 3

To find out the localising power of the skin.

Work in pairs, one person acting as the subject.

1 Subject: close your eyes.

2 Partner: touch the skin on the back of the subject's hand with the point of a fine felt pen.

3 Subject: without looking try to touch the skin in exactly the same place with the point of another pen.

How far apart are the two dots on the skin?

What does this tell us about your sense of touch?

Investigation 4

Reading with your fingers

Blind people read by means of 'braille', which was invented in France by Louis Braille. In braille each letter of the alphabet is represented by a character consisting of one to six dots embossed on thick paper.

1 Obtain a braille exercise card.

2 Without looking, put the tip of your finger on one of the characters. Do this with your non-writing hand.

How many dots is the character composed of?

Draw the arrangement of the dots on a piece of paper.

3 Now take your finger off the character and look at the dots.

Is your drawing of the dots correct?

4 Repeat the above with other characters, and try using different fingers.

Do your attempts at feeling the characters correctly improve with practice?

Are certain fingers better at it than others?

Which finger do you find the best? What do you think makes some fingers better at it than others?

5 Some blind people use another system called 'moon' which was developed by an Englishman called Dr Moon.

Carry out the above experiment on a sheet of 'moon'.

Are the characters easier to tell apart than the braille ones?

Investigation 5

An interesting aspect of our temperature sense

1 Obtain three bowls: the first should contain ice-cold water, the second hot water, and the third water at room temperature.

2 Place your left hand in the cold water, and your right hand in the hot water for one minute.

3 When the minute is up, place both hands in the room temperature water

What does each hand feel like?

What does this tell us about our temperature sense?

Investigation 6

Looking at the tongue

1 Look at the top side of your tongue in a mirror: if necessary shine a torch into your mouth, so as to show it up more clearly.

Can you see your taste buds?

How many different kinds of taste buds can you see?

2 Swallow your saliva, and dry your tongue with a tissue.

3 Place a lump of sugar on your dry tongue.

Can you taste it?

4 Let your tongue get wet with saliva, and then place the sugar lump on your tongue again.

Can you taste it now?

What conclusions do you draw from this experiment about the way we taste things?

Investigation 7

Which parts of your tongue respond to different tastes?

All the materials used in this investigation must be clean, and the experiment should be carried out under close supervision.

Work in pairs, one person acting as the subject.

1 Your teacher will give you four small beakers containing respectively a dilute solution of sugar, salt, acetic acid and quinine (a bitter substance). Each beaker should have a small paintbrush with it.

2 Subject: swallow your saliva and dry your tongue with a tissue, then stick your tongue out as far as you can.

3 Partner: draw the tongue in outline on a piece of paper.

4 With the brush put a little sugar solution onto different parts of the subject's tongue and note whether or not the subject can taste it.

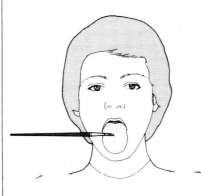

5 Put a cross in your drawing of the tongue to indicate the areas sensitive to sugar.

6 Subject: wash your mouth out with water, and dry your tongue again.

7 Repeat the experiment with the salt, acetic acid and quinine.

Indicate in your drawing of the tongue whereabouts each substance can be tasted.

What conclusions do you draw from this experiment about our sense of taste?

Assignments

1 What kind of receptors are stimulated when you:
 a) move a hair on the back of your hand,
 b) cut your finger with a knife,
 c) read braille,
 d) put clothes on,
 e) place some food in your mouth?

2 What is meant by sensory adaptation? Give one example of it, and explain why it is useful.

3 Suggest a reason why our sense of pain is useful.

4 Explain the difference between referred pain and phantom pain, and try to explain each.

5 a) Why do you think it is difficult to taste things when you have a cold?
 b) It is said that dogs can taste their food as it passes down the gullet (oesophagus). How could you find out if this is true?

6 A large number of volunteers were tested to find out the lowest (minimum) skin temperature that causes the sensation of pain. The results are shown below.
 a) Explain in words exactly what the graph shows.
 b) What conclusions can you draw from the results?

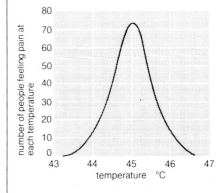

A look at the eye

Most people depend more on their eyesight than on any other sense, for it tells us so much about the world around us. The organ of sight is the eye.

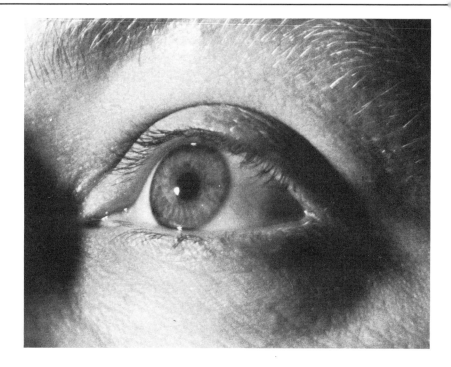

Figure 1 The human eye.

The external structure of the eye

Each eye consists of an **eyeball** which is held in a socket in the skull (Figure 2). The eyeball is surrounded by a thick coat (Investigation 1). This is transparent at the front, and white and opaque at the sides and back, it is called the white of the eye or **sclera**. The transparent front part is called the **cornea** and it is covered by a very thin membrane called the **conjunctiva** which is really part of the skin. Occasionally this gets infected and inflamed, resulting in a disease called **conjunctivitis** or **pink-eye**. Though very sensitive to touch, the cornea does not contain any blood vessels: it gets all its nourishment from the fluid inside the eyeball.

Beneath the cornea in the centre of the eye is what looks like a black hole: this is called the **pupil** and it leads to the inside of the eye. The pupil is surrounded by the **iris** which is the coloured part of the eye.

The eyeball is held in place, and moved, by six muscles. A large **optic nerve** runs from the back of the eye to the brain. When you look at something, messages are sent off in this nerve and as a result you see the object.

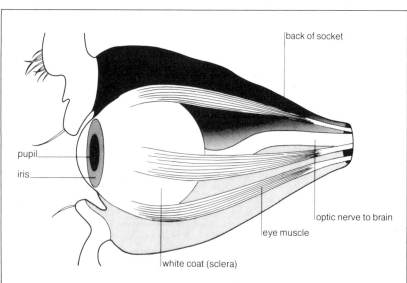

Figure 2 The eyeball in its socket. Only three of the six eye muscles are shown. The remaining ones are on the other side.

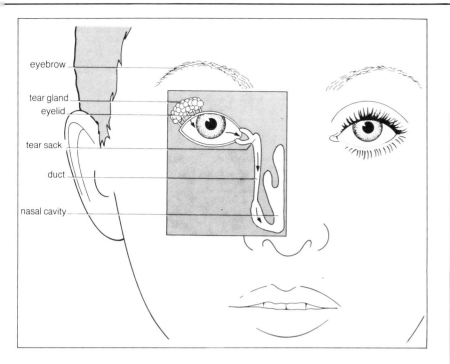

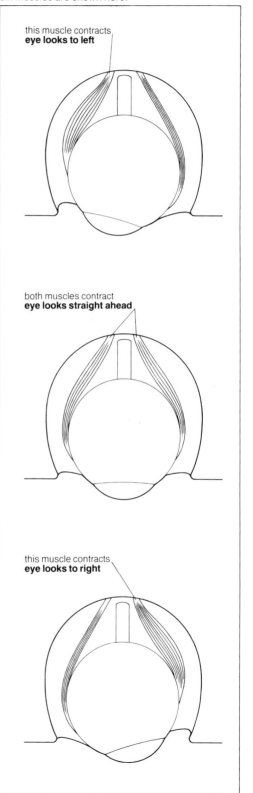

Figure 3 This diagram shows the main structures that keep the eye clean. The arrows indicate the way tears normally flow.

Figure 4 These diagrams show how the external eye muscles move the eye in its socket. Only two of the six muscles are shown here.

How is the eye protected?

The eye is a precision instrument and must be protected and kept clean. The main structures involved in this are shown in Figure 3. The surface of the eye is kept moist by a lubricating fluid produced by the **tear gland**. When you blink, the fluid is spread over the surface of the eyeball by the eyelids which act rather like windscreen wipers. The fluid contains a substance called **lysozyme** which kills germs. Surplus fluid drains into the duct in the corner of the eye and trickles down to the nasal cavity. When a person cries, so much fluid is produced that it cannot be drained away, and so it rolls down the cheeks as tears.

As well as spreading fluid over the eyes, the eyelids protect them. Thus we blink when something passes close by, and the eyelashes help to stop sweat and dirt running into the eyes. Infection of the eyelids can cause a **stye** in which the edge of the lid becomes red and sore.

Some animals, such as reptiles and birds, have a transparent shutter which can slide sideways over the eye. This **third eyelid** protects the surface of the eye while allowing it to see at the same time.

How do we move our eyes?

The muscles which move the eyes run from the sides of the eyeball to the back of the socket. Between them, these muscles can rotate the eyes in various planes within their sockets. The action of two of the muscles is shown in Figure 4.

Human beings can only move both eyes together. However, certain reptiles, e.g. the chameleon, can move their eyes independently of one another. What are the advantages and disadvantages of being able to do this?

Inside the eye

Figure 5 shows the inside of the eye. The two most important structures are the **lens** and the **retina**.

The lens is like a transparent balloon. It is held in position by fine ligaments which run from its edge to the surrounding **ciliary body**. The eyeball in front of and behind the lens is filled with a transparent jelly-like fluid called the **humour**.

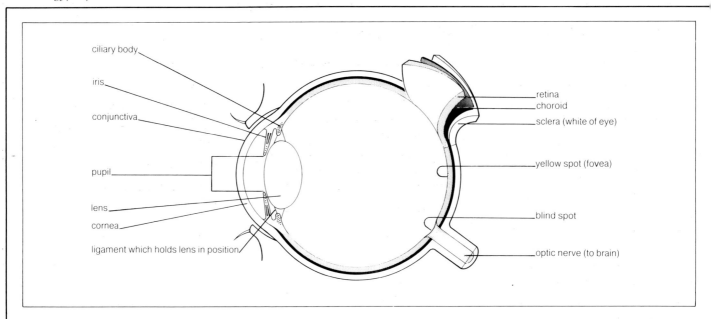

Figure 5 The internal structure of the human eye.

The retina is a delicate layer of tissue lining the inside of the eyeball. It contains millions of sensory cells which are sensitive to light entering the eye through the pupil: this is how the eye sees things. The part of the retina responsible for seeing things most clearly is a small area right in the middle, immediately behind the lens. This is called the **yellow spot** or **fovea**.

Beneath the retina is a layer of black tissue called the **choroid**. This absorbs the light and prevents it being reflected back out of the eye. There are numerous blood vessels in the choroid which supply the retina with oxygen and food substances.

The point where the optic nerve leaves the eye is called the **blind spot**. It's the only part of the back of the eye without sensory cells, and so it is incapable of seeing (Investigation 2).

Controlling the amount of light that enters the eye

If a bright light is shone in your eye, a reflex action occurs (Investigation 3). The pupil gets smaller and this prevents too much light getting into the eye.

The opposite happens in the dark: the pupil gets larger, so more light can enter the eye.

These changes in the size of the pupil are brought about by the iris. In the iris there are muscles which can make it either constrict or open up (Figure 6). This is bound up with the way the eye works, which is the subject of the next Topic.

Figure 6 Here you can see how the pupil responds to a dim and bright light.

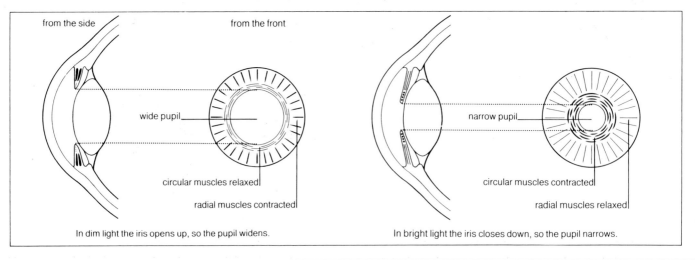

In dim light the iris opens up, so the pupil widens.   In bright light the iris closes down, so the pupil narrows.

Investigation 1

Looking at the eye

1 Look at one of your eyes in a mirror.

 Which structures in Figures 1 and 2 can you see?

2 Move your eyes up and down and from side to side so as to see the 'whites' of your eyes.

 What structures can you see ramifying over the 'whites'?

3 Examine an animal's eye which has been obtained from the butcher.

 Name the structures you see now which you couldn't see before.

4 With scissors cut a hole in the side of the eyeball.

 What comes out of the hole?

 What effect does making the hole have on the shape of the eyeball?

5 Find the lens and the retina.

 What other structures in Figure 5 can you see?

6 Cut the lens out of the eye.

 What does the lens feel like?

Investigation 3

The pupil reflex

Work in pairs, one person acting as subject.

1 The subject should close his eyes for ten seconds and then open them.

 What happens to his pupils when he opens his eyes?

2 Shine a torch in the subject's eye, and watch the pupil.

 What happens to the pupil?

3 The subject should look at an object in the distance and then near at hand.

 What happens to his pupils when he does this?

Investigation 2

Demonstrating the blind spot

1 Look at the picture below: hold it about 10 cm from your eyes.

2 Close your left eye, and look at the house with your right eye.

3 Slowly move the picture away from your eyes, keeping your right eye focused on the house all the time

 What happens to the ghost as you move the picture away from you?

 How would you explain this?

4 Repeat the experiment with both eyes open.

 What happens this time?

 How would you explain the difference?

Assignments

optic nerve	protecting the cornea
tear gland	sensitivity to light
external eye muscle	sending messages to brain
retina	moving eyeball
iris	preventing too much light entering eye

1 Each of the structures listed above in the left hand column is responsible for doing one of the jobs listed in the right hand column. Write them down in the correct pairs.

2 The pictures below show what a person's eyes look like if he moves them in certain ways. By means of diagrams like the ones in Figure 4, explain how each position is brought about.

3 Explain briefly what causes pink-eye and a stye.

4 If you look at something a long way off and then near at hand, your pupil gets smaller.

 a) How is this change brought about?
 b) Why do you think it happens?

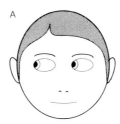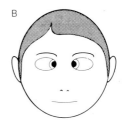

How does the eye work?

When you take a photograph, light enters the camera and is focused by a lens onto a light-sensitive film at the back. The eye works in the same kind of way.

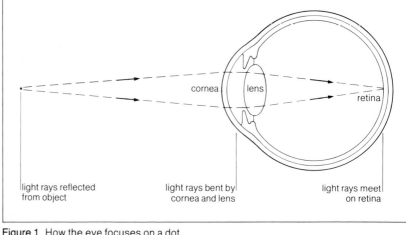

Figure 1 How the eye focuses on a dot.

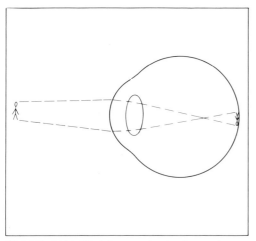

Figure 2 A simplified diagram showing how an image is turned upside down by the lens in the eye. The same thing is done by the lens in a camera.

Figure 3 How the eye keeps an object in focus as it gets closer.

What happens when we look at something?

Suppose you are looking at a dot in the distance. Light rays are reflected from it and enter your eye (Figure 1). The light rays are bent inwards as they pass through the cornea and lens, so they meet on the retina at the back of the eye. Here they produce an accurate image of the dot. *For the image to be clear and in focus the light rays must meet exactly on the retina.* The bending of light rays in this way is called **refraction**, and it plays a very important part in giving us good eyesight.

Seeing things the right way up

Suppose you are looking at a person. Figure 2 shows how light rays reflected from the person pass through your eye. Notice that the rays coming from the head cross those that come from the feet. The result is that the image is upside-down on the retina.

Why, then, don't we see everything upside down? The answer is that the brain comes to the rescue and turns the picture the right way up for us. An experiment has been done in which a man was given a special pair of spectacles which made him see everything upside down. After a while his brain made the necessary correction and he began to see things the right way up again.

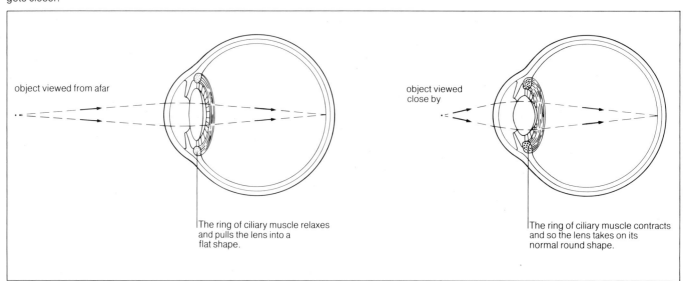

object viewed from afar

The ring of ciliary muscle relaxes and pulls the lens into a flat shape.

object viewed close by

The ring of ciliary muscle contracts and so the lens takes on its normal round shape.

Seeing things close at hand

Suppose a distant object comes much closer. If the eye did not adjust in some way, the light rays would meet *behind* the retina and the dot would be out of focus. However, the eye does adjust: the lens becomes rounder. This bends the light rays more, so that they meet on the retina as before (Figure 3).

This adjustment of the eye for looking at near objects is called **accommodation** and it is made possible by the fact that the lens is soft and can change its shape. The change of shape of the lens is brought about by the ring of ciliary muscles. When the muscle relaxes, the ciliary ring springs outwards and pulls the lens into a flattened shape. When the muscle contracts, the ciliary ring moves inwards, releasing the tension on the lens which consequently becomes rounder.

In old people the lens sometimes hardens, so accommodation becomes difficult and they find it hard to see things close at hand. This is why grandma holds her book a long way away when she's reading. However, the defect can be remedied by wearing glasses.

Much more serious is when the lens becomes opaque and won't let light through at all. This is known as a **cataract**. The only remedy is to take out the lens in an operation and replace it with an artificial one.

Defects of the eye

1 Short-sighted people
A short-sighted person can focus on things close to but not a long way off. This is due to the eyeball being too long, or the lens too strong, with the result that the light rays meet in front of the retina.

Short-sightedness is corrected by wearing glasses which bend the light rays outwards before they reach the eye (Figure 4A).

2 Long-sighted people
A long-sighted person can focus on things a long way off but not close to.

This is due to the eyeball being too short, or the lens too weak, with the result that the light rays meet behind the retina. This condition is corrected by wearing glasses which bend the light rays inwards before they get to the eye (Figure 4B). The same kind of glasses are worn by old people whose lenses have hardened.

Some people have a defect of the eye called an **astigmatism**. This is caused by the cornea and/or lens being unevenly curved, so the light rays meet on the retina in one plane but not in another. This, too, can be corrected by wearing glasses.

Various tests can be carried out to find out how good your eyesight is (Investigation 1). They are normally performed by an optician.

Seeing in depth

If you look at a solid object such as a book through only one eye, it looks flat. However, if you look at it through both eyes, it appears to have depth. In other words you see it in **three dimensions**.

So seeing things in depth depends on using both eyes. Each eye sees a slightly different aspect of the same object. In our brain the two images are combined to give us a single three dimensional view of the object.

As well as making the world look more interesting, this helps us to judge distances. For example, if you are driving along a road and there are two cars in front of you, you know roughly how far apart they are.

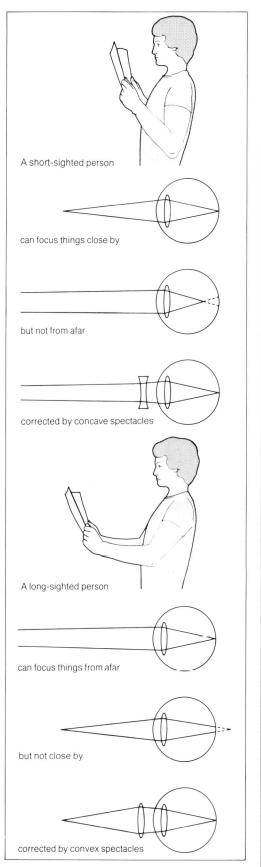

A short-sighted person

can focus things close by

but not from afar

corrected by concave spectacles

A long-sighted person

can focus things from afar

but not close by

corrected by convex spectacles

Figure 4 Short sight and long sight and how they can be corrected by wearing the right kind of glasses.

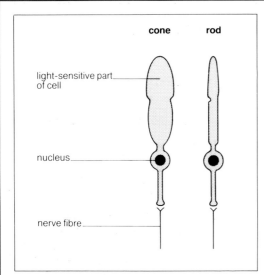

Figure 5 The two kinds of sensory cell found in the retina. The cones are used when it is light, the rods when it is dark or gloomy.

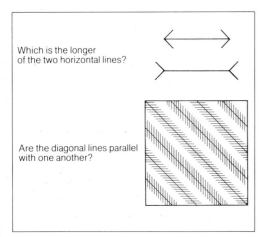

Which is the longer of the two horizontal lines?

Are the diagonal lines parallel with one another?

Figure 6 Is seeing always believing? Here are two well known optical illusions.

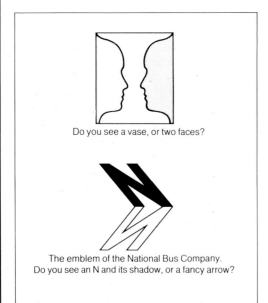

Do you see a vase, or two faces?

The emblem of the National Bus Company. Do you see an N and its shadow, or a fancy arrow?

Figure 7 What do you see?

What does the retina do?

If you look straight at an object, and then look at it out of the corner of your eye, you will find that its appearance changes. From being clear and sharp it becomes indistinct and fuzzy. Also it is difficult to tell what colour it is.

How can we explain this difference? The retina is divided into two parts. When we look straight at something, we are using the central part immediately behind the lens: the **yellow spot** or **fovea**. This contains sensory cells called **cones** which detect things clearly and in colour. When, however, we look at something out of the corner of the eye, we are using the part of the retina further out towards the edge of the eye. This contains sensory cells of a different kind, called **rods**. They detect things less clearly and in black and white (Figure 5).

Seeing in the dark

From what's just been said you might have got the idea that the edge of the retina isn't much use. However, it is good at seeing things in dim light (Investigation 2). You can prove this by looking at a faint star: it's much easier to see it out of the corner of your eye than by looking straight at it.

The reason for this is that the rods work better than the cones in the dark. They contain a substance called **visual purple** which is broken down by very small amounts of light, This causes them to send impulses to the brain. No sooner has the visual purple been broken down than it's re-made and is ready to be used again. This happens very quickly and is the reason why a flickering light can sometimes be detected out of the corner of the eye.

Visual purple is made from vitamin A which is particularly abundant in carrots. A person who hasn't been getting enough vitamin A in his food has difficulty seeing in the dark, a condition called **night blindness**.

Have you noticed that when you go into a gloomy room from bright sunlight you can't see anything at first but gradually things become visible? This is known as **dark adaptation**. The explanation of it is that all the visual purple has been broken down by the sunlight and it takes a while for the rods to re-make it and start working.

How do we see colours?

The ability of the eye to detect colours is due to the cones in the central part of the retina. The colour of an object is determined by the wavelength of light reflected from it. Scientists believe there are different kinds of cones for detecting different wavelengths and therefore different colours.

Not all animals can see colours. Dogs for example see everything in black and white. Certain people are completely colour-blind and cannot make out colours at all. Others cannot tell the difference between red and green, and this can cause difficulty when looking at warning signs. Various tests can be carried out to find out if a person is colour-blind (Investigation 3).

Is seeing believing?

What we see depends not only on our eyes, but also on our brain. We have already seen that the brain turns images the right way up. What else does it do?

Briefly it does two things: it fills in gaps, and sometimes it distorts things. The way the brain fills in gaps is dealt with on page 244. The way it distorts things is shown by **optical illusions** (Figure 6). The image registered by the eye is accurate, but the brain plays a trick on us and makes the image misleading.

To some extent this is because we are used to seeing certain patterns, and are baffled by anything unusual. In other words we tend to see what we *want* to see, or are used to seeing. Figure 7 illustrates this nicely. Can you think of any other examples of this?

Investigation 1

How good is your eyesight?

1 Hang a card on the wall on which there are two parallel lines one millimetre apart.

2 Gradually back away from the card until the two parallel lines appear as one, then stop.

 How far are you from the card?

 Compare your distance with that of other people in the class.

3 Hang an eyesight test card on the wall.

4 Stand facing the card six metres away.

 How many lines can you read?

 At a distance of six metres, a person with normal eyesight should be able to read as far down as line 6.

5 Repeat the eyesight test on each eye separately.

 Can you see better out of one eye than the other?

How does your eyesight compare with others in your class?

Investigation 2

Seeing in the dark

Do this experiment in a dimly lit room.

1 Stare into a bright light for five minutes: a bench lamp will do.

2 Turn the light off.

3 Look straight ahead. Your teacher has placed a certain object at the front of the room.

 Write down the name of the object as soon as you can tell what it is.

 Tell your teacher when you have done this.

 How long did it take you to recognise the object?

 Compare your time with that of others in your class.

 What was happening in your eye:

 a) while you were staring at the bright light,
 b) while you were in darkness afterwards?

Investigation 3

Seeing colours

1 Obtain two cards, one red and the other green.

2 Look at the two cards out of the corner of your eye.

 Can you tell which colour is which?

 How would you explain your observation?

3 Obtain a set of colour-blindness test cards. On each card there are numerous coloured dots. People with normal colour vision can make out certain numbers, whereas colour-blind people can't.

4 Test your eyes with the cards, following the instructions carefully.

 Can you see colours normally or are you colour-blind?

 If you are colour-blind, are you totally colour-blind or are you colour-blind only to red and green?

 How many students in your class, if any, are (a) totally colour-blind, and (b) red-green colour-blind?

People who are red-green colour-blind say that they have no difficulty telling whether the traffic lights are red or green. How would you explain this?

Assignments

1 The picture below shows a person wearing 'half-moon' spectacles. What sort of eye defect do you think he has, and why are these particular spectacles useful to him?

2 What are the advantages of having two eyes rather than only one?

3 Explain the reason for each of the following:

 a) When you go into a cinema from bright sunshine, you cannot see the seats at first, but gradually they become visible.
 b) If you are trying to see a faint star in the night sky, it is better to look slightly to one side of it rather than straight at it.
 c) When it is getting dark at night, it is impossible to make out the colours of cars on the road.
 d) If you look at a cinema screen out of the corner of your eye, you can see it flickering.

 e) If both your eyes are open and you press the side of one of your eyeballs, you see double.

4 Nocturnal animals, i.e. animals which sleep during the day and come out at night, tend to have wide pupils and lots of rods in their retinas. Suggest a reason for this.

5 With a piece of straight-edged paper cover the top half of the following phrase:

HAPPY BIRTHDAY

Can you read it?

Now cover the bottom half of the phrase. Can you read it now? Explain the difference.

The ear and hearing

The ear does two jobs: it enables us to hear and it also helps us to keep our balance. In this Topic we will look at the structure of the ear and the way it works in hearing.

The structure of the ear

People tend to think of the ear as just a flap on the side of the head, but there's much more to it than that as you can see in Figure 1. The flap is simply a device for catching sounds and directing them into the hole just in front. The flap itself is called the **pinna** and it contains gristle to keep it stiff.

The hole leads into a short tube called the **external ear channel**. The skin lining the first part of the channel is hairy and secretes wax which catches germs and dust, preventing them from getting into the ear.

Stretched across the inner end of the channel is a tough membrane, the **eardrum**. On the other side of the eardrum is a chamber called the middle ear. This contains three tiny bones called the **ear ossicles**: because of their shapes they are called the **hammer** (malleus), **anvil** (incus) and **stirrup** (stapes). They run from the inner side of the eardrum to a membrane covering a small hole on the other side of the middle ear chamber. This is called the **oval window**, and it leads to the inner ear.

The inner ear consists of a series of chambers and canals filled with fluid. It is made up of two parts which, though connected, do quite different jobs. The two parts are:

1 The hearing apparatus

This consists of a tube called the **cochlea** which is coiled like a snail's shell. Inside the cochlea there are sensory cells which are connected to the brain by the auditory nerve.

Figure 1 The structure of the human ear, slightly simplified.

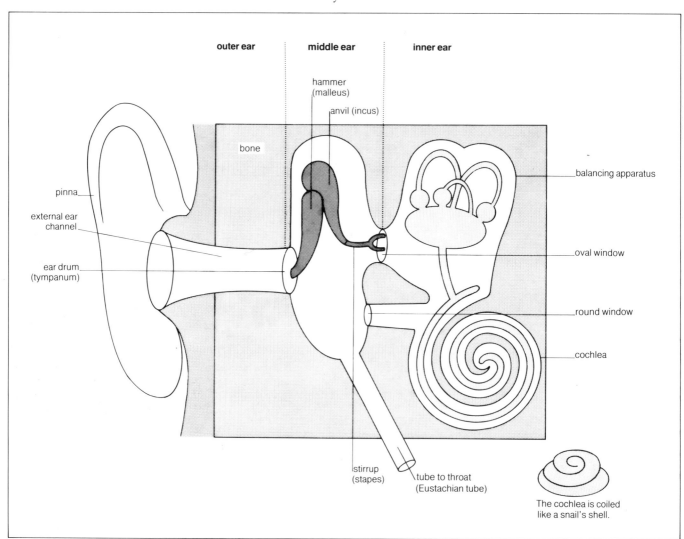

The cochlea is coiled like a snail's shell.

2 *The balancing apparatus*
This is dealt with in the next Topic.

The whole ear is embedded in the temporal bone which forms the side of the skull just above the pinna.

Why do our ears go pop?

Suppose the pressure in the middle ear chamber was to suddenly increase. This would make the eardrum bulge outwards. Not only would this be painful, but it might even burst the eardrum. So it's important that the pressure in the middle ear should be more or less the same as the atmospheric pressure outside the eardrum (Figure 2). The structure that makes this possible is the **Eustachian tube**. This connects the middle ear with the throat and it allows air to get in and out of the middle ear.

If you go up in an aeroplane, the atmospheric pressure outside the eardrum falls. As a result, the eardrum bulges outwards. You can rectify this by, for example, yawning or swallowing: this opens the Eustachian tube and so equalises the pressure on the two sides of the eardrum. The result is that the eardrum springs back into its normal position, making your ears pop.

The reverse happens when the aeroplane comes down to land, or if you go down a deep mine in a lift. In this case the atmospheric pressure increases causing your eardrum to bulge inwards. You can rectify this by swallowing or yawning as before.

The Eustachian tube is, therefore, an important part of the ear. The trouble is that if you have a heavy cold germs may get up it into the middle ear. The Eustachian tube may then become blocked, and pressure may build up in the middle ear causing earache.

How does the ear hear?

Suppose you are walking down the street and there's a loud bang. The noise sets off vibrations or sound waves which travel through the air in all directions, rather like ripples in a pond when you throw a stone in it. Within a fraction of a second the sound waves reach your ear, and the pinna directs them into the external ear channel.

Being rather small and pressed back against the side of the head, the human pinna is not much good at catching sound waves. However, the much larger pinna of an alsatian dog or a fox is more efficient, particularly if it's pricked up. Of course the human pinna can be improved by putting your hand behind it.

The sound waves now pass along the external ear channel to the eardrum. When they hit the eardrum they make it vibrate. This in turn moves the ear ossicles backwards and forwards, causing the membrane covering the oval window to vibrate. The vibrations of this membrane then move the fluid in the cochlea. This stimulates the sensory cells which send off messages in the auditory nerve to the brain.

Getting used to sounds

If you are subjected to a continuous or repetitive noise, you soon get used to it – provided of course that it isn't *too* loud. In fact after a time you may stop hearing it altogether. We call this **adaptation**.

There are two possible explanations of adaptation. One is that after a time the sensory cells in the cochlea stop sending messages to the brain, even though they are still being stimulated. Another is that the messages go on being sent, but the brain takes no notice of them. Either way after a time the sound is no longer heard. This means that we do not constantly respond to background noises that do not really matter.

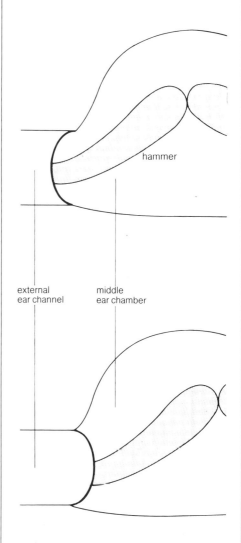

If the air pressure in the middle ear is greater than the pressure outside, the ear drum bulges outwards.

hammer

external ear channel

middle ear chamber

If the air pressure in the middle ear is less than the pressure outside, the ear drum bulges inwards.

Figure 2 These diagrams show what happens if the air pressure is not the same on the two sides of the ear drum.

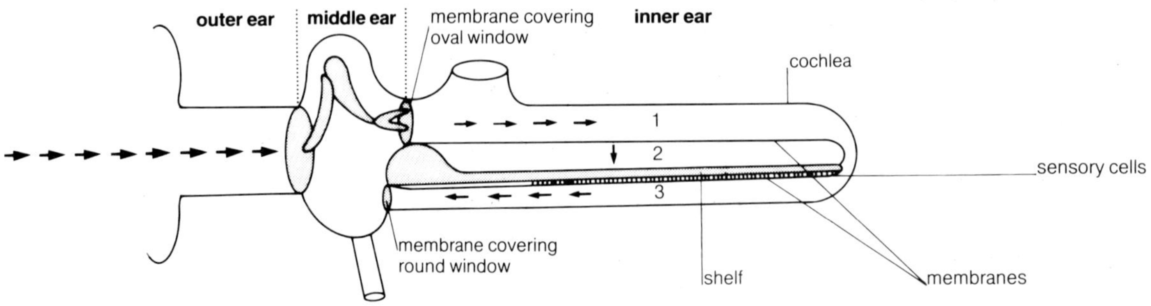

Figure 3 In this diagram the arrows show the passage of a sound wave through the ear. The cochlea is shown straightened out and its three canals are numbered 1, 2 and 3.

A closer look at the cochlea

Look at Figure 3. You will see that the cochlea is divided lengthways into three canals. These are numbered 1, 2 and 3 in the diagram. They are separated from each other by delicate membranes which can move up and down. When the membrane covering the oval window is pushed inwards by the stirrup, the fluid in the first canal is moved. This pushes the first membrane downwards, which moves the fluid in the second canal. This, in turn, pushes the second membrane downwards, which moves the fluid in the third canal. This finally presses on the membrane covering the round window, causing it to bulge outwards into the middle ear chamber.

How do these movements cause messages to be sent off in the auditory nerve? Figure 3 shows the sensory cells in the cochlea. You will see that they are attached to the membrane which separates the second and third canals. When this membrane moves up and down, it distorts the sensory cells, and as a result they send off messages in the nerve.

Loud and soft sounds

If you play a note on a piano, its loudness depends on how hard you hit the key. This in turn determines the distance through which the wire moves up and down when it vibrates: this is called the **amplitude**.

The loudness of a sound is registered by the ear in the same way. Soft sounds vibrate the eardrum, and hence the cochlea membrane, only slightly, whereas loud sounds cause much greater vibrations.

There are many different ways of testing one's ears to find out how sensitive they are. Investigation 1 is a simple method.

High and low notes

With a piano you make different notes by hitting different keys. The actual note, or pitch, depends on how rapidly the wire vibrates: this is called the **frequency**. Wires that vibrate at high frequency give high notes, whereas those that vibrate at low frequency give low notes.

Although the details are different, the ear works in the same kind of way. It has been found that the cochlea membrane is made of thousands of parallel fibres which run across it. The fibres towards the base of the cochlea are short and stiff: they vibrate very rapidly and are sensitive to high notes. In contrast, the fibres towards the apex are long and more flexible: they vibrate more slowly and are sensitive to low notes.

What causes deafness?

There are several different types of deafness, depending on which part of the ear is affected.

People sometimes become temporarily deaf because they produce too much hard wax, which consequently blocks up the external ear channel. This is easily removed by the doctor syringing out the ears with warm water.

An explosion, or a blow on the side of the head, may rupture the eardrum, causing deafness. However, the eardrum usually heals quite quickly and then the person gets his hearing back.

A much more serious type of deafness is caused by connective tissue growing into the middle ear chamber. This prevents the ear ossicles moving, in much the same way as a piston may become seized up with rust. If nothing is done about it, the person may become permanently deaf. However, the person's hearing can sometimes be improved by wearing a hearing aid which amplifies the sound waves, and in severe cases an operation may prove helpful.

There are other causes of deafness. For example, it may be caused by damage to the cochlea. If a person is subjected to a repeated loud sound of a particular pitch, the sensory cells may become damaged, making him deaf to that particular note. It's said that some pop singers have become deaf to certain notes because of this.

How can we tell where a sound comes from?

Normally when you hear a sound, you know where it comes from. This is because we have two ears, one on each side of the head. Suppose you hear a noise from the right. Sound waves reach the right ear a fraction of a second before they reach the left ear (Figure 4). So impulses are sent to the brain from the right ear slightly before they are sent from the left ear. From this the brain knows that the sound must have come from the right.

Now suppose you can hear a faint buzz and you want to find where it's coming from. With two ears you can compare the loudness of the sound on each side of the head, and this will guide you towards the source of the sound (Investigation 2).

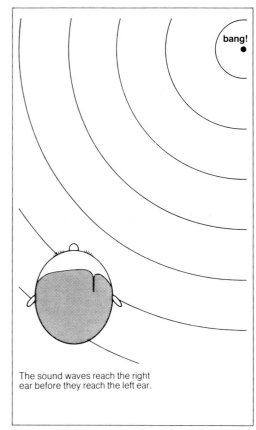

The sound waves reach the right ear before they reach the left ear.

Figure 4 Where did that bang come from?

Investigation 1

How well can you hear?

This experiment must be done in a quiet room.
Work in pairs, one person acting as subject.

1 The subject should sit down and plug one ear with cotton wool.

2 Hold a watch close to the subject's ear and gradually move it away until he can no longer hear it.

At what distance does the subject cease to hear the watch?

3 Hold the watch out of hearing range, then gradually move it towards the subject's ear until he can just hear it.

At what distance does the subject start to hear the watch?

Are the two distances which you have measured the same?
If not, which one is the furthest?
How would you explain the difference?

Investigation 2

Finding an object by sound

This experiment is best done as a class with one person acting as subject.

1 Blindfold the subject outside the room.

2 Place a ticking clock somewhere in the room.

3 Bring the subject in and ask him to find the clock.
(Someone should stand close to the subject to prevent him bumping into the furniture.)

4 Watch the subject's head as he goes about this task.

In what way do you think the movements of his head help him to find the clock?

What do you think he would do if one of his ears was plugged with cotton wool? Find out by repeating the experiment with one of the subject's ears blocked.

Assignments

1 If a person ruptures his eardrum he finds it difficult to hear until it has healed. Why is this?

2 Why do your ears go pop when you go up in an aeroplane?

3 What jobs are done by each of the following:
(a) the pinna, (b) the ear ossicles, (c) the round window, (d) the Eustachian tube?

4 People who are subjected day after day to a very loud noise of a particular pitch, may eventually become permanently deaf to all sounds of that pitch.

a) What is meant by the word pitch?
b) What do you think might cause this kind of deafness?

5 It is claimed that having two ears enables us to tell where a sound comes from. Devise an experiment to find out if two ears are really needed for this.

How do we keep our balance?

Figure 1 shows an ice-skater in action. How does she manage to stay upright and stop herself falling over?

Figure 1 An ice-skater in action.

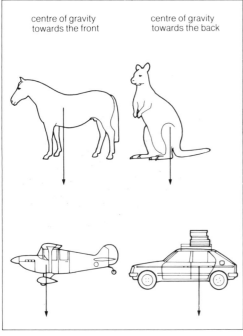

Figure 2 The centre of gravity is the point through which the weight (mass) of a body acts. It is indicated by the arrows in these diagrams.

Centre of gravity

Suppose you balance an empty tray on your finger, so that it is horizontal, like this:

Your finger marks the point where the weight or mass of the tray is concentrated. This point is called the **centre of gravity**. *The centre of gravity is the point through which the mass of a body acts.*

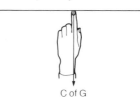

The mass of an empty tray is spread out evenly all over, so the centre of gravity is in the centre. But suppose there's a cup at one end. This will make the tray heavier at that end, so the centre of gravity will be shifted in that direction.

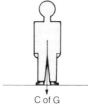

So the position of the centre of gravity of an object depends on how the mass is distributed within it. Figure 2 shows the position of the centre of gravity in a few well-known objects.

Stability

Imagine a model man standing in an upright position. The centre of gravity is immediately above the part of his body on which the model stands, i.e. the feet:

With the centre of gravity in this position, the model stays upright and doesn't topple over: we say that it is **stable**.

Now suppose we tilt the model slightly to one side, like this:

On letting go, the model returns to its original position: in other words it remains stable.

An object is said to be stable if it returns to its original position when displaced.

But suppose we tilt the model a bit more, like this:

C of G

This time, instead of returning to its original position when we let go, it becomes **unstable** and falls over.

Why is the model stable when we tilt it slightly, but unstable when we tilt it more? The answer lies with the centre of gravity. Look again at the previous two diagrams. With a slight tilt, the centre of gravity still falls within the area of the body on which it was standing. But with a larger tilt, the centre of gravity falls outside this area. *When displaced, a body becomes unstable and is liable to fall over if its centre of gravity falls outside the area on which it was standing.*

It is different with people

Now suppose our model is a real man and he leans over as in the last diagram. His centre of gravity now falls outside the area of his body on which he stands. However, in contrast to the model, he can stop himself falling over. This is because various muscles, particularly those in the legs, tighten up, and the body moves in such a way as to bring the centre of gravity back into its original position. So the body, having become unstable, becomes stable again.

This comes into many things we do. Even in a simple action like walking, the body constantly becomes unstable for a moment, only to regain stability immediately afterwards.

What makes the ice-skater in Figure 1 so skilful? It's because she can let her body become highly unstable, for a long period, and then bring it back into a stable position.

How do we keep stable?

Stability is maintained by a number of reflexes which are set off by the stimulation of certain sense organs. Let's look at these in turn.

1 *Eyes*
The eyes are more important in balance than you may think (Investigation 1). By looking at fixed objects such as the skyline and the sides of buildings, you become aware of the horizontal and vertical planes. This helps you to keep your body in the right position, and it explains why it's difficult to keep your balance in the dark.

2 *Pressure receptors*
If you're standing to attention and you lean forward, you can feel the extra pressure on the front of your feet. This makes you lean back again, so you don't fall forward. The feeling comes from receptors in the skin which are sensitive to pressure and they pass messages to the brain (Investigation 3).

3 *Stretch receptors*
When you lean forward you can also feel tension in the muscles at the back of your leg. All our muscles have special receptors inside them which are stimulated by being stretched. If a muscle is stretched, as often happens when the body becomes unstable, the body responds in such a way that the stretching is relieved.

4 *Ears*
Our ears contain a special balancing apparatus which is shown in Figure 3. It consists of two main parts: the **semicircular canals** and the **ear sack**.

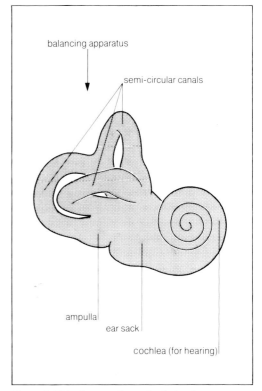

balancing apparatus

semi-circular canals

ampulla

ear sack

cochlea (for hearing)

Figure 3 This diagram shows the balancing apparatus in the ear.

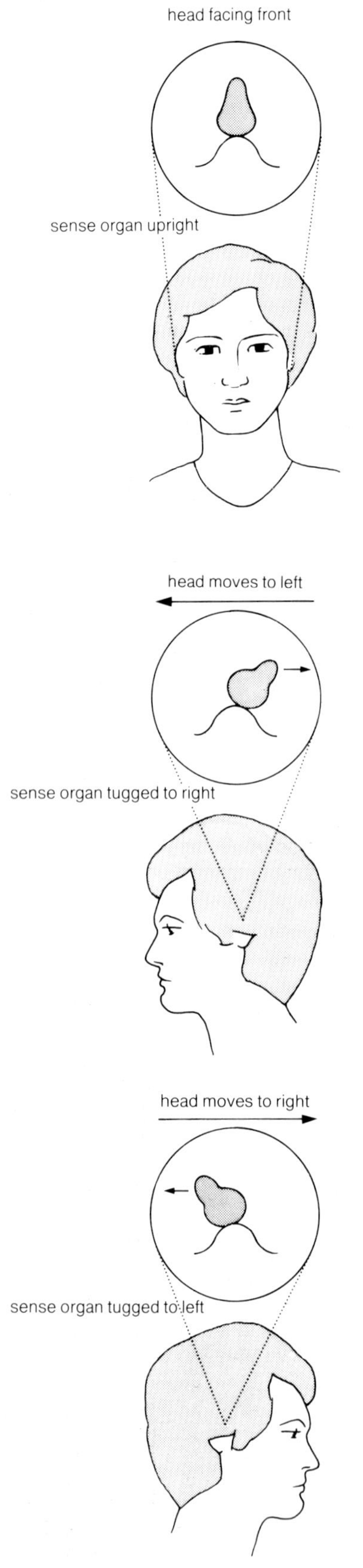

head facing front

sense organ upright

head moves to left

sense organ tugged to right

head moves to right

sense organ tugged to left

Figure 4 These diagrams show what happens to our semicircular canal organs when we turn our head suddenly.

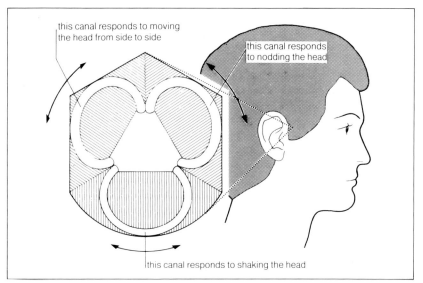

this canal responds to moving the head from side to side

this canal responds to nodding the head

this canal responds to shaking the head

Figure 5 The three semicircular canals are at right angles to each other, so movement of the head in any plane can be detected by the sense organs.

The semicircular canals

If you nod or shake your head you can feel your head moving. What's more you know if it's moving quickly or slowly, or if it changes speed. The organs which tell you this are the semicircular canals.

The semicircular canals are filled with fluid, and each one has a little swelling at one end called an **ampulla**. The ampulla contains a sense organ which sticks into it like a little finger. If you suddenly move your head in the same plane as the canal, the fluid tugs on the sense organ (Figure 4). As a result it sends messages to the brain, telling it that the head has moved.

We have three semicircular canals in each ear, and they are situated at right angles to each other. Each one is sensitive to movement in a different plane: one of them responds when you shake your head, another when you nod, and the third when you move it from side to side (Figure 5).

If you are spun round at a constant speed, as, for example, when you go on a roundabout, the fluid in the semicircular canal stays still relative to your head (Investigation 2). However, when you stop, it goes on swirling round and round for a while. This stimulates the receptors and makes you feel dizzy.

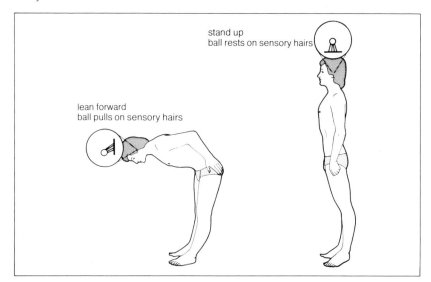

stand up
ball rests on sensory hairs

lean forward
ball pulls on sensory hairs

Figure 6 These diagrams show what happens to the organs in the ear sack when we change the position of our head.

The ear sack

Even when your eyes are closed, you know the position of your head, i.e. whether it's vertical or horizontal. We get this information from the ear sack just beneath the semicircular canals (see Figure 3).

The ear sack is filled with fluid, and it contains a tiny ball of chalk which is attached to a group of sensory cells. If your head is upright, the ball sits neatly on top of the sensory cells. However, if your head is bent forward, the ball pulls on the sensory cells causing them to send messages to the brain (Figure 6).

Investigation 1

The importance of the eyes in balance

1 Stand up with your eyes open.

2 Raise one leg off the floor.

Do you find it easy to stand on one foot?

3 Now repeat the above with your eyes closed.

Do you find it harder or easier to stand on one foot now?

4 Stand up with your feet together and your arms at your sides. Look straight ahead. Note the extent to which you sway from side to side.

5 Now close your eyes and continue to stand as before.

Are your swaying movements greater or less than before?

Why do you think the swaying movements occur?

What part is played by the eyes in balance?

6 Sit down and close your eyes.

7 Place the heel of one foot on the toes of the other.

Is it easier to do this when you're looking at your feet?

8 Put your finger on the end of your nose.

Can you do this more accurately with your eyes open?

What do you think these experiments tell us about balance?

Investigation 2

To see how the semicircular canals work

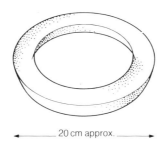

◄——— 20 cm approx. ———►

1 Obtain a circular trough of the kind shown in the illustration: a circular cooking mould does nicely.

2 Half fill the trough with water and place it on the turntable of a record player.

3 Float a match on the water, and wait for it to be still: the match will help you to see which way the water moves.

4 Rotate the turntable quickly through a quarter turn.
What happens to the water?

5 Spin the turntable slowly at a constant speed.

What happens to the water (a) when you start spinning, and (b) once the spinning has got underway?

6 Stop the turntable suddenly.

What happens to the water after the turntable has stopped?

What conclusions can you draw about the way the receptors in the semicircular canals are stimulated when the head moves?

Investigation 3

The part played by pressure and stretch receptors in balance

1 Stand up with your feet together. Whereabouts do you feel pressure on your feet?

2 Lean forward as far as you can. Where do you feel most of the pressure now?

3 Now lean backwards. Where do you feel the pressure now?

4 Lean forward again. Which muscles feel tense? What movements stop you falling over?

What does this experiment tell us about balance?

Assignments

1 When you are sitting in a train, which sense organs tell you that you are moving? Explain your answer fully.

2 Why do we need *three* semicircular canals in each ear, rather than only one?

3 When you wake up in the morning you know where your leg is in relation to the rest of your body, i.e. whether it is straight or bent or in a particular position. What sort of receptors give us this information, and what part do they play in helping us to keep our balance?

4 Why do you feel dizzy after you have been on a roundabout in a funfair?

5 Describe an experiment which could be done to find out how important the eyes are in enabling human beings to walk straight.

Introducing the skeleton

Man, like many other animals, possesses a skeleton which forms a framework inside the body. The next three Topics are about the skeleton and its functions.

Figure 1 The main parts of the human skeleton.

The parts of the skeleton

You can discover a lot about the human skeleton just by looking at it (Investigation 1). It is divided into two main parts (Figure 1).

1 Axial skeleton

Structures which lie in the centre of the body, namely, the skull, backbone (vertebral column) and rib cage. The ribs run from the backbone to the breastbone (sternum).

2 Appendicular skeleton

Structures which lie on either side of the body, namely, the limb girdles (shoulders and hips) and the limbs (arms and legs). The limbs are attached to the girdles, the arms to the shoulder girdle and the legs to the hip girdle.

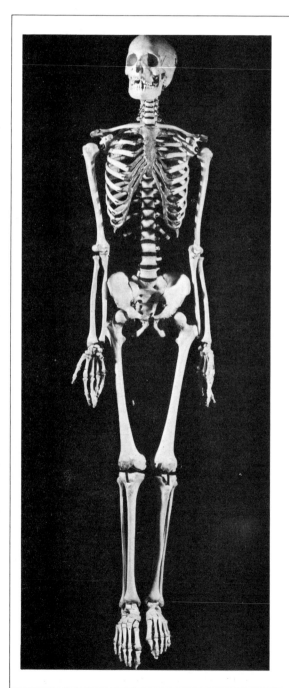

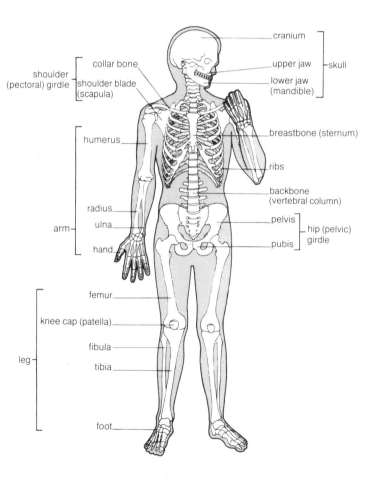

The main part of the shoulder girdle is the shoulder blade (scapula). We have two shoulder blades, one on either side of the rib cage and they are completely separate from each other. In contrast the two sides of the hip girdle are fused together.

The various bones that make up the skeleton are connected in such a way that they can move, or **articulate**, with one another. The places where this happens are called **joints**. The bones are held together by **ligaments**: these are tough elastic strands which run from one bone to another across the joints. The most mobile parts of the skeleton are of course the arms and legs (limbs) and in particular the hands and feet (Figure 2).

What does the skeleton do?

The skeleton does four main jobs:

1 **It supports the body**, giving it shape and form. Without it, the whole body would collapse.
2 **It protects the soft organs**. Thus the cranium protects the brain, and the vertebrae protect the spinal cord (Figure 3). The ribs and breastbone protect the lungs and heart, and the pelvis shields the reproductive organs.
3 **It makes blood cells**. Red blood cells, and certain kinds of white blood cells, are made *inside* certain bones.
4 **It brings about movement**. In doing this the skeleton works with muscles which are attached to it.

What is the skeleton made of?

The skeleton is made mainly of **bone**, which is hard because it contains minerals. The main mineral is calcium, and this is why growing children need plenty of this element in their food. If you take the calcium out of a bone by treating it with an acid, it becomes soft like rubber (Investigation 2).

Although bone looks dead, it is really very much alive. It is a living tissue containing blood vessels and nerves, and special bone cells which make new bone and repair it when damaged.

Between the bones there is a softer material called **cartilage** (gristle). This acts like a shock absorber, preventing the bones from jarring when we move around. In this respect the cartilage discs between the vertebrae are especially important, and they also help to make the backbone flexible.

The skeletons of different vertebrates vary in the amount of cartilage they contain. In sharks and their relatives the entire skeleton is made of cartilage.

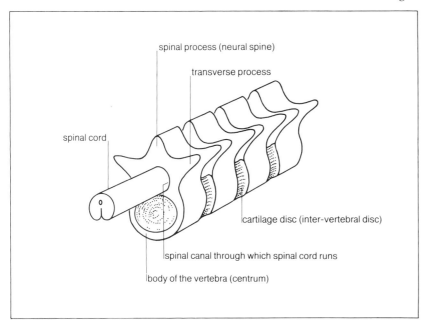

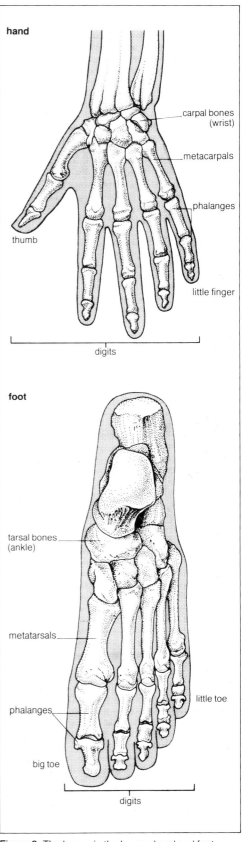

Figure 2 The bones in the human hand and foot.

Figure 3 The backbone helps to support the body and it also protects the spinal cord.

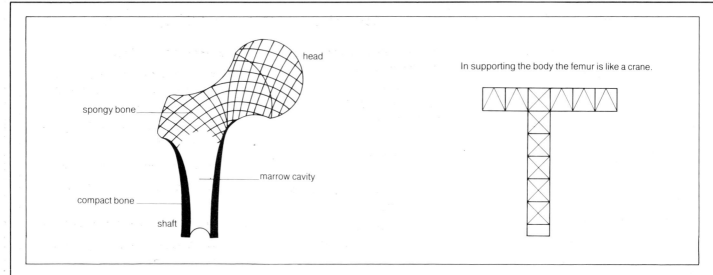

In supporting the body the femur is like a crane.

Figure 4 The structure of a typical bone can be seen in this diagram of the femur. The femur has to bear the mass of the body and so it must be strong. This is achieved mainly by the spongy bone: the fibres form a frame like the metal lattice in a crane.

Inside a bone

If you saw a bone right down the middle you can see its inside (Investigation 3). Figure 4 shows the structure of a bone such as the femur. The outer part consists of dense **compact bone**. Beneath this at the end of the bone there is a criss-cross network of bony fibres called **spongy bone**. In the centre there is a cavity filled with a soft substance called **marrow**. Yellow marrow consists mainly of fat. Red marrow, which is found in certain bones such as the pelvis and ribs, is where blood cells are made.

The skeleton of other vertebrates

Man is a peculiar mammal in that he walks on two legs: he is known as a **biped**. Most other mammals walk on all fours and are called **quadrupeds**. You might expect this to make a lot of difference to the skeleton. However, if you compare the human skeleton with that of a quadruped such as a rabbit, you find they are really very similar (Figure 5). The same bones occur in each, though their individual shapes are different. Other vertebrates have basically similar skeletons too, though there are individual variations. These can usually be related to the animal's way of life and how it moves (Investigation 4).

Figure 5 The skeleton of a rabbit.

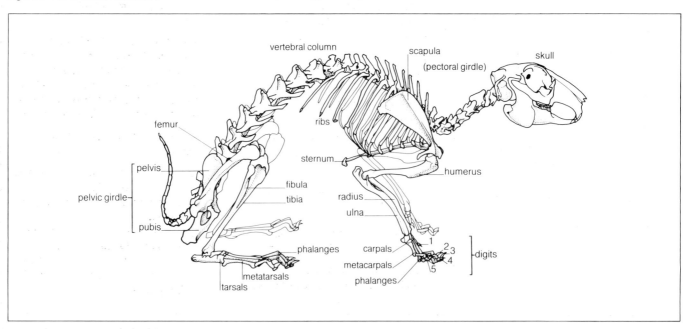

Investigation 1

Looking at the human skeleton

1 Examine a human skeleton.

Which structures shown in Figure 1 can you see?

Which structures:
a) hold the body up?
b) are involved in locomotion?
c) protect the soft organs?

2 Compare the shoulder and hip bones.

What do they have in common?

In what ways do they differ?

Can you explain the differences?

3 Examine the hand and foot in detail.

Which structures shown in Figure 2 can you see?

In what ways are they suited to their jobs?

4 Examine a vertebra in detail.

Which structures shown in Figure 3 can you see?
What are the functions of the vertebrae?

Investigation 2

To find the effect of taking the calcium out of a bone

1 Your teacher will give you a bone which has had all the flesh and marrow removed.

2 Obtain a 3 per cent solution of hydrochloric acid in which some salt has been dissolved.

3 Put the bone in the acid, and leave it for several days.

4 After several days lift the bone out of the acid with forceps, and wash it in water.

5 Dry the bone with a cloth.

Can you bend the bone?
What effect has the acid had on it?

What is the function of calcium in our bodies?

Investigation 3

Looking inside a bone

1 Your teacher will give you a fresh bone which has been sawn in half down the middle.

Which structures shown in Figure 4 can you see?

What does the bone marrow feel like?

What do you think the marrow is made of?

How could you test your suggestion?

2 Now look at a dry bone which has been cut in half.

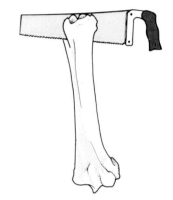

Where is the spongy bone and what job does it do?

Investigation 4

Looking at other skeletons

1 Look at the skeleton of a four-legged mammal such as a rabbit.

Write down five ways in which it differs from the human skeleton, apart from its size.
How would you explain the differences in terms of what the skeleton has to do?

2 Examine the skeletons of other vertebrates such as a bird, frog and fish.

In each case relate the structure of the skeleton to the kind of life which the animal leads.

Assignments

1 Each of the words in the left hand column is related to one of the words in the right hand column. Write them down in the correct pairs.

tarsals   hip
rib   wrist
pelvis   ankle
carpals   chest
femur   leg

2 Which of the structures in the left hand column in the previous question:

a) are important in locomotion,
b) help us to write,
c) protect the lungs,
d) are part of the axial skeleton,
e) play a part in raising the arm?

3 What is the common name for cartilage, mandible, patella, vertebral column, scapula?

4 Most dogs enjoy the marrow part of a bone. Why is the marrow good for them?

5 Someone has said that from the mechanical point of view the human backbone is like a skyscraper. However, someone else claims that it is more like the leaning tower of Pisa. Who do you think is right, and why?

6 Explain the reason for each of the following:

a) It is easy to slice through the skull of a shark with a knife.
b) Ligaments stretch when you pull them hard.
c) A bone which is treated with acid eventually becomes soft.
d) The head of a limb bone such as the femur contains a network of bony fibres.

How do we move?

One of the most important functions of the skeleton is to support the body and enable it to move. In doing this it works with the muscles.

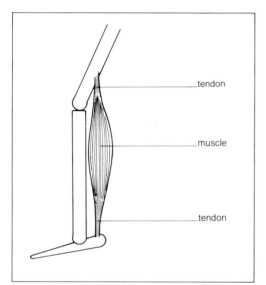

Figure 2 A muscle is attached to the bones of the skeleton by a tough tendon at each end.

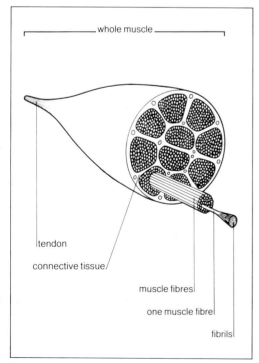

Figure 3 A whole muscle is made of muscle fibres, which in turn are composed of very fine strands called fibrils.

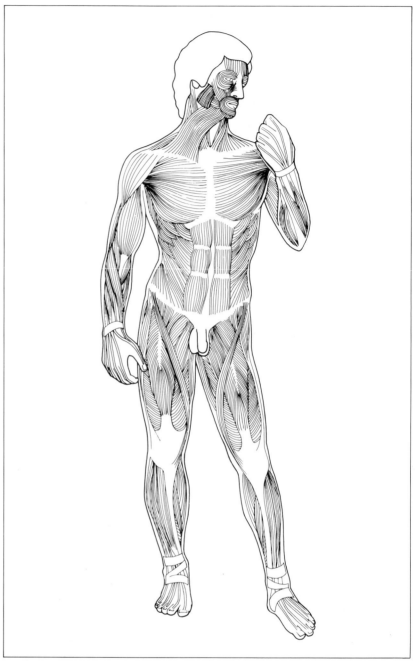

Figure 1 The main superficial muscles of a human.

Muscles and the skeleton

The entire skeleton is covered with muscles (Figure 1). This kind of skeleton in which the bones are situated *inside* the muscles is called an **endoskeleton** and it is characteristic of all vertebrates. If you buy a leg of lamb from the butcher, you can see that the meat (muscle) completely envelops the bone.

A structure like the arm or leg contains numerous muscles which move it in different directions (Investigation 1). Each muscle is attached to the skeleton at both ends by a **tendon** (Figure 2). The tendons are very tough and don't stretch much when they are pulled.

Each muscle is composed of hundreds of **muscle fibres** enclosed within a connective tissue envelope. With a pair of needles you can tease out the fibres (Investigation 2). Scientists have discovered that they are made up of even finer strands called **fibrils** (Figure 3).

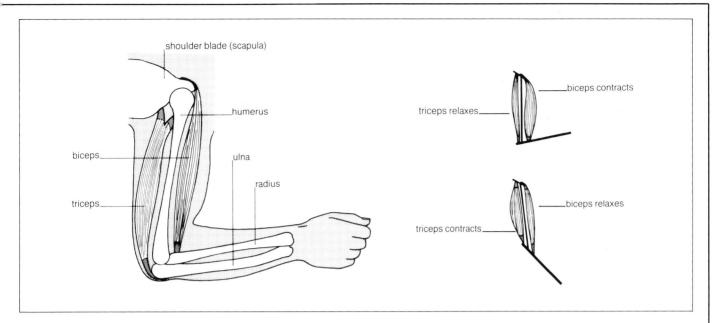

Figure 4 The biceps and triceps muscles move the arm at the elbow joint.

Every muscle has a nerve going to it. When it gets into the muscle, the nerve splits up into branches which supply the individual muscle fibres. The point where the nerve joins the muscle fibre is called the **nerve-muscle junction**.

When a message reaches the end of the nerve, it crosses the nerve-muscle junction and then spreads along the muscle fibres making them **contract**. If, for some reason, the nerve isn't working, or if the nerve-muscle junctions are blocked, the muscle cannot contract and becomes paralysed. Certain drugs block the nerve muscle junctions: an example is curare which South American natives used to put on their arrowheads to paralyse the animals they were hunting.

How do muscles move the skeleton?

To illustrate this, let's consider the arm. Two main muscles move the arm: the **biceps** bends it and the **triceps** straightens it (Figure 4). These two muscles produce opposite effects so they must not contract at the same time, otherwise the arm won't move at all. The nervous system ensures that this never happens. Each muscle has its own nerve supply, so that when messages are sent to the biceps, telling it to contract, they stop being sent to the triceps, and vice-versa.

Muscles such as the biceps which bend a limb are called **flexors**; those like the triceps which straighten it are called **extensors**. Of course we have flexors and extensors in our legs as well as our arms, and they play an important part in walking and running.

Joints

The structure of a joint is shown in Figure 5. It is enclosed within a tough **capsule**. Immediately beneath the capsule is a thin **synovial membrane** which secretes a fluid into the space inside. This **synovial fluid** serves as a lubricant enabling the two bones to slide smoothly against each other. It's like the oil between the moving parts of a machine.

The ends of the two bones are made of cartilage. Being comparatively soft, this prevents jarring when the two bones move against each other.

Joints are weak points in the skeleton, and it is important that they should be protected. The knees are particularly vulnerable because of their exposed position and complicated structure. To protect them, they are covered by a small bone called the **knee cap**, and in front of this there is another cavity filled with synovial fluid which serves as a cushioning device.

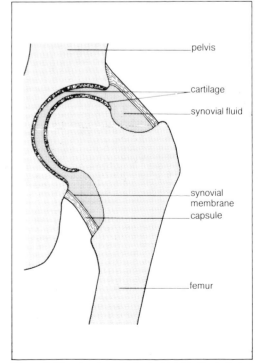

Figure 5 The structure of a typical joint is illustrated here by the hip joint between the femur and the pelvis.

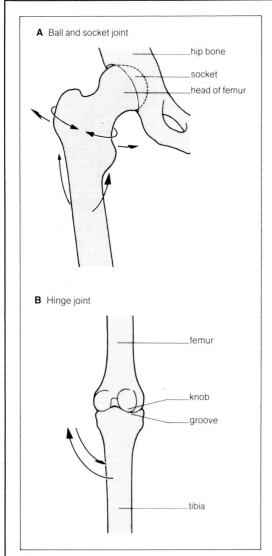

A Ball and socket joint

hip bone
socket
head of femur

B Hinge joint

femur
knob
groove
tibia

Figure 6 The two main kinds of joint found in the human body.

Figure 7 The three different kinds of lever found in the human body.

If you move your leg around, you will notice that at the hip you can move it in any direction, whereas at the knee you can only move it backwards and forwards. This difference is due to the kinds of joints which are found in these two places (Investigation 3).

The hip joint consists of a ball at the top of the femur which fits into a cup-like socket in the pelvis. This is called a **ball and socket joint**, and it allows movement in any plane (Figure 6A). However, the knee joint is constructed differently: it consists of two knobs at the bottom end of the femur which fit into two grooves at the top of the tibia. This is called a **hinge joint**, and it allows movement in only one plane (Figure 6B).

The arm works on the same principle as the leg. What kind of joint do you think we have at the shoulder, and the elbow?

Bones as levers

Suppose you are trying to force open the lid of a box with an iron bar like this:

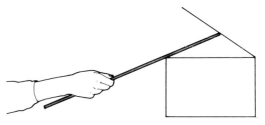

You are using the iron bar as a **lever**. *A lever is a bar which is turned about a fixed point.* The fixed point is called the **fulcrum**. In the lever illustrated above, there is a **load** (the lid) on one side of the fulcrum, and a force or **effort** is being applied by your hand on the other side:

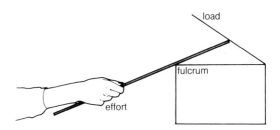

load
fulcrum
effort

Now some of our bones work as levers. There are three kinds of lever, which differ in the position of the fulcrum relative to the effort and load. All three are found in the human skeleton, and Figure 7 gives some examples.

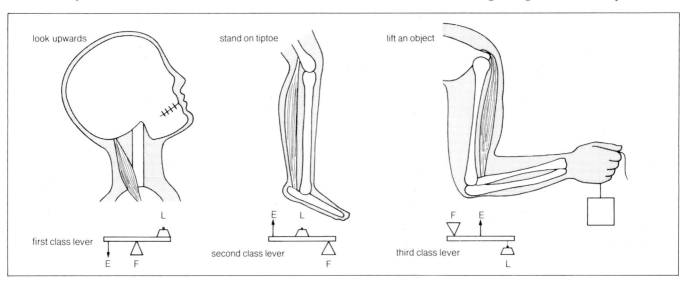

look upwards

first class lever

E   F

stand on tiptoe

E   L

second class lever

F

lift an object

F   E

third class lever

L

Investigation 1

Looking at muscles in relation to the skeleton

Your teacher will provide you with a pig's trotter which has been obtained from the butcher.

1 Remove as much of the skin as possible.

2 Pull the various muscles one by one.

 What movement does each one produce?

3 Cut away one of the muscles from its neighbours, and follow it down to the bone.

 Where do the muscles and bones lie in relation to each other?

4 Observe the tendon by which the muscle is attached to the bone.

5 Feel the tendon and test its strength by pulling it.

 How does it differ from the muscle?

6 How is the tendon joined to the muscle?

Meat that comes from the *end* of a muscle is often tough and gristly.

Why do you think this is?

Investigation 2

Finding out about the structure of a muscle

1 Your teacher will give you a small piece of muscle from the leg of a frog or rat.

2 Put the piece of muscle on a microscope slide.

3 With needles tease out the muscle fibres, and spread them out on the slide.

4 Add a drop of salt solution (0.75 per cent), and cover with a coverslip.

5 Observe under the microscope (low power).

 Can you see the individual muscle fibres?
 What do they look like?
 Can you see anything inside them?
 At a guess, how many fibres do you think the muscle possesses altogether?

Investigation 3

Looking at joints

1 Move your arm about at the shoulder and elbow.

 How much freedom of movement is there at each of these joints?

2 Move your leg about at the hip and knee.

 How much freedom of movement is there at each of these joints?

3 Look at examples of the above joints obtained from the butcher.

 Move the bones so as to see what kind of movement occurs at the joint.

 How much freedom of movement is there?

 How does the amount of freedom of movement fit in with the structure of the joint?

4 Examine the structure of the joint.

 Which structures in Figure 5 can you see?
 What enables the two bones to move smoothly against each other?

Assignments

What job does each of the following structures do:
 a) tendons,
 b) synovial fluid,
 c) nerve-muscle junction,
 d) spongy bone,
 e) inter-vertebral discs?

2 The following table gives the maximum speeds of four different animals in kilometres per hour:

cheetah 70
greyhound 64
racehorse 64
man 29

Suggest reasons why man has the slowest speed of the animals listed.

3 Why is it important that tendons should not stretch when a muscle contracts?

4 The diagram, right, shows some of the muscles, bones and nerves in the human arm.

 a) What will happen to the position of the forearm if muscle X contracts?
 b) What happens to muscle X when muscle Y contracts?
 c) What happens to muscle X when messages travel down nerve 1?
 d) When messages are travelling down nerve 1, what happens to messages in nerve 2?
 e) If the distance AB is 2 cm and AC is 30 cm, what effort must be exerted by muscle X to lift a bucket weighing 20 kg?
 f) If muscle X was attached to the forearm bone at point D, would it require more or less effort to raise the same load? Explain your answer.

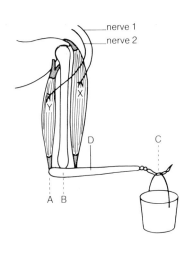

Aches, pains and broken bones

*In this Topic
we will look at some
of the things that can go wrong
with our skeleton
and muscles.*

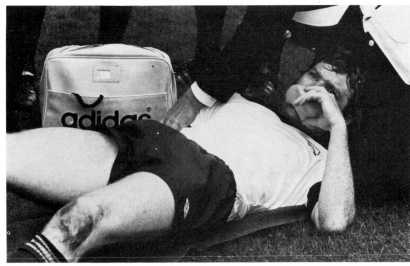

Figure 1 An injured footballer being taken off the pitch.

Broken bones

A broken bone is called a **fracture**. There are many different kinds of fractures, some more serious than others (Figure 2).

Suppose you break your arm. The first thing the hospital does is to take an X-ray. A machine sends a beam of X-rays through your arm: the rays pass through the skin and muscles, but not through the bones. Behind your arm is a photographic film which goes dark everywhere except where the bones are. So an X-ray photograph will show up the bones in the body, and any breaks can be seen clearly (Figure 3). In hospital X-ray pictures are taken by a specially trained person called a **radiographer**.

How does a broken bone mend?

A bone mends in three main stages (Figure 4):

1 When the bone is fractured, blood vessels are broken, so much bleeding occurs. The blood congeals around the fracture forming a clot. This may press on the tissues causing a lot of pain.
2 Bone cells multiply and move into the blood clot where they lay down new bone tissue. In this way the two separated parts of the bone become joined together again. In the mending process, a ring of new bone tissue is formed round the fracture, so the mended bone is slightly thicker in the region of the fracture – rather like the joint which a plumber makes when he connects two lengths of pipe.
3 The new bone is now re-modelled: any unwanted bits are broken down

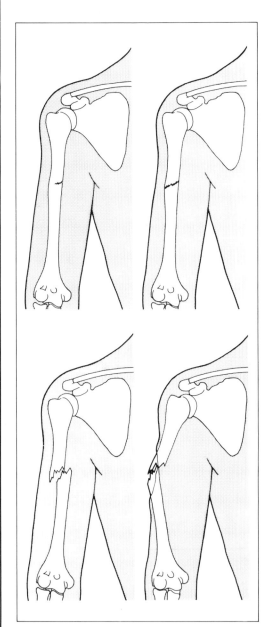

Figure 2 Four different ways that the humerus can be fractured.

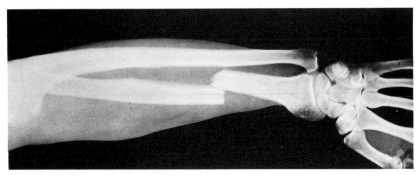

Figure 3 An X-ray of a fractured arm.

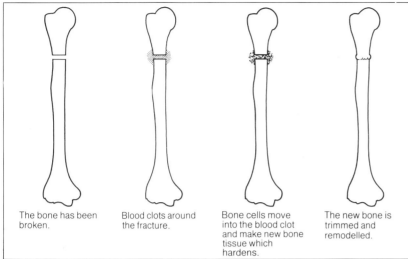

The bone has been broken.

Blood clots around the fracture.

Bone cells move into the blood clot and make new bone tissue which hardens.

The new bone is trimmed and remodelled.

Figure 4 How a fracture mends.

Figure 5 This injured miner is put in splints before being moved.

and reabsorbed, so the final mend is almost undetectable in an X-ray.

For a broken bone to heal neatly, the two ends must be correctly positioned, which means that the arm must be kept still. This is achieved by putting the arm in plaster or holding it in position with a splint (Figure 5).

A severe fracture may take many months to heal, much longer than it takes other tissues such as skin. This is because it takes a long time for bone to grow and harden.

What is a slipped disc?

The cartilage discs between the vertebrae in the backbone are made up of two parts: the outer part is hard and fibrous, whereas the middle part is soft and rubbery.

Now these discs have to carry a heavy load. Sometimes the strain is so great that the outer part of the disc splits open, and the rubbery material bulges out (Figure 6). This may press on a nerve, causing a lot of pain. So the disc doesn't really *slip*: it bursts.

Whereabouts the pain occurs depends on which part of the backbone is affected. If the disc is towards the top of the backbone, the person gets neckache and armache; if it's in the middle, he gets backache; and if it's at the bottom, he gets legache (sciatica):

A person with a slipped disc in his back wears a special corset which holds the vertebrae still and thereby relieves the pain. If the disc is in his neck, he wears a special collar. If the pain is very bad, an operation may have to be performed in which the protruding part of the disc is removed.

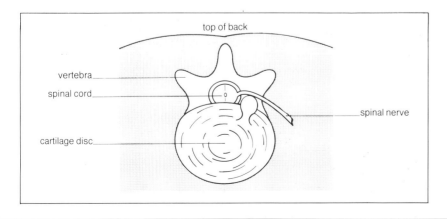

top of back

vertebra

spinal cord

spinal nerve

cartilage disc

Figure 6 A slipped disc. The cartilage disc has burst and a rubbery bulge sticks out of it as shown.

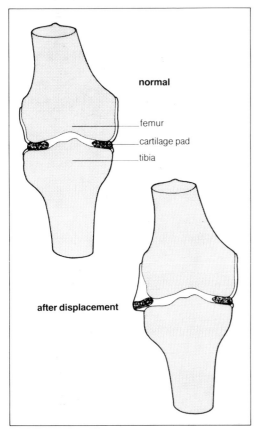

Figure 7 The knee joint has a pair of cartilage pads which sometimes get displaced.

Figure 8 An X-ray of a dislocated hip joint.

Faulty joints

All sorts of things can go wrong with our joints. One of the most common mishaps is to wrench one of them, thereby tearing a ligament or tendon. This is called a **sprain**. A sprained ankle may be caused by suddenly twisting the foot inwards, which tears the ligament on the outer side. The same kind of thing can happen in the wrist.

You sometimes hear of footballers who need to have a cartilage removed from their knee. Since the knee has to bear a considerable strain, the knee joint contains a pair of cartilage pads, which serve as extra shock absorbers. Now occasionally one of these pads becomes loose and gets pushed out of place (Figure 7). This can be an awful nuisance as well as painful, and on occasions it may 'lock' the knee joint completely so that no movement is possible, highly embarrassing for a footballer if it happens in the middle of a game! The only answer is to remove the cartilage in an operation.

Sometimes a joint becomes swollen and painful because its lining gets inflamed and produces too much synovial fluid. This tends to happen in joints which are used a lot, particularly the knee and elbow. Tennis players often suffer from it, and it's called **tennis elbow**.

Just in front of the knee cap is a small sack containing synovial fluid. This, too, can become inflamed, particularly in people who kneel a lot: it's called water on the knee. Usually these conditions are put right by bandaging the joint and resting it.

Sometimes a person is involved in an accident in which the upper arm bone is forced out of the shoulder socket; in fact with some people this can happen remarkably easily. The doctor can usually put the arm back by moving it about in a certain way.

Occasionally a baby is born with the head of the femur outside its socket. This is called a **dislocated hip** (Figure 8). The doctor puts this right by moving the legs about in such a way as to bring the head of the femur back into its socket. The child is then put in plaster with its legs pushed far apart for many months. This may run in families. Nowadays, a simple test is carried out immediately after birth on *all* babies to make sure their hips are all right.

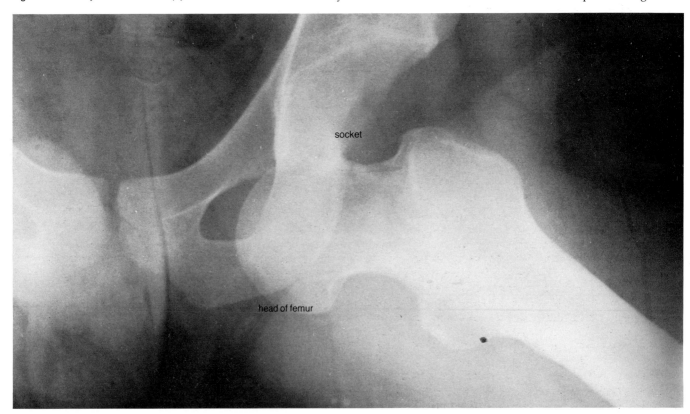

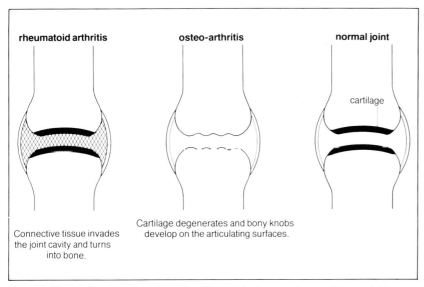

Figure 9 These diagrams show the main difference between osteo- and rheumatoid arthritis.

Arthritis

Many people complain of **arthritis**: the joints swell up and hurt, and movement is difficult. There are two kinds of arthritis (Figure 9):

1 Osteo-arthritis

This occurs mainly in elderly people and is due to wear and tear of the joints. The cartilage gradually breaks down so the joints lose their shock absorbers, and the bones no longer move smoothly against each other.

2 Rheumatoid arthritis

In this case connective tissue grows into the joints and eventually hardens, so the two bones become fused together, making movement impossible. This kind of arthritis tends to run in families and can start at any age.

Arthritis is a painful and crippling disease, but a lot can be done about it these days. For example, it is possible for the head of the femur to be replaced by a stainless steel 'ball', and the socket to be replaced by a plastic 'cup', so the person is given an artificial hip joint.

Muscle troubles

Many people suffer from aches and pains in their muscles, particularly as they get older. The general name for this is **rheumatism** or **lumbago**.

Doctors aren't certain what causes rheumatism, but it may be caused by inflammation of the connective tissue in the muscle: the tissue swells up and presses on the nerve endings and blood vessels, preventing blood flowing through the muscle and thus causing pain.

Rheumatism tends to be brought on by cold and damp, and there's no doubt that warmth and massage can bring relief. Otherwise, not much can be done about it. It's one of the main reasons why people miss work: in England and Wales over five million days are lost each year as a result of rheumatism and backache, costing the country about £50 million a year.

Everyone gets **cramp** from time to time. This is caused by a muscle spasm: the muscle suddenly contracts so powerfully that it hurts. Cramp is brought on by cold, or by using a muscle a great deal. **Stitch** is a type of cramp which occurs in the abdominal muscles, usually after a hard bout of exercise.

Finally, people sometimes tear a muscle or tendon in an accident. In severe cases the muscle or tendon may be torn right across (Figure 10).

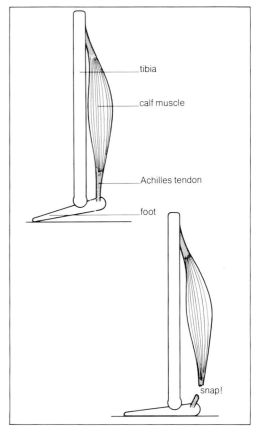

Figure 10 This person has torn his Achilles tendon, with the result that he can no longer stand on tiptoe.

Assignments

1 Why is a broken arm usually put in plaster?

2 'My back's killing me'. Give two possible things that might be wrong with this person's back.

3 Why do footballers sometimes have a cartilage removed?

4 Explain each of the following: sprained ankle, water on the knee, dislocated hip, cramp, rheumatism.

5 People sometimes suffer from a painful knee because they spend so much time kneeling. What do you think causes this?

6 In an X-ray why do the bones show up but not the skin, connective tissue, blood vessels and nerves?

Chemical messengers

Nerves provide one way by which messages can be sent from one part of the body to another. However, there is another way, and that is by means of glands.

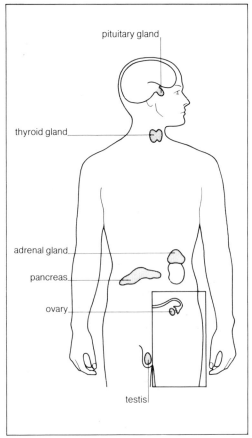

Figure 2 The main ductless glands in the human body.

What are glands?

There are many glands in the body, and their job is usually to produce, or **secrete**, some kind of useful substance.

Many of our glands shed their secretion into a tube or duct which carries it to wherever it's needed. For example, the salivary glands secrete saliva into the mouth cavity via the salivary ducts (see page 137).

In contrast, we also possess a number of glands which shed their secretion not into a duct but into the bloodstream (Figure 1). These are known as **ductless glands**, and the substances they produce are called **hormones**.

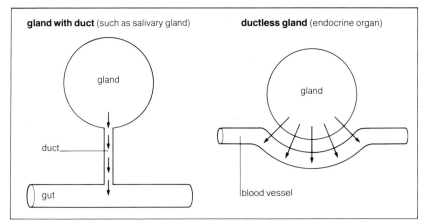

Figure 1 The two kinds of gland found in the body. The arrows show what happens to the substance (secretion) which the gland produces.

What do hormones do?

After being shed into the bloodstream, the hormone is carried to all parts of the body. It then produces an effect on certain organs. What the hormone is really doing is to carry a message from one part of the body to another, telling it to respond in a particular way. For this reason hormones are sometimes called **chemical messengers**.

Scientists have investigated the functions of different hormones by injecting them into animals such as mice, and noting the effects. Another approach has been to find out what happens when a particular gland is removed from the body.

The main ductless glands in the human body are shown in Figure 2. Each gland secretes one or more hormones, and their effects are summed up in Table 1.

Gland	Hormone	Function
Thyroid	Thyroxine	Controls the metabolic rate
Adrenals	Adrenaline	Prepares the body for action
Pancreas	Insulin	Regulates the amount of sugar in the blood
Ovaries	Female sex hormones	Control sexual development
Testes	Male sex hormones	Control sexual development
Pituitary	Growth hormone	Speeds up growth
	Thyroid-stimulating hormone	Stimulates the thyroid gland to secrete thyroxine
	Gonad-stimulating hormone	Stimulate the gonads (ovaries and testes) to secrete sex hormones

Table 1 Summary of the human body's main hormone-producing glands and their secretions.

Most hormones produce their effects rather slowly. However, there is one hormone that acts very quickly: this is **adrenaline**.

Adrenaline, the emergency hormone

Have you ever had that sinking feeling just before an important game of football, or an examination? The whole body tenses up, the heart beats faster and we feel alert and ready for action. This effect is brought about by adrenaline.

Adrenaline is secreted by the **adrenal glands**, which are situated close to the kidneys. As with other ductless glands, the hormone passes straight into the bloodstream and is then carried all round the body. The cells respond to it by using up more oxygen and releasing more energy. At the same time the heart beats more quickly and blood is diverted from the less important organs to the really important ones such as the muscles and brain (Figure 3). The overall effect is to prepare the body for an emergency. For this reason adrenaline has been described as the fight or flight hormone.

How are the glands controlled?

It's obviously important that the right quantity of hormones should be produced at all times. What tells a gland how much hormone it should secrete? The answer in many cases is the **pituitary gland**. This produces hormones which stimulate other glands to produce their secretions. For example, one of the pituitary hormones stimulates the thyroid gland to

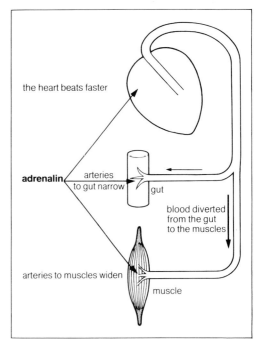

Figure 3 This diagram shows how the hormone adrenalin affects the circulation in an emergency.

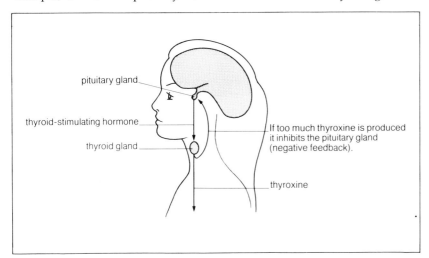

Figure 4 This diagram shows how the activity of the thyroid gland is controlled by the pituitary.

secrete thyroxine, and others stimulate the gonads (ovaries and testes) to secrete sex hormones. Because it controls the other glands, the pituitary is sometimes called the **master gland**. It's rather like the conductor of an orchestra, telling each gland when to be active.

The pituitary works in a very neat way. Suppose the thyroid is producing too much thyroxine, the thyroxine itself tells the pituitary to secrete *less* thyroid-stimulating hormone. The amount of thyroxine will then automatically fall (Figure 4). This mechanism is called **negative feedback** and it plays an important part in controlling many biological processes. The general name which we give to this kind of control process is **homeostasis**. Homeostasis literally means 'staying the same' and it ensures that the conditions inside the body remain constant.

Despite this kind of control mechanism, things sometimes go wrong and either too much or too little of a hormone is produced. This can have serious consequences and an example of it is given on page 293.

Assignments

1 Ductless glands such as the thyroid contain a large number of capillaries which are located close to the cells. Why do you think this is?

2 If necessary use the index to answer this question. The pancreas is made up of two parts: part of it secretes insulin and part of it secretes pancreatic juice.

 a) Which part is functioning as a ductless gland?
 b) What is insulin and what effect does it have in the body?
 c) What does pancreatic juice contain?
 d) Where does pancreatic juice go?

3 In Table 1, which hormones:

 a) make a person more active,
 b) cause the male to start producing sperms,
 c) cause a sinking feeling in the stomach,
 d) cause the liver to convert glucose into glycogen,
 e) cause breasts to develop in the female?

4 Nerves and ductless glands both provide a way of sending messages from one part of the body to another.

 Write down four differences between the two systems.

How do plants respond to stimuli?

The leaf or stem of a plant does not normally react when you touch it. This might suggest that plants do not respond to stimuli. However they do respond, though more slowly than animals, and in a different way.

Growth responses

Plants don't respond to stimuli by means of muscles. Instead they normally respond by growing in a particular direction. Such growth responses are called **tropisms**. They are much slower and longer-lasting than the responses given by animals.

Plants respond to three main kinds of stimuli: light, gravity and touch. Let's look at each in turn.

Light

Look at Figure 1. This shows the effect of lighting some cress seedlings from one side. The seedlings have bent over towards the light. Most plants respond to light in this way, and it ensures that the leaves get plenty of light for photosynthesis. A growth response to light is called **phototropism**. A structure such as a shoot which grows *towards* light is said to be *positively phototropic*.

How is this response brought about? A simple experiment helps us to see how (Investigation 1). A shoot has its tip covered with a little tinfoil cap, and is then lit from one side. Instead of bending towards the light, it grows straight up (Figure 2).

It seems that normally the tip receives the light stimulus, but the bending itself occurs *behind* the tip. This suggests that some kind of message is sent from the tip to the part of the shoot a little further back.

On page 292 some experiments are described which show that a shoot is made to grow by a hormone called **auxin** produced in the tip: the auxin passes down the shoot causing the cells behind the tip to expand. We can explain the shoot's response to light by suggesting that when the shoot is lit from one side more auxin gathers on the dark side than on the light side, so the dark side grows faster.

An experiment can be done to test this idea (Figure 3). The tip of a shoot is cut off and placed on a block of agar jelly which is divided by a thin partition into two halves. The tip is then lit from the right. After a while, the light is turned off and the agar block is placed on the top of the cut shoot. The result is that the shoot bends over to the right.

How can we explain this result? It seems that more auxin from the tip gets into the left hand side of the agar block than into the right hand side. So when it is placed on the cut shoot it causes more growth on the left hand side.

These and other experiments all point to the same conclusion: *lighting a shoot from one side causes more auxin to be present on the dark side than on the light side and this makes the shoot bend towards the light* (Figure 4).

Figure 1 These cress seedlings were lit for several days from the right hand side.

Figure 2 An experiment to find out if covering the tip of a shoot affects its response to light.

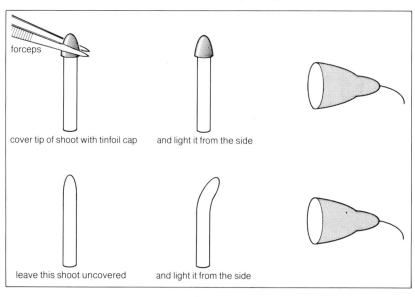

forceps

cover tip of shoot with tinfoil cap   and light it from the side

leave this shoot uncovered   and light it from the side

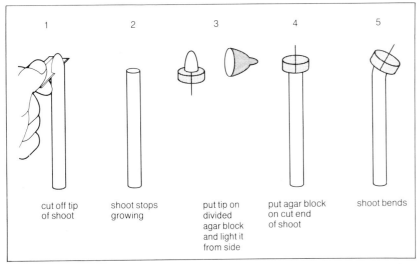

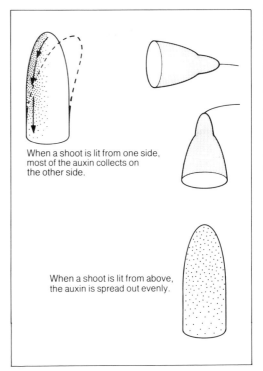

Figure 4 The direction of light affects the distribution of auxin inside the shoot.

Figure 3 An experiment to find out if a hormone is responsible for making a shoot bend towards light.

What about roots – how do they respond to light? Experiments indicate that the roots of most plants don't respond to light at all. However, some roots grow *away* from light, that is they are *negatively phototropic*.

Gravity

Look at Figure 5. This shows what happens if you place a newly germinated broad bean seedling in a horizontal position in the dark; it is put in the dark so as to avoid any effects caused by light. The result is that the shoot bends upwards and the root downwards. This is a growth response to gravity, and it is called **geotropism**. If a structure grows towards gravity, we say it is *positively geotropic*; if it grows away from gravity, we say it is *negatively geotropic*. Whatever way up a seed is, the shoot always grows upwards and the root downwards (Investigation 2). This means that if you plant some seeds, you need not worry which way up they are: nature will always make sure that the shoots and roots grow in the right direction.

Now look at Figure 6. Marks are made at equal intervals along the straight shoot and root of a seedling which is growing vertically. The seedling is then placed in a horizontal position. As the shoot bends upwards, the marks on the lower side gradually get further apart. This suggests that growth occurs

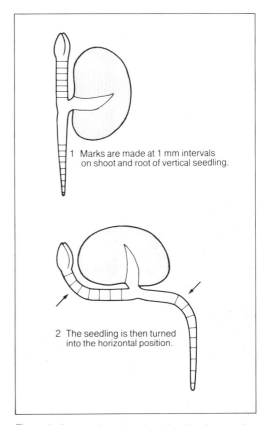

Figure 6 An experiment to show that the shoot and root respond to gravity by growing more quickly on one side than the other. The way the lines get pushed apart suggests that growth takes place in the positions marked by the arrows.

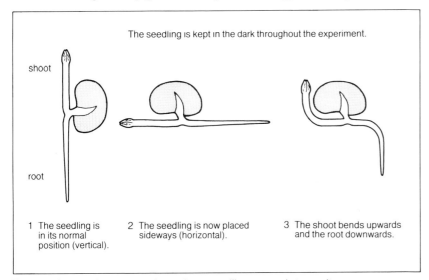

Figure 5 An experiment to see how a bean seedling responds to gravity.

Figure 7 The stems of this wisteria plant are twisting around a supporting branch.

more quickly on the lower side than on the upper side. However, in the root the marks on the *upper* side get further apart, which suggests that in this case growth occurs more quickly on the upper side than on the lower side. In both instances growth takes place behind the tip where the cells are actively lengthening.

How is this response brought about? One possible explanation is that the auxin which is produced at the tip of the shoot sinks under gravity to the lower side causing it to grow faster on that side; however, in the root it produces the opposite effect: it causes it to grow more *slowly* on the lower side.

Scientists believe that this is the correct explanation for the shoot. However, the root's response is thought to involve another hormone which *slows down* growth on the lower side.

Look again at Figure 5. Imagine you were to turn the seedling round so that the shoot points downwards and the root upwards. You would expect the shoot and root to change their direction of growth, the shoot bending upwards and the root downwards. This is precisely what happens. Experiments of this sort can be done with an instrument called a **klinostat** (Investigation 3). The results suggest that however much you change the seedling's position, the shoot and root always grow in the right direction.

Touch

Many people grow sweet peas in their gardens. This plant has a floppy stem, so it can't support itself. However, it can grow upwards, clinging to poles or trellises by means of short side branches called **tendrils**. A tendril is really a modified leaf: when it touches an object it winds itself round it. The side of the tendril which is touching the object grows more slowly than the other side, so it is made to bend round it.

There are many other climbing plants besides sweet peas. Some of them have tendrils; in others, such as runner beans and bindweed, the stem winds itself round firm objects such as poles, drainpipes or the stems of other plants (Figure 7).

Some specialised plants respond remarkably quickly to touch. For example, the leaves of the sensitive mimosa plant fold inwards when you touch them, (see page 10).

Quick responses are also given by **carnivorous plants**, which feed on small animals such as insects and spiders. An example is the Venus fly-trap, a tropical plant which is often a curiosity in greenhouses. This has a leaf which is divided into two halves by a hinge down the middle. The two 'half-leaves' are set at an angle to each other like an open book (Figure 8). If an insect such as a fly lands on the surface of the leaf, it sets off a response in which the two sides of the leaf suddenly close up together trapping its prey. Rigid spines round the edge of the leaf prevent the insect getting out. Thus imprisoned, the insect's body is broken down by digestive juices which are secreted by special cells in the leaf. It takes up to a week for the insect to be digested and absorbed, after which the leaf re-opens.

This response is rather like an animal reflex. However, the plant has no nerves or muscles, so the mechanism is quite different. No one knows for certain how it is brought about.

Other responses

Plants respond to water, chemicals and temperature. Thus roots tend to grow *towards* soil which is well watered, contains the right chemicals, and is reasonably warm; and they grow away from poor soil which does not have these qualities.

For flowering to occur, plants need to be given a certain amount of light beforehand. This kind of response is called **photoperiodism**. And seeds will not usually germinate unless they are chilled beforehand, which ensures that they do not germinate until after the winter.

Figure 8 The Venus flytrap is an unusual plant in that it responds quickly to the stimulus of touch.

Investigation 1

To find which part of a shoot responds to light

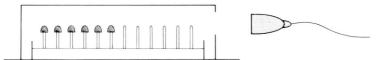

1 Obtain a dish containing about 12 seedlings of e.g. oat, wheat or barley. The shoots should be at least 10 mm long.

2 Make six little 'caps' out of aluminium foil.

3 Put a cap on six of the shoots so that it covers the tip, as shown in the illustration.

4 Light the dish from one side, as shown in the illustration, and leave it for several days.

How does the appearance of the covered seedlings differ from the others?

What conclusions can you draw about how shoots respond to light?

Investigation 2

To find if a seedling responds to gravity

1 Obtain six broad bean seedlings whose roots are just visible.

2 With pins attach them to a sheet of cork in various positions as shown in the illustration.

3 Put the cork in a small aquarium tank with a little water at the bottom, and put a sheet of glass on top.

4 Cover the tank with an upturned cardboard box so as to keep the seedlings dark.

5 After about a week observe the seedlings and sketch their appearance.

What conclusions do you draw regarding the way the shoot and root respond to gravity?

Investigation 3

Experiments with a klinostat

A klinostat is a small cylindrical chamber which can slowly rotate. The chamber contains a piece of cork to which young seedlings can be pinned.

1 Obtain two young broad bean seedlings whose roots are about 1 cm long.

2 Pin one of the seedlings to the cork in the klinostat as shown in the left hand illustration.

3 Pin the other one to a piece of cork in

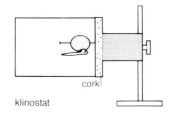

klinostat

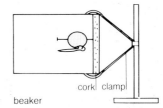

beaker

a beaker as shown in the right hand illustration. This is your control.

4 Put some moist cotton wool in the klinostat chamber and the beaker so as to keep the seedlings moist.

5 Cover the klinostat and beaker with

an upturned cardboard box, and leave the klinostat running for several days.

6 After several days, observe the two seedlings.

How do they differ in appearance? Explain your observations.

Assignments

1 Describe an experiment which you would do to find out if a bean root responds to light coming from one side.

2 Mr Lewis spends the morning in the garden. He plants some seeds and puts many of them in the soil upside down.

Does this matter? Explain your answer.

3 A small quantity of auxin is painted onto the shoot and root of a seedling in the positions shown by the arrows in Figure 6. How do you imagine the seedling will appear after 48-hours?

4 A young bean seedling is placed in a klinostat in the position shown in the diagram below. The seedling is then rotated slowly for two days. Draw what you suppose would be the shape of the seedling at the end of the two-day period.

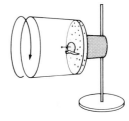

5 Describe one experiment which in your opinion provides the strongest evidence that a hormone is involved in making a shoot bend towards light.

Continuation
of life

*African elephants,
old and young, photographed
in Amboseli National Park, Kenya.
In the next group of topics we
shall see how organisms
grow and reproduce.*

How do living things grow?

Growth is the increase in size which takes place as an organism develops. It ensures that the organism is the right size to survive in its environment.

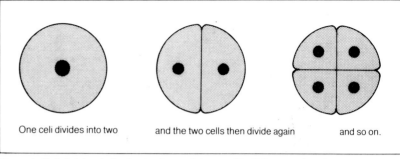

Figure 1 Cell division is the basis of growth in a multicellular organism.

One cell divides into two   and the two cells then divide again   and so on.

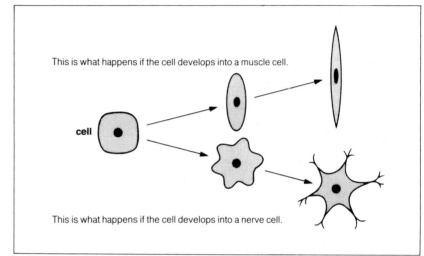

This is what happens if the cell develops into a muscle cell.

cell

This is what happens if the cell develops into a nerve cell.

Figure 2 A cell may change its shape and form and develop into a particular kind of cell with a specific function.

Cell division

Most animals and plants are composed of many cells. However, they usually start off as a single cell, the fertilised egg. This divides into two cells, then into four, eight, sixteen and so on. This process of cell division is the basis of growth.

Figure 1 shows what happens when a cell divides. The nucleus divides first, and then the rest of the cell divides across the middle. At first the daughter cells are smaller than the original parent cell, but they soon grow to full size. For this to happen they must take in food substances to provide the necessary materials and energy. This is why a growing organism needs plenty of food.

Eventually the cells change their shape and form and turn into particular types of cell, depending on their position in the body. In the human body, for example, a cell might develop into a smooth muscle cell if it happens to be in the wall of the gut, or into a brain cell if it's in the head (Figure 2). The process by which cells become specialised like this is called **differentiation**, and it plays a vital part in the construction of the full-grown adult organism.

Normally once a cell has become specialised in this way it does not divide any more.

Growth in humans

In a growing child cell division takes place in all parts of the body. As a result, the child gets steadily larger. However, different parts of the body grow at different rates: this is because cell division occurs more quickly in some places than in others. For example the head grows quickly in the early stages of development and then slows down, whereas the legs and arms grow slowly at first and then speed up later (Figure 3).

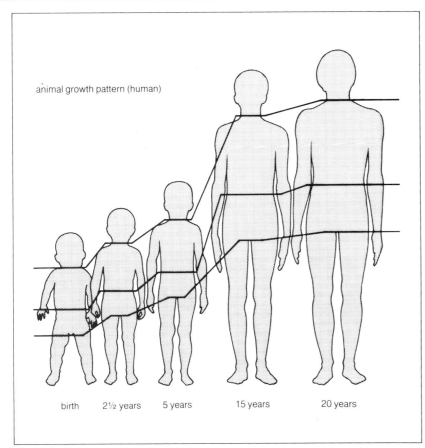

Figure 3 These diagrams show how different parts of the body increase in size during the first twenty years of a person's life.

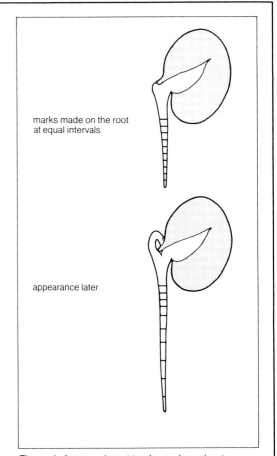

Figure 4 An experiment to show whereabouts growth occurs in a young bean root.

Eventually no more cells are added to the body, and so the person stops growing. In humans this happens at the age of about eighteen, though in most other animals it occurs much sooner (Investigation 1).

Certain cells need to be replaced, so these go on dividing throughout life: they include the cells at the base of the epidermis in the skin, the cells in the bone marrow from which blood cells are formed, and the cells from which eggs and sperms are formed. Cells will also start dividing again when the body is cut or damaged, and the new tissues so formed help to patch up the wound and heal it.

Growth in plants

We have seen that in a growing child cell division takes place all over the body. This is true of most animals. In a young plant, however, cell division is restricted to certain regions called **meristems**. The main meristems are at the tip of the shoot and root (Investigation 2). If you make marks on a young shoot or root, you can see exactly where growth is occurring (Figure 4).

Consider a growing shoot (Figure 5). In the tip, cells are continually dividing. These young cells draw in water by osmosis and expand. This has the effect of lengthening the shoot, and helps it to thrust its way upwards. The same kind of thing happens in the roots as well, and it helps them to push their way down into the soil. So in plants growth is achieved not just by cell division, but by cell expansion as well (Investigation 3).

While the cells are expanding, they differentiate into specialised tissues according to their position in the plant and the task which they have to perform. For example, most of the cells in the shoot develop into packing tissue, in certain regions the cells develop into transport tissues (xylem and phloem).

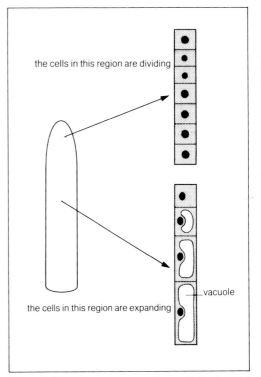

Figure 5 This diagram shows where cell division and expansion take place in a growing shoot. The same applies to the root.

How does a stem get thicker?

The kind of growth just described makes the stem get longer. However, as you know, stems get thicker too. Think of a tree, for instance, whose trunk gets thicker year by year. As this occurs *after* the stem has increased in length, we call it **secondary growth**.

This is how secondary growth occurs. At the same time as the stem is getting longer, a layer of cells develops inside it called the **cambium**. If you look at a cross-section of the stem, the cambium appears as a ring of cells situated between the xylem and phloem (Figure 6). Now the cambium cells are able to divide long after the other cells have stopped doing so. They divide to form new xylem tissue (wood) towards the inside, and new phloem tissue towards the outside. In fact far more xylem tissue is formed than phloem, so the amount of wood in the stem increases greatly.

Meanwhile another layer of dividing cells is formed just under the surface of the stem. This is called the **cork cambium**, and its cells divide to form the hard corky part of the bark.

Secondary growth takes place in the summer, but it stops in the winter. If you cut a tree down, you can see rings of wood corresponding to each year's secondary growth: they are called **annual rings**, and by counting them you can tell the age of the tree. You can do the same with individual twigs (Investigation 4).

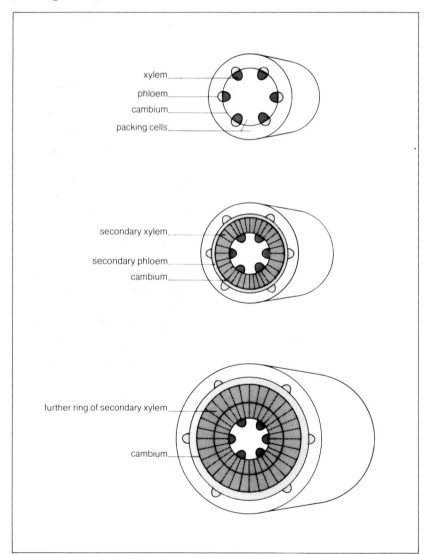

Figure 6   Secondary growth enables a stem to get wider year by year.

Investigation 1

Measuring the growth of an animal

If you have a young pet, such as a kitten or puppy you can carry out this investigation at home. Alternatively you can do it in the laboratory, using a small mammal such as a mouse or gerbil.

1 With a ruler measure the animal's length from the tip of its nose to the *base* of its tail.

2 Weigh the animal and find its mass.

3 Repeat steps 1 and 2 at regular intervals (at least twice a week) until growth appears to stop.

4 Plot your results on a piece of graph paper, putting length and mass on the vertical axis, and time on the horizontal axis.

Did the length and mass stop increasing at the same time?
If not, can you explain the reason?

Investigation 3

Examining the inside of a young root

1 Look at a prepared longitudinal section of a young root under the microscope.

2 Observe the cells just behind the tip.

Draw one of the cells in outline to show its shape.

What were these cells doing when the root was alive?

3 Now look at the cells further back.

Draw one of them to show its shape.

How did the cells come to be this shape?

How does the change in shape help the root to grow?

Investigation 2

To find where growth takes place in a shoot

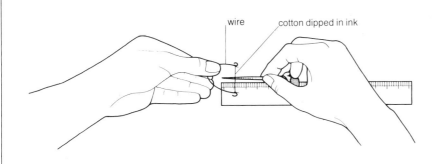

wire   cotton dipped in ink

1 Obtain a broad bean seedling with a root at least 2 cm long.

2 With Indian ink, make a series of marks along the length of the root 1 mm apart. Use the special 'pen' shown in the illustration.

3 Pin the seedling to a piece of cork with the root pointing downwards.

4 Put the cork in a jar with a little water in the bottom to keep it moist, and cover it with a sheet of glass.

5 After several days examine the seedling.

Are the marks still the same distance apart?
Where does growth occur in the root?

Investigation 4

Looking at secondary growth in a plant

1 Obtain prepared transverse sections of a series of twigs of different ages.

Alternatively you can cut your own sections as instructed on page 207. If you do this, stain the sections by putting them in a watch glass full of acidified phloroglucin. This will show up the woody part of the twig by staining it red.

2 Put the sections against a light background and if necessary look at them through a hand lens (magnifying glass).
How do they differ in appearance?
Can you tell how old each one is?

3 Look at the cut end of a series of older stems or branches.

How do they differ in appearance?

Can you tell how old they are?

Assignments

1 Fill in the missing words in this passage:

In order for an organism to grow, its cells ____ (a) ____. This process requires ____ (b) ____ which comes from the organism's ____ (c) ____. Later on the cells ____ (d) ____ into different types of cells depending on their ____ (e) ____ in the body.

2 Give three ways in which growth in animals differs from growth in higher plants.

3 Explain briefly how a growing stem increases in length, and how later it increases in width.

4 With a ruler measure in millimetres the width of the head and the length of the legs in the diagrams in Figure 3. Plot the results on a sheet of graph paper so the curves can be compared.

a) Which grows more quickly between the ages of 5 and 15 years, the head or the legs?
b) By how many times does one grow faster than the other?

5 A scientist sows a large number of seeds all at the same time, and he wants to measure the rate of growth of the seedlings. Here are three methods which he might use:

a) He measures the *heights* of fifty plants every day and takes the average.
b) He digs up five plants every day and estimates their *mass* by weighing them.
c) He digs up five plants every day and dries them by heating them in a hot oven until all traces of water have been driven off. He then weighs them, thereby obtaining their *dry mass*.

Write down the advantages and disadvantages of each method.

How is growth controlled?

What makes an organism grow at a certain rate, and to a particular size? In this Topic we will look into this question, and also see what happens when growth goes wrong.

How is growth controlled in plants?

If you cut off the tip of a shoot, the part behind will stop growing. However, if you put the tip back, the shoot will start growing again (Figure 1). There seems to be something in the tip which stimulates growth to take place (Investigation 1).

What is it? Figure 2 shows an experiment which was done to find the answer. The tip of a shoot was cut off and placed on a small block of agar jelly. Deprived of its tip, the shoot stopped growing. After a few hours the agar block was placed on the cut end of the shoot. The result was that the shoot started growing again.

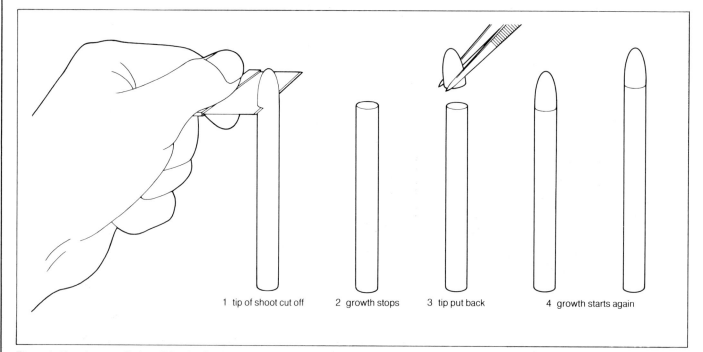

1 tip of shoot cut off   2 growth stops   3 tip put back   4 growth starts again

Figure 1 Experiment to find out if the tip of a shoot is needed for growth to occur.

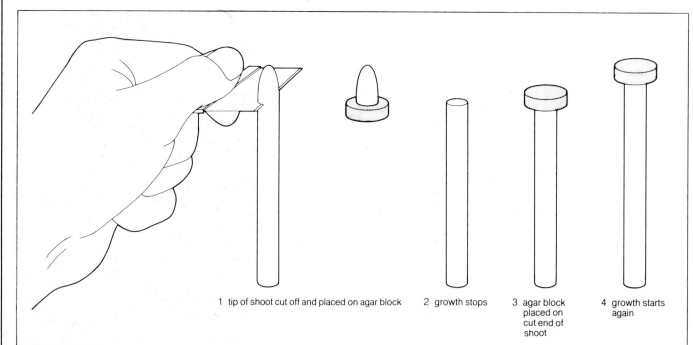

1 tip of shoot cut off and placed on agar block   2 growth stops   3 agar block placed on cut end of shoot   4 growth starts again

Figure 2 Experiment to find out if a hormone produced by the tip makes the shoot grow.

This experiment suggests that the tip of the shoot produces a chemical substance which passes down the shoot, making it grow. Scientists have managed to isolate this substance and have christened it **auxin**.

Auxin is produced at the tip of the shoot and then it slowly diffuses down to the roots producing various effects on the plant as it flows along. It therefore functions as a hormone, rather like those found in animals (see page 280).

Auxin doesn't always *stimulate* growth; sometimes it *stops* it. For example, as it passes down the stem, it tends to *prevent* side branches growing out, thus making the plant tall and straight. If you cut the apical bud off such a plant, the flow of auxin stops and side branches will then develop

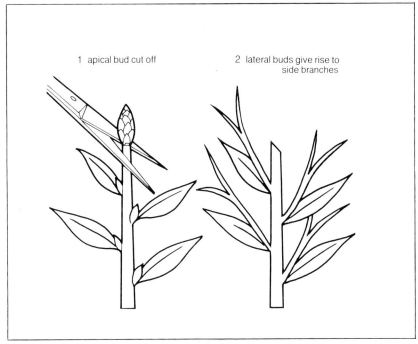

1 apical bud cut off   2 lateral buds give rise to side branches

Figure 3 This picture shows what happens if you cut the apical bud off a plant.

(Investigation 2). Gardeners sometimes cut the tops off plants to make them more bushy (Figure 3): this is the secret behind making a good thick hedge.

Since auxin was isolated, scientists have discovered other hormones which play an important part in plant growth. Nowadays these substances, or very similar ones, are manufactured in chemical factories; they are known as **growth substances** and are much used in gardening and horticulture. For example a substance similar to auxin is used for helping cuttings to 'take' (see page 332): the cut stem is dipped in the substance and this stimulates roots to grow out from it (Investigation 3).

How is growth controlled in animals?

In man a **growth hormone** is produced by the pituitary gland at the base of the brain, and this speeds up growth. If too little of this hormone is produced during childhood, the person remains short and becomes a dwarf. On the other hand, if too much is produced, the person may grow into a giant (Figure 4).

Another hormone which helps children to grow is **thyroxine**: this is produced by the thyroid gland in the neck (see page 280).

Hormones control growth in other animals too. For example, scientists have carried out experiments which show that in insects growth is brought about by a hormone produced by certain cells in the brain.

Figure 4 A pituitary giant and dwarf side by side.

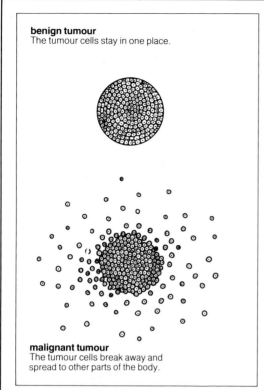

benign tumour
The tumour cells stay in one place.

malignant tumour
The tumour cells break away and
spread to other parts of the body.

Figure 5 Two kinds of tumour which can occur in
man and other animals.

When growth goes wrong

We have seen that growth takes place by cell division, and that in animals there comes a time when it stops. Some kind of control process prevents the cells dividing any more.

On occasions this control process may break down in some part of the body and the cells start dividing again. This results in the formation of a disorganised mass of simple cells which don't perform a useful function. Doctors call this a **growth** or **tumour**.

There are two kinds of tumour: **benign** and **malignant** (Figure 5). A benign tumour stays in one place and does not harm the surrounding tissues except by pressing on them. In contrast a malignant tumour spreads: cells become detached from it and are carried by the lymph or blood to other parts of the body where they invade and destroy the tissues. Tumours of this kind are known as **cancer** (carcinoma).

Cancer is second only to heart disease as a cause of death in Britain, and people are very frightened of it. However, there are many cases of people being cured of it and better methods of diagnosis and treatment are constantly being devised. The surest remedy is for a surgeon to remove the tumour before it has a chance to spread, but cancer can also be treated by drugs and radio-active rays (radiotherapy). Such treatment kills the tumour cells or at least stops them dividing (Figure 6).

The success of the treatment depends partly on how soon the tumour is discovered. A person who has a complaint that won't go away should go to the doctor. A persistent cough, chronic indigestion, a lump under the skin, bleeding from the anus – any of these *might* be a sign of cancer if they don't clear up. A quick test can often be carried out in a hospital to see if it's serious. For example, a chest X-ray will show up cancer of the lung, and a doctor can find out if a woman has cancer of the womb by taking a smear from the neck of the womb: this is called a cervical smear. He then looks at it under the microscope.

What causes cancer? No one knows for certain, though there are many theories and a great deal of research is going on into this question at the present time. You *cannot* catch it from other people, but there is no doubt that it can be brought on by environmental hazards such as atomic radiation, asbestos dust and smoking.

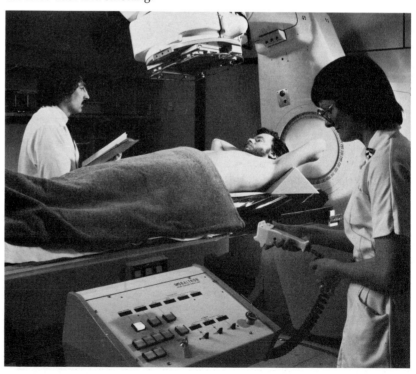

Figure 6 A patient being positioned on the treatment
couch before receiving radiation therapy from a
computer controlled machine.

Investigation 1

To find the effect on growth of cutting off the tip of a shoot

1 Obtain a dish containing about 12 seedlings of e.g. oat, wheat or barley. The shoots should be at least 10 mm long.

2 With small scissors cut the tip off 6 of the shoots: make your cut not more than 4 mm behind the tip.

3 Leave the dish in a uniformly lit place for several days.

4 After several days, observe the seedlings.

How does the appearance of the decapitated seedlings differ from the others?

What conclusions can you draw about how growth is controlled in the shoots?

Investigation 2

To find the effect of removing the apical bud from a plant

1 Obtain two potted plants which do not have any side-branches.

2 Cut off the apical bud at the top of the stem from one of the plants, but not from the other one.

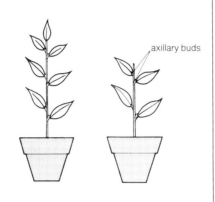

axillary buds

3 Observe the two plants at intervals during the next few weeks.

How do they differ in appearance?

What effect has been produced by removing the apical bud?

Investigation 3

To find the effect of a growth substance on cuttings

1 Cut off two healthy side-branches from a mature geranium plant. Make the cut just below a 'node' (see page 123).

2 Obtain some rooting hormone from a gardening shop and make up a solution of it.

3 Pour the solution into a small beaker to a depth of not more than 2 cm.

4 Pour distilled water into a second beaker to the same depth.

5 Stand one of the geranium cuttings in the beaker of hormone, and the other cutting in the beaker of water. Leave them side by side in a warm, evenly lit place and observe them at intervals during the next week or so.

Do the cuttings produce roots from the cut stem?

If so, which one produces them first?

What conclusions would you draw from this experiment?

Assignments

1 Give the name of one hormone which controls growth in plants and one which controls growth in animals. In each case say where the hormone comes from.

2 Explain the reason behind each of the following:

a) A gardener cuts the tops off his cypress trees so as to make a thick hedge.
b) He dips his cuttings in 'rooting powder' before he sticks them in the soil.

3 The diagram on the right shows an experiment which a scientist carried out on a growing seedling. The tip of the shoot was cut off and a thin piece of metal placed between the tip and the rest of the shoot.

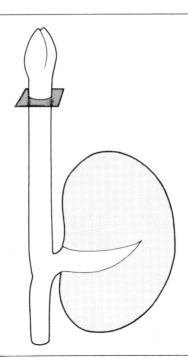

a) What do you think the scientist was trying to prove?
b) What should his control be?
c) What do you think the effect would be if the piece of metal extended only half way across the shoot?

4 What is the difference between a benign and a malignant tumour? Write down three ways in which a person with a malignant tumour can be treated.

Eggs, sperms and sexual development

Sexual reproduction involves the union of an egg and a sperm. As a result, a new individual comes into being. In this Topic we will look at eggs and sperms and see when and how they are formed.

The structure of eggs and sperms

A human **sperm** is shown in Figure 1. It is extremely small and is shaped like a tadpole with a head and tail (Investigation 1). The tail flaps from side to side, enabling it to swim. The sperm consists of only one cell, and the nucleus is in its head.

A human **egg** is shown in Figure 2. It, too, is a single cell but it is much larger than the sperm. It is shaped like a round ball and is surrounded by a thin membrane and a layer of jelly. The nucleus is situated towards the centre.

The nuclei of the sperm and egg contain chromosomes, thread-like bodies which carry genes. The genes are responsible for passing on the parents' characteristics to the offspring.

How are eggs and sperms made?

Sperms are made in the **testes** of the male (Figure 3). Eggs are made in the **ovaries** of the female (Figure 4). If you look at thin sections of the ovary and testis under the microscope you can see the eggs and sperms developing (Investigation 2 and 3). Collectively eggs and sperms are known as **gametes**, and the ovaries and testes are called **gonads**.

What happens to eggs and sperms?

If left on their own, eggs and sperms simply die. However, if a sperm gets close to an egg, it bumps into it repeatedly and sooner or later it penetrates it. The nuclei of the sperm and egg then join together. The process by which a sperm fuses with an egg is called **fertilisation** (Figure 5). The fertilised egg is called a **zygote**.

Sexual development

A new-born baby has a complete set of sex organs. However, the testes of a baby boy are not yet able to make sperms, and the ovaries of a baby girl can't produce eggs although thousands of *immature* eggs are already present, ready and waiting.

Between the ages of about twelve and fourteen, the sex organs suddenly become active: the testes start making sperms, and the ovaries start producing eggs. This change constitutes **puberty**, and only when a person reaches this stage is he or she capable of producing children. The time when puberty occurs varies from person to person: it usually occurs slightly earlier in girls than in boys, and interestingly it occurs earlier now than it did about fifty years ago. Why do you think this is?

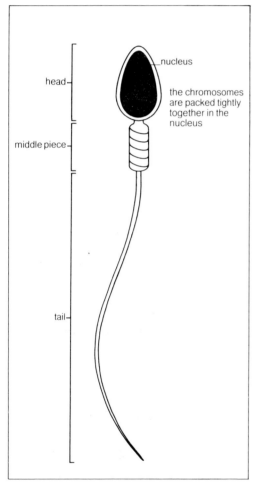

Figure 1 A human sperm. The head is about 3 micrometres wide and the whole sperm including the tail is about 60 micrometres long.

Figure 2 A human egg with a sperm alongside to show their relative sizes. The egg has a diameter of about a tenth of a millimetre: that is about forty times wider than the head of the sperm. Its nucleus is much larger than the sperm's nucleus and so the chromosomes are more spread out.

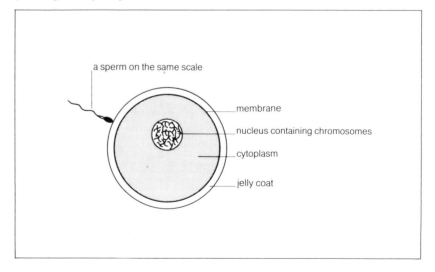

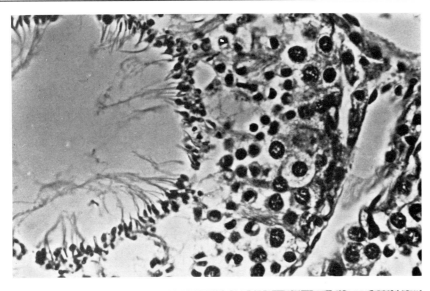

Figure 3 A section of a testis as it appears under the miscroscope.Notice the sperm tails. They are hanging into one of the many tubules of which the testis is composed.

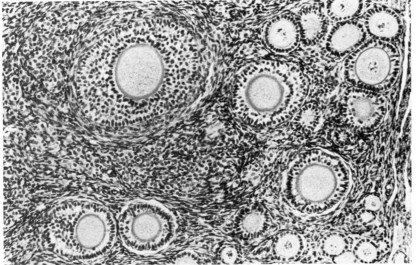

Figure 4 A section of an ovary as it appears under the microscope. The round objects are immature eggs.

Puberty is brought on by **sex hormones** which start being produced by the ovaries and testes themselves. The male sex hormones are called **androgens** and the female ones are called **oestrogens**. These in turn are activated by hormones from the pituitary gland at the base of the brain.

The sex hormones bring about other changes as well. For example, in boys

Figure 5 These diagrams show how fertilisation occurs in the human.

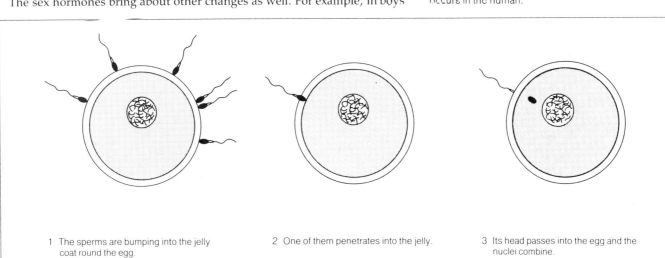

1 The sperms are bumping into the jelly coat round the egg.

2 One of them penetrates into the jelly.

3 Its head passes into the egg and the nuclei combine.

the voice breaks and hair starts growing on the legs, chest and face. A boy who has been through puberty soon finds himself having to shave. However, if a male is castrated, i.e. has his testes removed, before he reaches puberty, these changes do not take place. In medieval times choirboys were sometimes castrated to prevent their voices breaking. Such a person is called a eunuch, and in the choir they were known as castrati.

In the female, the breasts start developing, and fat is laid down in the thighs, giving the curves characteristic of the female body.

The various changes which have just been described constitute the person's **secondary sexual characteristics**. At this stage boys and girls both become more interested in the opposite sex.

Puberty comes on quite suddenly. If a boy is in the habit of masturbating, he finds that semen is produced when he reaches a climax. If he does not masturbate, he may have a 'wet dream': semen gradually builds up in his body and is discharged spontaneously while he is asleep.

The female starts having 'periods' which are characterised by bleeding from the vagina. This bleeding is caused by the lining of the womb (uterus) breaking down, and it is known as **menstruation**. Menstrual periods tend to occur irregularly at first, but eventually they take place at fairly regular intervals of about twenty-eight days. Generally the bleeding goes on for about five days. Some women don't feel any ill effects at these times, but others feel tense and under the weather for several days beforehand. The reason why menstruation takes place is explained on page 300.

The changes that occur at puberty are perfectly normal, but people who are worried about them should ask for help from their parents or a counsellor at school.

Growing old and sexual decline

A female will go on producing an egg every month until she reaches the age of forty-five to fifty. Her ovaries then stop producing eggs, and after that she is no longer able to become pregnant. This is brought about by changes occurring in her hormones, and it may make her feel tired and run-down for some months. It is called the **menopause** or 'change of life'.

Men do not go through a change of this kind. Normally a man goes on producing sperms until well into his seventies. However, he may experience a gradual decline in his desire for sexual activity.

The various changes which have been discussed in this Topic are summarised in Table 1.

Table 1   Summary of sexual development of the human male and female.

	Male (♂)	Female (♀)
AT BIRTH	Testes have descended into scrotal sac but they do not make sperms yet.	Ovaries containing immature eggs present in abdomen but they do not produce eggs yet.
12–14 years **PUBERTY**	PITUITARY GLAND ↓ gonad-stimulating hormones ↓ TESTES ↓ male sex hormones (androgens) ↓ Testes start producing sperms. Secondary sexual characters develop, e.g. growth of body hair and breaking of voice.	PITUITARY GLAND ↓ gonad-stimulating hormones ↓ OVARIES ↓ female sex hormones (oestrogens) ↓ Ovaries start producing eggs. Secondary sexual characters develop, e.g. growth of breasts and laying down of fat in thighs.
45–50		MENOPAUSE ('change of life') Ovaries stop producing eggs
70–75	Testes stop making sperms	

Investigation 1

Looking at sperms

Your teacher will do the first three steps in this investigation.

1 Obtain a male rat which has just been killed for dissection.

2 Cut open the scrotal sac.

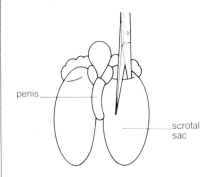

3 Cut into the testis and suck up a little of the milky fluid into a pipette.

4 Put two drops of the fluid on a slide, and add one drop of salt solution (0.9 per cent).

5 Put on a coverslip.

6 Observe under the microscope: low power first, then high power.

Can you see any sperms?

Which structures shown in Figure 1 can you see?

Are the sperms moving?

If not, suggest reasons for this.

Investigation 2

Looking inside the testis

1 Obtain a prepared slide of a mature testis of a mammal (a thin section which has been cut, stained and mounted on a slide).

2 Examine it under the low power of the microscope.

3 Locate sperms with their long tails.

Which structures shown in Figure 3 can you see?

Where are the sperms made?

How do you think they are formed?

Investigation 3

Looking inside the ovary

1 Obtain a prepared slide of a mature ovary of a mammal (a thin section which has been cut, stained and mounted on a slide).

2 Examine it under the low power of the microscope.

3 Locate the eggs. You will find them towards the edge of the ovary.

Which structures shown in Figure 4 can you see?

Assignments

1 Write down five ways in which a human sperm differs from an egg. Far more sperms are produced than eggs: why do you think this is?

2 Which of the following are associated with the ovary, which with the testis, and which with both the ovary and testis: oestrogen, sex hormones, wet dream, androgen, pituitary gland?

3 Briefly explain each of the following terms:

   a) puberty
   b) sex hormone,
   c) menstruation,
   d) secondary sexual characteristics,
   e) menopause.

4 In a certain town in Canada there is a horizontal line by the door of buses. If the top of the passenger's head comes below this line, he is only charged half fare. The bus company has had to raise the level of this line twice during the last thirty years. Why do you think they have had to do this?

5 Professor J. M. Tanner has estimated the relative rates of growth of the brain, the body in general, and the reproductive organs in human beings. His findings are shown in the graph below. Explain what the graph shows, and then suggest reasons why the three curves are different.

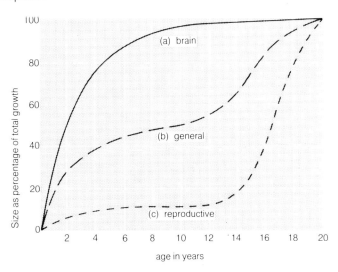

The menstrual cycle

Approximately once a month from puberty to the menopause the human female experiences a menstrual period. This is part of a cycle of events which occurs inside her body.

What happens during the cycle?

The cycle starts with menstruation for which reason it is known as the **menstrual cycle**. During menstruation the lining of the uterus breaks down and passes out through the vagina, with much bleeding. This is what's meant by 'having a period'.

Thousands of immature eggs are present in the ovaries. Immediately after menstruation one of these eggs starts developing (Figure 1). As it develops, it becomes enclosed in a protective structure called a **Graafian follicle** which gradually gets larger and becomes hollow. About two weeks after the beginning of menstruation, the follicle moves to the edge of the ovary and the mature egg pops out of it into the oviduct. This process is called **ovulation**.

While the follicle has been developing in the ovary, the lining of the uterus has gradually been healing and building itself up again, so that when ovulation occurs it is ready to receive a fertilised egg, should one become available.

When the egg is shed from the ovary, the follicle stays behind and develops into a solid object called the **yellow body** (corpus luteum). We shall see what its job is presently. Meanwhile the lining of the uterus continues to develop: it thickens and numerous blood vessels grow into it. About two weeks after ovulation, the yellow body withers away and at the same time the lining of the uterus breaks down and menstruation occurs. The whole cycle then begins all over again.

Figure 1 shows how the changes which occur in the ovary and the uterus fit in with each other.

How is the menstrual cycle controlled?

The menstrual cycle is controlled by hormones. The pituitary gland at the base of the brain produces a hormone which causes the follicle to develop in the ovary. The ovary in turn produces another hormone (**oestrogen**) which causes the lining of the uterus to repair itself after menstruation.

At the time of ovulation the pituitary gland starts producing a second hormone which causes the follicle to turn into the yellow body in the ovary. The yellow body in turn produces another hormone called **progesterone** which causes the lining of the uterus to become thicker and full of blood vessels.

We can sum up by saying that oestrogen and progesterone repair the uterus after menstruation and prepare it for receiving an embryo. If the egg does not get fertilised, the two hormones stop being produced: as a result the lining of the uterus breaks down and menstruation occurs.

What happens to the menstrual cycle during pregnancy?

If a woman conceives and becomes pregnant, her menstrual periods stop until after the baby has been born. In fact the sign that the woman is pregnant is that she will miss her usual 'period'.

What causes menstruation to stop like this? The presence of an embryo in the uterus causes the yellow body to stay in the ovary and go on producing progesterone. As a result, the lining of the uterus remains intact, and continues to thicken and build itself up. So progesterone *prevents* menstruation. At the same time it stops any further eggs being produced by the ovaries.

Breeding seasons

In the human female the sexual cycle just described goes on all the time and the female can become pregnant at any time of the year. This is true of many other animals too, including rats, mice and rabbits. However, some animals have a special breeding season when they are said to be on heat. Only at these times do their ovaries produce eggs. For example, dogs breed in the early spring, cats in the spring and autumn, and hamsters any time between March and October.

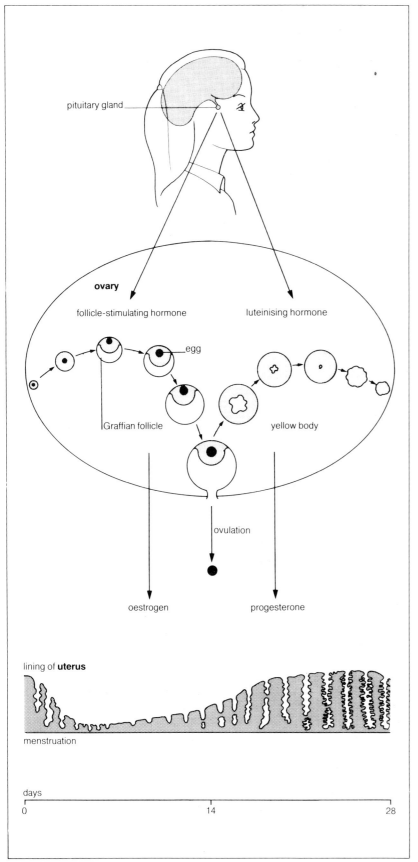

Figure 1 This diagram summarises the main things that occur during the menstrual cycle of the female.

1 How does the menstrual cycle get its name?

2 Explain what happens to the lining of the uterus when:

   a) a Graafian follicle develops in the ovary,
   b) the yellow body (corpus luteum) degenerates,
   c) the Graafian follicle changes into a yellow body,
   d) ovary-stimulating hormones are being produced by the pituitary gland,
   e) ovulation occurs.

3 This question is about the hormones which control the menstrual cycle.

   a) Name the hormones which prepare the uterus for pregnancy.
   b) Where is each hormone produced?
   c) How do the hormones get to the uterus from the organ which produces them?
   d) At what stage in the menstrual cycle is each hormone *most* active?
   e) At what stage in the cycle are both hormones *least* active?

4 Why is it important that a female's menstrual periods should stop when she is pregnant? What causes them to stop?

5 The following graph shows how the body temperature of a human female changed in the course of her menstrual cycle.

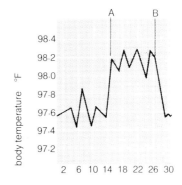

Suggest reasons why her temperature went up at point A on the graph, and down again at point B.

Bringing eggs and sperms together

In this Topic we will see how sperms and eggs are brought together. This will involve studying the reproductive system.

Reproductive system of the male

The reproductive system of the human male is shown in Figure 1. Many of the structures you can see for yourself by dissecting a mammal such as a rat (Investigation 1). The **testes** are suspended in the scrotal sac just behind the penis. Their job is to manufacture sperms. Because they are positioned outside the main body cavity, they are slightly cooler than the rest of the body. This is important because the testes make sperms more rapidly in cool conditions. During development of a baby boy, the testes start off in the abdominal cavity, but later they move down into the scrotal sac. Normally this has happened by the time the baby is born, but occasionally one or other testis does not come down until after birth.

Each testis is made up of a large number of narrow **sperm tubules** where the sperms are made. If they were placed end to end, these tubules would be over 500 metres long, that's long enough to go right round a football pitch. This gives the testes a high production rate.

As the sperms are produced, they move into a coiled tube called the **epididymis** where they are stored. The epididymis lies alongside the testis in the scrotal sac and it leads to the **sperm duct**. This is connected with the **urethra** which runs down the centre of the **penis**. The head of the penis, known as the **glans**, is highly sensitive and is protected by the **foreskin**. The foreskin is sometimes removed in the operation known as **circumcision**: this may be done for religious reasons (for example all Jews normally have it done), or because the foreskin is too tight. The operation is carried out at an early age, and there is no evidence that it is in any way harmful.

Figure 1 The reproductive system of the human male.

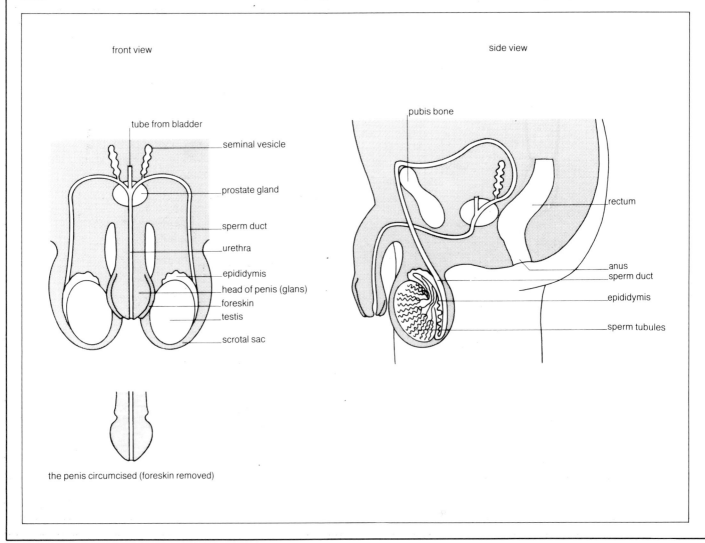

front view

tube from bladder — seminal vesicle — prostate gland — sperm duct — urethra — epididymis — head of penis (glans) — foreskin — testis — scrotal sac

the penis circumcised (foreskin removed)

side view

pubis bone — rectum — anus — sperm duct — epididymis — sperm tubules

Various glands open into the sperm duct and urethra. These include the **prostate gland** which is wrapped round the top of the urethra like a scarf. The glands produce a fluid which keeps the sperms alive and helps them to swim vigorously. This fluid, together with the sperms themselves, make up **semen**. If ejaculation does not occur, the sperms simply die and disintegrate.

If you look at Figure 1 you will see that the bladder and sperm ducts both open into the urethra. However, urine and semen never pass down the urethra at the same time. This is because it's impossible to urinate when the penis is erect, and it's impossible to ejaculate when the penis is limp.

The reproductive system of the female

The female's reproductive system is shown in Figure 2, and as with the male the various structures can be seen by dissecting a mammal such as the rat (Investigation 2). There are two **ovaries**, one on either side of the abdomen. Once every 28 days or so one or other of the ovaries produces an egg which is shed into the **oviduct**. The egg then moves slowly down the oviduct towards the **uterus**. If it isn't fertilised within a day or so it will die.

The uterus, or womb, is where the baby develops. Below it is the **vagina**. The lining of the uterus and vagina secrete a lot of mucus.

In Figure 2 you will see that the vagina opens to the outside quite separately from the tube that carries urine from the bladder. Close to the urinary opening is a small protuberance called the **clitoris**. This is the female's equivalent of the penis and it can become erect during sexual excitement.

Figure 2 The reproductive system of the human female.

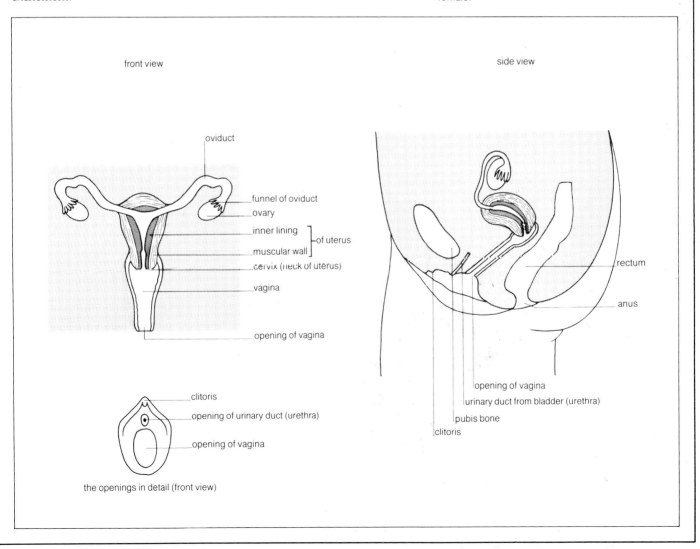

front view

oviduct

funnel of oviduct
ovary
inner lining
} of uterus
muscular wall
cervix (neck of uterus)
vagina

opening of vagina

clitoris
opening of urinary duct (urethra)
opening of vagina

the openings in detail (front view)

side view

rectum

anus

opening of vagina
urinary duct from bladder (urethra)
pubis bone
clitoris

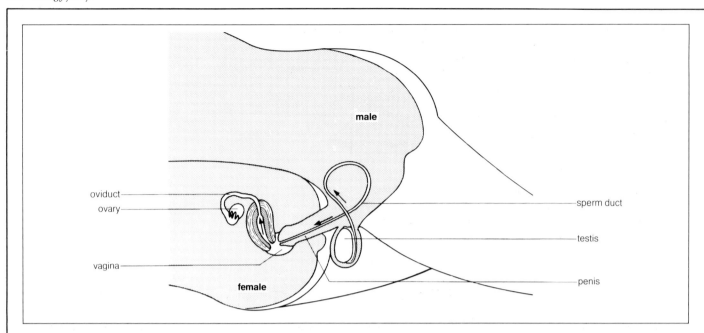

Figure 3 In intercourse sperms pass from the male to the female as indicated by the arrows in this diagram.

Intercourse

For sperms to get to the egg, **intercourse** (**copulation**) must take place.

First the penis of the male becomes stiff and hard. This is called an **erection**, and it's brought about by the blood pressure increasing inside the penis. The male then puts his erect penis into the vagina. Drops of fluid, secreted by the male's glands, emerge from the tip of the penis and serve as a lubricant, as does the mucus lining the vagina.

The male then moves his penis rhythmically inside the vagina. The head of the penis is very sensitive, and repeated stimulation of it culminates in **ejaculation**: this is a reflex in which the semen is expelled from the urethra with considerable force. It's brought about by a series of contractions which sweep down the sperm ducts and along the urethra.

Ejaculation is accompanied by a pleasurable feeling called an **orgasm**. The female may experience an orgasm too, though it usually takes longer for her to reach this point. It's brought about by stimulation of the clitoris.

Normally the male produces about 4 cm³ of semen: that's about a teaspoonful. This may not seem much but it can contain as many as 500 million sperms. Once deposited in the vagina, the sperms swim through the mucus lining the uterus and up the oviducts (Figure 3).

Figure 4 After the egg has been fertilised it develops into a little ball of cells which becomes implanted in the lining of the uterus.

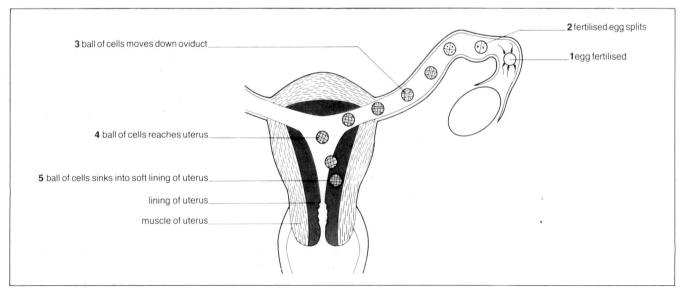

If there's an egg in the oviduct, one of the sperms may bump into it and fertilise it. Normally this happens within a few hours of intercourse. If there is no egg present, the sperms may stay alive in the oviduct for two or three days and if an egg is shed from the ovary within this time, it may get fertilised. Although it's only a short distance from the vagina to the top of the oviduct, the sperms are very small and the journey is not an easy one. Many never reach the egg. The reason why so many are produced is to raise the chance of one of them getting through.

What happens to the fertilised egg?

After fertilisation, the egg divides up into a little ball of cells which moves down the oviduct to the uterus (Figure 4). It then sinks into the soft lining of the uterus, a process called **implantation**. If this happens, we say that the woman has conceived. She is now pregnant.

Investigation 1

Looking at the male reproductive system

1 Your teacher will give you a male rat which has been dissected so as to show the reproductive system.

2 Find the penis and scrotal sacs.

3 Cut into one of the scrotal sacs and locate the testis (See Investigation 1, p. 299).

What is the coiled tube lying alongside the testis?

Where does it lead?

Which other structures shown in Figure 1 can you see in your dissected rat?

4 Feel the hard bone covering the top end of the urethra.

What is this bone? (see page 268)

What do you think it is for?

Investigation 2

Looking at the female reproductive system

1 Your teacher will give you a female rat which has been dissected so as to show the reproductive system.

2 Find the ovaries. These are small round organs on either side of the abdominal cavity.

3 Locate the oviduct and uterus.

The rat differs from the human in having a Y-shaped uterus as shown in the illustration below.

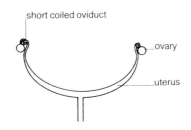

short coiled oviduct

ovary

uterus

What other structures shown in Figure 2 can you see in your dissected rat?

4 Feel the hard bone just posterior to uterus.

What is this bone? (see page 268)

What do you think it is for?

5 Find the opening of the vagina to the exterior.

There is another small opening close to the vaginal opening: what is it?

Assignments

1 Why is it important that a very large number of sperms should be present in the semen?

2 Why is it an advantage for the testes to be situated in the scrotal sac outside the main body cavity? Can you think of any disadvantages? What do you think the consequences would be if the testes were located *inside* the body cavity?

3 The diagram below shows a transverse section through a penis.

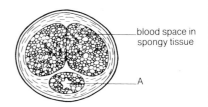

blood space in spongy tissue

A

a) What is the name of the structure labelled A?
b) Give two functions of this structure.
c) At what particular times does it perform each of these functions?
d) What is the function of the spongy tissue?
e) Why is this function important?

4 Which structures in the female are equivalent to these structures in the male:

(a) penis, (b) testes, (c) sperm ducts, (d) urethra?

In each case say in what respect the structures are equivalent.

Pregnancy and birth

*If an egg gets fertilised
it divides into a ball of cells
which sinks into the lining of the
uterus. The woman is then
pregnant.*

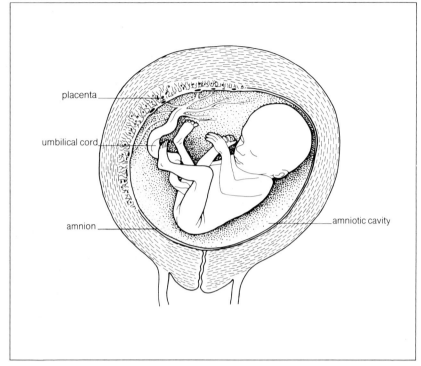

Figure 1 The foetus 14 weeks after the beginning of pregnancy.

What happens in the uterus?

The ball of cells is an **embryo**. An embryo is simply an early stage in the development of an organism from the egg to the adult. In the next two months the cells multiply and differentiate, and gradually the embryo grows into a **foetus**. The foetus looks like a miniature human being (Figure 1).

The private pond of the foetus

As the embryo develops it becomes surrounded by a thin membrane called the **amnion**. This membrane encloses a cavity called the **amniotic cavity** and it's filled with a watery **amniotic fluid**.

As development goes on, the amniotic cavity expands like a balloon until eventually it fills the entire uterus. The foetus floats in the middle of it, in a kind of 'private pond'. The amniotic fluid cushions the foetus, protecting it from being bumped and damaged as the mother moves around.

How is the foetus kept alive?

Attached to the belly of the foetus is a tough strand called the **umbilical cord**. This is its lifeline, bringing it all the things it needs such as oxygen and food substances, and taking carbon dioxide and excretory waste away from it.

The umbilical cord runs to a structure called the **placenta**, which is attached to the lining of the uterus (Figure 2). The placenta is shaped like a plate, and it has numerous finger-like **villi** which stick into blood spaces in the wall of the uterus. The mother's blood circulates through these spaces, and the villi contain blood capillaries which are connected to the foetus by an artery and vein in the umbilical cord. The barrier separating the foetus's blood from the mother's blood is very thin. As the foetus's blood flows through the placenta it picks up oxygen and dissolved food substances from the mother's blood. At the same time it sheds carbon dioxide and excretory waste (urea) *into* the mother's blood.

So the placenta supplies the foetus with everything it needs. In addition it produces the hormones oestrogen and progesterone which stop menstrua-

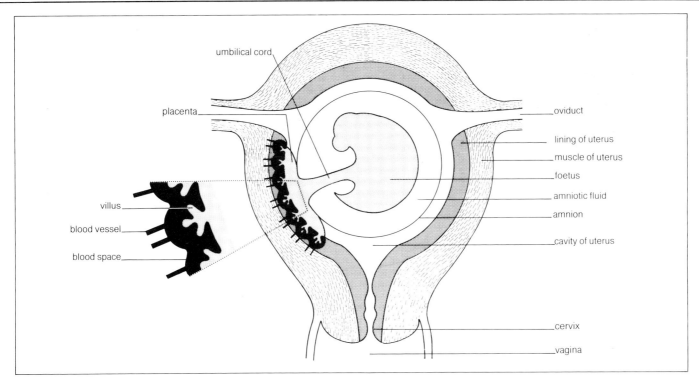

Figure 2A This diagram shows the uterus of a pregnant woman.

tion and prevent any further eggs being produced by the ovaries. These hormones are also produced by the ovaries themselves (see page 300).

Growth of the foetus

By the end of the third month the foetus is fully formed right down to the fingers and toes. It now grows until it fills the uterus. Meanwhile the uterus expands greatly in order to accommodate the foetus, and its wall becomes thicker and more muscular in readiness for birth. As the baby grows it gets more active: it moves its arms and legs and by the end of the fourth month it may kick quite hard.

Care during pregnancy

During pregnancy it's important that the woman should eat the right kind of food and not do anything which might injure her baby. From time to time she visits an **ante-natal clinic** where she is examined by a doctor to make sure that everything is progressing normally, and given advice on how to prepare for the birth of her baby.

There are times during pregnancy when the woman may feel rather unwell, and the doctor may give her some medicine to help. However, doctors are cautious about prescribing new drugs, however thoroughly they have been tested beforehand. In the 1960s a number of pregnant women in Britain were given a drug called thalidomide to help them get to sleep: as a result some of them gave birth to babies with severe deformities, such as no arms or legs.

This was a tragedy which no one wants to see repeated. Even smoking can affect the health of the foetus. The point is that the placenta is very good at supplying the foetus with the things it needs, but by the same token it may supply it with things it does *not* need.

Certain germs may also get across the placenta and harm the baby. Such is the case with the virus that causes German measles, and this is why girls who have not had this disease are always immunised against it. The germs that cause venereal diseases can also get across the placenta and damage the baby.

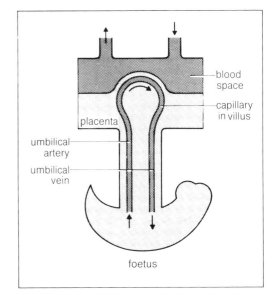

Figure 2B Schematic diagram showing the relationship between the blood of the foetus and mother. Although the two bloodstreams come very close to each other, they never mix.

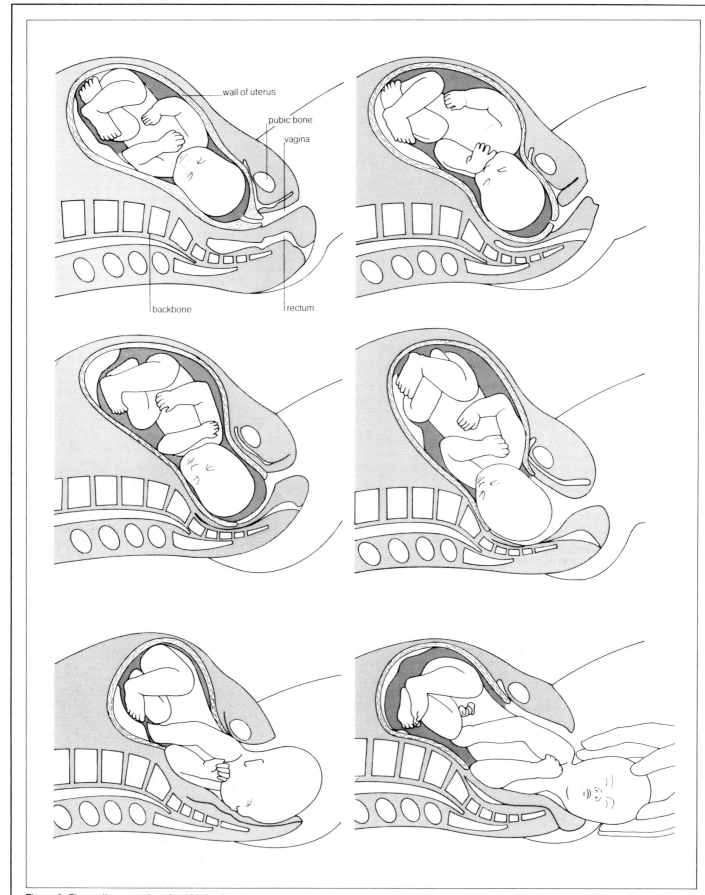

Figure 3 These diagrams show how birth takes place.

Birth

After about nine months the uterus begins to undergo occasional contractions which become steadily more frequent and powerful. This is called **labour**, and it's the first sign that the baby is about to be born. At about this time the amnion bursts like a balloon and the amniotic fluid escapes through the vagina: this is called the 'breaking of the waters'. Soon afterwards the uterus starts contracting so powerfully that the foetus is pushed through the vagina. Usually the baby is positioned in such a way that it comes out head first (Figure 3). However, babies are sometimes born feet first or bottom first.

Once the baby has come out, it starts to breathe. If it doesn't start breathing of its own accord, the doctor or nurse may give it a tap on the backside causing it to take a quick breath in. The umbilical cord, no longer needed, is tied and cut: you can see it in Figure 4 – the scar becomes the person's navel or 'belly button'. Meanwhile the placenta comes away from the wall of the uterus and passes out through the vagina. This is called the **afterbirth**. The average mass of a newborn baby is just over 3 kg. The birth, or delivery as it's called by doctors and nurses, is now complete.

Birth is brought about by a change in the amounts of hormones produced by the placenta, and also by a hormone which starts being produced by the pituitary gland. This is called **oxytocin** and it brings about the contractions of the uterus.

A woman may have her baby either in hospital or at home, depending on circumstances. Most babies are born quite easily. Sometimes, however, the baby needs to be helped out with forceps: these are like a large pair of tongs and are used to gently grasp the baby's head as it emerges. In particularly difficult cases the medical staff may feel it's best to give the mother an anaesthetic and remove the baby by cutting open the wall of the abdomen and uterus. This operation is known as a **Caesarian section** because Julius Caesar is believed to have been born this way.

Sometimes birth occurs before the ninth month, and the baby is **premature**. If the baby weighs more than about 2 kg it will probably survive, though it may have to be kept warm in an incubator and given a special oxygen supply until it is mature enough to support itself.

Figure 4 This baby has just been delivered and has taken its first breath in. Notice the umbilical cord on the right.

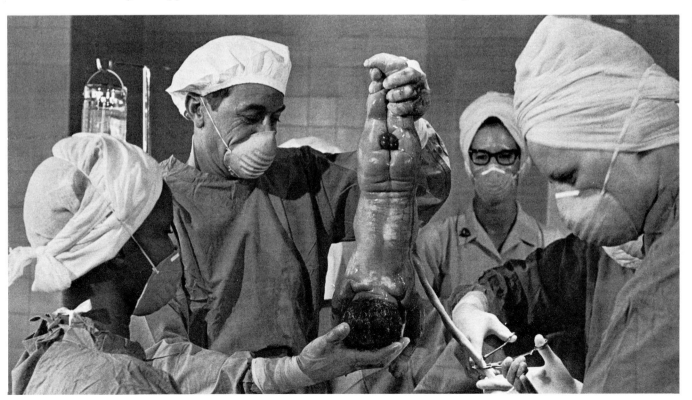

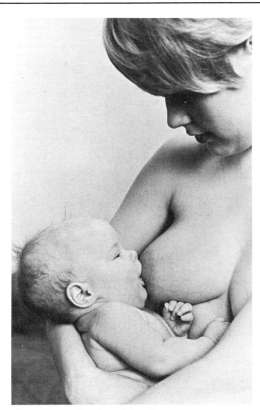

Figure 5 A mother breast-feeding her baby.

Occasionally things go wrong at an early stage of pregnancy and the foetus is expelled from the uterus. If it's not already dead, it dies almost immediately afterwards. This is called a **miscarriage**. It is, of course, very distressing for a woman who is looking forward to the birth of her baby. However, it does not mean that she cannot become pregnant again, and during a subsequent pregnancy the doctor will give her special help to prevent it happening again.

Caring for the newborn baby

During pregnancy the mother's breasts enlarge and the **mammary glands** inside them get ready to secrete **milk**. Soon after birth the baby starts to suck its mother's nipples (Figure 5). This stimulates the breasts to release the milk and make more.

Milk is the baby's food for the first few months of life. Of course it doesn't *have* to get milk from its mother: it can be fed equally well from a bottle with a teat. However, there's a lot to be said for breast-feeding: the mother's milk is a natural food delivered at just the right temperature and it allows a close contact between the mother and her baby. It's also cheaper. It's not possible for the baby to take in solid food at this stage, for it has no teeth to chew it with and in any case its digestive system would be unable to cope with solids.

Mothers wrap their babies up well, particularly in cold weather. This is necessary because the baby's ability to control its body temperature takes time to develop. Whereas adults adjust to the cold by shivering and making more heat, the baby is not yet able to do this efficiently.

These are just a few aspects of looking after one's baby. For several weeks after the birth a district nurse or health visitor will visit the mother at home and give her advice on looking after her baby, and after six weeks the mother will go back to the clinic for a post-natal examination to make sure that her own health is satisfactory. In most areas she will go on being visited until the child starts going to school at the age of five.

Twins and multiple births

In the human, only one embryo usually develops in the uterus at a time. However, two may sometimes be present together, each with its own placenta and umbilical cord. These are known as **twins**.

There are two kinds of twins: identical and non-identical (Figure 6). Non-identical twins arise if two eggs are shed from the ovaries at the same time, and both are fertilised. Although the two babies will be born together, they don't have the same genes and will be no more alike than brothers or sisters. Twins of this sort are known as fraternal twins and they may be of different sexes.

Identical twins arise in a quite different way. A single egg is shed from one of the ovaries and fertilised in the usual way. It then splits into two cells, *each* of which develops into an embryo. The two cells have exactly the same genes, and so the two babies will be exactly alike and will be the same sex.

Sometimes the two cells into which the egg splits do not separate completely. The result is that the two embryos are joined together at some point, resulting in **Siamese twins**. If the connection is not too extensive, the two babies can be separated by an operation; but if the internal organs are intimately connected, it's very difficult to separate them and it is unlikely that both will survive.

Occasionally three or more eggs are produced by the ovaries at the same time, resulting in triplets, quadruplets or even quintuplets. Such **multiple births** tend to occur in women who have been given a fertility drug to help them get pregnant. As the foetuses grow, it becomes more and more difficult for the mother's uterus to contain them, and so she gives birth early, often around the seventh month. Very rarely do all the babies survive.

Although multiple births are unusual in man, they are quite usual in other mammals such as dogs, cats and mice. The offspring constitute a **litter**.

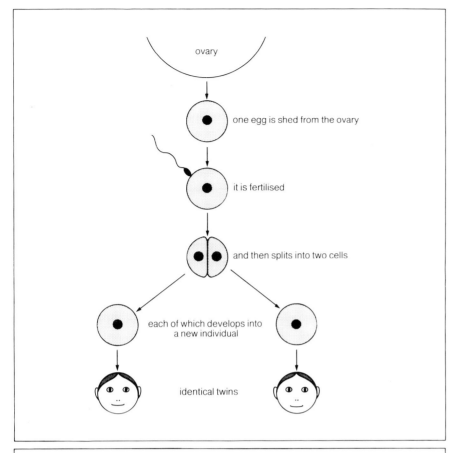

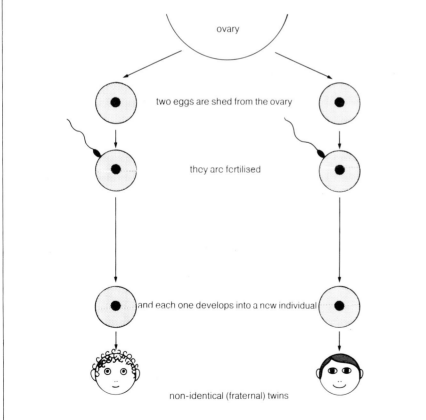

Figure 6 How twins occur.

Assignments

1 What functions are performed by (a) the muscle in the wall of the uterus, (b) the amniotic fluid, (c) the umbilical cord, and (d) the mammary glands?

2 Name five jobs which are carried out by the placenta. What is it about the structure of the placenta which makes it ideally suited to do these jobs?

3 Babies are usually born head first. What advantages are there in being born this way? What changes take place in the baby's body soon after it is born?

4 Some of the food requirements of an adult woman with a body mass of 55 kg, and of the same woman in an advanced state of pregnancy are given in the following table:

	non-pregnant	pregnant
Energy	9200 kJ	10 700 kJ
Protein	29 g	38 g
Vitamin A	750 μg	750 μg
Vitamin D	2.5 μg	10 μg
Vitamin C	30 mg	30 mg
Calcium	0.5 g	1.2 g
Iron	28 mg	28 mg

a) Suggest one reason why the woman requires more energy when she is pregnant.

b) Name two kinds of food from which she is likely to obtain most of this energy.

c) Suggest one reason why she requires extra protein when she is pregnant.

d) Why do you think she needs extra vitamin D and calcium when she is pregnant?

e) What are the *percentage* increases in the amount of energy, protein, vitamin D and calcium which she needs when she is pregnant?

f) Which substances in the table do not need to be increased during pregnancy?

g) Why do you think those particular substances do not need to be increased?

Sex without pregnancy

We normally associate sex with having babies. However, there are various circumstances in which sex does not lead to pregnancy.

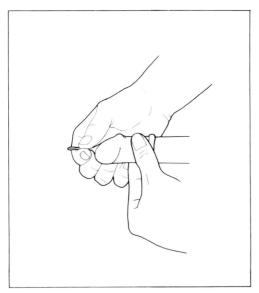

Figure 1 The sheath is here seen being slipped onto the erect penis.

Some couples are unable to have children

One of the commonest causes of this is that either the male cannot produce sperms or the female cannot produce eggs. The person is said to be **sterile**. Sometimes the male does have sperms in his semen but there simply aren't enough of them to ensure fertilisation.

A couple who find that they can't produce a baby should discuss their problem with a doctor. The doctor can arrange for them to see a specialist who may be able to find out what's wrong.

A lot can be done to help childless couples these days. For example a woman who isn't producing eggs can be treated with a **fertility drug**. This is a hormone preparation and it stimulates her ovaries to start working. In the past this treatment has sometimes proved more successful than either the patient or the doctor bargained for: the woman has become pregnant and produced quadruplets or even quintuplets!

Recently, scientists have succeeded in fertilising human eggs *outside* the body. The egg is taken out of the oviduct in an operation and fertilised by sperm in a test tube. The fertilised egg is then put into the uterus to develop. Although this research is at an early stage, it may prove to be a way of helping childless couples in the future.

Preventing pregnancy

Some couples do not wish to have children at a particular time, and they achieve this by means of **contraception**.

Contraception is any procedure which prevents conception. There are many different methods, and some of them involve the use of an artificial device called a **contraceptive**. The methods fall into three groups: (1) those that stop sperms reaching the egg; (2) those that stop eggs being produced, and (3) those that stop the fertilised egg developing in the womb.

Now let's look at each method in turn, and you can decide for yourself which group each one fits into.

Withdrawal

A commonly used way of avoiding conception is for the male to withdraw his penis from the vagina just before he ejaculates. However, the lubrication fluid which comes from the penis before ejaculation may contain some sperms, and so this is an extremely unreliable method of contraception.

The rhythm method

A woman will only become pregnant if there is an egg in her body to be fertilised. Now there is a certain length of time in the sexual cycle when no eggs are available: this is round about the time of menstruation. If she has intercourse at this time, it is unlikely that she will become pregnant. We call this period of time the **safe period**.

The trouble is that the length of the safe period varies. It depends on how long the female's eggs can stay alive in her oviducts before being fertilised, and on how long the male's sperms can stay alive in her body before an egg becomes available for fertilisation. It also depends on when exactly she ovulates, which can vary a lot. In some cases the safe period may last as long as three weeks, but in other cases it may only last a few days. The rhythm method is therefore unreliable, unless carried out under the expert guidance of a doctor.

The sheath

This is a type of contraceptive, worn by the male, which stops sperms getting into the female's vagina. Known as the condom or 'French letter', it consists of a rubber sheath, shaped like the finger of a glove, which fits over the erect penis (Figure 1). A little bag-like extension at the end catches the semen when ejaculation occurs. The rubber is very thin and is usually coated with a lubricant.

The sheath is a very reliable contraceptive if used properly, and is available from chemist shops and vending machines in public lavatories.

The cap or diaphragm

This type of contraceptive also forms a barrier to sperm, but it is worn by the female. It consists of a dome-shaped piece of rubber with a metal spring round the edge. It comes in various sizes, and the correct one can be prescribed by a doctor. It fits over the bottom end of the uterus and the woman is taught how to put it in herself (Figure 2).

The cap is a pretty good contraceptive. However, it may slip out of place if it hasn't been fitted properly, and sperms may get round it. It's best to use it with a **spermicide** and leave it in for at least six hours after intercourse.

Spermicides

Spermicides are substances which kill sperms. They are available as sprays, creams or tablets (pessaries). The female should put the spermicide as far up her vagina as possible, not more than ten minutes before intercourse (Figure 3). The tablets dissolve in her normal vaginal secretions, forming a kind of foam.

On their own spermicides are not very effective because sperms sometimes manage to get through them. However, when used with the diaphragm or sheath, they can be very effective.

The pill

This is a tablet which is taken by the female. It is known as the **oral contraceptive** and it prevents any eggs being produced by the ovaries. One tablet has to be taken daily throughout the sexual cycle except for about a week round about the time of menstruation. Although the tablets prevent eggs being produced, the woman's menstrual periods still take place at the usual times. The tablets have to be prescribed by a doctor and are available on the National Health Service.

How does the pill work? During pregnancy the ovaries and placenta produce certain hormones which prevent any further eggs being produced by the ovaries (see page 300). The pill contains chemical substances identical with these hormones.

The pill is very effective and there are normally no unpleasant side effects. Since it came onto the market in the 1950s it has done more than any other contraceptive to reduce the number of unwanted births. If a woman who has been on the pill stops taking it, she will start producing eggs again, and can then become pregnant in the normal way.

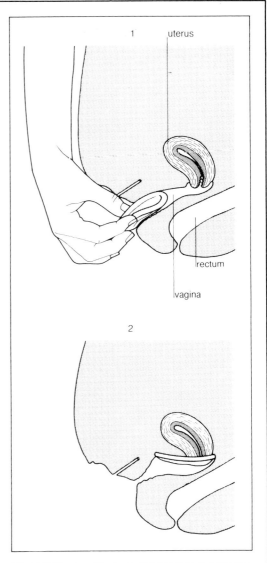

Figure 2 The cap (diaphragm) being inserted into the vagina. Note how it fits over the bottom end of the uterus.

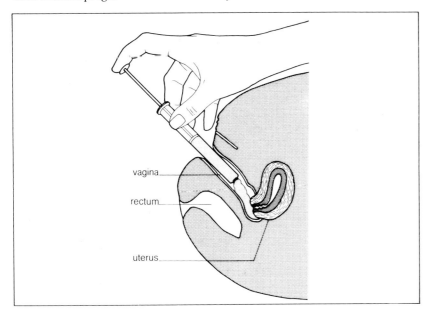

Figure 3 A spermicidal cream being put into the vagina by means of a syringe.

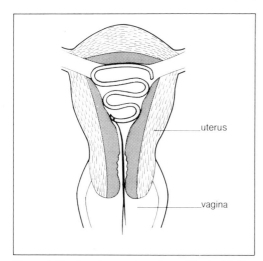

Figure 4 The loop is seen here in position inside the uterus. The strings enable it to be taken out easily by the doctor. It is one of the most widely used intra-uterine devices.

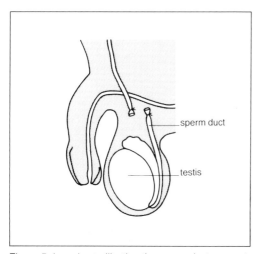

Figure 5 In male sterilisation the sperm ducts are cut and tied, as shown here in side view.

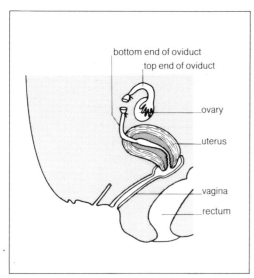

Figure 6 In female sterilisation the oviducts are cut and tied, as shown here in side view.

Intra-uterine devices

These devices, called IUDs for short, are made of plastic or metal and are put into the uterus by a doctor who simply pushes them up through the vagina. They come in all sorts of shapes: one of the most widely used is shown in Figure 4.

IUDs prevent the embryo becoming implanted in the lining of the uterus. No-one knows exactly how they achieve this but they certainly stop the female becoming pregnant. They don't normally cause any discomfort, and are a reliable method of contraception. If a woman with an IUD wants to become pregnant, the IUD can be taken out by a doctor.

Sterilisation

In this method the person has an operation which prevents conception permanently. It can be carried out on either the male or the female. In the male the surgeon ties and cuts the sperm ducts (Figure 5). This prevents sperms getting to the urethra from the testes: the man can still ejaculate, but his semen will not contain any sperms.

In the female the surgeon ties and cuts the oviducts (Figure 6). This stops any sperms getting up the oviducts and so her eggs cannot be fertilised.

Sterilisation is the most complete method of contraception. The operation is simple and quick and there are no unpleasant consequences. Men sometimes fear that sterilisation might reduce their masculinity or change their personality in some way, but this is not true. The only snag is that once you've been sterilised, there is no possibility of having any children in the future. So it's no use having second thoughts afterwards!

Abortion

Despite the various contraceptive methods described above, women often become pregnant when they don't want to. The only way to avoid giving birth to the baby is to destroy the foetus in the womb. This is known as **abortion.**

There are various ways of carrying out an abortion, and it should always be done in a clinic or hospital by a qualified doctor. It is extremely dangerous for it to be done any other way. The procedure gets more and more difficult as the foetus gets older, and most doctors consider that an abortion should not be carried out after about five months.

In the United Kingdom and many other countries abortion is only allowed if the doctor considers that by continuing the pregnancy the woman's health is at risk. Some people think that the law should be changed to make abortion more easily available. Many moral and ethical issues are involved in this difficult question.

Family planning and birth control

All the procedures discussed in this Topic are methods of **birth control**. Birth control is extremely important, particularly in over-populated countries where there isn't enough food to go round. Every couple should consider how many children they can support, and in the light of this should use whatever birth control methods are necessary and in keeping with their beliefs and conscience. This is known as **family planning**. In Britain and many other countries there are family planning clinics where people can get advice about birth control methods.

Masturbation

Normally an orgasm is achieved by the stimulation which accompanies the rhythmical movements of the penis inside the vagina. However, it can also be achieved by stimulating the penis or the clitoris with, for example, the hand. This is called **masturbation.** It is not in the least harmful and indeed can give considerable relief at times when intercourse is impossible. What *can* be harmful is the feeling of guilt which people sometimes have about it.

Homosexuality

Homosexuality is having sexual feelings towards members of one's own sex, in contrast to heterosexuality which is having sexual feelings towards members of the opposite sex. Homosexual relationships can occur between men or between women. The two partners may reach an orgasm by, for example, masturbating each other or, in the case of male homosexuals, by one partner putting his penis into the anus of the other, a practice known as anal intercourse. However by no means all homosexuals engage in physical contact.

Homosexual relationships occur quite often in places like prisons and boarding schools where people of the same sex are cooped up together. Such relationships develop mainly because of frustration and are not usually permanent or harmful. The people involved are generally able to form relationships with the opposite sex when the circumstances allow. However, there are people who are *only* capable of relating sexually to persons of their own sex. Such people are usually as normal in appearance and behaviour as heterosexuals, though some male homosexuals may behave in an effeminate manner.

Some individuals are sexually attracted to members of their own sex *and* members of the opposite sex. They are known as bisexual.

There is nothing wrong with being a homosexual, though a great deal of harm is sometimes done by feelings of guilt and isolation which may accompany it. If a person finds that he or she has homosexual feelings and is worried about it, it's better to talk it over with parents, a counsellor or a trusted friend than to bottle it up. Many different types of professional help are available to those who want it.

Assignments

1 What does the word contraception mean?

The sheath is one of the most reliable contraceptives. What do we mean by the word 'reliable' in this context? Why is the sheath so reliable, and what disadvantages do you think it might have?

2 What are the advantages and possible disadvantages of the 'pill' as a method of contraception?

3 What is meant by the 'safe period'?

Below is shown what normally happens on certain days during a typical 28-day menstrual cycle:

Days 1 – 5   Lining of uterus disintegrates (menstruation)

Day 14   Egg released from ovary (ovulation)

The egg cell can be fertilised for up to three days after ovulation. Sperms can live in the female's uterus and oviducts for as long as eight days though they normally retain the capacity to fertilise an egg for only one or two days.

a) On which days could the egg be fertilised?
b) On which days could intercourse result in fertilisation?
c) Why is the 'safe period' an unreliable method of birth control?
d) There are cases of women becoming pregnant having had intercourse just before, or just after, menstruation. Suggest possible reasons for this.

4 A man and his wife find that they are not managing to produce any children though they are having intercourse regularly and are not using any method of contraception. Suggest five possible reasons for their lack of success.

5 The graph below shows the number of pregnancies in girls under 16 in England and Wales each year from 1948 to 1968 inclusive: Suggest reasons why there has been a steady rise in the number of pregnancies during this time.

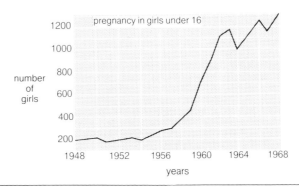

Sex and disease

Sex involves close contact between two people. It is therefore an easy way of passing germs from one person to another.

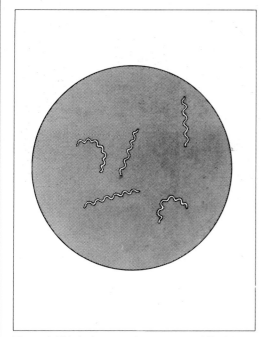

Figure 1 This is the germ that causes syphilis. It is a spiral-shaped bacterium called a spirochaete.

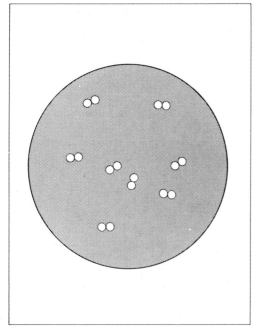

Figure 2 This is the germ that causes gonorrhoea. It is a ball-shaped bacterium called a coccus, and they normally occur in pairs.

Venereal diseases

Any infectious disease is liable to be passed from one person to another during sexual intercourse. However, certain diseases are normally *only* passed this way. These are called **venereal diseases** or **VD** for short. The germs which cause them cannot survive even for a short time outside the body, so it's impossible to pick them up from such objects as towels or lavatory seats.

Venereal diseases can be very serious, in some cases deadly. The trouble is that in the early stages the symptoms may be very slight, and the person may not know that there's anything wrong. For this reason venereal diseases are sometimes called 'hidden diseases'. They occur in all sections of society, and are just as common amongst homosexuals as heterosexuals.

We will now look briefly at a few of the more common diseases.

Syphilis

This is caused by a certain kind of bacterium (Figure 1). The first sign of the disease usually appears about two to four weeks after intercourse, and consists of a sore on, or near, the genital organs: usually just inside the vagina in the female, and on the end of the penis (the glans) in the male. The sore doesn't hurt and it usually lasts for only a week or two at the most, so the person may not notice it. The germs then move to other parts of the body, causing a mild fever and rashes on the skin. After a few weeks these symptoms disappear, and the person appears to recover. However, the germs are still in the bloodstream and eventually they may attack the brain, making the person go blind and insane. This final stage of the disease may not occur until many years after the original infection.

If a pregnant woman has syphilis, the germs are likely to pass across the placenta into the baby's bloodstream. As a result the baby may be born dead ('stillborn'), or it may be born with the disease and become crippled with it later. Nowadays all pregnant women have their blood tested to make sure that it does not contain any syphilis germs.

Gonorrhoea

This is also caused by a bacterium, though of a different kind from the one that causes syphilis (Figure 2). The first sign of the disease usually appears a few days after intercourse, and consists of a burning sensation when urinating. This happens because the tube down which the urine passes, the urethra, becomes inflamed. There may also be a yellowish discharge from the reproductive opening: this is easily seen in the male, but the female may not notice it because it is mixed with her normal secretions. Sometimes the germs spread to other parts of the body, and the person may feel ill and get swollen and painful joints. Gonorrhoea can cause sterility, making it impossible to have children.

If a pregnant woman gets gonorrhoea, her baby may become infected as it passes through the vagina during birth. As a result it may develop very sore eyes which, if untreated, can lead quickly to blindness.

Other sexual diseases

A number of other diseases can be passed from one person to another by intercourse or close sexual contact. Some of them can be spread in other ways as well, so strictly speaking they aren't *venereal* diseases. However, they are treated in venereal clinics. Here are a few of them:

Urethritis

In this disease the urethra gets inflamed, much as it does in gonorrhoea, causing pain and irritation when urinating. The kind of urethritis associated with sex is caused by a virus, and in rare cases it may lead to another condition called Reiter's disease which is characterised by aching joints (arthritis), painful feet and sore eyes (conjunctivitis).

Herpes
This is also caused by a virus – the same kind that causes chicken pox. The symptoms include sores on the genital organs, usually inside the vagina in the female and on the head of the penis (the glans) in the male.

Thrush
This is caused by a fungus which may make the mouth cavity and genital organs sore and inflamed. In the female there is a white or yellowish discharge from the vagina.

Trichomoniasis
This is caused by a protozoan. The symptoms may be hardly noticeable, but sometimes it causes discomfort when urinating and in the female there may be a discharge from the vagina.

Hepatitis
In this disease the liver gets infected. It has been found that viral hepatitis (a type of hepatitis caused by a virus) can be passed from one person to another by sexual contact.

How to avoid venereal diseases

The only *sure* way of avoiding venereal disease is to make certain that one's sexual partner hasn't got it. Obviously this is not easy, but having casual sexual relations with all sorts of different people is asking for trouble.

The wearing of a contraceptive sheath by the male helps to prevent most venereal diseases, but it doesn't give complete protection. Venereal diseases are much more common now than they were thirty years ago, and the reason may be that the sheath has been replaced by 'the pill' as the main means of contraception. The pill allows people to have sex without the risk of pregnancy, but it gives no protection against venereal disease.

Curing venereal diseases

Syphilis and gonorrhoea can be cured with antibiotics such as penicillin *provided thorough treatment is carried out at an early enough stage*. The trouble is that often people don't realise they have got the disease, and so they do nothing about it until it's too late.

Virus diseases such as urethritis and hepatitis are more difficult to cure, because they don't respond to antibiotics. This means that the patient must wait for them to clear up of their own accord. In some cases this may take many months or even years, and often the disease disappears for a while and then comes back again.

Most large general hospitals have a special clinic where people can be examined to find out if they have got any venereal diseases (Figure 3). The examination takes only a short time, and includes a urine and blood test. When the results are known treatment can, if necessary, be given. All this is confidential and free of charge and it does not have to be done through the person's own doctor: if someone suspects that he has caught a venereal disease, all he has to do is to report direct to the hospital. If it turns out that he *has* got the disease, he will be urged to give the names of anyone with whom he has had sexual intercourse in the last few weeks. A social worker will then visit them and suggest that they should attend the clinic too.

In this way venereal diseases are kept under control and prevented from spreading through our society.

Figure 3 A person visiting the local VD clinic for a check-up.

Assignments

1 Why is syphilis described as a 'hidden disease'?

2 Why is syphilis considered to be a very serious disease?

3 Why is gonorrhoea easier to cure than herpes?

4 The graph below shows the number of new cases of gonorrhoea seen in VD clinics each year from 1936 to 1970:

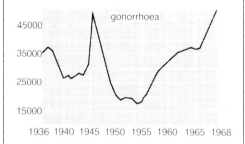

Suggest reasons why the number of cases:

a) rose rapidly in the first half of the 1940s,

b) fell rapidly in the last half of the 1940s,

c) rose steadily from the mid-1950s onwards.

Sex in flowering plants

Sexual reproduction occurs in plants as well as animals. In higher plants the part of the plant responsible for this is the flower.

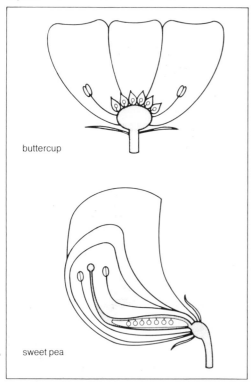

Figure 1 The flower contains the plant's reproductive organs.

Figure 3 Side views of the buttercup and sweet pea flowers to show the kind of variety that one finds in flower structure. In both cases, the flower has been sliced down the middle so as to show the internal structure. In reality there are far more stamens than are shown here, particularly in the buttercup.

The structure of flowers

The basic structure of a flower can be seen in plants like cherry, plum and hawthorn (Investigation 1). This kind of flower is illustrated in Figure 2.

The flower is made up of a series of rings of structures. The outermost ring consists of several small green leaf-like **sepals**; then come the **petals** which are often brightly coloured; next come the **stamens** which look rather like pins; and finally in the centre there is a club-shaped **carpel**. All these structures are situated at the end of a stalk which is slightly swollen to form the **receptacle**.

At the base of each petal you will see an area which is slightly thicker than

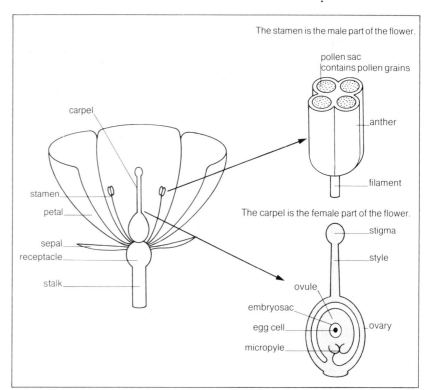

Figure 2 The structure of a typical flower.

the rest: this is called the **nectary** and it produces a sugary liquid called **nectar**.

The stamens constitute the male part of the flower. Each one has a knob at the top: this is called the **anther**, and it contains four **pollen sacs** in which pollen grains are formed. The pollen grains are equivalent to an animal's sperms. The rest of the stamen is known as the **filament**.

The carpel constitutes the female part of the flower. It consists of three parts: a slightly swollen **stigma** at the top, then a slender stalk called the **style**, and a swollen **ovary** at the bottom. Inside the ovary there is a small body called the **ovule** which is attached to the wall of the ovary by a short stalk. The ovule contains a little bag called the **embryosac** and inside this is an **egg cell**. There is a small hole in the wall of the ovule called the **micropyle**. You need a microscope to see the embryosac and egg cell.

Variations on the theme

You have only to look at a few flowers to realise that they aren't all exactly like the one just described although they all follow the same basic plan (Investigation 2).

Figure 3 shows two flowers to illustrate the kind of variation that one finds. Notice that the buttercup has numerous carpels each containing one ovule; in contrast, the sweat pea has a single carpel containing a row of ovules.

You will also notice that these two flowers differ in shape. In the buttercup the petals are all identical as you go round the flower. Such a flower is

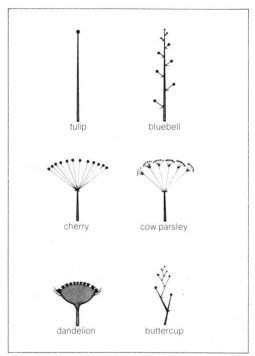

Figure 4 There are two ways of describing the shapes of flowers.

Under **radial symmetry**:
You can cut this flower in any of these planes to get two halves which are mirror images of each other.
Examples: buttercup, cherry, hawthorn

Under **bilateral symmetry**:
You can cut this flower only in this plane to get two halves which are mirror images of each other.
Examples: sweet pea, snapdragon, orchid

described as **radially symmetrical**. However, in the sweet pea the petals at the top and bottom are different from those on either side. This kind of flower is described as **bilaterally symmetrical**.

The difference between radially and bilaterally symmetrical flowers is illustrated in Figure 4.

Flowers vary in many other ways too. For example, the petals may be joined together to form a kind of 'trumpet' in, for example, the foxglove, and in other flowers the sepals are often missing.

How are flowers arranged on the stem?

There is a great deal of variety in the way flowers are arranged on the main stem of the plant, and on how many flowers there are altogether (Investigation 3). Some common arrangements are shown in Figure 5. The collection of flowers on the shoot is known as the **inflorescence**.

Pollen and pollination

Pollen grains are very small bodies like specks of dust. Their job is the same as an animal's sperms: to fertilise the eggs.

The pollen grains develop inside the anthers. When the anther is mature, it splits open and the pollen grains are released. They are then conveyed to another flower of the same species, and if one of them gets onto a stigma, it sticks to it. The process by which the pollen grains are conveyed from the anthers to the stigma is called **pollination** (Figure 6).

Fertilisation

Once a pollen grain has landed on a stigma, it sends out a snake-like outgrowth called a **pollen tube**. This grows into the stigma and down the style. It is attracted by sugar in the stigma and nourished by substances in the tissues of the style. Towards the tip of the pollen tube there is a **male nucleus** which is equivalent to the nucleus in the head of an animal's sperm.

Having reached the ovary, the pollen tube pushes its way into the ovule, usually through the micropyle. The tip of the pollen tube now grows towards the egg cell in the centre of the ovule. Finally the male nucleus fuses with the egg cell. This is **fertilisation**, and is equivalent to the fertilisation of an egg by a sperm in an animal.

The fertilised egg now divides up into a ball of cells which becomes an **embryo**. This remains in the centre of the ovule, and becomes surrounded by a special tissue called the **endosperm** which supplies it with food.

Meanwhile the ovule itself becomes the **seed** and the wall around it hardens to form the tough seed coat. While this is happening the ovary develops into a **fruit**. So the seed becomes surrounded by a fruit: cut open any fruit and you will normally find seeds inside it.

Finally water is drawn out of the seeds so they become very dry. They then become dormant, and in this state they can survive bad conditions such as drought and cold.

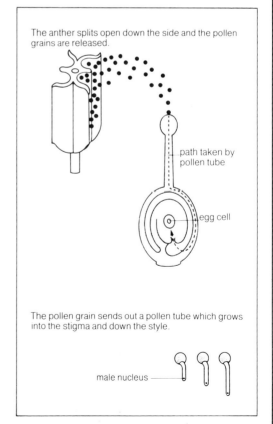

Figure 5 These are some of the ways the flowers may be arranged on the stem of a plant. Notice that the dandelion 'flower' is not just one flower but a whole mass of very small flowers or florets sitting on top of the swollen end of the stem.

The anther splits open down the side and the pollen grains are released.

path taken by pollen tube

egg cell

The pollen grain sends out a pollen tube which grows into the stigma and down the style.

male nucleus

Figure 6 In pollination pollen grains are transferred from an anther to a stigma. A pollen tube then grows down the style to the ovary where fertilisation occurs.

Figure 7 Pollen being scattered from two hazel catkins.

Figure 8 A bee collecting nectar from a flower.

The flower has now done its job, so the sepals shrivel up and the petals fall off, leaving the fruit with the seeds inside.

Different methods of pollination

Pollination is normally carried out either by wind or by insects (Investigation 4). It can also be carried out artificially by man.

Wind-pollinated plants include grasses, hazel and willow. The familiar hazel catkins and pussy willows are clumps of very small male flowers whose pollen is blown about by the wind. The pendulous way they hang down and are shaken by even a slight gust of wind ensures that the pollen is scattered over a wide area (Figure 7).

Insect-pollinated plants include buttercups, roses, wallflowers and many others. Insects such as bees visit the flowers to feed on the nectar. As the insect pokes its head into the flower, its hairy body gets covered with pollen (Figure 8). When the insect visits other flowers, some of the pollen gets onto the stigmas, thereby pollinating them.

Experiments have shown that insects such as bees are attracted to flowers by their colour, shape and smell. Some flowers have gone to great lengths to entice the insect and make sure that it gets covered with pollen. For example, in orchids some of the petals form a kind of platform which the insect can land on. Marks on the petal guide the insect to the nectar, like the landing lights on an airport runway.

Some flowers are constructed in such a way that when the insect lands on it and pushes its head in, the stamens are jerked so that the insect's body gets well and truly covered with pollen.

One of the most interesting cases is an orchid whose flower looks like a female wasp. When a male wasp sees the flower, it tries to copulate with it and gets covered with pollen. It then tries the same thing with another flower, pollinating it in the process.

Wind and insect pollination are quite different and require different adaptations on the part of the flowers. You can often tell whether a particular flower is pollinated by wind or insects just by looking at it. Table 1 sums up the main differences between them.

Cross-pollination and self-pollination

By now you must be wondering why flowers go to such lengths to spread their pollen. Why not let the pollen fall onto a stigma in the *same* flower? Actually this does sometimes happen, and we call it **self-pollination**. But it isn't good for the species. **Cross-pollination**, in which the pollen is transferred to another flower, creates variety and is really much better (see page 348).

So the various mechanisms which we have been discussing are really ways of making certain that cross-pollination takes place. There are other devices too. For example, in some flowers the stamens ripen before the carpels, so the pollen grains will have been dispersed by the time the carpels are ready. In other flowers the carpels ripen before the stamens. And some flowers are exclusively male or female and these may be found on separate plants. This is true of holly, for example.

Wind-pollinated flowers	Insect-pollinated flowers
1 Generally small	Generally larger
2 Petals green or dull coloured	Petals often brightly coloured
3 Do not produce nectar	Petals have nectaries which produce nectar
4 Flower hangs down for easy shaking	Flower faces upwards
5 Stamens and stigma hang out of the ring of petals	Stamens and stigma inside the ring of petals
6 Large number of pollen grains produced	Smaller number of pollen grains produced
7 Pollen grains very light with smooth surface	Pollen grains heavier with spikes for sticking to insect
8 Stigma has feathery branches for catching pollen	Stigma is like pinhead and lacks branches

Table 1 Summary of the main differences between typical wind-pollinated and insect-pollinated flowers.

Investigation 1

Looking at the structure of a basic flower

1 Obtain a flower from e.g. a cherry or hawthorn tree.

2 Identify the sepals, petals, stamens and carpels.

3 Remove some of the petals so as to see the inside of the flower more clearly.

   Which structures shown in Figure 2 can you see?

4 Make an accurate drawing of the flower and label its parts.

5 Pull off a sepal, petal, stamen and carpel, and lay them on a piece of paper.

6 Examine each one under a hand lens. Draw them in outline.

7 Cut open the carpel.

   Can you see an ovule inside it?

8 Cut open an anther.

   Can you see pollen grains inside?

Investigation 2

Looking at other flowers

1 Obtain up to six different kinds of flowers.

2 Examine each one carefully and write down how it differs from the flower which you looked at in Investigation 1.

3 Make a list of the ways in which all six flowers are similar to the flower in Investigation 1.

Suggest reasons why each kind of flower has its own characteristic shape and form.

Investigation 3

Looking at how flowers are arranged on the stem

1 Obtain up to six different kinds of flowering plant.

2 Make a simple but accurate sketch of each plant, showing how the flowers are arranged on the stem.

3 In each case compare the arrangement of the flowers with the diagrams in Figure 5.

   Which diagram does it resemble most closely?

Suggest why each arrangement should be useful to the plant.

Investigation 4

Exploring the differences between wind and insect pollinated flowers

1 Your teacher will give you one wind-pollinated and one insect-pollinated flower.

   Which do you think is which?

   How do you know?

2 Look carefully at each flower.

   Which features listed in Table 1 does it possess? Does it have any other adaptations for pollination besides the ones listed in the Table?

3 Examine the insect pollinated flower in more detail.

   What sort of insect do you suppose pollinates it?

   Give reasons for your suggestion.

4 If possible look at flowers being visited by insects. In each case observe what the insect does, and note any special adaptations which the flower has for being pollinated by that particular insect.

Assignments

1 Each of the words in the left hand column is related to one of the words in the right hand column. Write them down in the correct pairs.

   ovule   colour
   petal   egg cell
   pollen   sugar
   nectary   sperm
   leaflet   sepal

2 Explain the difference between pollination and fertilisation.

   Why do plants generally produce very large numbers of pollen grains?

3 The flower shown diagrammatically below is pollinated by wind:

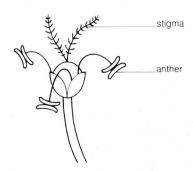

   What special features can you detect which suit it to this method of pollination?

4 Primroses are pollinated by bees. Some primrose flowers have their anthers high up and their stigmas low down, whilst others have their stigmas high up and their anthers low down, as shown in the illustration below:

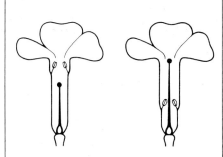

   In what way might this arrangement favour cross-pollination?

Seeds

Flowering plants produce seeds. The seeds survive summer drought and winter cold, and in the right conditions they give rise to new plants.

How long can seeds survive?

In 1933 a Japanese botanist found some lotus seeds in a dried-out lake bed in Manchuria. He sent them to the Royal Botanic Gardens at Kew. Scientists at Kew put the seeds on moist blotting paper, and they sprouted into new plants. These seeds were later found to be over a thousand years old.

More recently some viable lupin seeds were found buried in frozen soil in Canada. These were estimated to be over ten thousand years old.

Not all seeds can survive for as long as this. Many survive for up to a hundred years, others for about ten, and some last for only a few days.

Many seeds can withstand very bad weather. The seeds of desert plants can stand up to long periods of drought, sprouting into new plants as soon as it rains. And the seeds of Arctic plants can survive extremely cold winters.

How do seeds do this? No-one knows for certain. In its dried-out state, and protected within the seed coat, the embryo goes to sleep – it becomes **dormant**. In this state it *appears* to be dead, but when conditions become suitable it bursts into life.

The structure of seeds

If you split a bean pod open you will see the seeds (beans) inside (Investigation 1). Each seed is attached to the pod by a short **seed stalk**, and it is surrounded by a tough **seed coat**. (Figure 1). A black scar marks the position of the seed stalk. Just above this is a tiny hole, the **micropyle**.

Inside the seed coat is the **embryo**. This consists of a baby shoot (the **plumule**) and a baby root (the **radicle**). Attached to the embryo is a pair of thick wing-like structures called the **seed leaves** or **cotyledons**. They contain starch, and they feed the embryo when it starts to grow into a new plant.

What happens when a seed produces a new plant?

If you put some seeds in a moist place, you can see what happens when they produce new plants (Investigation 2). The process is called **germination**. Stages in the germination of the broad bean are shown in Figure 2, and are explored in Investigation 3.

First the seed takes up water, mainly through the micropyle. This makes it swell. As a result, the seed coat bursts open, and the young root and shoot grow out. The root grows downwards, and the shoot upwards. The shoot is bent like a hook: this protects its delicate tip as it pushes its way up through the soil.

The tip of the root is protected by a mass of loosely packed cells called the **root cap**. This prevents it being damaged as it grows down into the soil. The root gives off side-branches which help to anchor the young plant and absorb water and minerals from the soil. Slender root hairs increase the surface area for absorption.

The shoot eventually breaks through the surface of the soil. It then straightens, and the first green leaves open out. We now call the young plant a **seedling**.

For the embryo to grow like this, food is needed. Where does it come from? To begin with it comes from the cotyledons which remain inside the seed coat beneath the soil. Starch in the cotyledons is turned into soluble sugar: this is then transported to the tips of the shoot and root where growth takes place.

Once the seedling has formed its first green leaves, it can make its own food by photosynthesis. It is then self-supporting: the cotyledons are no longer needed, and so they wither away.

Different kinds of germination

Not all seeds germinate like the broad bean. For instance, in the sunflower the seed leaves do not stay beneath the soil. Instead they are lifted out of the soil with the growing shoot. When they reach the light, they turn green and

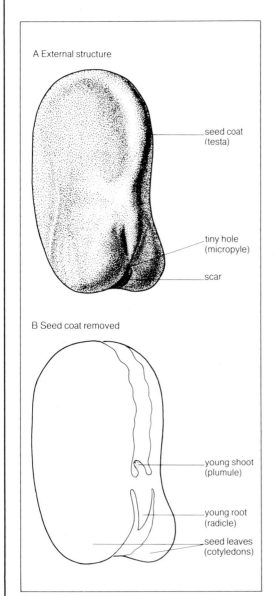

A External structure

seed coat (testa)

tiny hole (micropyle)

scar

B Seed coat removed

young shoot (plumule)

young root (radicle)

seed leaves (cotyledons)

Figure 1 The structure of a broad bean seed.

start feeding the seedling by photosynthesis. Until this happens the embryo is fed by a special tissue called the **endosperm** which is present in the seed.

Wheat germinates like the broad bean, but the shoot instead of being hooked points straight up. Its delicate tip is protected by a sheath which we call the **coleoptile**. When the first leaves open out, they break through the coleoptile which then falls off. Wheat is a type of grass, and all grasses have a coleoptile.

What conditions are needed for germination?

How annoying it is when you plant seeds in the garden and they don't grow. This is because the seed must have the right conditions in order to germinate. We can find out what these conditions are by trying to germinate seeds in different conditions (Investigation 4). From these experiments we can draw the following conclusions:

1 **Water** is essential for germination: without it seeds cannot swell up and burst open, and the embryo cannot grow.

2 A supply of **oxygen** is needed. This enables seeds to respire so they have plenty of energy for germination.

Figure 2 Germination of a broad bean seed.

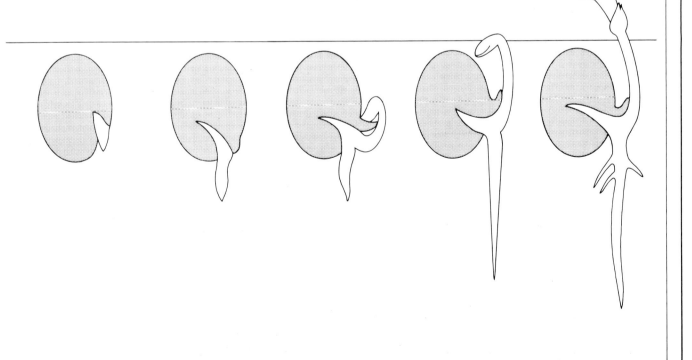

The root starts growing out about 3-4 days after soaking the seed.

The root grows downwards into the soil.

The shoot grows out and starts growing upwards. It is hook-shaped to prevent damage to the tip.

The root continues to grow downwards and the shoot upwards.

The shoot breaks through the surface of the soil and straightens. The first leaves open out. Side branches grow out from the main root. The seedling looks like this about 12-14 days after soaking the seed.

3 A **suitable temperature** is required. This varies with different plants. Usually seeds won't germinate when the temperature is below 0–5°C or above 45–50°C.

4 The effect of **light** is variable. Most seeds don't mind if it is light or dark. However some germinate only in the dark. Others require light: The amount needed may be very small – one quick flash is enough in some cases. Once the young shoot begins to grow above the soil, light is needed for the leaves to make chlorophyll and start photosynthesising.

Man's use of seeds

Seeds contain a store of food for feeding the new plants as they develop. This makes them a good source of food for man. A very important group of plants in this respect are **cereals** such as wheat and barley. From the seeds of these plants we get some of our most useful food including bread, cornflakes and porridge.

When wheat and barley are fully grown, the seeds, or **grains** as they are called, are clustered together at the tops of the stems. Each grain is enclosed in a tough covering called the **husk** which is the fruit. **Combine harvesters**, so-called because they *combine* two actions in one, cut the wheat stems *and* separate the grain from the husks. After harvesting, the grain is stored in **granaries.**

The structure of a wheat grain is shown in Figure 3. It is surrounded by a coat known as the **bran**. Inside is the embryo and a mass of endosperm tissue.

Wheat is used for manufacturing flour from which, of course, bread is made. Usually the bran and embryo are removed from the grain, leaving only the endosperm. This consists almost entirely of starch. When it is ground up it gives us **white flour**.

Sometimes the entire grain, including the bran, is ground up. This gives us **brown flour** from which wholemeal bread is made. The brown colour is due to a pigment in the bran. You have probably been told that wholemeal bread is better for you than white bread. This is because the whole wheat grain contains not only starch but other useful substances as well. These include vitamin B$_1$ and cellulose. The cellulose makes wholemeal bread coarser than white and provides roughage in the diet.

Wheat grain, and hence flour, contains a protein called **gluten** which makes dough sticky. When dough is put in the oven its stickiness causes it to hold in gas, so it rises. Gluten is therefore important in baking bread.

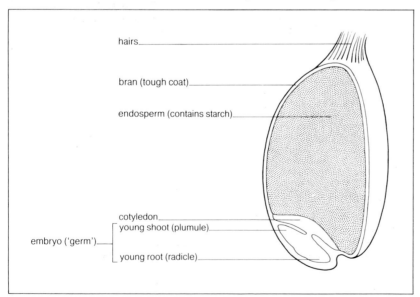

Figure 3 A wheat seed sliced down the middle so as to show its internal structure.

Investigation 1

Looking at seeds

1 Split open a bean pod, and notice the row of beans inside. The beans are the seeds, and the pod is the fruit.

2 Examine the seeds of various plants. Can you explain their similarities and differences?

3 Look at the outside of a broad bean seed. Notice the structures shown in Figure 1A.

4 Take a broad bean seed which has been soaked in water, and cut it in two as shown in the following illustration. Can you see the structures shown in Figure 1B?

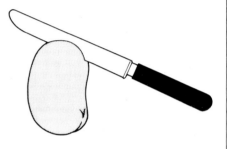

5 Pipette a drop of iodine onto one of the cotyledons inside the seed. What colour does the cotyledon turn? What does this tell us?

Investigation 2

How do different seeds germinate?

1 Lay sheets of blotting paper in the bottom of a series of dishes.

2 Moisten the blotting paper with water.

3 In each dish sprinkle some seeds of different plants, e.g. mustard, cress, wheat, radish, etc.

4 Observe differences in the way the seeds germinate.

How do the seedlings obtain food?

Investigation 3

Watching seeds germinating

1 Put some water in the bottom of a jar.

2 Roll up a piece of blotting paper, and put it in the jar as shown in the illustration. Tilt the jar so the blotting paper is thoroughly wetted: this will make it stick to the side of the jar.

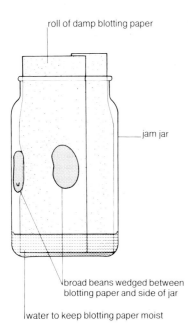

roll of damp blotting paper

jam jar

broad beans wedged between blotting paper and side of jar

water to keep blotting paper moist

3 Push several broad bean seeds between the blotting paper and the side of the jar. Be careful they don't fall into the water at the bottom.

4 Observe the broad beans at intervals over the next ten days or so, and watch stages in germination.

Do your observations agree with the diagrams in Figure 2?

Why is the shoot hook-shaped to begin with?

Where does the seedling get its food from?

What do you think makes the shoot grow upwards and the root downwards?

Investigation 4

To find out the conditions needed for germination

1 Push some cotton wool into the bottom of five large test tubes.

2 Pour a little water into four of the test tubes, so as to moisten the cotton wool. Leave the other one dry.

3 Sprinkle some cress seeds onto the cotton wool in each test tube.

4 Set up the test tubes as shown in the illustration.

5 Observe the test tubes at intervals during the next few days.

In which tubes does germination take place, and *not* take place?

What conclusions do you draw as regards the conditions needed for germination?

Which tube serves as the control in this investigation?

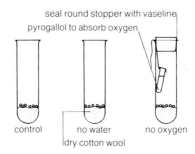
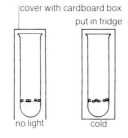

seal round stopper with vaseline
pyrogallol to absorb oxygen

cover with cardboard box
put in fridge

control | no water / dry cotton wool | no oxygen | no light | cold

Assignments

1 Peas were placed in a retort flask which was set up as shown in the diagram below. The flask was then left for two days:

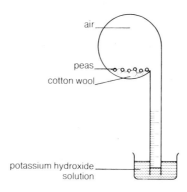

air

peas

cotton wool

potassium hydroxide solution

a) What would you expect to have happened to the level of the potassium hydroxide solution after two days?

b) What would have happened to the composition of the air in the flask?

c) What conclusion would you draw from the result of this experiment?

d) By means of diagrams show what controls are needed in this experiment.

2 Seeds which are planted too deep in the soil won't germinate. Suggest *two* possible reasons for this. Describe an experiment which you would carry out to test *one* of your suggestions.

3 The graph below shows how the dry mass of a germinating seed (and seedling) changes from the moment germination starts. (The dry mass is estimated by getting rid of all traces of water from the plant and then weighing it.)

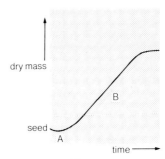

dry mass

seed

A

B

time

a) How do you think this experiment was actually carried out?

b) Explain what is happening at points A and B on the graph.

Fruits and dispersal

We usually think of a fruit as something soft, juicy and good to eat. In this Topic we will see the reason for this, and we shall understand why fruits are important to plants.

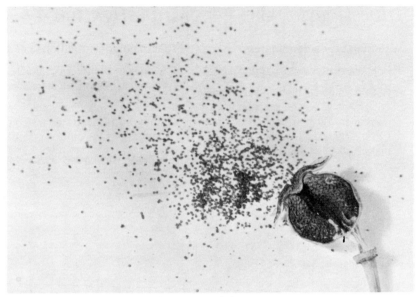

Figure 1 Seeds being dispersed from a poppy fruit.

What are fruits?

The fruit is the part of the plant which surrounds the seed or seeds. It is formed from the ovary in the flower. The number of seeds in a fruit depends on how many ovules were present in the ovary to start off with.

The whole purpose of the fruit is to help disperse the seeds, ensuring that they are spread over as wide an area as possible (Figure 1). The way a particular fruit helps dispersal depends on what kind of fruit it is. Looking at fruits as a whole, there are two main kinds: **fleshy** and **dry**.

Fleshy fruits

A fleshy fruit is one in which the main part of the fruit, formed from the ovary, is soft and juicy. The seeds, which are usually hard, are located somewhere inside (Investigation 1). Examples are plums and tomatoes (Figure 2).

Fleshy fruits usually taste good and are a useful food. They are therefore eaten by animals, particularly birds. The soft part of the fruit is digested, but the seeds, protected by their hard coat, resist the action of the animal's digestive juices and pass out with its faeces. They may be deposited a long way from where they were formed, and in this way the seeds get widely dispersed. Fleshy fruits are often highly coloured so as to attract animals – think of bright red cherries for example.

There is another kind of fleshy fruit in which the soft juicy part is formed, not from the ovary, but from the receptacle of the flower. As such fruits are formed from a part of the flower other than the ovary, they are called **false fleshy fruits**. Examples are apple and strawberry (Figure 3).

Man cultivates fruits on a large scale. However, the food substances inside them tend to be very dilute because so much water is present. For example, strawberries consist of over 90 per cent water and their main food value lies in the fact that they happen to contain quite a lot of iron.

Dry fruits

A dry fruit is one in which the fruit is relatively hard and dry. There are many different kinds, but one of the best known is the **pod**, characteristic of the pea and bean family (Investigation 2). The 'peas' and 'beans' themselves are the seeds (Figure 4).

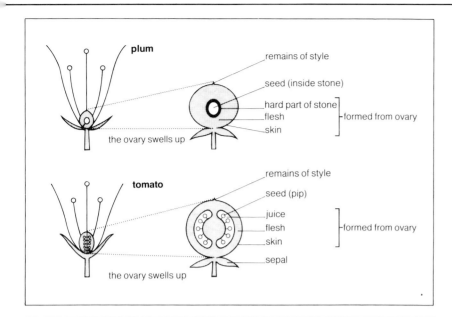

Figure 2 In fleshy fruits such as the plum and tomato, the ovary swells up and its wall becomes soft and succulent. The plum contains a single seed, which is inside the 'stone', whereas the tomato contains a large number of seeds (the 'pips').

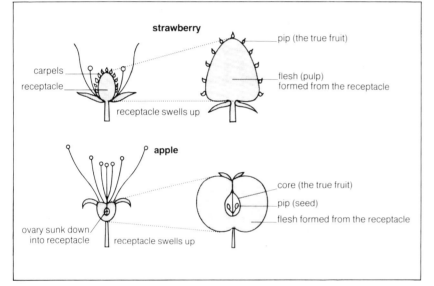

Figure 3 In the strawberry and apple, the pips are formed from the ovary of the original flower and so they represent the true fruit. The soft part of the strawberry and apple are formed from the receptacle, which swells up after fertilisation has occurred.

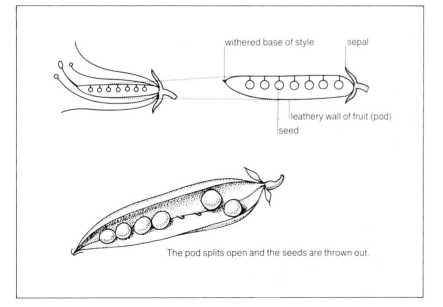

Figure 4 A dry fruit of the kind formed by pea and bean plants. The peas and beans themselves are the seeds.

Figure 5 The fruit of a sycamore tree has 'wings' which enable it to whirl through the air for quite long distances.

Figure 6 Fruit of the coco-de-mer palm from the Seychelles.

Pods have a leathery skin. When the skin dries out the pod splits open with such force that the seeds are scattered over a fairly wide area. Fruits which throw out their seeds by splitting open are known as **dehiscent fruits**.

There are many other kinds of dry fruits, all with special methods of dispersal. For example, the poppy fruit in Figure 1 is like a pepper pot and the seeds are scattered when the plant is shaken by the wind. In contrast the fruits of goose-grass are covered with hooks which cling to the fur of animals and to people's clothes. Some fruits are light and have 'parachutes' or 'wings' which enable them to be carried by the wind (Figure 5); and fruits like the coconut contain air spaces which enable them to float in water. Floating seeds can obviously be larger than air-borne seeds, and the coconut in Figure 6 is in fact the largest seed known. Nuts such as acorns and beechnuts are dispersed by untidy animals such as squirrels which collect them and accidentally drop them.

Investigation 1

Looking at fleshy fruits

1 Obtain two tomatoes. These are the fruits of the tomato plant.

What are the small leaf-like structures at the end?

2 Cut one tomato transversely, and the other longitudinally.

How are the pips (seeds) arranged inside the fruit?

Which structures in Figure 2 can you see?

3 Look at fruits growing on a tomato plant.

What were they formed from originally?

How are the seeds dispersed?

How is the fruit adapted for this kind of dispersal?

4 Examine other examples of fleshy fruits.

Find out where the seeds are, and how they are arranged.

5 Examine an apple.

Where are the seeds in this fruit?

Which structures in Figure 3 can you see?

Investigation 2

Looking at dry fruits

1 Obtain a pea or bean pod.

The pod is the fruit of the pea or bean plant. It is also found in other plants belonging to the same family, e.g. gorse, broom, laburnum and lupins.

2 Open the pod by splitting it down the side.

How are the seeds arranged inside the pod?

Which structures in Figure 4 can you see?

3 Look at pods that are still attached to the whole plant.

What were they formed from originally?

How are the seeds dispersed?

How is the pod adapted for this kind of dispersal?

4 Examine other examples of dry fruits.

In each case locate the seeds and try to explain how they are dispersed.

Assignments

1 Give one example of each of the following:

a) a fleshy fruit formed from the ovary of the flower,

b) a false fleshy fruit formed from the receptacle,

c) a dry fruit whose seeds (but not the fruit itself) are eaten by man,

d) a dry fruit whose seeds *and* fruit are eaten by man,

e) a winged fruit which is dispersed by wind.

2 The seeds found inside fleshy fruits generally have a hard seed coat. Suggest two functions of the seed coat.

3 In a fleshy fruit, food substances such as carbohydrate are highly concentrated inside the seeds, but very dilute in the fleshy part of the fruit. Explain the difference.

4 Below are shown the fruits of two different plants.

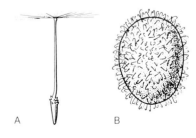

A   B

In each case explain how you think the fruit is dispersed.

Vegetative reproduction

In this Topic we shall see how plants survive from one year to the next by means other than seeds, and how this gives them an alternative method of reproduction.

Perennating organs

You may have noticed that many garden plants die down in the autumn, but the following spring they grow up again in the same place. This is because they form special storage organs during the summer which they fill up with food such as starch. The organ remains dormant in the soil after the rest of the plant has died, and the next year a new plant grows out of it (Figure 1).

Such organs enable plants to carry on from one year to the next, and so they are called **perennating organs**. (The word perennating means 'through the year'.)

There are many examples of perennating organs, and they can be formed from different parts of the plant. Let's take an example to illustrate what happens (Investigation 1). As you know, plants such as onions and daffodils form **bulbs**. The bulb consists of a mass of closely packed swollen leaves which rest on a short thick stem. If you plant a daffodil bulb in the autumn, it will stay alive in the soil till the spring. It then sends up a shoot, using food from its swollen leaves to support it. The green leaves of the new plant then make sugars which are moved back into the bulb for storage until the following year. Meanwhile the plant produces a flower which then dies, leaving the bulb in the soil (Figure 2).

Figure 1 These diagrams show how a storage organ enables a plant to survive the winter and come up again the following spring.

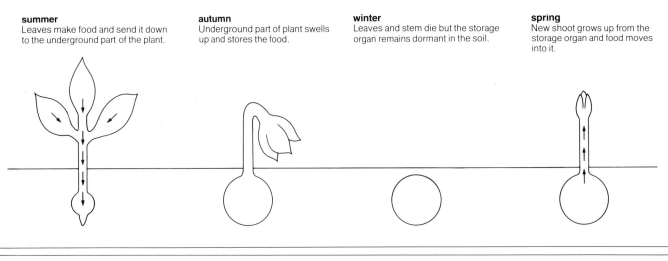

summer
Leaves make food and send it down to the underground part of the plant.

autumn
Underground part of plant swells up and stores the food.

winter
Leaves and stem die but the storage organ remains dormant in the soil.

spring
New shoot grows up from the storage organ and food moves into it.

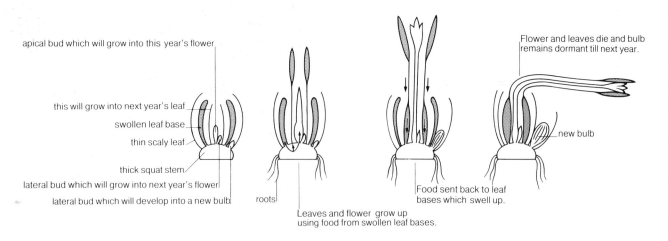

apical bud which will grow into this year's flower

this will grow into next year's leaf

swollen leaf base

thin scaly leaf

thick squat stem

lateral bud which will grow into next year's flower

lateral bud which will develop into a new bulb

roots

Leaves and flower grow up using food from swollen leaf bases.

Food sent back to leaf bases which swell up.

Flower and leaves die and bulb remains dormant till next year.

new bulb

Figure 2 These diagrams show how a bulb enables a plant such as a daffodil to survive from year to year.

These perennating organs enable some plants to grow up year after year. Such plants are called **perennials**. However, some last for only two years: they pass through one winter by means of a perennating organ and then die. These are called **biennials**, and an example is the carrot. Normally flowering occurs in the second year.

Vegetative reproduction

Consider a potato plant. This forms **stem tubers** (the potatoes) which are swollen underground stems (Investigation 2). Now a single plant produces not just one tuber, but many, perhaps five or six altogether (Figure 3). These tubers rest in the soil during the winter, and the following summer each one gives rise to a new plant. This is therefore a method of reproduction as well as a way of getting through the winter. It is known as **vegetative reproduction**.

Other plants, too, can reproduce by means of their perennating organs. A bulb, for example, may sprout a new bulb from the side during the growing season. When a gardener digs up his daffodil bulbs after the plants have died down, he will break off any such new bulbs and plant them separately the following year.

Other methods of vegetative reproduction

Vegetative reproduction does not necessarily involve the formation of perennating organs. Many plants reproduce vegetatively by other means. This generally involves the plant sending out a side branch from which roots grow down into the soil.

Figure 3 These diagrams show how a potato plant survives from year to year.

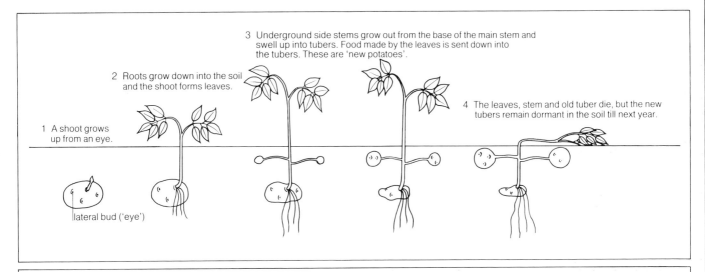

1 A shoot grows up from an eye.

lateral bud ('eye')

2 Roots grow down into the soil and the shoot forms leaves.

3 Underground side stems grow out from the base of the main stem and swell up into tubers. Food made by the leaves is sent down into the tubers. These are 'new potatoes'.

4 The leaves, stem and old tuber die, but the new tubers remain dormant in the soil till next year.

Side stem (runner) grows out from the base of the main stem.

New plants develop from the lateral buds of the runner.

Figure 4 How plants produce runners. The photograph shows a creeping buttercup with two young plants growing from the runner.

For example, some plants form **runners**. A runner is a branch of the main stem which lengthens and creeps along the surface of the ground. Roots grow down from it at intervals as shown in Figure 4. Eventually the parts of the runner in between the roots wither away, so a row of new plants is produced. Strawberry plants spread in this way.

Figure 5 This common pot plant known as *Chlorophytum* has formed several stolons.

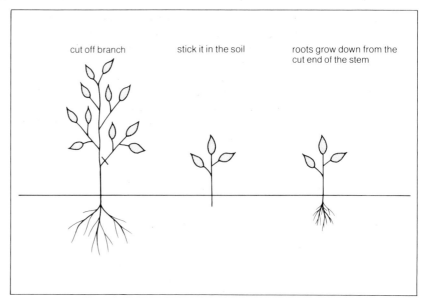

cut off branch   stick it in the soil   roots grow down from the cut end of the stem

Figure 6 Taking a cutting.

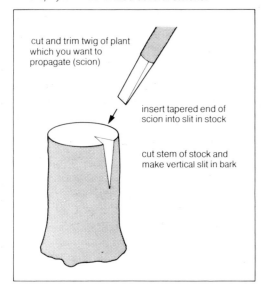

cut and trim twig of plant which you want to propagate (scion)

insert tapered end of scion into slit in stock

cut stem of stock and make vertical slit in bark

Figure 7 Grafting one plant onto another. This particular kind of graft is called a crown graft.

Figure 8 A gardener binds the scion and stock together after carrying out a crown graft.

Gardeners sometimes produce new shrubs by a process called **layering**. A young branch of a shrub is bent down and pressed into the soil. If necessary a brick can be used to hold it down. Hopefully, roots will grow out from the covered part of the branch. Once the roots are established, the branch can be cut so a new plant is produced. Some plants do this naturally, sending out a drooping branch which touches the ground and then roots itself. An example is shown in Figure 5. Rooting side branches of this kind are called **stolons**.

People often reproduce their favourite plants by taking **cuttings** (Investigation 3). To do this you cut off a healthy young branch, preferably just below a node, and remove most of its leaves. You then stick the cut end into some good soil. With luck, roots will grow out, and the cutting becomes established as a new plant (Figure 6).

Grafting

This is a special method of reproducing trees and is used a lot by gardeners. It involves placing the cut stems of two plants in contact with each other so that the tissues join together and become continuous. Figure 7 shows one way of doing this.

A twig (the scion) is cut off the tree you want to reproduce, and it's grafted onto the cut stem (the stock) of another type of tree. Once the two cut surfaces have been brought together, the two plants should be bound with tape or raffia and the joint covered with wax to prevent evaporation and stop microbes getting into it (Figure 8).

Another way of reproducing trees is by **budding**. In this case you cut a T-shaped slit in the bark of the stock and you cut out a bud from another plant. You then insert the bud into the slit. As with grafting, the two are then tied together and protected with wax.

Budding enables a large number of new plants to be grown on a single rootstock. Each bud grows into an individual plant and they will all be the same genetically, like identical twins. So if a gardener wants to produce a large number of, say, apple trees or rose bushes of a particular kind, this is a good way of doing it.

Reproducing without sex

The various mechanisms which have just been described are all examples of organisms reproducing without sex: the organism produces new individuals without the help of a sexual partner. We call this **asexual reproduction**.

Asexual reproduction is carried out by many organisms, plants and animals alike: the formation of potato tubers, fission of *Amoeba*, and budding of *Hydra* are all examples. If you look up asexual reproduction in the index you will be able to find other examples.

The offspring formed by asexual reproduction have exactly the same genes as the parent. They therefore resemble the parent in every detail. This is one of the main advantages of producing new plants vegetatively, for one can be certain that they will be just as good as the parent plant – provided of course that they are reared in the same conditions.

Investigation 1

Looking at bulbs

1 With a knife slice an onion bulb in two down the middle.

Which structures in Figure 2 can you see?

Why are the inner leaves so thick?

2 Look at another bulb which is beginning to sprout into a new plant.

What is the new shoot growing from?

Where is the shoot getting its food from?

What will happen to the bulb when the new plant is full-grown?

Investigation 2

Looking at potatoes

1 Examine a potato tuber.

Which structures in Figure 3 can you see?

The 'eyes' are small lateral buds.

2 Look at a potato tuber which has been left for some weeks in a warm place and is 'sprouting'.

What structures are the new shoots growing from?

How does the tuber differ in appearance, and the way it feels, from the previous one?

Where are the new shoots getting their food from?

What will happen to this tuber eventually?

3 Look at a complete potato plant which has been carefully dug up.

Can you see the old tuber from which it grew?

How many new tubers has it formed?

Investigation 3

Taking cuttings

1 Fill a pot with good soil: the soil should contain plenty of humus and be well watered.

2 Cut off a healthy side-branch from a mature geranium plant.

Make the cut just below a 'node' (see page 102).

3 Remove most of the leaves. This is to prevent it losing too much water.

4 With a stick make a hole in the soil and put the stem of the cutting into it.

5 Press the soil firmly round the cutting.

6 Place the pot in a warm, well-lit place, and observe it at intervals during the next few weeks.

If you wish, you can dip the cut end of the stem into some hormone rooting powder before you stick it in the soil.

Assignments

1 Why are perennating organs also described as storage organs?

2 What, if anything, is wrong with each of these remarks?

  a) Potato tubers are formed at the ends of the roots.

  b) Grafting is better than budding because it produces more new plants.

  c) Daffodils reproduce asexually by means of bulbs.

  d) In a carrot plant flowering does not occur until the third year.

  e) In a bulb the bud in the centre gives rise to a new bulb.

3 Potato tubers which are suitable for planting are called 'seed potatoes'. When a gardener plants seed potatoes he usually rubs off all but about two of the 'eyes'. Why do you think he does this?

4 What are the advantages to a gardener of propagating a plant by vegetative means?

5 When taking a cutting it is advisable:

  a) not to take a shoot which has a flower on it,

  b) to cut off some of the leaves before you plant it.

Give a possible reason for each of the above.

Trees and shrubs through the year

Many plants die down in the autumn and grow up the following year. However, trees and shrubs keep going all the time.

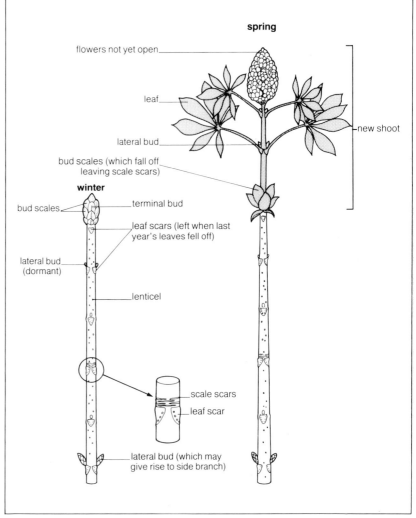

Figure 1 A horse-chestnut twig in winter and spring.

How do trees and shrubs survive?

Think of a tree such as a horse-chestnut. How does it manage to survive the winter? Here are three reasons:

1 It has a thick layer of corky **bark** which protects its trunk and branches, and helps to keep the tissues underneath warm.
2 The horse-chestnut is **deciduous** and drops its leaves in the autumn. As a result far less water evaporates from the tree than would otherwise be the case. This is useful because if the ground freezes the roots can't take up water from the soil. A plant whose roots are in frozen soil is as short of water as a plant living in a dry desert.
3 As winter sets in, the tree becomes **dormant**: it goes to sleep, as it were, and no further growth or activity takes place. It remains in this state until the following spring.

Twigs

If you look at a twig of, say, a horse-chestnut tree in winter, you will find that it looks pretty dead (Investigation): the leaves have dropped off, leaving only a series of scars where they were attached. The buds are closed and dormant, and there is no sign of life. However, in the following spring, things begin to happen: the terminal bud at the end opens up and a leafy shoot grows out (Figure 1). After a few weeks a group of flowers is formed, and later on seeds

Figure 2 A horse-chestnut twig in the autumn showing several fruits.

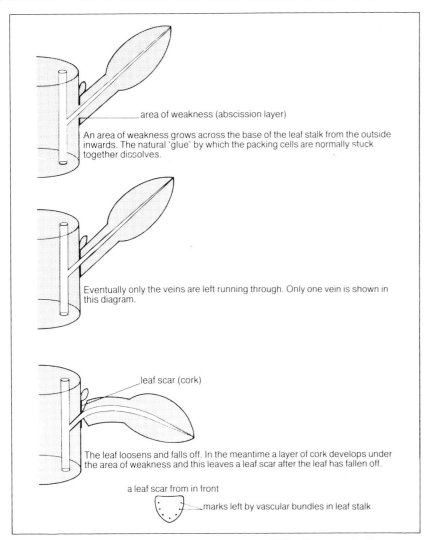

area of weakness (abscission layer)

An area of weakness grows across the base of the leaf stalk from the outside inwards. The natural 'glue' by which the packing cells are normally stuck together dissolves.

Eventually only the veins are left running through. Only one vein is shown in this diagram.

leaf scar (cork)

The leaf loosens and falls off. In the meantime a layer of cork develops under the area of weakness and this leaves a leaf scar after the leaf has fallen off.

a leaf scar from in front

marks left by vascular bundles in leaf stalk

Figure 3 These diagrams show what happens when a leaf falls off a tree or shrub.

are produced. In the autumn the fruits appear: the horse-chestnut fruit contains the familiar 'conker' which is actually the seed (Figure 2).

What makes the leaves fall off?

As winter approaches a layer of cells grows across the leaf stalk at the point where it's attached to the main stem or branch: the cells form a partition, leaving only the veins (vascular bundles) running through (Figure 3). This creates a region of weakness at the base of the stalk, and eventually it breaks and the leaf falls off. A layer of cork then grows over the cut surface, creating a **leaf scar** in which the marks left by the veins can be seen (see Figure 2).

Autumn tints

Why do leaves change colour in autumn? You may remember that leaves contain yellow and orange pigments (xanthophyll and carotene) in addition to the green pigment chlorophyll; however there is normally so much chlorophyll in the leaf that these other pigments don't show up. Towards the end of the summer the chlorophyll breaks down and disappears, but the other pigments remain. So the leaves change from green to yellow or orange. In addition, the leaves of some plants start making red and purple pigments (anthocyanins) at this time, and this makes their leaves particularly beautiful.

Investigation

Looking at horse-chestnut twigs

1 In spring look at a horse-chestnut twig whose buds are still closed.

   Which structures in Figure 1 can you see?

2 Stand the twig in a jar of water in a warm, well-lit place, and leave it.

3 Watch it at intervals during the next few weeks.

4 Observe the way the terminal bud opens and gives rise to a new leafy shoot.

5 Later in the spring look at a tree whose shoots have formed flowers.

6 In the autumn look at a tree which bears fruits.

   What part of the flower are the fruits formed from?

Assignments

1 Explain each of the following:

   (a) bud scale, (b) leaf scar,
   (c) scale scar, (d) lenticel.

2 In a winter twig, what may each of the following give rise to:

   (a) terminal bud, (b) lateral bud?

3 Why do leaves change colour from green to yellow in the autumn?

4 In the autumn a frost may cause the leaves to fall sooner than would otherwise be the case.

   Explain why this is. What other weather conditions, besides frost, may hasten the falling of leaves?

Chromosomes, genes and cell division

The nuclei of all living cells contain chromosomes. In this Topic we shall see what chromosomes are and what they do.

When a cell is resting the chromosomes are long, thin and thread-like,

but when the cell is about to divide the chromosomes become shorter and fatter.

Figure 1 Chromosomes as they appear at different stages in the life of a cell.

Figure 2 The full set of chromosomes of a human male as seen under the microscope just before the cell divides.

Figure 3 In this chromosome the genes are shown as white discs. The genes at particular positions (1, 2, 3 etc.) generally control specific characteristics.

What are chromosomes?

If you stain cells with certain dyes, the chromosomes show up under the microscope (Investigation). However, if you look at an ordinary *resting* cell, you can't see the chromosomes very well. This is because they are very thin, like fine pieces of thread. To see them clearly you must look at the cell when it's about to divide: at this stage they get shorter and fatter and this makes them show up distinctly (Figure 1).

In Figure 1 notice that the chromosomes can be arranged in pairs according to their sizes. Scientists have found that in virtually all cells, both animal and plant, the chromosomes are in pairs like this. The two chromosomes belonging to a pair look exactly alike, and we call them **homologous chromosomes**: homologous comes from Greek and means 'agreeing'.

The total number of chromosomes in the cell varies from one type of organism to another. The cell in Figure 1 has only ten (five pairs). However, man has 46 (23 pairs). You can see them in Figure 2. Can you count them, and can you identify the different pairs?

What are genes?

Chromosomes contain **genes** which determine the individual's characteristics, such as eye colour and nose shape. The genes are arranged along the length of each chromosome like a string of beads (Figure 3). Often a given characteristic is controlled by a pair of genes, which occur at a particular position on a pair of homologous chromosomes. Such genes are known as **alleles**.

How does a gene cause a particular characteristic to develop? Here is a very simple example to illustrate what is thought to happen. Suppose that brown eyes are caused by the presence of a certain pigment in the iris: the body makes this pigment because one of the genes tells it to. The gene exerts its action by making the cells produce a specific enzyme, and this in turn causes the production of the pigment.

What happens when cells divide?

Cells divide in two different ways, by **mitosis** and **meiosis**. Mitosis is the kind of cell division that occurs during growth and asexual reproduction. Meiosis, on the other hand, takes place during the formation of eggs and sperms. The chromosomes behave differently in these two types of cell division, as we shall now see.

Mitosis

Figure 4 shows what happens to the chromosomes during mitosis. The parent cell has four chromosomes: two long ones and two short ones. Notice that the two daughter cells have exactly the same number and kinds of chromosomes as the parent cell: two long and two short.

If you study Figure 4, you will see how this is achieved. Before the cell starts to divide, each chromosome produces an exact copy or replica of itself. The original chromosome and its replica are called **chromatids**, and they are held together by a structure called the **centromere**. The chromatids now line up across the middle of the cell. Then they part company and move to opposite ends of the cell. Finally the cell splits across the middle, and the chromatids become the chromosomes of the daughter cells.

Do you agree that the chromosomes behave in such a way that the

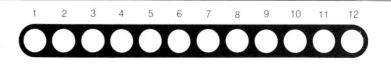

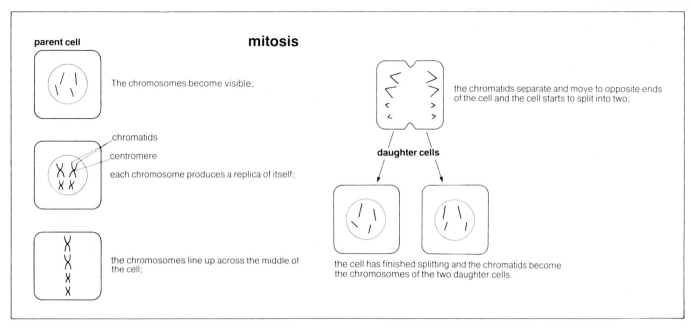

daughter cells are bound to have the same number and kinds of chromosomes as the original cell? This means that as an organism grows, all the new cells will have the same chromosome make-up. And it ensures that when an organism reproduces asexually its offspring are identical to the parent as far as their chromosomes and genes are concerned.

Meiosis

Figure 5 shows what happens during meiosis. The parent cell has four chromosomes as before. However, in this case each daughter cell contains only *two* chromosomes: one long one and one short one. In other words the daughter cells contain *half* the original number of chromosomes.

If you study Figure 5, you will see how this comes about. The chromosomes form chromatids as in mitosis. However, they line up across the middle of the cell in a different way. In this case homologous chromosomes *come together*, and then move away from each other to opposite ends of the cell which then splits in two. There now follows a second cell division in

Figure 4 These diagrams show what happens to the chromosomes when a cell divides by mitosis.

Figure 5 These diagrams show what happens to the chromosomes when a cell divides by meiosis.

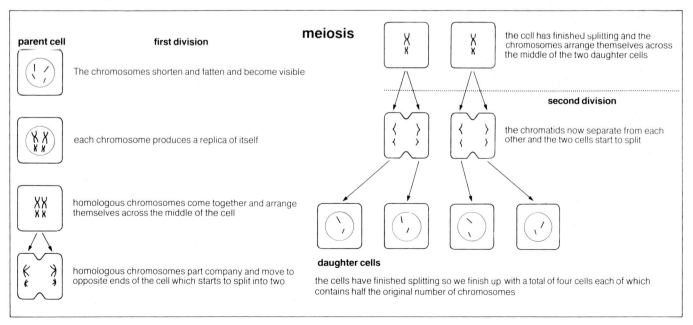

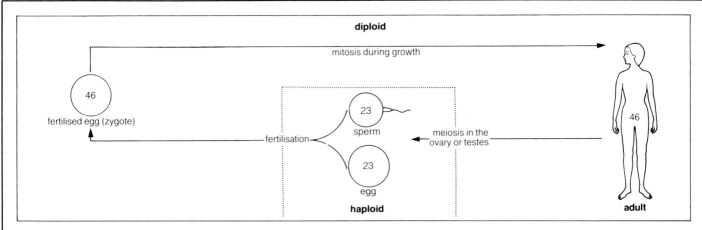

Figure 6 The human life cycle. The figures refer to the number of chromosomes in the cells. Cells which have the full number of chromosomes, in this case 46, are described as diploid. Those that have only half the full number (i.e. the eggs and sperms) are described as haploid.

which the chromatids part company. So in meiosis there are two cell divisions, one after the other, and the chromosomes behave in such a way that the daughter cells have half the number of chromosomes present in the original parent cell.

Why is it necessary for eggs and sperms to be formed this way? The answer lies in what they have to do, i.e. join together in the act of fertilisation (Figure 6). When a sperm fertilises an egg, their nuclei fuse together and the full number of chromosomes is restored.

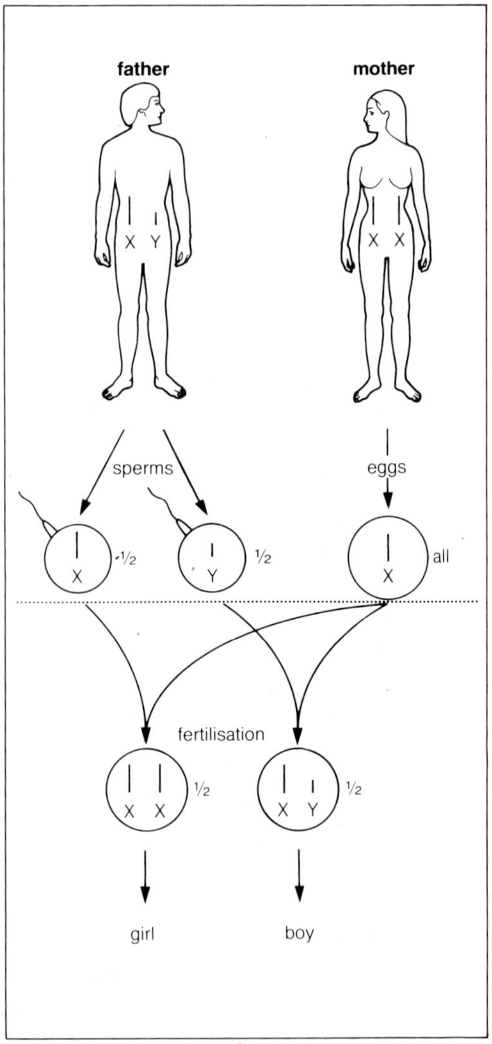

A boy or a girl?

What determines whether the fertilised egg develops into a boy or a girl? Well, it depends on what kind of chromosomes the egg contains.

The cells of an adult human contain 46 chromosomes (23 pairs). One pair are called **sex chromosomes** because they determine the individual's sex. There are two types of sex chromosome: a long one known as the **X** chromosome, and a short one known as a **Y** chromosome. Males contain an **X** and a **Y** chromosome, whereas females contain two **X** chromosomes.

Now the sperms which a male produces in his testes contain only one of these two chromosomes, either an **X** or a **Y** . This is because they are formed by meiosis. In fact, of all the sperms formed, half will be **X** and half **Y**. On the other hand, all the eggs which the female produces in her ovaries will contain an **X** chromosome. This is shown in the top part of Figure 7.

Now when fertilisation occurs, the egg may be fertilised by either an **X** sperm or a **Y** sperm. In fact if fertilisation is random, as it's believed to be, there is an equal chance of either happening. If an **X** sperm fertilises the egg, the zygote will contain two **X** chromosomes and this will develop into a female. On the other hand if a **Y** sperm fertilises the egg, the zygote will contain an **X** and a **Y** chromosome and will develop into a male. This is shown in the bottom part of Figure 7.

More about genes

We have seen that a chromosome is made up of a string of genes which determine the individual's characteristics. An enormous amount of research has been carried out on genes to find out what they consist of and how they work.

Genes consist of a chemical substance called **deoxyribonucleic acid**, or **DNA** for short. In the early 1950s two scientists at Cambridge, James Watson and Francis Crick, discovered the structure of the DNA molecule: they did this by working out the positions of various atoms in the molecule.

Watson and Crick found the DNA molecule to be like a twisted ladder or, more technically, a **double helix** (Figure 8). The rungs of the ladder, they discovered, are made up of pairs of organic bases. There are four such bases altogether, and they are known by their initial letters **A**, **C**, **T** and **G**. The bases fit together as shown in Figure 8: **A** always pairs with **T**, and **C** with **G**.

Figure 7 A boy or a girl? It all depends on the sex chromosomes.

So we have a series of pairs of bases along the length of the DNA molecule. Now the *order* in which the bases are arranged is variable, and this is how the genes exert their effects in the organism. For example, the order of base-pairs for producing, say, brown eyes, will be different from the order that produces green eyes, and so on. We can sum it up like this: DNA contains a set of coded instructions which tell the organism how to develop. These instructions form the **genetic code**.

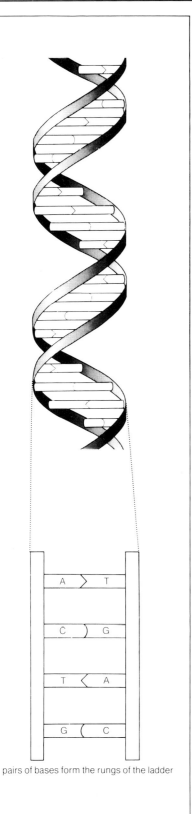

Figure 8 The DNA molecule is like a twisted ladder. It contains the genetic code. The code is contained in the sequence of the pairs of bases which occur along the length of the molecule.

pairs of bases form the rungs of the ladder

Investigation

Looking at chromosomes

Obtain a prepared longitudinal section of a root tip. Alternatively make your own slide like this:

1 Cut off the end of a young root about 5 mm back from the tip.

2 Place the root tip in a watch-glass of acidified acetic orcein stain.

3 Put the watch-glass on a warm hotplate for five minutes.

4 Place the root tip on a slide with a drop of the acetic orcein stain.

5 Break the root tip up with a needle so as to spread the cells out as much as possible.

6 Put on a coverslip, cover it with blotting paper and press down on it gently.

7 Look at your slide under the microscope: low power first, then high power.

Can you see chromosomes in any of the cells?

How are the chromosomes arranged in the different cells?

In each case try to decide if the cell is resting, about to divide, in the middle of dividing, or has just finished dividing.

Assignments

1 In a resting (non-dividing) cell the different chromosomes cannot be distinguished from each other, but in a cell which is about to divide they can be. Explain the difference.

2 What is the difference between a chromosome and a gene?

3 Explain the meaning of each of the following:

(a) homologous chromosomes,
(b) chromatids, (c) gene, (d) DNA.

4 Look at the cell shown below and then answer the questions beneath it.

a) How many chromosomes are there altogether?

b) How many pairs of homologous chromosomes are there?

c) If this cell divided by mitosis, how many chromosomes would there be in each daughter cell?

d) If the cell divided by meiosis, how many chromosomes would each daughter cell contain?

e) If the cell divided by meiosis how many daughter cells would be formed?

5 In a human being how many chromosomes are present in:

a) a brain cell,
b) a sperm cell in the testis,
c) an egg which has just been produced in the ovary,
d) a skin cell,
e) a fertilised egg.

6 Mr and Mrs Cross have three children, all boys. They are sure that their next child will be a girl. Do you agree? Give the reason for your answer.

Heredity

Why do children look like their parents? The answer lies in the way genes are passed from parents to their children. This is the science of heredity or genetics.

Crossing plants

Suppose we have a bed of plants some of which have red flowers, and others white. We take some pollen from a red flower and place it on the stigmas of a white flower: in this way we cross the two plants. When the seeds develop, we sow them in the soil.

In time new plants grow up from the seeds, and we find they all have red flowers:

Parents: red × white
↓
Offspring: all red

How can we explain this? Look at Figure 1. Each parent plant possesses in all its cells a *pair* of genes which control flower-colour. The red-flowered parent contains two genes which make the flowers red: we can call them **RR**, the white-flowered plant contains two genes which make the flowers white: we can call them **rr**.

Now the pollen grains and egg cells (i.e. the gametes) contain only one of these genes. Each pollen grain or egg cell produced by the red-flowered parent contains one **R** gene; and each pollen grain or egg cell formed by the white-flowered plant contains one **r** gene.

When fertilisation takes place, the **R** and **r** genes are brought together; so each of the offspring contains one **R** gene and one **r** gene. We can call it **Rr**.

Now the offspring all have red flowers even though they contain an **r** gene. We can explain this by saying that the **R** gene somehow suppresses the **r** gene, so it can't exert its effect: putting it another way, it is **dominant** over the **r** gene. That's why we have written it with a capital letter. The **r** gene on the other hand is **recessive**, and so it is represented by a small letter.

The genes controlling flower colour are located on a pair of homologous chromosomes, one gene on each chromosome (see page 336). The pollen grains and egg cells are formed by meiosis, in which the homologous chromosomes get separated from each other. This is why the pollen grains and egg cells contain only one of these genes instead of the normal two.

Another plant cross

The red-flowered offspring in the previous experiment belong to the **first filial generation (F1)**. Now suppose we take two of these plants and cross them with each other. Or alternatively we might self pollinate one of them. The resulting seeds are then planted, and the new plants grow up and bear flowers. They belong to the **second filial generation (F2)**.

This time we get a mixture of red-flowered and white-flowered plants. On counting each, we find that roughly three-quarters of them are red, and the remaining quarter white. In other words they are in a ratio of 3 to 1:

Parents: red × red
↓
Offspring: ¾ red ¼ white

How can we explain this? Look at Figure 2. Each parent plant contains an **R** and an **r** gene (**Rr**) as we have already seen. Now the pollen grains and egg cells produced by these plants contain *either* an **R** gene *or* an **r** gene. In fact there should be equal numbers of each type of gamete (**R** and **r**).

Fertilisation is completely random, and it's sheer chance as to which kind of pollen grain fertilises which kind of egg cell. There are three possible ways the genes might come together: two **R** genes might combine, giving **RR**; an **R** gene might combine with an **r** gene, giving **Rr**. Both these combinations will, of course, produce red flowers because, as we saw earlier, the **R** gene is dominant to the **r** gene. Alternatively two **r** genes might combine, giving **rr** which will produce white flowers.

Figure 2 shows how these combinations are brought about. If you look at the checkerboard at the bottom of the diagram, you will see that three-

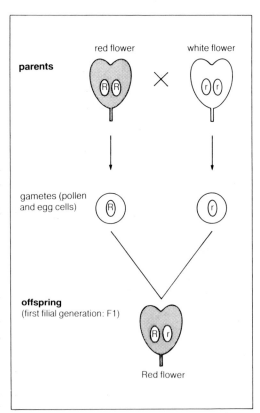

Figure 1 The result of crossing two plants with red and white flowers. The genes are indicated by letters: R, red; r, white. The sausage-shaped objects surrounding the genes are the chromosomes.

quarters of the offspring contain at least one **R** gene and will therefore have red flowers. The remaining quarter are all **rr** with white flowers. These are the proportions which we would *expect*, and they are confirmed when the actual cross is carried out.

Doing a back cross

Plainly there are two kinds of red-flowered plants: those that have two **R** genes (**RR**), and those that have an **R** and an **r** gene (**Rr**). Both look exactly alike, so you can't tell which is which just by looking at them. How, then, could you tell if a given plant is **RR** or **Rr**?

One way would be to cross it with one of the original white-flowered plants. This is called a **back cross**. If the red-flowered plant is **RR**, the offspring will all be red-flowered (as in Figure 1). On the other hand if it's **Rr**, we would expect to get a mixture of red-flowered and white-flowered plants in roughly equal proportions (Figure 3).

Producing plants with the same flower colour

Suppose you are a market gardener and you find that your customers want mainly white-flowered plants. How could you produce nothing but white-flowered plants?

The answer would be to cross two white-flowered plants with each other, or to self-pollinate one of them. You know that these plants must be **rr**, so the offspring are bound to be **rr** too, that is white just like the parents. And if you cross two of these offspring with each other, or self-pollinate one of them, their offspring will also be white **rr**, and so on down the generations. This is called **breeding true**, and it gives us a **pure line**. In a pure line all the individuals have the same genes with respect to a particular characteristic.

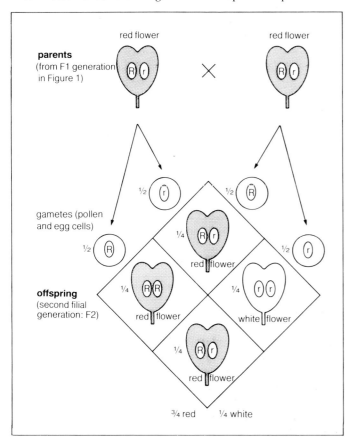

Figure 2 The result of crossing two red flowered plants from the offspring in Figure 1, or self-pollinating one of them.

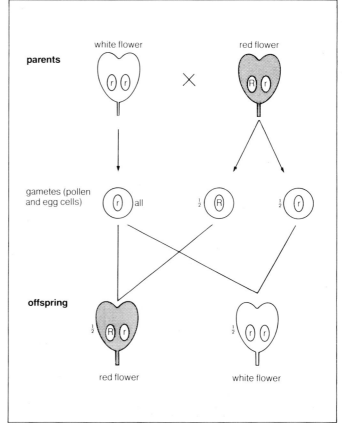

Figure 3 The result of a back-cross between a red flowered and white flowered plant. Symbols as in Figure 1.

Figure 4 Gregor Mendel, the Austrian monk who discovered the rules of genetics.

Suppose your customers wanted only red-flowered plants. In that case you would have to cross two red **RR** plants with each other, or self-pollinate one of them. You'd need to make sure that these plants were **RR** and not **Rr**, because if both of them were **Rr** you would get some white plants amongst the offspring.

Some technical terms

Looking back over the crosses which we have just considered, there are two ways of describing a plant. We can describe it in terms of its outward appearance, e.g. red-flowered or white-flowered. This is known as its **phenotype**. Alternatively we can describe it in terms of the genes that it contains, e.g. **RR**, **Rr** or **rr**. This is known as its **genotype**. When the genotype for a particular characteristic consists of two identical genes, for example **RR** or **rr**, we say that the organism is **homozygous**. If the two genes are both dominant, the organism is **homozygous dominant**; if they are both recessive, it is **homozygous recessive**.

When the genotype consists of two *contrasting* genes, for example **Rr**, we say that the organism is **heterozygous**. For example, the red-flowered offspring in Figure 1 would be described as heterozygous for flower colour.

Genes such as **R** and **r** which control the same characteristic but produce different effects are known as **alleles**. Alleles occur at the same positions on a pair of homologous chromosomes (see page 336). The term allele is commonly used by geneticists, but in this book I will keep to the word gene.

The rules of genetics

On the basis of the crosses described above we can make four general statements:

1 An organism's characteristics are passed down from one generation to the next by definite objects called genes.
2 The genes normally exist in pairs, one of which may be dominant to the other.
3 In a gamete only one of the two genes is present.
4 If the dominant and recessive genes are present together in an individual, it is the dominant one which produces an effect.

These are the basic rules of genetics. They were first discovered by an Austrian monk called Gregor Mendel (1822–1884) (Figure 4). He did experiments with pea plants, studying the inheritance of such characteristics as the colour of the flowers, height of the plant, texture of the seeds, and so on.

Since Mendel's day the same rules have been found to apply to other plants and also to animals. For example, an American scientist called T.H. Morgan studied inheritance in the fruit-fly *Drosophila*. This little insect has a number of clear-cut characteristics such as the colour of its eyes and the size of its wings, and it breeds quickly. It is therefore ideal for studying heredity (Investigation 3). In more recent years scientists have found that certain human characteristics are inherited in the same kind of way.

Human genetics

Try rolling your tongue longways into a U-shape without the help of your lips (Figure 5). Some people can do this, others can't. Tongue-rolling is caused by a dominant gene which we can call **T**. People who can roll their tongue are either homozygous dominant (**TT**) or heterozygous (**Tt**). People who can't roll their tongues are homozygous recessive (**tt**).

What happens if a non-roller mates with a heterozygous roller? The answer is given in Figure 6. Half the children should be rollers, and half non-rollers. Of course human beings don't produce large numbers of offspring like plants and fruit-flies, so it doesn't mean much to put it that way. It is more useful to say that there is an *equal chance* of any given child which they produce turning out to be a roller or non-roller.

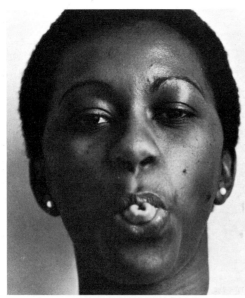

Figure 5 This girl can roll her tongue.

What happens if two heterozygous rollers mate? If you look at Figure 7 you will see the answer: there is a three to one chance that any given child which they produce will be a roller.

What will be the outcome of matings between (1) two homozygous rollers, (2) two non-rollers, (3) a homozygous roller and a non-roller?

We have seen that a heterozygous individual can roll his tongue – indeed he can roll it just as well as a homozygous dominant individual. However, he possesses the 'non-rolling' recessive gene (**t**) which he may pass on to his children. He is therefore described as a **carrier** of the recessive gene.

Inherited diseases

It doesn't matter if you can't roll your tongue. However, there are similar cases of inheritance where the genes produce harmful effects. Such genes may be either dominant or recessive. For example, there is a very serious defect of the pancreas called cystic fibrosis: this is caused by a recessive gene which is passed on just like the gene for non-rolling in Figure 6.

Certain inherited conditions are associated with the sex chromosomes (see page 338). For example, red-green colour blindness is caused by a recessive gene which is located on the X chromosome. Bleeder's disease (haemophilia) is similar.

If a couple give birth to a child with an inherited disease, or if there is a history of a particular disease in either of their families, their doctor can arrange for them to see a **genetic counsellor**. The genetic counsellor will try to work out the chance of their next child being born with the disease. To do this he will need to know the parents' pedigrees.

In the case of a recessive disease, a heterozygous person will be healthy because the 'normal' gene is dominant to the gene for the disease. However, he is a carrier of the recessive gene and may hand it on to his children.

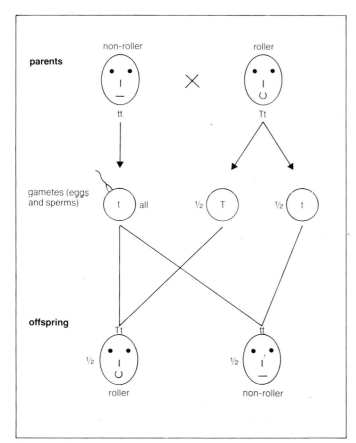

Figure 6 The inheritance of tongue-rolling in man. The diagram shows the result of a non-roller mating with a heterozygous roller. Gene for tongue rolling: T; gene for non-rolling: t. T is dominant to t.

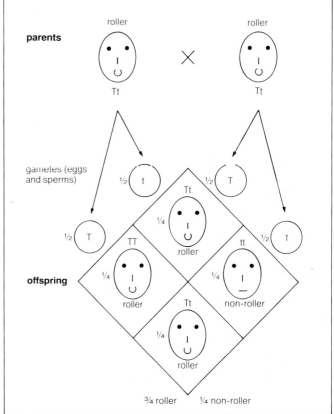

Figure 7 The result of a mating between two heterozygous tongue rollers.

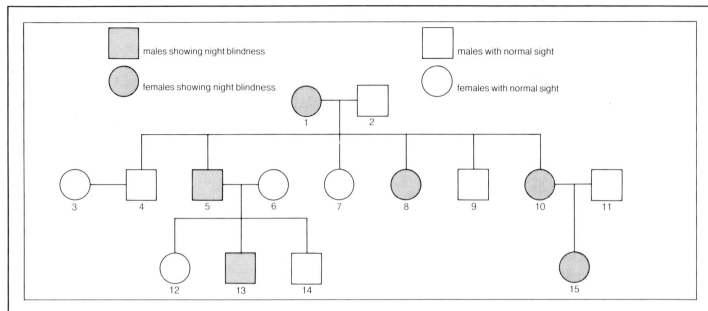

Figure 8 Family pedigree showing the inheritance of night blindness, a condition in which it's difficult to see in dim light. The condition is controlled by a single pair of genes, the gene for night blindness being dominant.

Figure 9 Two members of the Habsburg family showing the famous 'Habsburg lip'. Left: Felippe IV of Spain 1605–1665. Right: Ferdinand I 1793–1875.

Pedigrees

A pedigree is an individual's ancestral line of descent with respect to a particular characteristic or **trait**. It involves tracing back his or her history through the parents, grandparents and so on. A pedigree can be established for any kind of organism whose ancestors are known.

A human pedigree for a particular trait is shown in Figure 8. Males are represented by squares, females by circles. Individuals showing the trait are represented by a filled-in square or circle; those not showing it are represented by an open square or circle.

Having built up a chart like this, it's possible to work out the genotypes, or *possible* genotypes, of the various individuals. From this it may be possible to work out the chance of the trait appearing in the next generation.

It's amazing how some features persist in a family. A famous example of this is the drooping lower lip of the Habsburg family (Figure 9). By looking at family portraits, this feature can be traced back through several centuries. It's thought to have been caused by a single dominant gene.

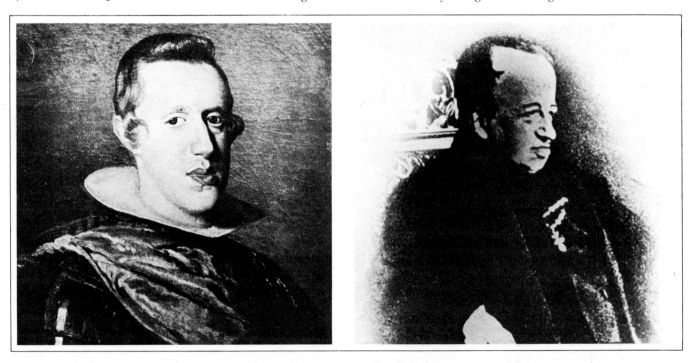

Do genes always show dominance?

Look at Figure 10. Here a red-flowered plant is crossed with a white-flowered plant as in the first experiment described in this Topic (Figure 1). However, instead of getting red-flowered plants in the next generation, the offspring are pink!

In this case the **R** gene doesn't completely suppress the **r** gene: we say that it is only *partially* dominant over it. So when the flowers develop, the **r** gene is able to produce a slight effect, resulting in a pink colour.

This kind of thing occurs in both plants and animals. An animal example is found in shorthorn cattle: if you mate a dark-haired bull with a white-haired cow, the calves have a mixture of dark and white hairs giving them a light red coat known as **roan**.

The inheritance of blood groups

You may remember that everyone's blood belongs to a particular group known as *A, B, AB* or *O*. Now the particular group that a person belongs to depends on whether he possesses certain genes. There are three genes altogether, but a given individual can only have two of them.

To belong to group *A*, you must have either two **A** genes (**AA**) or an **A** gene with an **O** gene (**AO**). To belong to group *B*, you must have either two **B** genes (**BB**) or a **B** gene with an **O** gene (**BO**). To belong to group *AB*, you must have the **A** and **B** genes together. And finally, to belong to group *O*, you must have two **O** genes (**OO**). Neither the **A** nor the **B** genes are dominant over each other. However, both are dominant to the **O** gene.

Suppose a man belonging to group *A* marries a woman belonging to group *O*, and they have a child. What blood group will the child belong to? The answer depends on whether the husband's genotype is **AA** or **AO**. If it's **AA**, then the child must belong to group *A*. On the other hand, if the husband's genotype is **AO**, there's an equal chance of the child belonging to group *A* or group *O*. If you are uncertain about this have a look at Figure 11.

Blood groups are sometimes used in court cases. For example, Mrs Green claims that Mr White is the father of her child. Their bloods are tested, and it turns out that Mrs Green belongs to group *B*, the child to group *AB* and Mr White to group *O*. This shows that Mr White could not possibly be the father of the child. Not all cases are as clear-cut as this!

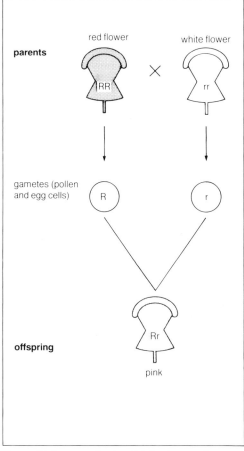

Figure 10 An example of partial dominance: the gene for red flowers **R** is only partly dominant over the gene for white flowers **r**.

Figure 11 These diagrams show the way the *ABO* blood groups are inherited in man.

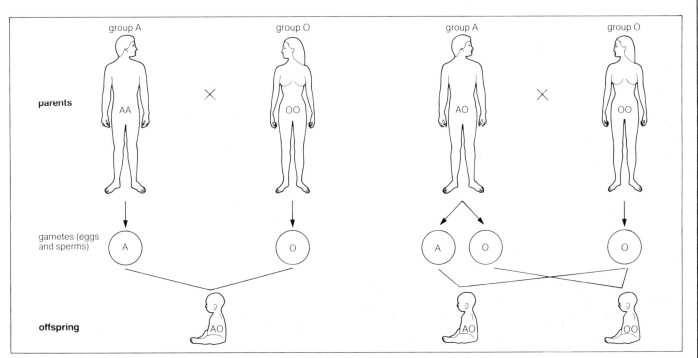

Investigation 1

How are genes sorted out in heredity?

A simple situation

1 Obtain two beakers.

2 In one beaker place 50 black beads.

3 In the other beaker place 25 black beads and 25 white ones, and stir them up thoroughly.

4 Have two empty beakers in front of you, side by side.

5 Close your eyes and take a bead from each beaker. Then look at them. If they are both black, put them in the left hand beaker, if one is black and the other white, put them in the right hand beaker.

6 When you have used up all the beads, count the number of pairs of beads in each beaker.

How many *pairs* of black beads are there?

How many *pairs* of black and white beads are there?

What are the proportions of each combination?

In this exercise, what do the beads represent?

What do the beakers represent?

Why did you have to close your eyes when taking beads?

This exercise resembles the kind of thing that happens in one of the crosses illustrated in Figures 1–3. Which one?

In what way does the exercise differ from what really happens?

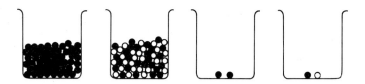

Investigation 2

How are genes sorted out in heredity?

A more complex situation

1 Obtain two beakers.

2 In each beaker place a mixture of black and white beads. 25 of each colour. Stir the beads up thoroughly in each beaker.

3 Have three empty beakers in front of you, side by side.

4 Close your eyes and take a bead from each beaker. Then look at them. If they are both black put them in the left hand beaker; if they are both white, put them in the right hand beaker; if one is black and the other white, put them in the middle beaker.

5 When you have used up all the beads, count the number of *pairs* of beads in each beaker.

How many *pairs* of black beads are there?

How many *pairs* of white beads are there?

How many *pairs* of black and white beads are there?

What are the proportions of each?

In this exercise what do the beads represent?

What do the beakers represent?

Why did you have to close your eyes when taking the beads?

Which of the crosses illustrated in Figures 1–3 does this exercise resemble?

In what way does the exercise differ from what really happens?

Investigation 3

Inheritance in the fruit fly

1 Your teacher will give you a bottle containing male flies with short wings and another bottle containing female flies with long wings.

The bottles contain food which the fruit flies like, and a piece of rolled up filter paper for them to cling to.

2 Anaesthetise about 15 males and 10 females (see illustration). Be sure to keep them apart.

3 Put the anaesthetised flies in separate groups on a white tile.

4 With a paintbrush carefully place the flies one by one in a new bottle containing food.

5 Place the bottle on its side so the flies don't fall into the food.

6 When the flies have recovered, stand the bottle in an incubator at 25°C.

7 One week later, look at the bottle.

Are there any larvae present? The larvae should form pupae.

Can you see any pupae yet?

8 Now anaesthetise the parent flies and then kill them.

Why do you think this is necessary?

9 After a further week look at the bottle again.

Are there any adult flies present? Don't proceed any further until all the adults have emerged from the pupae.

10 Anaesthetise the adult flies and put them on a white tile.

Do they have long or short wings? Which condition is dominant and which recessive?
Explain your result by writing out a genetic chart like the one in Figure 1.

11 Carry out other crosses as instructed by your teacher.

Explain your results and summarise each cross you do with a genetic chart.

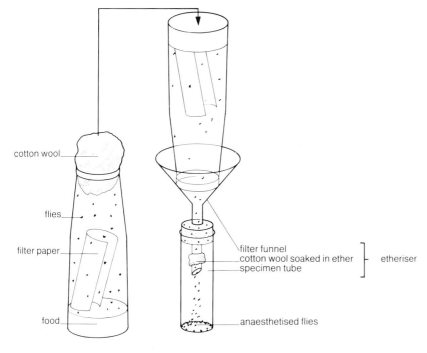

remove cotton wool and invert culture bottle over etheriser

cotton wool

flies

filter paper

food

filter funnel
cotton wool soaked in ether
specimen tube

etheriser

anaesthetised flies

Assignments

1 A black mouse mates with a brown mouse, and all the offspring are black.

Why are no brown offspring produced? Explain your answer fully.

2 If two of the black offspring from question 1 mate with each other, what kind of offspring would you expect and in what proportions? Draw a diagram to illustrate what happens.

3 In human beings the gene for brown eyes is dominant to the gene for blue eyes. A brown-eyed man marries a blue-eyed woman and they have five children. Three of the children have brown eyes and two have blue eyes. What are the genotypes of (a) the mother, (b) the father, and (c) the children? Explain how you arrive at your answer.

4 An albino is a person who has no pigment in the skin so he is very pale. This condition is caused by a recessive gene. An albino man marries a normal woman one of whose parents was an albino. How likely is it that their first child will be an albino? Give your reasons in full. Certain individuals in this family are 'carriers'. Which ones are carriers and what does this word mean?

5 Look at the pedigree in Figure 8 and then answer these questions about it:

a) Using B as the symbol for the night-blindness gene (dominant) and b for the normal gene (recessive), write down the possible genotypes of all the people in the chart.
b) Explain in words how you know the genotype of person 1.
c) How are persons 13 and 15 related to each other?
d) How do you know the genotypes of 13 and 15?
e) If 13 and 15 should marry, what is the chance that any of their children will be night-blind? Explain your answer.
f) If 14 and 15 marry, what is the chance of any of their children having night-blindness. Explain your answer.

Why do organisms vary?

If you look at a group of people such as those in Figure 1, you will notice that they are all different: they vary in height, the colour of their hair, shape of the face, and so on.

Figure 1 No two people are exactly alike.

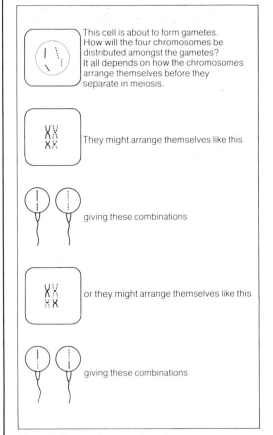

This cell is about to form gametes. How will the four chromosomes be distributed amongst the gametes? It all depends on how the chromosomes arrange themselves before they separate in meiosis.

They might arrange themselves like this

giving these combinations

or they might arrange themselves like this

giving these combinations

Figure 2 These diagrams show the different ways two pairs of chromosomes can be distributed amongst the gametes.

What is meant by variation?

If you measure the heights of a whole lot of people all of the same age, you will probably find that there is a steady gradation from very short to very tall (Investigation). The same kind of thing applies to weight, hair colour, intelligence, and many other features. This kind of variation, where there is gradual transition between the two extremes, is called **continuous variation**.

What causes this kind of variation?

The main reason why people vary in the way just described is that each individual possesses a different combination of genes. The particular combination of genes in *your* body is probably different from that of any other human being, past or present. In other words you are unique!

What causes this uniqueness? The answer is chance. Through the egg and sperm, your parents gave you a set of genes. However, the *particular* genes which you got from your mother, and the *particular* ones you got from your father, was simply the result of chance.

The reason for this lies partly in the way your parents' chromosomes behaved during meiosis when the egg and sperms (gametes) were formed. The genes are carried on the chromosomes, and during meiosis half the chromosomes go into one gamete and half into the other. Now it is sheer chance as to which particular chromosomes get into which gamete (Figure 2): it's rather like dealing out a pack of cards. Bearing in mind that each chromosome contains its own particular set of genes, you will realise that a great deal of variation can be brought about this way.

However, there's more to it than that. Do you remember how the chromosomes come together in meiosis (see page 337)? When this happens the chromosomes get wrapped round each other, and bits of them may break off and change places. In this way genes get shifted from one chromosome to the other. This is called **crossing-over**, and it brings about further variation in the offspring.

The two processes mentioned above will make all the gametes different with respect to the genes they contain. Now it's just chance as to which particular sperm fertilises an egg: in other words fertilisation is completely random. This provides yet another source of variation.

As a result of the mechanisms just described, every individual has a unique set of genes. This explains why brothers or sisters differ from one another, and why children differ from their parents, although they may resemble each other to a certain extent.

Another kind of variation

Do you remember tongue-rolling? This is the ability to roll your tongue into a U-shape (see page 342). Now a person can either roll his tongue, or not; there are no 'in-betweens'.

This is called **discontinuous variation**. There aren't many examples of it in humans, but in other organisms we see it a lot: for example the different coloured flowers and different kinds of fruit-flies mentioned in the Topic on heredity (page 340).

This kind of variation arises as a result of a process called **mutation**.

Mutation

Mutation is a sudden change in the genetic make-up of an organism.

Mutation sometimes leads to people being born with a defect such as an extra toe or a missing arm (Figure 3). In Britain about two per cent of babies are born with a defect of some kind. Sometimes the defect results in the child being physically or mentally handicapped.

Mutation occurs during meiosis when the eggs and sperms are being formed. There are two main kinds: **chromosome mutation** and **gene**

mutation. In a chromosome mutation a major change occurs in one or more of the chromosomes. For example the individual may lack a particular chromosome, or have an extra one. The condition known as Mongolism or Down's syndrome, for instance, is caused by the presence of an extra chromosome. Sometimes part of a chromosome gets snapped off and lost, or it may turn round the wrong way. Doctors can find out if a person has this kind of mutation by looking at his chromosomes under the microscope.

A gene mutation is caused by a chemical change occurring inside an individual gene. You cannot see this kind of mutation under the microscope, as it does not alter the appearance of the chromosome. Scientists have shown that when a gene mutation occurs, there is a change in the order of bases in the DNA molecule. The change may be very small indeed, but it may have a severe effect on the organism. An example is cystic fibrosis (see page 343).

What causes mutation? The answer is that nothing actually *causes* it. It just happens by chance from time to time. It can be greatly speeded up if the organism is exposed to radioactivity or certain chemical substances. This is why these things are considered dangerous. Many of the people who survived the two atom bombs which were dropped on Japan at the end of the Second World War received massive doses of atomic radiation. As a result they showed a high incidence of mutation, and many of their children were born with defects, far more than in a normal population.

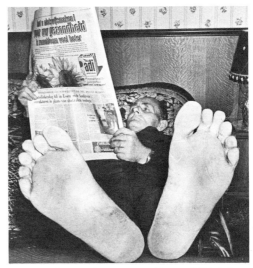

Figure 3 What is unusual about this man? This is an example of a mutation.

Variation and environment

Figure 4 shows a hydrangea plant. It's a common shrub in people's front gardens, and it has large clumps of flowers which are generally white, pink or blue. Now blue flowers develop only if the plant is grown in an acid soil. If the soil is chalky, pink or white flowers develop.

This is an example of variation being caused by the environment. Can you think of any other examples from plants or animals? Many of the differences between people, particularly in their behaviour, can be explained by the fact that they have been brought up in different environments. We still don't know for certain how important the environment is, compared with the genes, in making us different from one another. This particularly applies to features like intelligence and artistic ability.

Figure 4 A hydrangea in flower.

Investigation

Looking at examples of variation

1 Measure the height of each person in your class.

2 Divide the heights into 5 cm groups, starting with 120 cm and finishing up with 180 cm (ie. 120–125, 125–130, 130–135, etc.) and write down the groups in a list.

3 Work out how many people in your class fall into each group. Write the numbers alongside the groups in your list.

4 Construct a bar chart showing how height varies in your class.

Which group of heights do most people in your class belong to?

Which group of heights do the fewest people belong to?

Which group contains the largest number of people?

What is the average height in your class?

Suggest reasons why the members of your class should vary in height.

5 Make a list of other ways besides height in which the members of your class differ from each other.

Assignments

1 Two things happen during meiosis which help to make offspring different from one another. What are these two things?

2 In what sense is fertilisation a random process?

3 Explain carefully why (a) two brothers do not look alike, and (b) why identical twins *do* look alike.

4 Explain the difference between continuous and discontinuous variation.

5 What is meant by a mutation? Whereabouts do mutations occur, and what are their consequences?

Living things and their environment

This field, and the
nuclear power station
beyond, illustrate how man
can influence the natural world.
How living things relate to their
environment is discussed
in the next group
of topics.

Identifying and collecting organisms

There are so many animals and plants in the world. How do we begin to find out what they all are and where they live?

What is its name?

Suppose you have been out for a walk and you come back with the plants shown in Figure 1. How can you find out their names? One way might be to compare each one with pictures in a book. This is all right if it's a short book, but if it's a long one it can be a tedious business and it is difficult to know where to start.

It is much better to use a **key**. Keys are widely used by biologists to identify organisms quickly and accurately. A key is shown below. Use it to identify the plants in Figure 1. Use keys to identify other organisms, and try making a key of your own (Investigation 1).

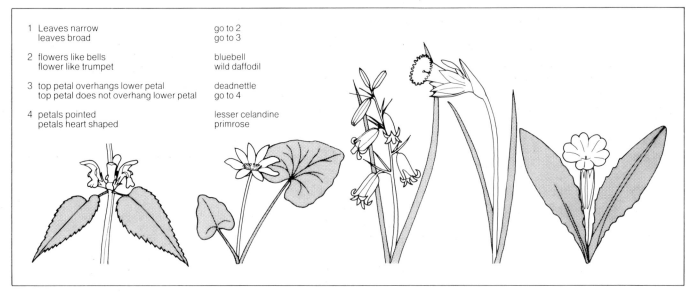

1	Leaves narrow	go to 2
	leaves broad	go to 3
2	flowers like bells	bluebell
	flower like trumpet	wild daffodil
3	top petal overhangs lower petal	deadnettle
	top petal does not overhang lower petal	go to 4
4	petals pointed	lesser celandine
	petals heart shaped	primrose

Figure 1 Five flowering plants. In each case the leaves are shaded.

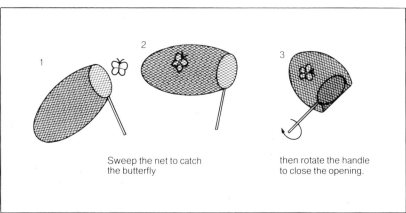

Sweep the net to catch the butterfly

then rotate the handle to close the opening.

Figure 2 How to use a butterfly net.

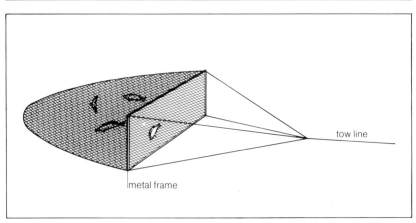

tow line

metal frame

Figure 3 A drag-net can be used for catching fish and other swimming animals in a pond.

How to collect organisms

Slow-moving animals like snails can be collected by hand. However, most animals are obtained by capturing them in some way. One of the most useful devices for doing this is a **net**. There are all sorts of nets, each suitable for a particular kind of animal. The main thing is that it should be large enough to contain the animal without damaging it, and the mesh should be fine enough to prevent the animal getting out (Figure 2).

Nets are useful for catching both water animals and land animals. To catch very small water animals, such as insect larvae and protozoans, it's best to use a fine net with a bottle at the end (see page 24). For larger animals, a drag-net may be needed. This consists of a bag attached to a rectangular frame. It is thrown into the water and then pulled towards the shore (Figure 3). A similar kind of net can also be dragged over land and is useful for collecting insects and other small animals from long grass or marshland.

Some animals are shy and timid, or only come out at night. These can only be caught by using some kind of **trap**. To catch insects which live on the ground a pitfall trap can be used (Figure 4). It consists of a jam jar which is sunk into the ground and covered with a stone. Insects can be enticed into the trap by putting bait inside: decaying meat will attract scavengers, and honey or jam will attract insects that normally visit flowers and feed on nectar.

Have you noticed that on a summer night swarms of flying insects may gather round a light? This fact is made use of in collecting moths. A moth trap is shown in Figure 5. It's best to use a mercury-vapour lamp because this gives out a lot of ultra-violet light to which insects are particularly attracted.

For small mammals such as mice and voles a mammal trap is used. This is a metal box, approximately 15 cm ×8 cm × 8 cm. At one end there is a door and at the other end there is a place for the bait, such as a piece of cheese. If a mouse enters the box and nibbles the bait, a mechanism is released which closes the door and traps the animal (Figure 6). It's rather like a mousetrap, but instead of killing the animal, it captures it without harming it.

These are just a few ways of collecting animals which you can try yourself (Investigation 2). For some animals, such as those living in the soil, special methods have to be used (see page 388).

How many are there?

Suppose you want to find out how many thistles grow in one field compared with another. How could you do this? One way would be to count all the thistles in each field. However, this would probably drive you mad. So what you do is to count the number of thistles in a series of small areas in each field, then work out the average. We call this process **sampling**. As many areas as possible are sampled, and they are selected at random so that the figures are not biased in favour of a particular result. Obviously if you were to deliberately choose areas where there were lots of thistles, the results could be totally misleading.

One of the simplest ways of sampling a habitat is to use a **quadrat**. This is a square frame made of wood or metal which is laid on the ground (Figure 7). Quadrats come in many different sizes, but for counting thistles in a field, a quadrat with a side of one metre would be suitable. You simply count up the number of thistles inside the frame, repeating the process in, say, ten different parts of the field. You then work out the average number of thistles per square metre (Investigation 3).

Let's imagine you find that Field A contains 4 thistles per square metre, whereas Field B contains 8. You then have to ask yourself whether these figures are caused by a difference in, for example, the soil in the two fields, or whether they are simply the result of chance. The answer isn't just a matter of opinion or of guessing; you can work out the extent to which chance comes into it by doing special statistical tests. These are beyond the scope of this book but they are very important in ecological studies.

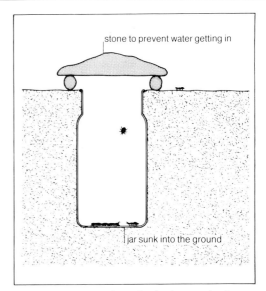

Figure 4 A pitfall trap for collecting insects that live on the ground.

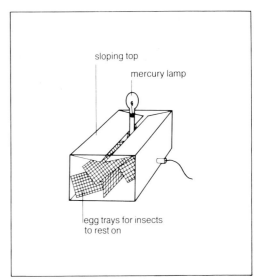

Figure 5 A moth trap.

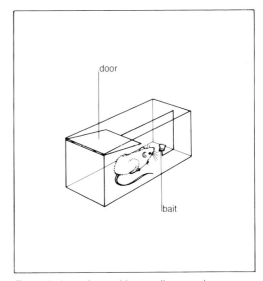

Figure 6 A trap for catching small mammals.

Figure 7 These students are using a quadrat to count weeds in a field.

What occurs where?

You may have noticed that in some places the kinds of plants in one area are different from those in another. In such cases it is useful to record exactly where each kind of organism occurs. You can do this by mapping the positions of each kind of plant in the whole field, but often it is simpler to make a **line transect**. A length of string or a plastic clothes-line is stretched across the area which you want to examine. You then record all the plants (and animals, too, if you wish) which occur on, or close to, the line (Investigation 4).

An example of a line transect is shown in Figure 8. Notice that some plants which occur at one end of the transect are completely absent at the other. Investigating the reason for such a difference is one of the most important aspects of ecology. This is taken up again in the next Topic.

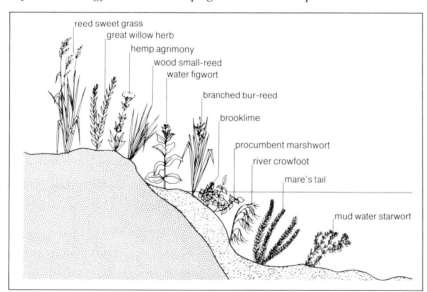

reed sweet grass
great willow herb
hemp agrimony
wood small-reed
water figwort
branched bur-reed
brooklime
procumbent marshwort
river crowfoot
mare's tail
mud water starwort

Figure 8 . A line transect through the edge of a ditch. Plants such as reeds which are found on the bank are not found in the water, and vice-versa.

Investigation 1

Making a key for identifying organisms

1 Your teacher will give you a collection of organisms, or pictures of organisms, together with their names.

2 Make a key, similar to the one on page 352, which would enable a person to find out the name of each organism.

3 Ask a friend to identify the organisms using your key.

4 If your friend runs into any difficulties, improve the wording of the key to make it clearer.

Investigation 2

Collecting and naming organisms

1 Use one of the methods described in this Topic to collect organisms from a habitat near your school or home.

2 If any of the organisms are small rapidly moving land animals such as insects, kill them by placing them in a killing bottle for a few minutes.

killing bottle

cotton wool soaked in chloroform

3 Examine each organism, using a hand lens (magnifying glass) or microscope where necessary.

4 Use the classification on pages 6–8 to find out what group each organism belongs to.

5 Use a simple key provided by your teacher to find out the name of each organism.

Investigation 3

To find out how many weeds there are in a field

1 Obtain a quadrat, 1 metre square.

2 Select a field, and decide what particular weed you wish to investigate.

3 Lay the quadrat on the ground, and count the number of weeds inside it. If some of the weeds are partly inside the quadrat and partly outside it, follow this rule:

For the top and left hand sides of the quadrat regard a weed as *inside* if any part of it, however small, falls inside the frame.

For the bottom and right hand sides of the quadrat, regard a weed as *outside* if any part of it, however small, falls outside the frame. Thus:

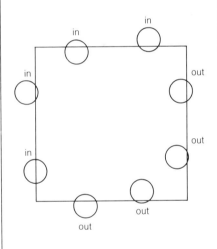

4 Repeat the above procedure with the quadrat in at least five different places, chosen at random.

5 Work out the average number of weeds per square metre in the field.

Do you think this is a good method of finding out how many weeds there are in the field? If not, why not?

What are the main reasons for any inaccuracies in the results?

What could be done to improve the method?

Investigation 4

Making a line transect

1 Obtain a plastic-covered clothes line which has been marked at one metre intervals.

2 Select a hedgerow which is not too thick but where there is plenty of ground vegetation.

3 Stretch the line across the ground at right angles to the hedge, like this:

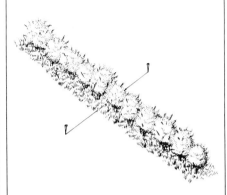

4 On a sheet of squared paper, record the plants which are touching the line as you go from one end to the other. If you don't know the name of a plant, make up a name or call it by a letter.

What you have done is to make a line transect of the habitat.

5 Make a diagram of the habitat, like the one in Figure 8.

Are any plants present on one side of the hedge but not on the other?

Suggest reasons for any differences you have observed.

Assignments

1 If you were designing a drag-net for catching fish, mention two qualities which it would need to have.

2 Your friend maintains that there are more minnows in his pond than there are in yours. What could you do to see if he is right?

3 How could you find out if there are more night-flying moths of a particular species in one area than another?

4 A scientist visits an uninhabited island and discovers the insects shown in the illustration below. Make up a name for each insect, and devise a key which would enable another visitor to the island to identify them.

What controls where organisms live?

The three-spined stickleback shown in Figure 1 lives only in streams and fresh-water ponds in temperate regions of the world. In this Topic we shall ask why certain organisms are confined to particular places.

Figure 1 A stickleback in its natural habitat.

What is a habitat?

The particular place where an organism lives is called its **habitat**. This may be a pond, hedgerow, wood, field, or even the body of another organism.

Every habitat has certain conditions which make it suitable for some organisms but not for others. These conditions make up the **environment**.

What is the environment?

An organism's environment consists of two parts: the physical environment and the biological (biotic) environment.

The **physical environment** includes temperature, light, humidity and so on. The precise features which make up an organism's physical environment depend on whether it lives on land or in water. For land organisms temperature and humidity are particularly important. For water-dwellers (aquatic organisms) the most important features are how salty the water is, and how rapidly it flows.

The **biological environment** is made up of all the other organisms in the habitat. One of the most important examples of an organism's biological environment is the presence of another organism which it feeds on.

We will now look more closely at some detailed examples of the physical and biological environments.

The physical environment

One of the most obvious components of the physical environment is light, and this can have a great influence on where different organisms occur. For instance, you have probably noticed that green plants only occur where there is light. In caves, they are completely absent, and even in dim places, such as under a beech tree, there are not very many.

Now let's look at another component of the physical environment: water. All living things need water and the amount of it in the environment can greatly affect the distribution of animals and plants. Some organisms lose water more quickly than others because it evaporates from the surface of the body at different rates. The rate of water loss depends on how dry the surrounding air is and on how waterproof the organism's surface is.

Let's take a plant to illustrate this. You can measure the rate at which water evaporates from a land plant by means of a **potometer**. Using this apparatus you can compare the rate of evaporation from the plant in different conditions, such as dry and humid air (Investigation 1)*.

From these experiments, we can draw the following conclusion: if the air is dry, the plant loses water more quickly than it does if the air is moist (humid). Water loss is also speeded up by air currents and warmth.

* What assumption is made in using a potometer to measure the rate of water loss from a plant?

Another thing you can do with a potometer is to compare the rate of water loss of two different kinds of plants taken from different habitats. From such experiments we discover that some plants lose water surprisingly slowly even in very dry air. This is because they have special ways of cutting down water loss. For example, their leaves may be covered by a thick waterproof cuticle, and they may have relatively few air pores (stomata). What's more, their leaves may be small so as to cut down the surface area across which water might evaporate. These kinds of plants can live in hot, dry places such as deserts. They are called **xerophytes**, and an example is the Joshua tree shown in Figure 2.

Animals have similar ways of reducing water loss. For example, reptiles such as lizards and snakes have a scaly, waterproof skin which enables them to live in hot, dry deserts (Figure 3). In contrast, amphibians such as frogs and newts have a thin, moist skin through which water readily evaporates. Such animals cannot survive long in dry conditions and are normally found only in damp places.

Figure 2 The leaves of this Joshua tree have special adaptations for reducing water loss enabling the plant to live in hot, dry deserts.

Figure 3 This lizard lives in hot places and has special adaptations for saving water.

Experiments can be done to find out how animals react to different conditions (Investigations 2 and 3). Usually they move around in such a way as to get into an environment where they are most likely to survive.

Comparing different habitats

Suppose you are interested in two habitats. You have noticed that each supports different kinds of animals and plants, and you suspect that this is because one is more humid than the other. Before going any further, you must find out exactly how humid each area is.

You can do this by means of a **porous pot atmometer** (Figure 4). This is like the potometer mentioned earlier, but instead of attaching a plant to one end you attach a porous pot. The sides of the porous pot are perforated by tiny holes through which water can evaporate, and the rate of evaporation will depend on the dryness of the air. By measuring the rate of evaporation in the two habitats, you can find out which is the more humid.

Suppose we find that one habitat is indeed more humid than the other. Can we be certain that this is why different kinds of organisms live in each habitat? No, because the two habitats may differ in other respects besides humidity. For example, one of them might be warmer or lighter than the other. This means that you must measure these other features as well. To measure the temperature you simply use a thermometer. To measure the light, you can use a light meter of the type that photographers use.

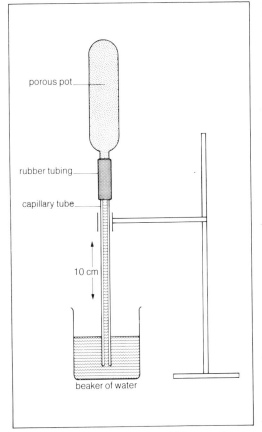

porous pot

rubber tubing

capillary tube

10 cm

beaker of water

Figure 4 A porous pot atmometer. This apparatus can be used to compare the humidity in different places.

Figure 5 A caterpillar of the privet hawk moth.

The biological environment

Look at the caterpillar in Figure 5. This feeds on privet leaves and occasionally on lilac and ash leaves. These particular plants form the caterpillar's biological environment. Without them the caterpillar could not survive. Organisms are always found close to their source of food: this is one of the most important aspects of their biological environment.

Now let's take a more specialised example. Bees normally occur only in places where there are flowers. Bees need flowers because they get nectar and pollen from them, so flowers are an essential part of the bees' biological environment. If all the flowers were to disappear from an area, the bees there would either die or have to move to another area.

It works the other way round too: the flowers need the bees for pollination, so bees are an important part of the plants' biological environment. Plants which depend on bees for pollination will only thrive and reproduce successfully in places where bees are present.

These are just two examples of the way one organism's distribution is related to that of another. Watch out for other examples as you study organisms in their natural surroundings.

Succession

Suppose you have an area of bare soil, and you leave it entirely alone. Soon various grasses and weeds appear, forming a covering of low vegetation. Then larger plants such as gorse and broom grow up, turning the area into a scrub. Next, shrubs such as hawthorn develop, and later on small trees such as birch and mountain ash grow up. Finally, large trees such as oak and beech become established (Figure 6).

There is thus a series of changes, one after the other. We call this a **succession**. The plants present at one stage alter the habitat in such a way that new species can move in. In other words each lot of plants *prepares* the habitat for the next lot. For example, oak and beech are able to move in at the end of the succession because the shrubs prepare the way for them: the shrubs shade the ground preventing other plants from developing, but oak and beech seedlings don't mind being shaded so they survive and become established.

Once these large trees have moved in, no further changes occur and we say that the habitat has reached a **climax**. The only things that can change it now are a forest fire, or man chopping down the trees.

While changes occur in the vegetation, changes also occur in the animals present in the habitat. For example, once the shrubs and trees grow up, birds and squirrels and many other animals can move in and make their homes in them.

Figure 6 A plant succession.

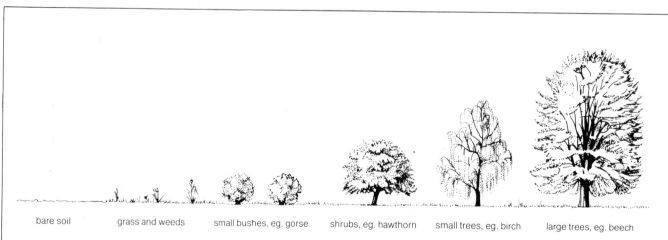

bare soil   grass and weeds   small bushes, eg. gorse   shrubs, eg. hawthorn   small trees, eg. birch   large trees, eg. beech

Investigation 1

Measuring the uptake of water by plants in different conditions

1 Obtain a leafy shoot of a tree such as sycamore.

2 With the cut end of the shoot under water, attach it to a capillary tube by means of a short length of rubber tubing.

3 Clamp the capillary tube to a stand, with the bottom end in a beaker of water as shown in the illustration.

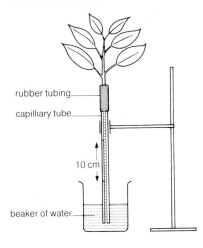

rubber tubing

capilliary tube

10 cm

beaker of water

4 Make two marks on the capillary tube 10 cm apart.

The apparatus which you have set up is called a potometer. You can use it to measure the rate at which the plant takes up water as follows:

5 Lift the capillary tube out of the beaker, touch the end of it with blotting paper, and then put it back. An air bubble will have been introduced into the capillary.

6 Time how long it takes for the air bubble to travel from the first to the second marks on the capillary tube.

7 When the air bubble has passed the second mark, push it out of the capillary tube into the beaker of water by squeezing the rubber tubing.

8 Repeat the experiment with plants in different conditions, and try it with different kinds of plants.

Explain your results.

Investigation 2

To find out how blowfly larvae react to light

Blowflies lay their eggs on dung, and the larvae live *inside* it.

How do you think the larvae would react to light: by moving towards it, or away from it?
You can find the answer as follows:

1 Obtain a sheet of white paper approximately 24 cm long.

2 Direct a lamp towards one end of the sheet of paper.

3 Switch the lamp off and darken the room.

4 Place about 6 blowfly larvae at the end of the sheet of paper where the lamp is.

5 Switch the lamp on and observe the blowfly larvae.

How do they react to your switching on the lamp?

What could you do to make sure their response is not caused by heat from the lamp?

Investigation 3

To find out how woodlice react to humidity

Woodlice normally live under logs and stones in damp places.

How do you think they would react to dry conditions?

You can find out by setting up a 'choice-chamber' containing a dry area and a moist area.

1 Set up the choice chamber as shown in the illustration below.

Anhydrous calcium chloride powder absorbs water, so the air on this side of the choice chamber will become very dry. In contrast, the air above the water will become relatively humid. Wait at least ten minutes to allow time for these conditions to develop.

2 Place about ten woodlice in the choice chamber.

3 Observe the woodlice at intervals during the next half hour or so.

Which end of the choice chamber do woodlice seem to prefer?

On what observation do you base your answer?

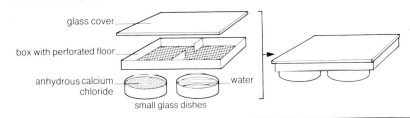

glass cover

box with perforated floor

anhydrous calcium chloride

water

small glass dishes

Assignments

1 Name one organism which forms the biological environment of each of the following:

(a) an adult mosquito, (b) a tapeworm, (c) a bumble bee, (d) a lion, (e) a tadpole.

2 A brick wall runs east-west. There are mosses on the north side but not on the south side.

Suggest two possible reasons for this. Describe experiments which you could do to find out which suggestion is correct.

3 A forester notices that in a certain area young sycamore seedlings die after reaching a height of about 6 cm, whereas in another area a short distance away sycamore seedlings grow successfully to a large size. Suggest possible reasons why the seedlings do badly in the first area.

Adaptation and survival

Figure 1 shows a caterpillar which looks like a twig. In this way the animal is camouflaged so it cannot be seen by its enemies. This is an example of adaptation.

Figure 1 This is not a twig but the caterpillar of the peppered moth.

What is meant by adaptation?

When we say that an organism is *adapted*, we mean that its appearance and structure make it well-suited to live in a particular habitat. The caterpillar in Figure 1 is a particularly striking example of adaptation. However, *all* living things are adapted, though it may not always be obvious (Investigation 1). Every organism must be adapted if it is to survive.

A detailed look at adaptation: the vertebrate limb

In Figure 2 you can see the forelimbs of four mammals including man. Notice how they differ. In each case, the limb is adapted to do a particular job. For example, in the bat the fingers are greatly lengthened to support the wing which is used for flying. In contrast the seal has short, flat fingers to support the flipper which is used in swimming.

Although the limbs illustrated in Figure 2 are all different and are used in different ways, they are all built on the same basic plan. This is illustrated in Figure 3: it is known as the **pentadactyl limb** – the word pentadactyl comes from Greek and literally means 'having five digits' (fingers or toes). This kind of limb is found in amphibians, reptiles, birds and mammals, though in different groups it is adapted to serve different purposes.

Structures which have the same fundamental design, though they may be used for different purposes, are described as **homologous**. The limbs of vertebrates provide a particularly good example of homologous structures, though there are many other examples (Investigation 2).

Does adaptation work?

When you look at an organism you can usually see certain features which *appear* to be useful adaptations. But how can we be sure that a particular feature actually helps the organism to survive?

To answer this let's look at an animal which is common in Britain: the peppered moth (Investigation 3). There are two forms of this moth: a white form and a black form. In the 1950s a scientist called Bernard Kettlewell studied the distribution of the peppered moth in Britain, and he found that in industrial areas such as Manchester and Birmingham the black form was the more common, whereas in non-industrial areas such as Cornwall and the north of Scotland the white form was the more common.

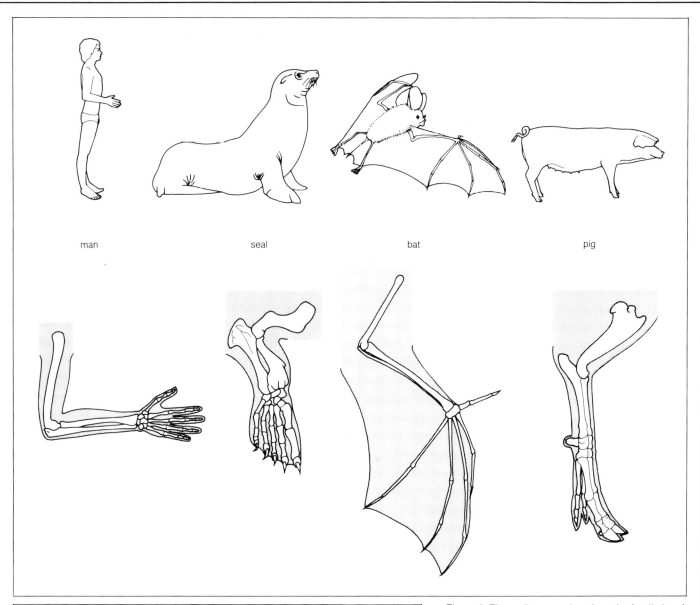

man   seal   bat   pig

Figure 2 These diagrams show how the forelimbs of four different mammals are adapted to carry out different functions.

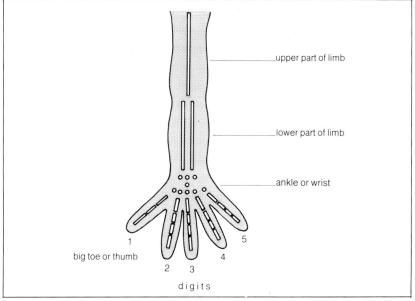

upper part of limb

lower part of limb

ankle or wrist

big toe or thumb

1   5

2   3   4

d i g i t s

Figure 3 The pentadactyl limb with its five digits. The legs of all land-dwelling vertebrates, including the mammals shown in Figure 2, are all variations on the basic theme.

Figure 4a (left) White and black peppered moths resting on a light coloured tree trunk.

Figure 4b (right) White and black peppered moths resting on a dark coloured tree trunk.

How can we explain these observations? The peppered moth rests on tree trunks and is fed on by thrushes which peck them off the trees. In non-industrial areas, the tree trunks are usually covered in lichens which are light in colour: against this background the white form of the moth is well camouflaged, whereas the black form shows up clearly (Figure 4a). As a result, the thrushes take mainly the black moths, and the white ones survive. However, in industrial areas the tree trunks are darkened by soot: here the black form of the moth is camouflaged, whereas the white form shows up (Figure 4b). The result is that in these areas the thrushes take mainly the white ones and the black ones survive.

We can sum this up by saying that in industrial areas the black colour is a useful adaptation. On the other hand, in non-industrial areas the white colour is a useful adaptation.

Living together as a means of survival

Many organisms are adapted to living inside, or sometimes on the surface of, another organism. The latter is called the host. Sometimes the host is harmed in some way, in which case we call the organism a **parasite**. The blood fluke and many fungi are examples of parasites. They feed on the host's tissues and damage them. Every parasite is specially adapted to live in or on its particular host, and to spread from one host to another. Look up some parasites in the index and find out in what ways each one is adapted.

Organisms which live in or on the bodies of other organisms don't always harm their hosts. For instance, mosses and lichens grow on the trunks and branches of trees. However, they don't feed on the tree's tissues and they don't damage it in any way.

In some cases the organism may actually help its host in some way. For example, the single-celled alga *Chlorella* which lives in the cells of the green hydra gives the hydra sugar which the alga makes by photosynthesis. In return the alga receives shelter and various other benefits from the hydra. This kind of relationship is called **symbiosis**. There is no doubt that it can help organisms to survive.

Investigation 1

Some examples of adaptation

1 Examine various organisms, or pictures of organisms, provided by your teacher.

2 Use the classification on pages 6–8 to find out what group each organism belongs to.

3 Examine each organism carefully, and write down one way in which it appears to be well adapted to its environment.

   In each case explain how you think the adaptation may help it to survive.

Investigation 3

Camouflage in the peppered moth

1 Set up a moth trap (see page 353).

   Leave it running overnight in a safe place out of doors. (A good place would be a flat roof in your school.)

2 Next day carefully take the egg trays out of the trap and examine the moths.

   Which ones are peppered moths?

   The proper name of this moth is *Biston betularia.*

   What kind are they: white or black – or both?

3 Put dead specimens of the white and black moth on sheets of light and dark paper. Against which background is each kind of moth camouflaged best?

4 Put the moth trap out again for several nights and examine the moths each day. Count the numbers of white and black ones.

   Which are the most common?

   Can you explain your observations in terms of the darkness of the tree trunks in your area?

Assignments

1 Give one example of how a *named* animal or plant is adapted to its environment, *excluding* any of the examples given in this Topic.

2 Write down *one* special adaptation which would be needed to enable:

   a) a small mammal to feed on the fruits from the top of a tree,
   b) a lizard to avoid being seen by birds in a sandy desert,
   c) a parasitic worm to live on the gills of a trout,
   d) the leaves of a plant to avoid being eaten by cows,
   e) a fish to be equally at home in fresh water and the sea.

3 In very hot, dry weather the grass in a lawn tends to dry up and go brown, whereas certain weeds such as dandelions are not affected by the weather.

   Suggest *two* ways by which these weeds might be able to withstand hot, dry weather better than grass.

4 The illustration below shows the appearance of a certain moth, viewed from above when at rest. Of what possible value might its shape and markings be?

   Describe further observations which could be carried out to test your suggestion.

5 Fifty black mice and fifty white mice were released into an area inhabited by a pair of owls. After four months the mice in the area were recaptured: only 38 black and 9 white mice remained.

   How would you explain this result?

   What further experiments could you do to find out if your explanation is correct?

   Suggest two ways in which owls are well adapted for hunting.

Investigation 2

An example of homology

1 Obtain preserved specimens of a locust, butterfly, beetle and housefly.

2 Examine the wings of these animals, the forewings first, and then the hindwings.

3 Draw the following table into your notebook:

4 In each box describe the shape and form of each wing, and say what it is used for.

How is the structure of each wing adapted to perform its function?

What other homologous structures can you see in these four insects?

	locust	butterfly	beetle	housefly
forewing				
hindwing				

The history of life

The world is full of millions of kinds of animals and plants, all adapted to their habitats. How did they get here? This is a question which scientists have thought about for hundreds of years.

Figure 1 Charles Darwin (1809–1882).

Creation or evolution?

There are two main theories to account for the existence of living things on this planet: **creation** and **evolution**.

Creation

This theory claims that living things were created by a supernatural being or God. Each kind of organism was given the necessary adaptations to enable it to cope with its environment. Some people believe that this happened only once. But others consider that it may have happened several times, earlier populations dying out and being succeeded by new ones. In any case it is assumed that the various species have remained unchanged through time, and are the same today as they were when they were first created.

Evolution

Supporters of this theory believe that the various organisms present in the world today are descended from simpler forms which inhabited our planet in an earlier age. It is thought that these simple ancestors gradually changed or *evolved* into the kinds of organisms which we see today.

This theory was first put forward by the 19th century naturalist Charles Darwin (Figure 1). In the 1830s Darwin sailed around the world in a ship called the *Beagle* (Figure 2). He visited many countries and islands and he studied the animals and plants there. He gradually became convinced that the various species which he observed had come into being by a process of slow and gradual evolution. In 1859 he published his famous book *The Origin of Species* in which he put forward evidence to support this idea. He also put forward a theory to explain *how* evolution may have taken place.

Evidence for evolution

Evolution is supported by many different kinds of evidence, some more convincing than others. Here we will concentrate on two particular lines of evidence.

Figure 2 *HMS Beagle*, the ship that was to take Darwin round the world, lying in the Catwater at Plymouth. The ship set sail on 27th December 1831.

1 *The structure of present-day organisms*

If you compare the structure of a group of animals such as the vertebrates, you find that they are all basically similar. For instance, if you look at the limbs of amphibians, reptiles, birds and mammals, you find that they are all based on the same design, namely that of the pentadactyl limb (see page 360). This suggests that they may all have developed from a common ancestor which lived long ago.

Structures which are found in different animals but have the same basic design are described as **homologous**. Most biologists believe that the existence of homologous structures provides strong evidence for evolution.

Here's another interesting observation. Certain snakes, notably pythons and their relatives have a pair of small claw-like structures about two-thirds of the way down the body, one on either side (Figure 3). They have no obvious function, and it has been suggested that they are the remains of a pair of legs which existed in an ancestor of the snakes millions of years ago. Structures which have no function today, but are thought to have been important in the past are called **vestigial structures**, and their existence would seem to support evolution.

2 *Animals and plants of the past*

Normally when an animal dies, its body decays. However, in the past dead animals have sometimes become buried in mud which later hardened to form **sedimentary rock**. Meanwhile tiny particles worked their way into the animal's bones. As a result, the bones gradually got turned into rock. These are known as **fossils**.

Over the years, thousands of fossils have been discovered. They range from a few isolated bones or fragments to almost complete skeletons. By carefully piecing the bones together, scientists have been able to work out the structure of the animals which used to live on our planet. The main conclusion is that these ancient animals were basically similar to those living today, but different in detail.

To illustrate this let's go back about 150 million years. At this time there were no mammals, and the world was dominated by dinosaurs such as the ones shown in Figure 4. Some of them were savage flesh-eaters, and others

Figure 3 It is possible that the 'claws' of this anaconda are the relics of a pair of hind legs which were possessed by the ancestors of snakes.

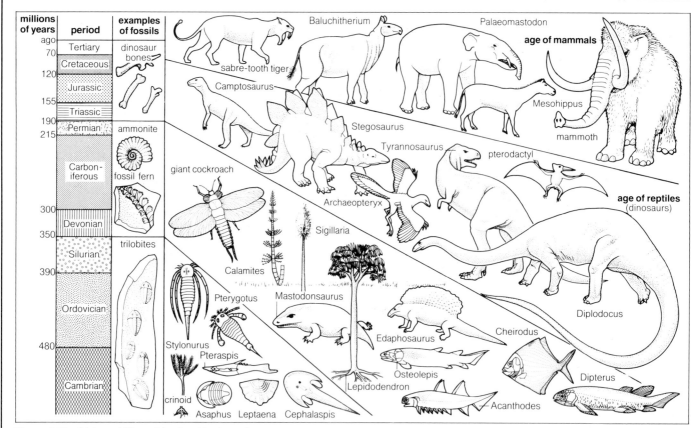

millions of years ago	period	examples of fossils
70	Tertiary	dinosaur bones
120	Cretaceous	
155	Jurassic	
190	Triassic	
215	Permian	ammonite
300	Carbon-iferous	fossil fern
350	Devonian	
390	Silurian	trilobites
480	Ordovician	
	Cambrian	

Figure 4 The history of life on this planet is shown by the fossils found in the rocks.

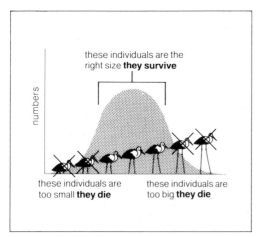

Figure 5 This simple diagram shows how natural selection ensures the survival of the fittest. The fittest individuals are those that are best adapted to the particular environment.

were docile vegetarians. Some of them grew extremely large: the one on the right hand side of Figure 4, for example, was about 30 metres long and the main part of its body was about the size of a two-storey house. It was much larger, and looked very different from, any land animal living today. However, careful studies of the bones of these dinosaurs tell us that they were reptiles and were basically similar to present-day crocodiles and lizards.

·Certain dinosaurs have been found whose skeletons were very similar to those of mammals, and this has led to the idea that these particular dinosaurs may have given rise to the first mammals. When sedimentary rock is formed it is laid down in layers, one above the other. The oldest layers are towards the bottom, and the most recent layers towards the top. In the sequence of fossils showing the change from reptiles to mammals, the most reptile-like fossils are found in lower layers than the mammal-like ones. This suggests that the reptiles came before the mammals and that there may have been a progression from one to the other.

This is just one change which may have taken place. Other fossils have been discovered which enable us to put the various animals and plants that used to inhabit the earth into a sequence like the one in Figure 4. The fossils come from different layers of sedimentary rock, and their approximate ages have been established by special chemical dating methods. All the organisms shown in Figure 4 died out eventually and became **extinct**. However, most biologists believe that the animals and plants which populate the world today are their descendants.

How has evolution taken place?

Darwin put forward an explanation of how evolution may have taken place which most biologists still believe to be correct. The explanation goes like this:

In any population of animals or plants there is usually fierce competition for food and an ever-present threat of being attacked by enemies. This creates a **struggle for existence** in which every individual is fighting desperately for survival. Now within a species there is considerable variation between

individuals, and some individuals are better adapted to the environment than others. For example, some may be particularly strong or good at running. These individuals will be more likely to survive than the others: this is sometimes called the **survival of the fittest**. These individuals are most likely to reproduce, and when they do so they may hand on their good qualities to their offspring.

This process is called **natural selection**. Nature, as it were, *selects* the fittest individuals and *rejects* the weaker ones (Figure 5). In this way, the species as a whole is constantly tending to improve and to change into something better.

We have seen that natural selection depends on variation. The kind of variation which is important in natural selection is *genetic* variation, particularly the sort that results from **mutation** (see page 348). Mutations are usually harmful, but occasionally a mutation may occur which makes an individual *better* adapted to its environment. This particular individual will win in the struggle for existence, and its useful adaptation will be passed to the offspring and will gradually spread through the population. In this way new kinds of organisms arise.

What evidence is there that natural selection really does take place? One of the clearest pieces of evidence comes from Kettlewell's studies on the peppered moth (see page 360). The black form of this moth is thought to have arisen as a result of a mutation in the 1840s. Thanks to the darkening of the trees from industrial smoke, this black form thrived in industrial areas. Its numbers increased until it eventually became the main type in those areas.

Another example of natural selection is provided by the way new kinds of germs arise. For example, new types of bacteria have been formed as a result of a mutation which makes them resistant to penicillin. In the same way new types of mosquito have arisen which are resistant to DDT. Obviously these new forms are at a great advantage, and so they spread quickly.

Artificial selection

We have seen that in bringing about evolution nature selects the fittest individuals and rejects the unfit. Now the same thing can be done by man. With a population of animals or plants at his disposal, a person can select those with good qualities and allow them to breed, whereas those with poor qualities can be killed or at any rate prevented from breeding. This is called **artificial selection** and it has been carried out by animal and plant breeders for centuries.

All our familiar breeds of farm animals, and domestic animals such as cats and dogs, have been produced this way (Figure 6). So have various varieties of garden plants, such as roses. By the same process, plant breeders have produced new varieties of crop plants which are better than the older ones. For example, we now have new varieties of wheat and rice which grow more quickly, give a higher yield of grain and are more resistant to disease. This is important in helping to meet the world's food problem, particularly in over-populated countries.

Figure 6 Three different breeds of dogs which have been produced by artificial selection.

Assignments

1 In what way do vestigial structures support the idea of evolution?

2 What are homologous structures? In what way do homologous structures provide evidence for evolution?

3 Rats are normally killed by a poison called warfarin. However, certain individuals are not affected by this poison, and they appear to be on the increase. How would you explain this?

4 In what respect is the breeding of dogs similar to, and different from, the process of evolution as put forward by Darwin?

5 Suppose the animals in Figure 5 are antelopes in an African Game Park. Suggest reasons why it is a disadvantage for them to be (a) very short, and (b) very tall.

Populations

People people everywhere! Figure 1 shows a beach in Britain on a fine day in the middle of summer. This Topic is about populations: how they grow and what happens if they get too big.

Figure 1 A crowded beach in Britain.

How do populations grow?

Suppose you introduce a few rabbits onto an unpopulated island. The rabbits reproduce and gradually the population increases. If you were to count the number of rabbits at intervals and plot them against time, you would find that the population rises as shown in Figure 2.

One of the most noticeable things about populations is that they increase very quickly. This is because the numbers go up by *multiplication*, like this:

$$2 \times 2 \times 2 \times 2 \times 2 \times 2$$
$$2 \rightarrow 4 \rightarrow 8 \rightarrow 16 \rightarrow 32 \rightarrow 64$$

In other words the total number *doubles* at regular intervals of time. This type of increase is described as **exponential**, and it is how populations grow, whether it is rabbits, flies or human beings.

Populations increase like this because new individuals are born at a faster rate than older ones die, in other words the **birth rate** is greater than the **death rate**.

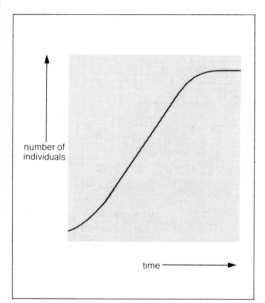

number of individuals

time ⟶

Figure 2 This graph shows how a population increases with time.

What stops a population growing?

Look again at the graph in Figure 2. Notice that once a certain number of individuals has been produced, the curve gradually flattens out: in other words the population growth slows down and levels off.

Why does this happen? In the case of our rabbits there are several possible reasons. Here are some of them:

1 The food (grass and other leafy plants) begins to run out, so some of the rabbits starve.
2 There are so many rabbits on the island that there is no room for any more burrows.
3 The rabbits are so overcrowded that diseases spread rapidly.
4 Being overcrowded the rabbits fight each other, resulting in many deaths.

These are the kinds of 'checks' which normally stop populations growing indefinitely. This may seem harsh and cruel, but it is a normal and important aspect of the balance of nature. In the case of the rabbits, the population would be kept at an even lower level if there were some animals on the island which ate rabbits: a few foxes perhaps (Figure 3), or a bird of prey like the hawk. Such **predators** normally prevent populations getting too large. In Europe and many other parts of the world rabbits are kept in check this way.

One of the main forces keeping down populations is **competition**. The main things that animals compete for are food and a place to make a home. Green plants compete mainly for light and also for water and nutrients in the soil.

Figure 3 Foxes feed on rabbits and help to keep their numbers down.

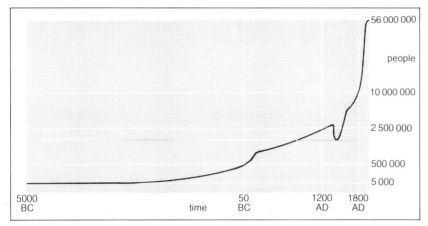

Figure 4 This graph shows how the population of Britain has grown since 5000 BC.

Controlling pests

Australia used not to have any rabbits, but some years ago some domestic rabbits escaped from their pen when it was swept away by a flood. These rabbits ran wild and bred at such a rate that parts of Australia soon became overrun with them and they did a great deal of damage to crops and gardens. Unfortunately there were not enough predators to keep them under control. Eventually a virus disease called **myxomatosis** was deliberately introduced to destroy them. This disease swept through the rabbit population very quickly, and their numbers dropped dramatically.

Myxomatosis shows us how man can reduce the population of a pest by introducing an organism which kills it. We call this **biological control**. The only alternative is to kill the pest with a poison: this is known as **chemical control** and an example is the killing of mosquitoes with insecticides such as DDT (see page 77).

Human population

Figure 4 shows the total number of people in Great Britain at various times during the last seven thousand years. Notice how the number has increased: in fact the graph looks rather like the one in Figure 2. In particular notice that the population has grown more during the last two hundred years than in the whole of the previous 5000 years!

Why has there been this recent explosive increase? In the old days a large number of people died of starvation and infectious diseases. Many died while they were still very young, in other words the **infant mortality** was high. However, in recent times there have been great improvements in food-production and in fighting disease. As a result no one starves any more, and far fewer people die of diseases. Infant mortality has been cut right down, and people live much longer than they used to.

We can sum it up this way: in Britain over the last few hundred years the death rate has fallen dramatically with the result that the population has shot up. Obviously this cannot go on for ever. Already many people feel that we have reached the point where there are too many people in Britain.

Because of this there has been a campaign in recent years to encourage young people to have smaller families. This involves **birth control**. Birth control certainly seems to be having an effect because the population of Britain is now beginning to level off; in fact in the last few years it has dropped slightly. Unfortunately this is *not* true of many developing countries where the population continues to rise at a horrifying rate. In fact the world population as a whole is increasing by about 80 million people each year, that's 9000 an hour, or 150 a minute. Someone has worked out that if this were to go on indefinitely, the whole of the earth's surface would be covered with people standing shoulder to shoulder in less than 600 years' time!

Assignments

1 Study Figure 2, then suggest two reasons why the population rises slowly to begin with and then speeds up.

2 If all its offspring survived, a single greenfly could produce 600 000 000 000 offspring in one season, with a mass of over 600 000 kg – roughly equivalent to 10 000 men. What prevents this happening?

3 It was suggested that in order to protect the deer in a nature reserve in the United States, the deer's predators (wolves and coyotes) should be killed. What do you think the consequences of doing this would be?

4 Study the graph in Figure 4 and then answer these questions.

   a) How do you think we know what the population of Britain was in 5000 BC?

   b) Suggest a possible reason why the population was rising around 50 BC.

   c) What do you think caused the sudden fall in the population midway between AD 1200 and 1800?

   d) Why has the population risen so quickly since AD 1800?

   e) Suggest one reason why the population appears to be levelling off now.

5 Suppose that a certain island becomes overrun with rabbits. Suggest three different ways by which the people living on the island might try to reduce the rabbit population. Briefly mention the advantages and disadvantages of each method.

Feeding relationships

In the natural world, animals feed on plants and on other animals. This is an essential part of the balance of nature.

Food chains

Suppose we put some weeds, tadpoles and a couple of water beetles into a jar and watch what happens (Investigation 1). We find that the tadpoles nibble at the weeds, and the water beetles eat the tadpoles. We can sum up the feeding relationship between the three organisms like this:

weeds → tadpoles → water beetles

We call this a **food chain**, and it is a basic feature of most habitats. Tadpoles feed only on plants and are therefore herbivores. In contrast, water beetles feed on other animals and are carnivores. In fact, the water beetle (and its larva) are extremely voracious: I have seen a larva get through over 20 tadpoles in an hour (Figure 1).

There are only three links in the food chain shown above. However, in a lake there might be some pike. These fish feed on water beetles amongst other things, so in the lake the food chain would be:

weeds → tadpoles → water beetles → pike

Producers and consumers

Let's think about this food chain in a bit more detail. The weeds make their own food by photosynthesis, and they get the necessary energy for doing this from the sun. Because they *make* food (i.e. manufacture organic substances), we call them **producers**. In contrast, the animals in the chain get their food by

Figure 1 The larva of the great diving beetle is one of the most savage carnivores found in ponds. It sinks its fang-like teeth into its prey and then sucks up its juices.

eating other organisms. For this reason we call them **consumers**. In this particular chain there are three consumers. The tadpoles are the first consumers, the water beetles are the second consumers, and the pike is the third consumer (Figure 2).

Now as you go along the food chain, the number of organisms which can be supported at each level gets less and less. Thus a given number of tadpoles will feed a relatively smaller number of beetles, and these beetles will feed a relatively smaller number of fish (Figure 3). This drop in numbers at each level in the food chain is called the **pyramid of numbers**. The reason why the numbers fall is that when an animal eats food, only a proportion of it gets into the structure of its body: the rest is broken down in respiration.

For the same reason there is also a drop in the total mass of living material at each level of a food chain. This is called the **pyramid of biomass**. On the other hand the size of the individual animals at each level tends to increase. This is because generally speaking carnivores are larger than their prey.

Food chains in the service of man

Going back to the lake, it is quite possible that a fisherman might catch one of the pike and have it for his supper. The food chain would then be:

weeds → tadpoles → water beetles → pike → man

From man's point of view, some of the most important food chains occur in the sea. In the surface waters where light can penetrate there are millions of

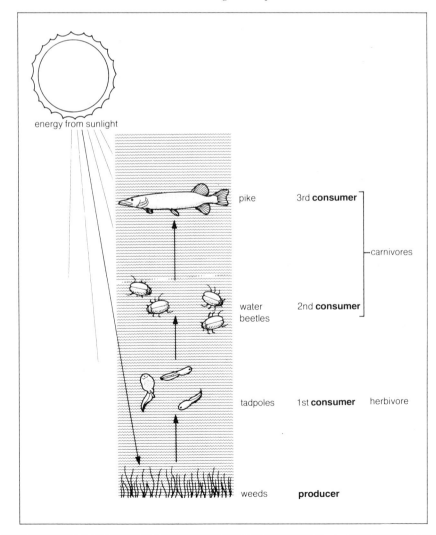

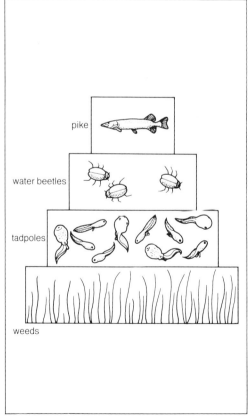

Figure 2 (left) The main steps in a typical food chain.

Figure 3 This diagram shows how the numbers of organisms decrease as you go along a food chain. It is called a pyramid of numbers. The total mass of living material also decreases at each level, and so does the total amount of energy contained in the bodies of the organisms.

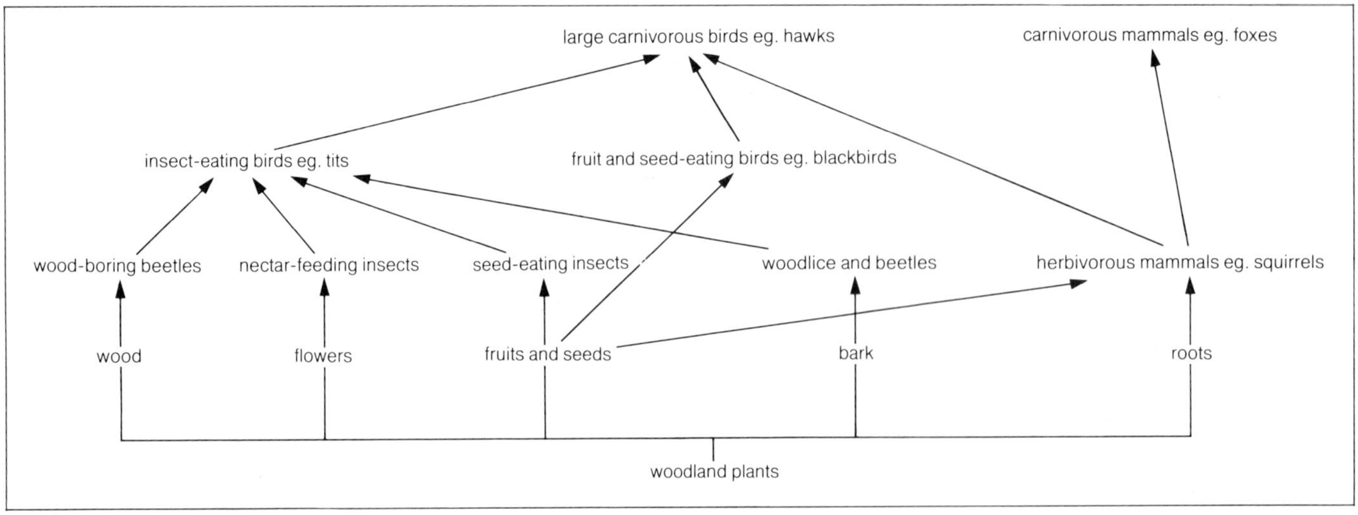

large carnivorous birds eg. hawks

carnivorous mammals eg. foxes

insect-eating birds eg. tits

fruit and seed-eating birds eg. blackbirds

wood-boring beetles

nectar-feeding insects

seed-eating insects

woodlice and beetles

herbivorous mammals eg. squirrels

wood

flowers

fruits and seeds

bark

roots

woodland plants

Figure 4 A food web in a wood.

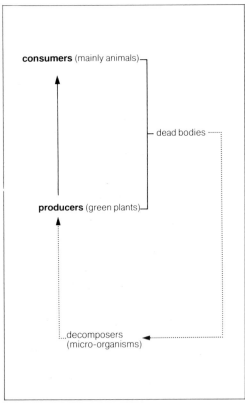

Figure 5 Decomposers enable the materials in the bodies of the producers and consumers to be used again.

consumers (mainly animals)

dead bodies

producers (green plants)

decomposers (micro-organisms)

microscopic algae (see page 44). These are plants and they provide food for numerous small animals. The plants and animals together constitute **plankton**. Plankton is extremely important because it provides food for fish. So a food chain in the sea would go like this:

plant plankton → animal plankton → fish → man

Certain important substances get concentrated as they pass along the chain. For example, vitamin D is synthesised by the plant plankton, and it then gets into the animal plankton. Eventually it gets into the fish which store and concentrate it in their livers. This is why cod liver oil is such a good source of this vitamin (see page 117).

On land, farming involves several important food chains. Here is one of them:

grass → sheep → man

Can you think of any others? And can you think of any circumstances in which the first consumer is eaten by another animal besides man?

Food webs

Let's go back to the jar with which we started this Topic. Suppose we remove the tadpoles, what will happen to the beetles? They will die, because we have taken away their only source of food.

However, if the tadpoles were to disappear from a pond or lake, the beetles would probably survive. This is because there would be other sources of food available to them. For example, it's quite likely that the pond would contain some small fish such as minnows which the water beetles could eat.

By finding out what all the organisms in a habitat feed on, you can build up a diagram summarising their feeding relationships (Investigation 2). This is called the **food web**. In a natural habitat such as a pond, it would be unusual for the organisms to be linked together in a simple chain. Food webs are much more common, and if the habitat contains a large number of different species, the web may be very complex. A comparatively simple food web is shown in Figure 4.

Decomposers

When the animals and plants in a habitat die their bodies decay. This is because they are broken down by bacteria and other microbes which feed on them. These microbes are called **decomposers**. As a result of their activities, simple substances are released from the dead bodies, and these can be used again by plants, i.e. by the producers (Figure 5). The decomposers thus play an important part in keeping life going in a habitat.

Ecosystems

We have seen that a habitat such as a pond or a wood contains three types of organisms: producers, consumers and decomposers. All three are influenced by physical features of the environment such as temperature and rainfall, and together they make up what biologists call an **ecosystem**. Every habitat has its own particular ecosystem consisting of a community of organisms. These interact in such a way that the ecosystem as a whole is maintained in a stable and harmonious manner.

Investigation 1

Observing a simple food chain

1 Collect some weeds, tadpoles and water beetles (*Dytiscus*). Either the adult water beetle or its larva will do. Keep the tadpoles and beetles in separate containers.

2 Obtain a jar: a large-size instant coffee jar will do nicely.

3 Fill the jar with water and put in a few weeds and about 10 tadpoles.

   Can you see the tadpoles feeding? What do they feed on?

4 Now put two water beetles (adults or larvae) into the jar with the tadpoles.

   What do the beetles feed on? Describe their method of feeding.

Investigation 2

Building up a food web

1 Set up an aquarium in your laboratory.

   Put into it as large a variety of animals from a local pond or stream as you can. Make the aquarium as natural as possible, and use water from the pond itself: be sure that you put plants in as well as animals.

2 Using a simple identification key find out the names of the animals and plants in the aquarium.

3 Observe the animals, and see if you can find out what each one feeds on. If necessary use books to help you.

4 Write down the names of:

   a) the producers,
   b) the herbivorous consumers,
   c) the carnivorous consumers.

   Which of the carnivorous consumers are *not* eaten by any other organism?

   What do you think would happen if you removed the consumers from the aquarium?

5 Construct a food web similar to the one in Figure 4, showing the feeding relationships of the animals in your aquarium.

Assignments

1 Fill in the missing organism in each of the following food chains:

   a) grass → ? → man
   b) grass → deer → ?
   c) algae → planktonic animals → ?
   d) lettuces → ? → fox
   e) aphid (greenfly) → ladybird → ?

2 The following is a food chain that ends up with man:

   plant → bee → man

   Explain precisely how plants provide food for bees, and how bees provide food for man.

3 Study the food web in Figure 4, then answer these questions:

   a) Give the names of two woodland plants.
   b) Give the name of one organism which is a second consumer only, and one which is both a second and a third consumer.
   c) Give one example of a nectar-feeding insect, and one example of a beetle that lives under the bark of trees.
   d) Give an example of a carnivorous bird other than the hawk, and of a carnivorous mammal other than the fox.
   e) This food chain does not include the leaves of the woodland plants. Write down a food chain which might lead from leaves.

4 In this Topic it is stated that as one proceeds along a food chain, each organism tends to be larger than the one before.

   a) Give an example of a food chain which illustrates this.
   b) Why do you think this is true?

   c) Give an example of a food chain which is an *exception* to this.

5 The following figures show the total mass of body material, measured as dry mass, from one square metre of grassland during one year:

plants	470.0 g
herbivores	0.6 g
carnivores	0.1 g

   Explain why the mass of body material decreases at each step of the food chain.

6 Energy from the sun passes through food chains. However, only a small proportion of the sun's energy gets into the bodies of the final consumers. What happens to the rest?

The wheel of life

One of the most important aspects of nature is that materials circulate. This means that they can be used over and over again.

The cycling of carbon

The air around us contains a small amount of carbon dioxide. This is constantly being absorbed by plants which use it for photosynthesis. The carbon dioxide diffuses into the leaves, and is built up into sugar and other complex carbon compounds.

Now when a plant is eaten by an animal the sugar gets into the cells in the animal's body. Here it is broken down into carbon dioxide and water to produce energy (respiration). As a result carbon dioxide is put back into the atmosphere.

When the animals and plants die they decay. In this process bacteria and other microbes feed on them. They too respire, and so once again carbon dioxide is put back into the atmosphere.

We can sum up by saying that *carbon dioxide is taken out of the atmosphere by photosynthesis, and put back into it by respiration and decay*. This is known as the **carbon cycle** and it is summarised in Figure 1.

If you look at this diagram you will see that carbon dioxide is also released into the atmosphere when fuels such as coal are burned (**combustion**). These fuels are formed by the fossilisation of dead plants. If it wasn't for man the carbon contained in these fuels would never be returned to the atmosphere, and the arrow leading to **fossil fuels** would be a dead end taking carbon out of the cycle for ever.

Fossil fuels include coal, oil and natural gas, and of course they give us energy when they are burned. Vast amounts have been laid down deep in the ground over millions of years. However, it's a slow process and at present we are using them up about one hundred thousand times faster than they are formed. Scientists estimate that at this rate we shall run out of them within a few hundred years unless we go over to other ways of getting energy.

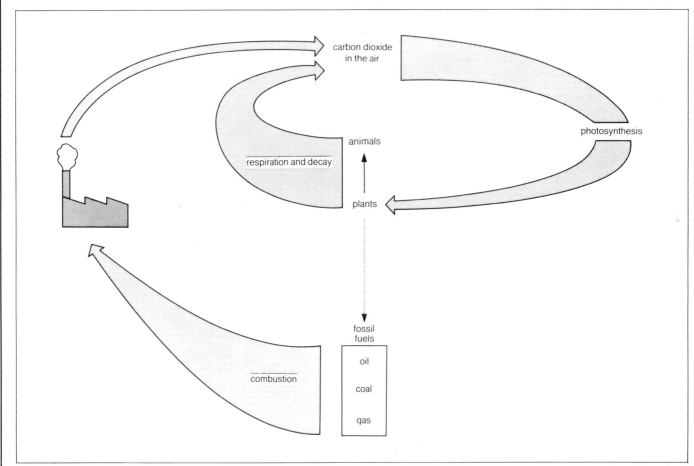

Figure 1 Summary of the carbon cycle.

The cycling of nitrogen

In the soil there are inorganic nitrogen compounds called nitrates which are dissolved in the soil water. These are absorbed by the roots of plants which then build them up into complex proteins.

Now think what happens when a plant is eaten by an animal. The nitrogen in the plant protein gets into the animal's body and becomes part of its protein.

When the animals and plants die they decay. The microbes which cause decay break the proteins down into ammonia. Ammonia is also formed from the animal's excreta. Now in the soil there are certain bacteria which turn ammonia into nitrates. They do this so as to get energy. The effect is to return nitrates to the soil, which can then be used again by plants. Because they enrich the soil in nitrates, they are known as **nitrifying bacteria**. You will now understand why compost and manure improve the soil for plant growth.

This circulation of nitrogen is known as the **nitrogen cycle** and it is illustrated in Figure 2. In this diagram you will see that certain bacteria release nitrogen from nitrates. Obviously they *lower* the nitrate content of the soil, and so we call them **denitrifying bacteria**.

Plants cannot make use of atmospheric nitrogen. However, certain bacteria can absorb nitrogen and build it up into protein. They are called **nitrogen-fixing bacteria**, and they are found free in the soil and also inside the roots of plants belonging to the pea and bean family. The nitrogen-fixing bacteria in the roots give some of their nitrogen compounds to the host plant, in return for protection. The relationship is therefore beneficial to both organisms, and is an example of symbiosis. Some of the nitrogen compounds are released into the soil which thus becomes fertilised (see page 181).

Assignments

1 Study the carbon cycle in Figure 1, and then answer these questions:

a) What is photosynthesis and how does it remove carbon dioxide from the atmosphere?

b) What is respiration, and how does it add carbon dioxide to the atmosphere?

c) Name two organic carbon compounds found in plants, and two found in animals.

d) Name the main kind of organisms that bring about decay, and explain how they put carbon dioxide into the atmosphere.

e) What happens chemically when coal is burned?

2 Farmers often plough clover and other leguminous plants into the soil. Why is this a good thing to do? Explain your answer fully.

3 Construct a diagram, similar to Figure 1, which summarises the way *oxygen* circulates in nature.

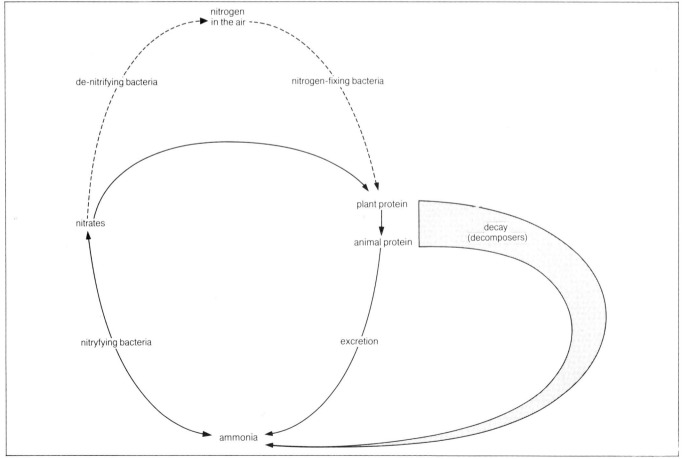

Figure 2 Summary of the nitrogen cycle.

Decay

*If it wasn't for
decay the world would
be piled high with dead bodies:
it is said that if no decay had occurred
since the reign of Elizabeth I, the
bodies of dead organisms
would cover the earth
to a depth of a
kilometre!*

Figure 1 Louis Pasteur (1822–1895) who discovered that there are microbes in the air. He is seen here in his laboratory.

dead material

bacteria land on it

they secrete digestive enzymes which break down the material into a liquid

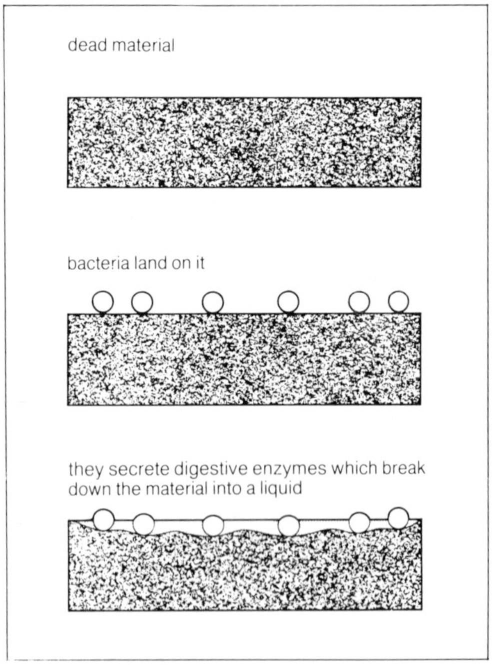

Figure 2 These diagrams summarise how decay is brought about.

What happens during decay?

When something decays – a dead animal if you like – it is gradually turned into a liquid. Three main things cause this to happen:

1 Immediately after death enzymes start breaking down the body. The organism literally digests itself. This may happen remarkably quickly: an animal's gut, for example, may become as thin as tissue paper within a day after death.
2 Various natural processes help to break the body up. Birds peck at it, worms wriggle through it, rain softens it, and alternate freezing and thawing breaks it up into pieces.
3 Certain **microbes**, mainly moulds and bacteria, feed on the dead remains and break it down. They are **decomposers**, and they are the main agents of decay.

Where do the microbes come from?

Decay microbes were discovered by the French scientist Louis Pasteur (Figure 1). He did an experiment similar to the one in Investigation 1. He showed that a substance such as broth goes bad because tiny organisms, too small to be seen with the naked eye, get into it from the surrounding air.

We now know that moulds and bacteria form **spores**. Countless millions of them float in the air around us, like specks of dust. When an animal or plant dies, it isn't long before some of these spores land on it. They then burst open, giving rise to new individuals. These grow and multiply, spreading quickly through the dead material. A teaspoonful of rotting vegetation may contain over a thousand million of them. However, you can only see them by looking at the material under a microscope (Investigation 2).

To feed on the dead material the microbes must first break it down into a liquid, just as we have to digest our food before we can absorb it. They do this by producing digestive enzymes which break down the solid material into a liquid (Figure 2). They then soak up all the nutrients they need. Soft remains like skin and flesh decay more quickly than hard structures like bone and wood (Investigation 3). A skeleton may remain intact for years after the rest of the body has decomposed.

Decay does not occur all at once: it happens step by step. A dead body is attacked by one kind of microbe first. This works its way in, and prepares the way for another organism and so on. As decay gets underway the rotting material may get warm, due to the heat given out by the millions of microbes. It may also smell unpleasant, due to certain gases being given off.

What conditions are needed for decay to occur?

A simple experiment can be done to answer this question. You simply put pieces of food in different conditions, and find out which ones decay (Investigation 4). Experiments of this sort tell us that the following conditions are needed for decay to occur:

1 There must be plenty of moisture
This is needed for the spores to germinate, and for the microbes to grow and multiply. If a dead body is kept dry it loses moisture and the skin shrinks, but it does not decay. This process is called **mummification**: the ancient Egyptians used it to preserve the bodies of their kings.

2 It must be warm enough
Most microbes which bring about decay thrive in a warm environment. If it is cold, decay is slowed down; and if it is freezing, it will not happen at all. Extinct mammoths, which died in Siberia thousands of years ago, have been dug out of the ice with their flesh intact: the intense cold had stopped their bodies decaying.

3 *Oxygen must be present*

The microbes which bring about decay need oxygen for respiration. If oxygen is lacking they respire without it, i.e. they respire anaerobically. The end-product of this is lactic acid (see page 162). The lactic acid *prevents* further decay from taking place. So when there is no oxygen present decay is incomplete.

4 *Chemicals must not be present which kill the microbes*

This sometimes happens when poisonous substances are discharged into lakes and rivers from factories. It also happens when a biologist puts a specimen in alcohol to preserve it. In the past it has happened when animals have fallen into tar pits: a tar pit is a lake full of an oily liquid. In California there are tar pits containing the undecayed skeletons of extinct sabre-toothed tigers which died there about a million years ago.

How can we make decay occur?

To bring about decay all we need to do is to put some dead material in a place where microbes can flourish. This is what a gardener does when he makes a **compost heap** (Figure 3).

Any rottable material can be used to make compost: old cabbages, potato peelings, tea leaves – you name it. In making a compost heap the following rules should be followed. Each has a scientific reason which you can think of for yourself.

1 Choose a place for your compost heap which is sheltered from the wind and sun.
2 Enclose the heap in some kind of container with sides and a rainproof lid.
3 Allow air to get underneath the heap.
4 As you pile up the dead material, put in a thin layer of manure or fertiliser every now and again. It's also a good idea to add a little lime occasionally.
5 Don't throw on sticks, plastic bags or materials tainted with oil, creosote or paint.
6 If you put on old cabbage stumps, bash them up with the back of an axe beforehand.
7 If you put in grass cuttings, mix them with bulkier material first.
8 Keep the heap moist, but don't saturate it.
9 Your heap should be neither too large nor too small. Ideally it should be about 1–2 metres high and 1 metre wide.
10 Turn the heap occasionally with a fork so the inside is moved to the outside, and the outside to the inside.

Why is decay important?

When microbes get to work on a dead body, they break its complex chemicals into simpler ones. Carbohydrates are broken down into carbon dioxide and water; proteins into simple nitrogen salts. These simple substances can then be absorbed by plants, which build them up again into the complex materials of their own bodies.

So decay puts back into the atmosphere and soil the chemicals which plants took out. It is why decaying matter in the form of manure or compost is so good for plants.

Now let's look at some other ways in which decay is useful.

Making cheese

The first step in making cheese is to curdle sour milk so that it forms curds and whey. The next step is to separate the curds from the whey. One way of doing this is to put the curdled milk in a muslin bag and squeeze the fluid whey out. The paste-like substance left behind is cheese.

At this stage the cheese is white and tasteless. It must now be **ripened**. This

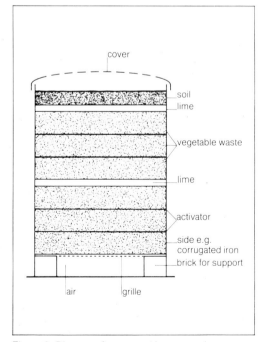

Figure 3 Diagram of a compost heap seen in sectional view.

Figure 4 Blue cheese is ripened by a mould which grows in it. This machine makes holes in the unripe cheese helping the mould to spread through it.

Figure 5 A greatly simplified diagram of a sewage works.

is where microbes come in. Some of them occur naturally in the cheese, others are added to it. They break the cheese down, softening it and giving it its characteristic smell and flavour.

There are many kinds of cheese, and each is ripened by particular microbes. In some cheeses the mould is visible as a network of threads. In others the microbes give off a gas which cannot get out: the result is that the cheese has cavities in it. Next time you are in a food shop, see if there are any cheeses that look like this.

Cheese should not be ripened for too long, otherwise the decay process goes too far and the cheese begins to liquefy and go very smelly.

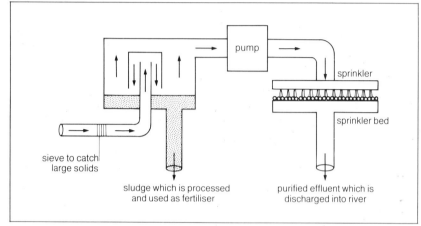

Decay gets rid of sewage

Sewage is got rid of in a sewage works (Figure 5). The process depends on decay.

First the sewage is pumped into a large tank. Here the solid matter sinks to the bottom, forming a sludge. This is broken down by anaerobic bacteria which give off the gas **methane** which can be used as a fuel, supplying power to run the sewage works. When the sludge has been broken down by these bacteria, it is dried and can be used as a fertiliser.

Meanwhile the liquid part of the sewage is pumped into a long pipe with holes in it. This sprinkles the liquid onto a bed of broken stones, called a **filter bed**. The stones are coated with a slimy film of aerobic bacteria. These bacteria break up any organic matter into simple substances like carbon dioxide and nitrogen salts. The liquid is collected from underneath the filter bed, treated with chlorine to kill any harmful bacteria, then discharged into a nearby river or lake.

Investigation 1

To find out where decay organisms come from

1 Make up some 'nutrient broth' as follows: put a broth tablet in a test tube and add 10 cm of distilled water to it. Boil it for several minutes so as to sterilise it.

2 Pour half the nutrient broth into each of the two test tubes, and set the test tubes up as shown on the right.

3 Sterilise both tubes by heating them in an autoclave or pressure cooker for 15 minutes, then let them cool.

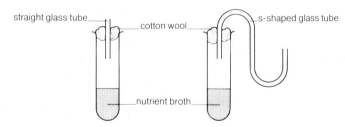

4 Observe the nutrient broth in the two test tubes at intervals during the next few days or weeks.

If the nutrient broth goes cloudy, it means that bacteria have got in and are turning it bad.

Which tube goes cloudy first?

How would you explain your observation?

Investigation 2

Looking at organisms which bring about decay

1 Pull off a *small* piece of mould from some stale bread with forceps. Put it on a slide. Add a drop of water, and cover it with a coverslip.

2 Pull off a piece of decaying earthworm with forceps, and put it on another slide. Add a drop of water and cover it with a coverslip.

3 Examine your slides under a microscope. You may see three types of organisms which help to bring about decay:

Bacteria	look like tiny dots
Moulds	look like fine threads
Roundworms	are slender with pointed ends

In what ways are these organisms suited to living on dead material?

Investigation 3

To find out how quickly different things decay

1 Fill a plant pot with soil.

2 Water the soil well.

3 Put different objects on the surface of the soil, e.g. a dead earthworm, dead insect, leaf, stick, bone, and piece of plastic or polythene.

4 Cover the pot with a polythene bag:

polythene bag

well-watered soil

thread for tying on polythene bag

plant pot

saucer

5 Put the pot in a warm place.

6 Examine the objects at intervals for the next 2–3 weeks.

How does the appearance of each object change during the 2–3-week period?

Explain your observations.

Why is it necessary to make sure there were no live earthworms in the soil?

Investigation 4

To find out what conditions are needed for decay to occur

1 Obtain four pieces of bread, about 1 cm square.

2 Moisten three of them with water; dry the other one thoroughly with a hair drier.

3 Sprinkle some dust on each piece of bread – this will almost certainly contain spores of moulds and bacteria.

4 Obtain four large test tubes. Label them A, B, C and D.

5 Put the pieces of moist bread in test tubes A, B and C. Put the dry piece in test tube D.

6 Set up the four test tubes like this:

A (moist bread): plug the test tube with cotton wool, and leave it in a warm place.
This is your control; the piece of bread is given all the conditions thought to be necessary for decay to occur.

B (moist bread): plug the test tube with cotton wool, and put it in a refrigerator.
This tests the effect of cold on decay.

C (moist bread): seal the test tube tightly with a stopper, and smear vaseline round the edge to make it airtight. Leave it in a warm place.
This tests the effect of lack of air on decay.

D (dry bread): plug the test tube with cotton wool, and leave it in a warm place.
This tests the effect of lack of moisture on decay.

7 Examine the four pieces of bread at intervals over the next 2–3 weeks.

What does each piece of bread look like at the end of a 2–3 week period?

Which pieces of bread have decayed, and which have not?

What conditions appear to be needed for decay to occur?

Is this a good experiment? Can you find faults with it, and how might you improve it?

Assignments

1 Decay is brought about mainly by microbes such as bacteria. However it is helped by several other agents. Name *five* such agents.

2 A body was found in a remote cave in the Sahara desert. Forensic experts estimated that it had been there for well over a hundred years. The skin, though dry and shrivelled, was still intact. Suggest reasons why it had not decayed.

3 When making a compost heap, why do you think it is a good idea to:

a) support it on bricks or large stones,
b) break up old cabbage stalks before you put them on the heap,
c) mix grass cuttings with bulkier materials when you put them on the heap,
d) add a thin layer of manure or fertiliser every now and again,
e) turn the heap occasionally with a fork?

4 The graph below shows how the temperature inside a compost heap changed after it had been set up.

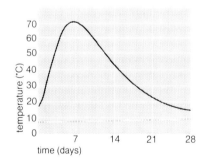

Explain why the temperature changed like this.

5 As a result of an accident a large amount of untreated sewage gets into a lake. Write a short letter to the local authority explaining the possible consequences if they don't do something about it quickly.

6 What part does decay play in (a) cheese making and (b) getting rid of sewage?

What is soil?

Soil is the surface of the earth's crust where plants have their roots and where many small animals make their homes. As the soil directly affects the growth of plants, it is of the utmost importance to man. Vast areas of it are used for growing crops.

How is soil formed?

Thousands of millions of years ago, the land was covered with bare rock. Gradually the surface of the rock was broken up by rain, wind, snow and frost into small particles. These particles were gradually piled up on top of the rock to form soil.

If you look at a cliff or a new motorway cutting, you will see that the soil is made up of layers. At the top is a dark layer where plants and other organisms live. We call this the **topsoil**: it is formed by surface weathering and the activities of the many organisms which live in it, and it contains the decaying remains of dead organisms.

Beneath the topsoil is a lighter-coloured layer of gravel, stones, clay and so on. This is called the **subsoil**. It contains the deeper roots of plants, but otherwise not much lives there.

Further down still is solid **rock**. This is non-porous and won't let rain through, so water tends to gather above it. The surface of this water is called the **water table**.

These three layers are shown diagrammatically in Figure 1. Their relative thicknesses, and the position of the water table, vary a great deal from place to place.

What does soil consist of?

Soil contains six main components: rock particles, soil water, humus, mineral salts, lime and air. We will now look at each in detail.

Rock particles

These vary in size. Depending on their size, they are classified into clay, silt, sand and gravel (Figure 2). Clay particles are so small that they can only be seen properly under the microscope. At the other extreme, gravel consists of small stones which can be separated from the rest of the soil by sieving.

The smaller soil particles can be separated from each other by shaking up a sample of soil with some water and letting it stand (Investigation 1). The sand sinks to the bottom, but the tiny clay particles remain suspended in the water above the sand.

Rock particles make up the framework of the soil, its 'skeleton' as it were. Both clay and sand are important in this respect. Clay holds onto water better than sand, thus making it sticky and helping to bind the rest of the soil

Figure 1 Sectional view of the earth's crust to show the different layers of soil and other materials.

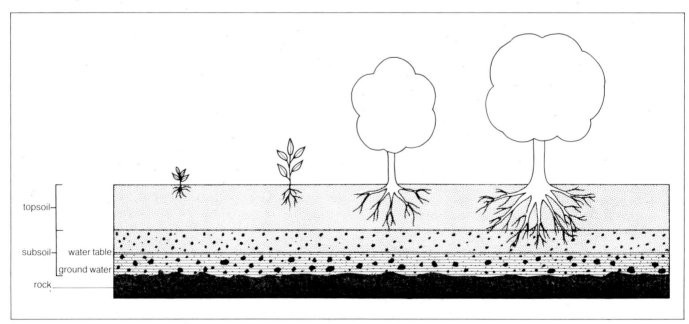

together. On the other hand, sand is looser and more easily penetrated by air and water.

Neither sand nor clay on their own make good soil. Good soil consists of a mixture of the two: this is called **loam**. Good gardening loam contains roughly twice as much sand as clay.

If you look at some good garden soil you will notice that the particles are stuck together in small clumps (Investigation 2). These are called **soil crumbs**: they make the soil coarser, helping air to get into it and water to drain through it.

The roots of plants grow down between the soil crumbs, gripping them as they do so. This gives plants a firm anchorage, which is why it is sometimes hard to pull them up.

Soil water

Soil particles are normally surrounded by a thin film of water. It is from these films that plant roots take up all the water they need. Unless the soil is excessively dry, these films are always present. What ensures that this is so? Let's answer this by thinking what happens after a heavy shower of rain (Figure 3).

The rain sinks down into the soil, wetting the soil particles. Eventually it reaches the water table. The roots of plants absorb water from the films surrounding the soil particles, and some water also evaporates from the surface of the soil. But as quickly as water is lost this way, more is drawn up from lower down. If you don't believe this, try doing Investigation 3.

The process which causes this is **capillary action**. It is the same process that causes water, or ink, to spread through a piece of blotting paper, or to rise in a narrow capillary tube.

For water to move through the soil like this, the soil particles must be the right size. If they are too large, water will sink straight through and will not be pulled up. Such soil is useless: not only does it fail to hold water, but useful nutrients are washed out of it as the water sinks through. This is called **leaching**. On the other hand, if the soil particles are too small and tightly packed, water cannot get through – it just stays on top or flows off the surface.

You can find out how much water is present in a sample of soil by doing Investigation 4. In good, well-watered soil the water should take up about a quarter of the total volume.

If there is very heavy rain and the drainage is poor, the soil may become

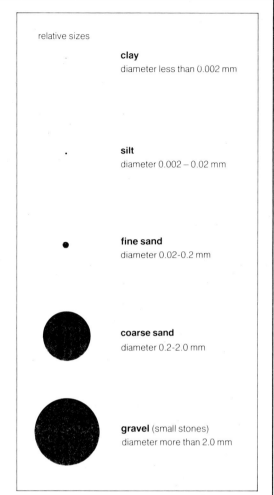

relative sizes

clay
diameter less than 0.002 mm

silt
diameter 0.002 – 0.02 mm

fine sand
diameter 0.02-0.2 mm

coarse sand
diameter 0.2-2.0 mm

gravel (small stones)
diameter more than 2.0 mm

Figure 2 The different kinds of particles which make up soil.

Figure 3 The water in the soil is constantly on the move, as shown in this diagram.

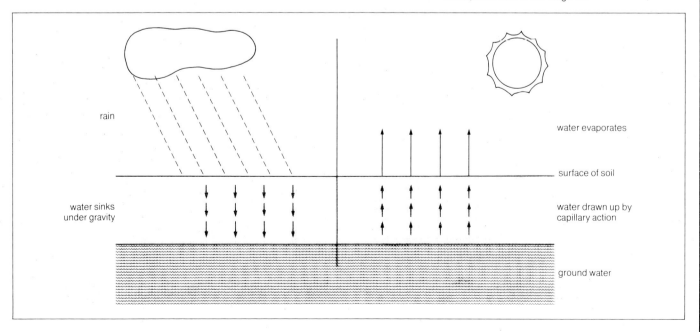

rain

water sinks under gravity

water evaporates

surface of soil

water drawn up by capillary action

ground water

full of water. It is then **waterlogged**. Waterlogged soil is short of oxygen, so roots cannot breathe properly. In swamps and bogs the soil is waterlogged all the time. Only certain kinds of plants will grow in these conditions. In waterlogged soil humus will not break down fully, so peat tends to accumulate.

Humus

When animals and plants die in the soil, their dead bodies gradually decay into a sticky jam-like liquid called **humus**. The layer of soil where most humus is found is the topsoil. Humus is black, so soil that contains a lot of it tends to be a dark colour. For the gardener one of the best sources of humus is compost (see page 377).

Humus makes the soil rich in nutrients which are needed for plant growth. It also forms a sticky coating round the soil particles, helping them to clump together into soil crumbs. Humus stores water and prevents valuable nutrients being washed out of the soil when it rains. It also helps to insulate the soil against extremes of heat and cold.

For humus to rot completely, oxygen is needed. If there is not enough oxygen, humus accumulates into a thick carpet of half-decayed material. This is **peat**.

You can find out how much humus is present in a sample of soil by doing Investigation 5. In good soil humus takes up about a tenth of the total volume.

Mineral salts

Dissolved in the soil water are various mineral salts. These provide plants with important elements such as nitrogen, phosphorus and potassium, and they are essential for growth.

Some mineral salts come from the rock which formed the soil. They may give the soil a particular colour: for example, the red soil of Devon contains a lot of iron salts. Other minerals, such as nitrogen salts, are formed when humus breaks down. This is why humus is so good for plants.

Lime

Lime comes from limestone, a type of rock which contains **chalk**. Chalk is calcium carbonate. All good soil contains a certain amount of this important chemical substance. You can find out if a sample of soil contains lime by doing a simple chemical test on it (Investigation 6).

Lime is important for three main reasons: (1) Calcium is one of the elements which all plants need for proper growth and development. (2) Lime helps soil particles to clump together into soil crumbs. (3) Calcium carbonate is an alkali, and this prevents the soil being too acidic: in gardeners' language it prevents the soil being 'sour'.

We can express how acidic or alkaline the soil is by a number called the pH. These numbers are arranged in a scale running from 0 to 14. A pH of 7.0 is neutral – neither acidic nor alkaline. A pH of less than 7.0 indicates acidity, and above 7.0 indicates alkalinity. You can do a simple test on samples of soil to find out their pH (Investigation 7).

Most plants grow best in soil which is round about neutral. However, some plants like alkaline soil, and others like acidic soil.

Soil air

In good soil there are plenty of spaces between the soil particles and crumbs. These spaces are filled with air. The oxygen in this air is needed for respiration by plant roots and the other organisms which live in the soil. Oxygen is also needed for the decay of humus – this is because the microbes responsible for decay are aerobic.

You can find out how much air is present in a sample of soil by the method

Figure 4 The White Horse at Westbury in Wiltshire. The turf has been removed in the shape of a horse, exposing the chalky soil which can be seen from miles away.

Figure 5 Heather and other moorland plants like living in acidic soil.

given in Investigation 8. In good soil about a quarter of the volume is taken up by air.

If the soil particles are too tightly packed together, or if the soil is waterlogged, the amount of oxygen will be lowered, and few organisms will be able to live there.

Different kinds of soil

You often hear people say that the soil in their garden is dreadful. There are many different types of soil, some good, some bad. Here are the main types:

Sandy soil

As the name implies, this kind of soil contains mainly sand. Sandy soil is loose, light and easy to dig: think how easy it is to dig sand on a beach. It contains plenty of air, and it drains well. However, it is a cold type of soil because it readily loses heat. It dries up quickly in hot weather, and useful chemicals are washed out of it when it rains. So sandy soil is not very fertile.

If you have a garden with sandy soil you should add humus to it. As well as putting goodness into the soil, the humus helps to bind the sand together. It also holds onto water when it rains.

You can prevent sandy soil from drying up by spreading a layer of peat or manure over the surface. This is called **mulching** and it also helps to keep the soil warm.

Clay soil

This kind of soil contains a lot of clay. It holds onto water and nutrients very well, so it tends to be rich in plant food. However, it is extremely heavy and difficult to dig, being sticky when wet and hard when dry. The soil particles are held together so tightly that there is little room for air in between. Rain, rather than draining through it, runs off the top.

If you have a lot of clay in your garden, you should add lime to it. This causes the particles of clay to clump together into soil crumbs. This breaks up the soil, getting air to it and draining it.

Chalky soil

This kind of soil contains a lot of lime. It is therefore very alkaline. As chalky soil comes from calcareous rock, it looks rather white. It is typical of the downs of Southern England: if the turf is stripped off, the white soil is visible from far away (Figure 4).

Chalky soil is usually rather clayey and therefore difficult to cultivate. If you have chalky soil in your garden, the best thing is to add humus to it. This makes it more acidic, and helps to break it up. Although most garden plants dislike a lot of chalk, many wild flowers thrive on it.

Peaty soil

This kind of soil contains a lot of peat. Although peat is useful, too much of it can make the soil acidic. Most plants dislike this kind of soil, though some like it. Moorland soil is very peaty, and heather is one of the few plants to grow well in such areas (Figure 5). If you have this sort of soil in your garden you should add plenty of lime to it.

The ideal soil

The best soil is a mixture of sand, clay, chalk and peat. This combines the advantages of all four. The disadvantages of each one on its own are counteracted by the presence of the others.

What should their proportions be? Good garden soil should contain roughly: 50 per cent sand, 30 per cent clay, 12 per cent humus, and 8 per cent lime.

Investigation 1

Separating the ingredients of soil

1 Quarter fill a large test tube with soil.

2 Add water until the tube is three-quarters full. Notice that air bubbles are given off: what does this tell you?

3 Put your hand over the open end of the tube, and shake well.

4 Put the tube in a rack, and let the soil settle. The heaviest constituents of the soil will sink to the bottom, the lighter ones will float at various levels.

   Does the appearance of your test tube agree with the illustration?

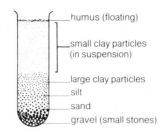

humus (floating)

small clay particles (in suspension)

large clay particles
silt
sand
gravel (small stones)

5 Repeat this experiment with different kinds of soil and compare the amounts of the different constituents in each.

Investigation 2

Looking at soil crumbs

1 Obtain some good garden topsoil. Look at it, and feel it with your fingers, then describe it as fully as you can.

2 Notice that the soil is composed of numerous small particles. Some of the particles may be separate, but others are clumped together into soil crumbs.

3 Put a soil crumb on a sheet of paper, and squash it with your finger. What do you think makes the particles stick together?

4 Look at the roots of a plant which has been pulled out of the soil. What do you think causes the soil to cling to the roots? How are the roots connected to the soil crumbs?

Soil crumbs make the soil better for plant growth – why?

Make a list of all the features of the soil which, directly or indirectly, help plants to grow in it.

Investigation 3

To see if water moves upwards through soil

1 Obtain a wide glass tube, about 20 cm long.

2 Plug one end of the tube with glass wool.

3 Hold the tube upright with the glass wool end downwards.

4 Fill the tube with dry soil above the glass wool.

5 Scatter some seeds on the surface of the soil at the top of the tube (mustard seeds will do).

6 Set the tube up so that the lower end is dipping into a dish of water. Note the time.

7 Observe the tube at intervals, watching to see if the water rises through the soil.

8 Note the time when the seeds start germinating.

   How long after setting up the apparatus do the seeds start to germinate?

   Do they germinate as soon as the water reaches them?

9 Repeat this experiment with different kinds of soil, and compare how long it takes for water to rise through them.

Investigation 4

To find out how much water there is in soil

1 Half fill a small crucible with soil.

2 Weigh the soil and crucible.

3 Put the soil and crucible in an oven at about 100 °C for at least 30 minutes. The water in the soil should evaporate.

4 Put the soil in a desiccator and let it cool down.

5 Re-weigh the soil and crucible.

   What is the difference in the mass of the soil before and after drying it in the oven?

   Do you agree that this is the mass of the water in the soil sample?

   What percentage of the soil is taken up by water?

6 Repeat this experiment with two soil samples, one taken after a period of dry weather, the other after heavy rain.

Investigation 6

To find out if soil contains lime

1 Put a little soil in a test tube.

2 Add a few drops of concentrated hydrochloric acid.

3 Do the contents of the test tube fizz? If they do there is lime in the soil. The fizzing results from the acid reacting with the lime (calcium carbonate).

What gas is given off in this reaction?

Is there any liquid left in the test tube after the fizzing has stopped?

Can you write a chemical equation for the reaction?

Why is lime useful in the soil?

Investigation 5

To find out how much humus there is in soil

1 Half fill a small crucible with soil.

2 Dry the soil by putting it in an oven at about 100 °C for at least 30 minutes.

3 Put the soil in a desiccator and let it cool down.

4 Weigh the dry soil and crucible.

5 Place the crucible of soil on a wire gauze on a tripod, and put a Bunsen burner underneath.

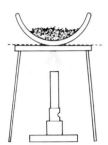

6 Light the Bunsen burner and heat the crucible with a strong flame for 15 – 20 minutes. The humus will burn up into carbon dioxide gas and water vapour.

7 Put the soil back in a desiccator and let it cool down.

8 Re-weigh the soil and crucible.

What is the difference in the mass of the dry soil before and after burning it?

Do you agree that this is the mass of the humus in the soil sample?

What percentage of the soil consists of humus?

How can you be certain that *all* the humus has been burned before you find the final mass of the soil sample?

How can you improve the experiment to make sure of this?

9 Repeat this experiment with different kinds of soil, and compare the percentages of humus in them.

Investigation 8

To find out how much air there is in soil

1 Obtain a measuring cylinder with a volume of 100 cm³.

2 Put soil into the cylinder up to the 50 cm³ mark.

3 Tap the cylinder on the bench to make sure that the soil is bedded down.

4 Run water slowly into the cylinder from a tap until it reaches the 100 cm³ mark.

5 Stir the soil and water gently so as to dislodge all the air bubbles from between the soil particles.

6 Note the new level of the water.

What is the difference between the new level of the water, and the original level?

Do you agree that this is the amount of air in the sample of soil?

What percentage of the soil consists of air?

Why is it important to plants that there should be air in the soil?

Investigation 7

To find out how acidic or alkaline soil is

1 Put a little soil in a test tube and cover it with distilled water.

2 Put your thumb over the end of the test tube, and shake it vigorously.

3 Obtain a piece of pH paper.

4 Dip the pH paper into the water. What colour does it go?

5 Compare the colour of the paper with the *colour code* supplied by the manufacturer.

6 What is the pH of the soil sample?

Is the soil sample acidic, alkaline or neutral?

7 Repeat this experiment with different kinds of soil and compare their pH's.

Assignments

1 Why is it better to water a garden in the evening rather than in the middle of the day? In what circumstances would it be all right to break this rule?

2 a) You put a potted plant on a saucer and pour some water onto the surface of the soil, but none of the water comes through. What has happened to it?

   b) Some people water their potted plants by standing the pots in a saucer of water. Will this work? Explain your answer fully.

3 What sort of soil:

   a) shifts easily beneath your feet,
   b) sticks to your shoes,
   c) looks very black,
   d) has a whitish appearance,
   e) is highly acidic?

4 On a certain mountain top the soil is only a few centimetres thick, whereas in a forest at the foot of the mountain the soil is about twenty metres thick. Suggest two reasons for the difference.

5 A student carried out an experiment to determine the percentage of water in a sample of soil. These are his results:

mass of crucible	10 g
mass of crucible plus damp soil	25 g
mass of crucible plus soil after drying	20 g

   a) How do you think the student dried the soil?
   b) What was the percentage of water in the soil?

Life in the soil

The soil provides a home for many different organisms and these can greatly affect its quality. Some organisms have beneficial effects on the soil, others are harmful.

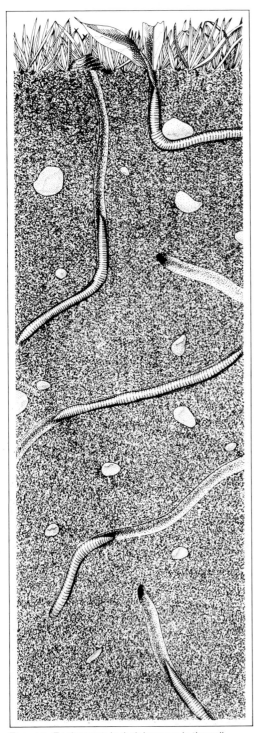

Figure 1 Earthworms in their burrows in the soil.

Collecting animals from soil

A few quite large animals live in the soil, for example earthworms and moles. However, most soil animals are small and you have to use special methods for collecting them.

One commonly used method depends on the fact that soil animals do not like light and move away from it. You obtain some soil and shine a light on it so as to drive out the animals (Investigation 1).

Having got the animals out of the soil, you can look at them in detail, find out what they are, and learn about their habits. Many of them live in the water surrounding the soil particles.

A helpful animal: the earthworm

There may be over two million earthworms in a hectare of good garden soil.

Like other soil animals earthworms don't like light, so during the day they live underground in their burrows (Figure 1). Here they eat soil, grinding it up in their guts. The undigested soil is then passed out through the anus onto the surface of the ground as **worm castings** (see page 56).

On warm damp nights worms come out of their burrows and feed on dead leaves which they pull down into the soil. If you go out into a garden with a torch, you can see the worms lying on the ground. Tread softly because the slightest vibration will make them jerk back into their burrows. In my garden I have counted over 30 worms in a square metre.

You can find out the effect which earthworms have on the soil by building a wormery (Investigation 2).

How do earthworms improve the soil?

Worms improve the soil in three main ways:

1 They turn it over
By constantly burrowing through the soil, they loosen it and mix it up. This helps to drain and aerate it, and it ensures that the various nutrients are evenly spread out. In this way worms do what a gardener does when he digs his garden. It has been estimated that in one hectare, worms may turn over as much as 50 tonnes of soil in a year.

2 They fertilise it
Worm castings contain nitrogenous waste which makes them very fertile. They also contain calcium carbonate which helps to make the soil less acidic. Worms also help to fertilise the soil by pulling leaves into it: once buried the leaves quickly decay and useful nutrients are released from them. The worms add further goodness to the soil when they themselves die and decay.

3 They make it finer
Having been ground up in the worm's gut, worm castings are very fine. Seeds get covered with this fine soil, which protects them and helps them to germinate successfully. And when the young roots emerge they can push their way easily through it.

Do worms do any harm? Their castings certainly look rather unsightly on a newly mowed lawn, and they can be a nuisance on golf courses. However, this is a small price to pay for all the good they do.

The mole

This amazing little animal can burrow over a hundred metres in a day. As it lives underground it is blind, and it has short strong forelegs for digging (Figure 2). It has a very large appetite and can eat its own mass in food in a day.

Its burrows serve as irrigation channels, helping to drain excess water from the soil. As it eats a lot of soil pests, it can be useful to gardeners and farmers. However, it also eats earthworms and makes unsightly mole hills on lawns, so many people regard it as more of a hindrance than a help.

The mole is one of a small number of mammals that make their homes in the soil. Can you think of any others?

Figure 2 A mole.

Some harmful soil animals

Unfortunately the soil contains a number of animals which are a nuisance to farmers and gardeners. Various insect larvae eat the roots of plants and do a lot of damage. These include 'wireworms', which are the larvae of a certain kind of beetle; and 'leather jackets' which are the larvae of the daddy long legs. Millipedes can also be harmful, and so can certain roundworms. These animals are shown in Figure 3.

Various poisons can be used to kill these pests. These are available as powders which can be dug into the soil before planting, or they can be dissolved in water and sprayed onto the soil after the plants have started growing.

Bacteria in the soil

In this Topic we have concentrated mainly on animals. However the soil also contains vast numbers of bacteria and other microbes which help the soil to support life. These bacteria bring about decay and enable carbon and nitrogen compounds to circulate in nature. Their action is explained on pages 374–5.

The effect of plants on the soil

The main effect which plants have on the soil is to hold it together. We can appreciate the importance of this by seeing what happens if the plants are suddenly removed from an area. Suppose a farmer puts too many sheep to graze on a hillside. As they run short of food, the sheep crop the grass so close to the soil that the grass dies. The soil, no longer bound together by the roots, starts getting loose. Rain and wind then wear it away, and soon the area becomes bare and incapable of supporting life.

This is called **erosion**, and one of its main causes is over-grazing of the kind just described. This rarely happens in countries like Britain where farming is of a high standard. However, it is all too common in many developing countries (Figure 4).

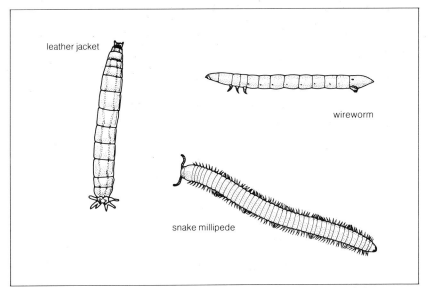

Figure 3 Three animals which are a nuisance in the soil. They eat the roots of plants.

leather jacket

wireworm

snake millipede

Figure 4 A severe case of soil erosion in East Java.

Investigation 1

Collecting small animals from the soil

Method 1

This method is used for collecting small arthropods such as centipedes and insect larvae.

1 Obtain a sample of good garden soil.

2 Set up the apparatus shown in the illustration, spreading the soil out on the perforated tray.

3 Switch the lamp on, and leave the apparatus for between one and three days. The light and heat from the lamp should drive the animals downwards out of the soil into the beaker.

4 Observe the contents of the beaker.

   Are there any animals in it?

   What kind of animals are they?

   Use a simple key to identify them.

Method 2

This method is used for collecting roundworms and other small animals that live in the soil water.

1 Obtain a sample of good garden soil.

2 Wrap the soil up in a bag made out of cheesecloth, and set it up as shown in the illustration.

3 Switch the lamp on, and leave the apparatus for several days. The light and heat should drive any small animals living in the soil water out of the bag.

4 Open the clip and run a little water from the funnel into the beaker.

5 With a pipette transfer a drop of the water to a microscope slide, and cover it with a coverslip.

6 Examine the slide under the microscope: low power first, then high power.

   Can you see any slender worms with pointed ends? These are roundworms (see pages 6–8).

   Can you see any other organisms on your slide? If so, try to identify them.

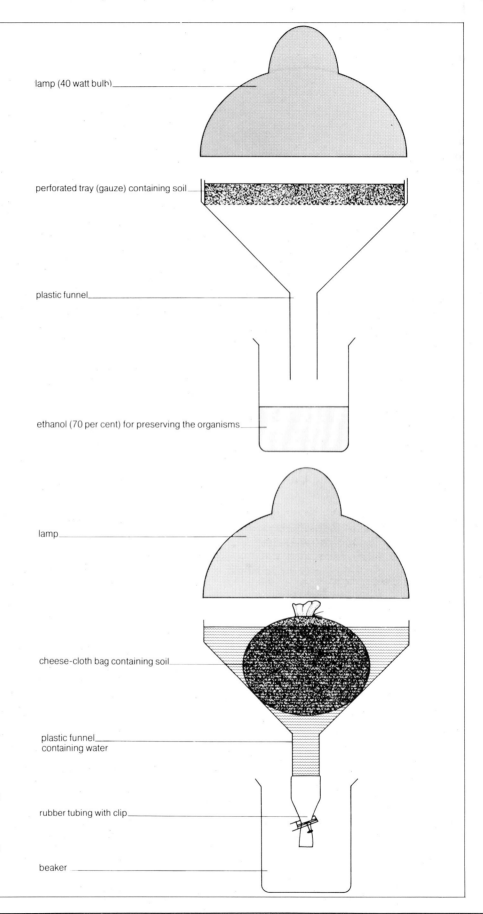

lamp (40 watt bulb)

perforated tray (gauze) containing soil

plastic funnel

ethanol (70 per cent) for preserving the organisms

lamp

cheese-cloth bag containing soil

plastic funnel containing water

rubber tubing with clip

beaker

Investigation 2

To find the effect of earthworms on the soil

1 Set up the wormery shown in the illustration.

2 Place about 10 worms on the surface of the soil.

3 Watch the worms burrowing.

4 Observe the wormery at intervals during the next week or so.

What happens to the leaves at the surface?

What happens to the two chalk-sand layers?

(Disturbance of these layers is an indication of the extent to which the worms are mixing up the soil.)

Make a list of the effects the worms bring about on the soil *which you can see for yourself.*

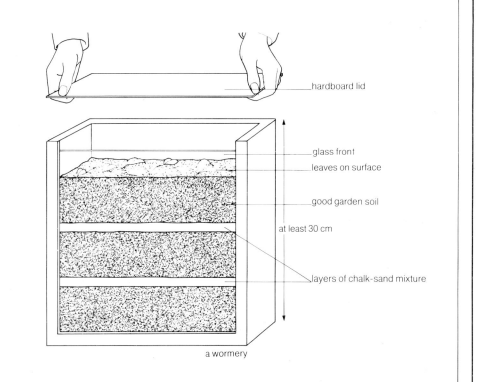

hardboard lid

glass front

leaves on surface

good garden soil

at least 30 cm

layers of chalk-sand mixture

a wormery

Assignments

1 Name two soil animals which are useful to man, and two which are harmful. Explain why each animal is useful or harmful.

2 The apparatus shown on the right was set up to find out if there were any microbes in a sample of soil:

After 48 hours the level of the water had risen in the left hand side of the U-tube and dropped in the right hand side.

a) Suggest an explanation for this result.
b) What should the control be in this experiment?

3 While digging the soil, a gardener notices that there are lots of earthworms in one part of his garden but none in another part. Suggest *one* possible reason for this. Describe an experiment which you could do to find out if your suggestion is right.

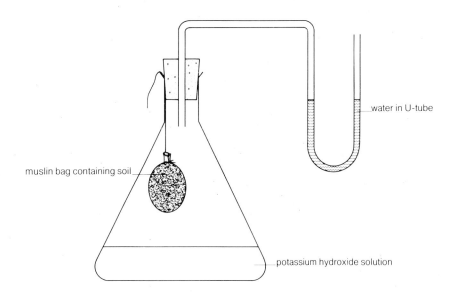

water in U-tube

muslin bag containing soil

potassium hydroxide solution

4 You suspect that worms prefer beech leaves to oak leaves. Describe an experiment which you could do to find out if you are right.

5 Why are (a) soil water and (b) soil air important to organisms which live in the soil?

What is it about the atmosphere surrounding the earth which makes it capable of supporting life? In this Topic we will look into this question.

Gas	Percentage volume
Nitrogen	78.1
Oxygen	20.9
Carbon dioxide	0.03
Other gases	0.97
Total	100.0

Table 1 Composition of the atmosphere (dry air) at sea level.

Figure 1 Photosynthesis releases oxygen into the atmosphere and removes carbon dioxide from it. Respiration and combustion do the opposite: they remove oxygen from the atmosphere and release carbon dioxide into it.

What does the atmosphere consist of?

The earth's atmosphere consists of three main gases: nitrogen, oxygen and carbon dioxide (Table 1). This is the air we breathe.

The proportions given in Table 1 are more or less the same all over the world: the air contains about the same amount of oxygen and carbon dioxide in England as it does in China. What is more, we believe the atmosphere has been like this for thousands, even millions of years: almost certainly it was the same in the age of the dinosaurs as it is today.

This is just as well, because a change in the amounts of oxygen and carbon dioxide in the atmosphere could make our air unfit for breathing. For example, suppose the amount of carbon dioxide was to go up and up, and oxygen down and down? You would feel drowsy, get a headache, become very hot and faint. Your brain would stop working properly and eventually you would die. This is why places where people live and work must be well ventilated. In a well ventilated room there is plenty of oxygen, and carbon dioxide does not build up to a harmful level.

How is the atmosphere kept constant?

When living things breathe (respire) they use up oxygen and give out carbon dioxide. However, when green plants photosynthesise they use up carbon dioxide and give out oxygen. In other words *respiration removes oxygen from the atmosphere and adds carbon dioxide to it, whereas photosynthesis removes carbon dioxide from the atmosphere and adds oxygen to it.*

In the world as a whole, respiration and photosynthesis are in balance with each other. The result is that the amounts of carbon dioxide and oxygen are kept constant (Figure 1).

Another process, besides respiration, takes oxygen out of the atmosphere and adds carbon dioxide to it. This is combustion (burning). What happens if we put some extra carbon dioxide into the air by, for example, burning coal in a factory or making a bonfire? If the amount isn't too great, plants simply respond by using it up more quickly. In this way the level of carbon dioxide in the atmosphere stays more or less the same.

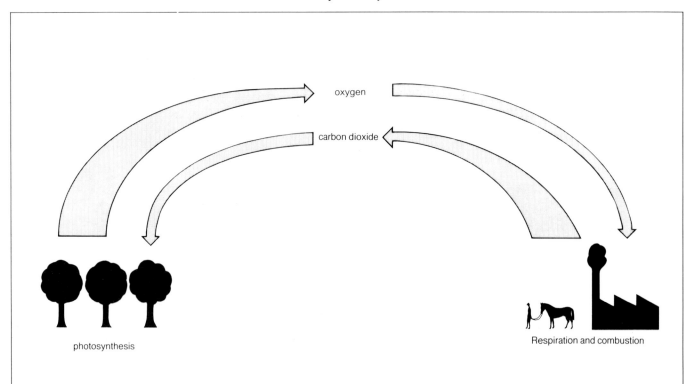

oxygen

carbon dioxide

photosynthesis

Respiration and combustion

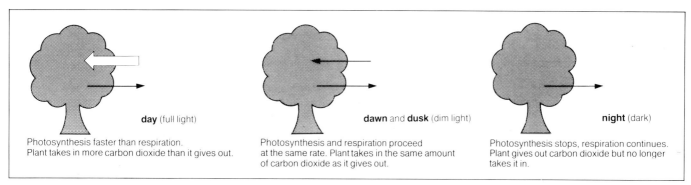

day (full light)

Photosynthesis faster than respiration.
Plant takes in more carbon dioxide than it gives out.

dawn and **dusk** (dim light)

Photosynthesis and respiration proceed
at the same rate. Plant takes in the same amount
of carbon dioxide as it gives out.

night (dark)

Photosynthesis stops, respiration continues.
Plant gives out carbon dioxide but no longer
takes it in.

A more detailed look at carbon dioxide

Try the Investigation. Your results should enable you to draw certain conclusions about the effect which plants, animals, and the two together, have on the amount of carbon dioxide in the atmosphere.

First, plants. Do you agree that they take up carbon dioxide in the light, but give it out in the dark? The carbon dioxide which they give out comes from their respiration.

Plants respire all the time, whether it is dark or light. However, in the full light of day they take up more carbon dioxide for photosynthesis than they give out in respiration. So when it is light they remove this gas from the atmosphere (Figure 2).

Animals, on the other hand, give out carbon dioxide all the time, whether it is light or dark. This comes from their respiration. So they are always adding carbon dioxide to the atmosphere.

What is the combined effect of animals and plants together? When it is dark they both give out carbon dioxide, with the result that the amount of carbon dioxide in the atmosphere goes up. But when it is light the animals give it out and the plants take it up, and the amount in the atmosphere falls (Figure 3).

The *overall* effect of animals and plants living together in a normal environment is to keep the level of carbon dioxide in the atmosphere more or less constant.

Figure 2 Whether a plant takes up, or gives out, carbon dioxide depends on how light it is. The arrows show the movement of carbon dioxide into and out of the tree.

Figure 3 Carbon dioxide is put into the atmosphere by animals all the time, and by plants at night. It is removed from the atmosphere by plants during the day.

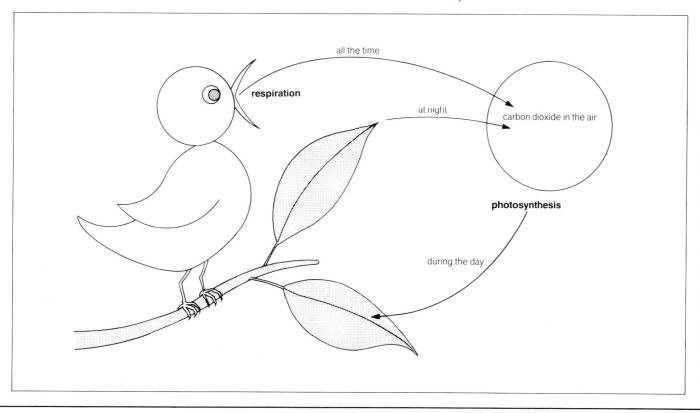

all the time

respiration

at night

carbon dioxide in the air

photosynthesis

during the day

Can man change the atmosphere?

What activities of man might raise the level of carbon dioxide, and lower the level of oxygen, in our atmosphere?

Man does two things which *might* cause this to happen:

1 He cuts down trees for timber and destroys vegetation to make room for towns.
2 He burns fuel, which uses up oxygen and produces carbon dioxide.

How could we find out if doing these things affects the amount of oxygen and carbon dioxide in the atmosphere? We would have to make measurements over many years of the amounts of these two gases in our atmosphere. This has been done in several parts of the world during the last hundred years. The results are very interesting.

In the case of oxygen, there has been no detectable change. This is because there is a vast supply of oxygen in our atmosphere which has been put there by plants over thousands of millions of years, and man in his short history has not used much of it. Even if we burned all the coal and other fuels that exist in the whole world, the oxygen content of the atmosphere would decrease by only a fraction of one per cent.

So we are not likely to run out of oxygen because we burn things. But might we run out of it because we destroy plants? This *could* happen, but it is unlikely. We would have to destroy an awful lot of plants to lower the amount of oxygen in the atmosphere, and even then any changes would take place very slowly.

What about carbon dioxide? Look at the graph in Figure 4. In the last hundred years or so the amount of carbon dioxide in the atmosphere has increased by about 12 per cent. The cause of this is not known for certain, but the burning of fossil fuels and farm waste is thought to have been at least partly responsible.

If carbon dioxide goes on increasing in the atmosphere, what might its effects be? It is unlikely to increase so much that our health would be affected. However, some scientists believe that it could bring about a change in climate by making the temperature go up. Why do you think this might happen and what could its consequences be?

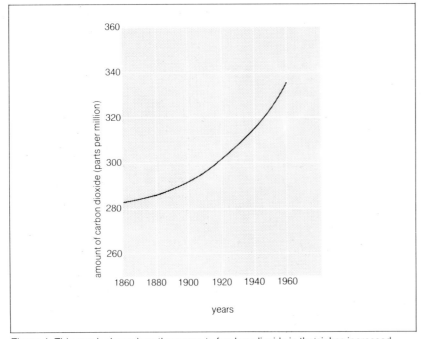

Figure 4 This graph shows how the amount of carbon dioxide in the air has increased during the last hundred years. The measurements were made in Vienna. Similar figures have been recorded in other parts of the world.

Investigation

How do organisms affect the amount of carbon dioxide in the atmosphere?

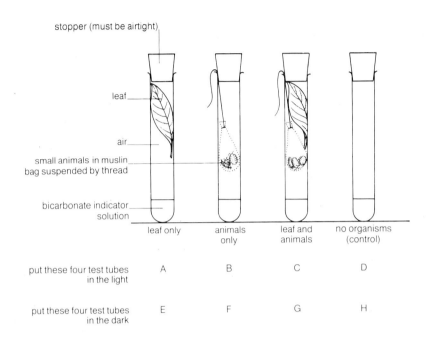

In this experiment we will make use of a bicarbonate indicator. This changes colour according to how much carbon dioxide is present.

1 Label eight test tubes A to H. Pour a little indicator into the bottom of each tube. Notice that the indicator is reddish-orange: this is its colour when it is in contact with ordinary atmospheric air.

2 Set up the eight test tubes as shown in the illustration. Seal all the tubes with a stopper so there is no chance of air getting in or out.

3 Put test tubes A to D in the light. (But do not use a lamp that is liable to heat them up.)

4 Put test tubes E to H in the dark. (A good way is to put them under a cardboard box.)

5 After about an hour give each test tube a quick shake.

Now compare the colour of the indicator in the test tubes. It is best to look at the colours against a plain white background.

The colour tells you how much carbon dioxide is present in the test tube compared with ordinary atmospheric air.

Yellow means there is more carbon dioxide than in atmospheric air.

Purple means there is less carbon dioxide than in atmospheric air.

Reddish-orange means there is the same amount of carbon dioxide as in atmospheric air.

Write down the colour of the indicator in each test tube.

For each test tube, say whether carbon dioxide has been added to, or removed from the air.

What is the effect of (a) plants, (b) animals, (c) plants plus animals, on the carbon dioxide content of the atmosphere?

Assignments

1 Why is it a good idea to open classroom windows whenever possible?

2 A scientist found that the average carbon dioxide content of the air in March was 0.02971 whereas in September it was 0.02905. Explain the difference.

3 Imagine there is a catastrophe in which all the plants of the world are suddenly destroyed. Describe in detail the effects which you think this might have.

4 It has been suggested that a suitable atmosphere might be maintained in a manned space capsule by having some plants inside. Do you think this is feasible? What problems might be encountered in putting it into practice?

5 At an agricultural research station, a group of scientists measured the amount of carbon dioxide in the air in the middle of a wheat field every three hours for 24 hours. Here are their results:

Time	Percentage of carbon dioxide in the air
24 (midnight)	0.042
3	0.037
6	0.031
9	0.029
12 (noon)	0.028
15	0.030
18	0.032
21	0.035
24 (midnight	0.042

a) Plot these results on graph paper.
b) Explain them as fully as you can.
c) How would you expect oxygen to change during the same period?

Man and the environment

The picture in Figure 1 shows an example of pollution. It is a scene that is all too common in industrialised countries today.

Figure 1 The air around this cement works is heavily polluted.

What is pollution?

Pollution is any process which leads to a harmful increase in the amount of a chemical substance in the environment. The harmful substance is called a **pollutant**.

It is often difficult to know for certain whether a particular substance is harmful or not. Its effects may not appear straight away, but only after a long period of time. Also a pollutant may affect some organisms more than others, For example, certain gases from factories may not affect man in the concentrations in which they normally occur, but they may damage plants. Some pollutants which are harmful to certain organisms may actually *help* others. For instance, a high concentration of carbon dioxide in the air, though harmful to animals, is useful to plants because it enables them to photosynthesise faster.

Water pollution

From time to time **oil** is spilled into the sea from a tanker or an off-shore oil rig. The oil forms a thick layer, or slick, which floats on the surface of the sea. The slick may then be carried by ocean currents to the coast where it is deposited on the shore.

The oil ruins the beaches for the local residents and holiday-makers. It also kills fish and sea birds. In the past, attempts have been made to get rid of oil slicks by spraying them with **detergents** which break the oil up into drops. The trouble is that the detergents are even more deadly than the oil and they kill many organisms that might have escaped the oil. Nowadays less destructive methods are used.

Another water pollutant is **sewage**. If untreated sewage is put into a river or lake, it is decomposed by bacteria which quickly multiply. The bacteria use up so much oxygen that there is not enough for the fish and other animals, which suffocate and die.

Finally various **chemical waste products** from factories are sometimes discharged into seas and rivers. They may be so concentrated that the fish are killed straight away. But sometimes they are taken up by the organisms at the beginning of a food chain, and then passed along the chain becoming more and more concentrated as they go from one stage to the next. This can be very dangerous. Some years ago over 60 people died in Japan from eating fish whose bodies contained mercury. The mercury had been discharged into the sea from a local factory and had then entered the food chains.

The insecticide DDT is known to get into food chains on land. It is thought to have a damaging effect on animal tissues, and may possibly be a danger to man. For this reason DDT has been banned in many countries. Here we have an example of a useful substance turning out to be a pollutant. Another example is dioxin, a powerful herbicide used for killing weeds in conifer plantations. This chemical, which was used by the Americans to defoliate the jungle during the Vietnam war, is now known to be highly dangerous to man as well as animals.

Various tests can be carried out to find out what particular chemicals affect animals and plants, and how tolerant organisms are to them (Investigation 1).

Air pollution

A widespread air pollutant is **smoke** from the burning of fossil fuel such as coal. The smoke contains particles of carbon which float through the air and settle on the surface of buildings and trees, turning them black (Figure 2). If breathed in, the particles irritate the breathing passages and can cause bronchitis. Simple methods of finding out how much smoke pollution there is in different places are given in Investigations 2 and 3.

Smoke contains two main gases: carbon dioxide and sulphur dioxide. The effects of carbon dioxide are discussed on page 390. Sulphur dioxide is a poisonous gas, but fortunately it is not normally produced in sufficient

Figure 2 Industrial smoke blackened Nottingham Castle until the original stone was barely recognisable. On the right, the same building after cleaning.

quantities to endanger man. However, it certainly affects plants, either killing them or reducing their yield. In industrial areas the concentration of sulphur dioxide in the air may reach 4 ppm.* Concentrations as low as 0.3 ppm may damage plants. In certain parts of North America the vegetation has been completely destroyed by sulphur dioxide from smelting works.

Lichens are particularly sensitive to sulphur dioxide and in polluted regions you do not find lichens growing on the tree trunks.

Sulphur dioxide reacts with water in the atmosphere to form sulphuric acid (Investigation 4). This aggravates bronchitis and other breathing complaints. The sulphuric acid is washed down into the soil by rain where it may make the soil highly acidic, thus affecting the growth of plants. It also eats into the surface of buildings, eroding the stone and brickwork (Figure 3).

Smog

Normally smoke from houses and factories goes straight up into the atmosphere and is blown away by wind and air currents. However, in certain conditions it stays close to the ground where it builds up to form **smog**. Smog is a mixture of smoke and fog, and it is caused by a layer of warm air developing above a region of colder air. The warm layer prevents the colder air from escaping. This is called **temperature inversion** (Figure 4).

Figure 3 Years of erosion by sulphuric acid in the rain led to the disfigurement of this statue at Hever Castle.

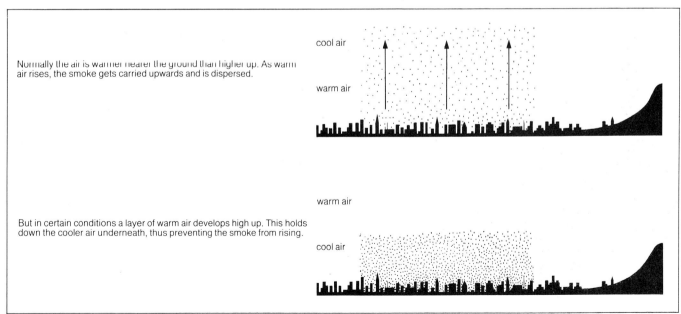

Normally the air is warmer nearer the ground than higher up. As warm air rises, the smoke gets carried upwards and is dispersed.

cool air

warm air

But in certain conditions a layer of warm air develops high up. This holds down the cooler air underneath, thus preventing the smoke from rising.

warm air

cool air

Figure 4 These diagrams show how smog can build up as a result of temperature inversion.

*ppm means parts per million. This is the number of cubic millimetres of sulphur dioxide in one million cubic millimetres (i.e. one cubic metre) of air.

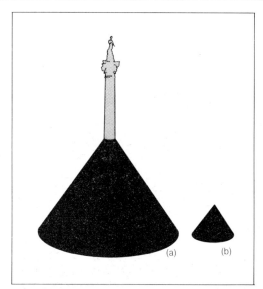

Figure 5 This picture shows what a month's fall of soot in central London would have looked like if swept into a pile in Trafalgar Square (a) in the early 1950s and (b) in the late 1970s. Nelson's Column is about 48 metres high, and the area of central London is approximately 120 square miles.

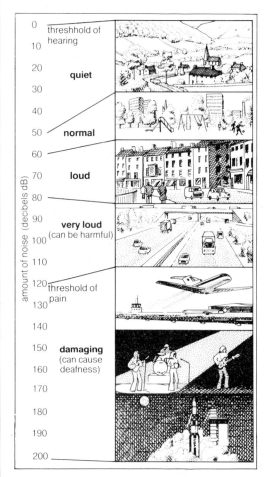

Figure 6 This illustration shows the noise scale as expressed in decibels, the standard unit of noise as measured with a sound meter.

Smog is not only unpleasant but it can be dangerous. In December 1952 there was a particularly bad smog in London. It lasted five days, and is said to have caused 4000 deaths. Street lamps had to be put on in the middle of the day, and in cinemas it was impossible to see the screen clearly from the back of the auditorium.

In 1956 a Clean Air Act was introduced by parliament making it illegal to burn coal and other smoky fuels in industrial areas except in special circumstances. This has led to a pronounced improvement in the atmosphere. In London the concentration of carbon particles and sulphur dioxide at street level was six times lower in 1977 than in 1950 (Figure 5).

Although smoky smogs have largely disappeared, there is another kind of smog which is still a problem. This is the type which you get in sunny places like Los Angeles. The Los Angeles smog is caused by the action of sunlight on the chemicals in motor vehicle exhaust, for which reason it is called **photochemical smog**. Temperature inversions are frequent in this area, and the result is a brown haze which makes people's eyes sting and can cause severe headaches. It also damages plants and is said to have reduced the yield of citrus fruits (oranges and lemons) in the Los Angeles area.

Land pollution

There are many examples of land that has been stripped of vegetation by industrial development and disposal of waste. The slag heaps from mines bear witness to the destructive effect which this can have on our environment.

Sometimes the tips from mines contain chemical substances which are poisonous to plants, and possibly to man too. For example, high levels of lead and cadmium have been found in the soil and crops in certain areas where mining used to be carried out. The trouble is that we just don't know the long-term effects which this kind of thing might have on people's health. However, many people feel that it is better to be safe than sorry and to err on the side of caution.

Have you ever thought what happens to all the rubbish which people throw away? Much of it will decay: this includes bits of food, potato peelings, tea leaves – in fact anything which can be broken down by microbes. Thanks to microbes, the chemicals in these materials can be set free and used again: in other words they are re-cycled in nature.

Other kinds of rubbish will not decay, because they are made of substances which microbes cannot live on. These include plastic, polythene and many other man-made materials. If you throw an apple core into a hedge, it will eventually decay, but if you throw a polythene bag it will remain there forever – unless of course some worthy citizen removes it.

Some man-made materials can of course be used again: for example paper, and scrap metal from used cars. But the majority cannot be re-used and must be got rid of somehow. Getting rid of this kind of rubbish is a major problem in modern society.

Noise pollution

At the start of this Topic we defined pollution as an increase in a harmful chemical substance in the environment. However, other things besides chemicals may be harmful, and these too can be regarded as pollutants. An example is **noise**.

Noise can be measured by a sound meter and is expressed in a unit called the decibel (dB) (Investigation 5). The quietest sound that the human ear can detect (zero decibels) is called the **threshold of hearing**. Figure 6 relates the loudness of sounds as measured in decibels to a series of everyday situations. Notice that a sound of 120 dB is on the **threshold of pain**: sounds louder than this actually hurt your ears and can give you a headache. If they go on continuously they may damage the sensory cells in the inner ear, causing permanent deafness. People who work close to machinery in foundries and mills are at a particular risk, and they normally wear ear-plugs.

Radiation pollution

Another type of pollution to which we are all exposed is **radiation** from radio-active materials. Radiation affects dividing cells, damaging their genes. It can cause leukaemia, which is cancer of the blood. It also causes mutations in the sex cells in the ovary and testis, and this can result in babies being born with deformities.

Most of the radiation to which we are exposed comes from the sun and outer space and is a natural part of our environment. However, man adds to this natural radiation by atomic explosions, nuclear power stations, and medical equipment such as X-ray machines. The total amount of radiation which an average person receives from all these man-made sources is about half the natural background radiation. This is not considered a hazard to health. However, people who work in places where radiation levels are particularly high, such as nuclear power stations, are obviously at greater risk than the rest of us, and so special precautions are taken to protect them.

Figure 7 *Left* Opencast coal mining at the Shipley Lake site near Ilkeston, Derbyshire.
Right The Shipley Country Park – the land was restored after the coalmining was finished.

Conservation

To *conserve* something means to protect it and keep it in a healthy condition. Applied to our environment, conservation means protecting the animals and plants from being harmed. This can be achieved in five main ways:

1 We must reduce pollution as much as possible, particularly those kinds which are liable to damage natural habitats and harm the organisms that live there. Areas which have been devastated by mining should be restored afterwards. Figure 7 shows that this is perfectly possible, though of course it costs money.

2 Animals that are killed for food, or any other purpose, must not be used up so quickly that their numbers start falling to the point that they may die out. This particularly applies to fish and whales.

3 Natural forests, particularly tropical forests and those containing rare trees such as the giant redwoods in California, should not be used extensively as a source of timber. It is better to rely on plantations in which the felled trees are replaced by new seedlings.

Figure 8 Burchell's zebra in Keyna.

4 One species of animal or plant should not be allowed to flourish at the expense of another, because this can upset the balance of nature. For example, killing off a predator may result in a large increase in the population of its prey. The prey then compete for food, and if they happen to be grass-eaters this can lead to over-grazing and soil erosion (see page 387).

In many countries certain areas have been set aside where the animals and plants are protected. In Britain these are known as **National Parks** and they contain an abundance of wildlife and natural scenery which everyone can enjoy. Because of man, most of the wild animals which used to roam the earth have long since died out. However, in East and Southern Africa there are large game parks with lion, giraffe, elephant and many other animals (Figure 8).

Investigation 1

Finding the effect of chemical pollutants on organisms

1 Set up a series of jars all containing the same amount of clean water.

2 Into each jar place a small selection of living organisms such as tadpoles, water fleas, insect larvae and so on.

3 Put a substance which you suspect may be a pollutant into each jar. You might try oil, detergent, paraffin, acid and so on.

4 Observe the behaviour of the organisms straight away and, if necessary, over a number of days.

Is the behaviour of any of the organisms abnormal?

Do any of them die? If so, which ones?

What conclusions do you draw?

Describe an experiment which you could do to find out the maximum concentration of a chemical substance a particular species of animal can tolerate without being harmed.

Investigation 2

To find out how much dust is deposited on outside walls

1 Cut off a short length of Sellotape about 8 cm long.

2 Place the sticky side of the Sellotape against a wall out of doors and press it gently.

3 Hold the Sellotape against a sheet of white paper and note how much dust it has picked up.

4 Repeat the experiment on other walls in different places, and compare the amounts of dust picked up.

Suggest reasons why some walls appear to be dirtier than others.

Investigation 3

To find out how much dust there is in the atmosphere

1 Obtain six slides which have been coated with a thin layer of agar jelly.

2 Place the slides in different places out of doors. Make sure they are in safe places where they won't be tampered with.

3 Leave the slides in position for a day or two.

4 Compare the amounts of dust deposited on the agar surface of the six slides.

5 Look at the slides under a lens or microscope to see how dense the dust particles are.

This investigation could form the basis of a large-scale survey of air pollution in your district. How would you carry out such a survey?

Investigation 4

Estimating the amount of pollution in rainwater

1 Obtain a jar about 15 cm tall, a funnel and filter paper.

2 Set them up as shown in the illustration.

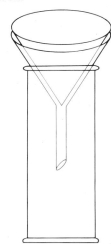

3 When it is raining, place the apparatus in a safe place out of doors and leave it there for at least 30 minutes.

4 Examine the filter paper for particles of dust which have been brought down by the rain.

5 Using universal indicator paper, measure the pH of the rainwater in the jar.

You will probably find that the pH is round about neutral.

However, if the air is badly polluted the pH may be acidic.

What is the acidity caused by?

If you have the chance, try doing this experiment in different parts of the country and compare the results.

Assignments

1 How did the word 'smog' get its name?

What causes smog to develop over a city?

2 Suggest one reason for each of the following:

a) London is a much cleaner city now than it was in the 1950s.
b) Discharging sewage into a river may kill the fish.
c) The Los Angeles smog is worse on bright days than on cloudy days.
d) The use of the insecticide DDT has been stopped in many countries.

3 Write out a chemical equation for the reaction which takes place when sulphuric acid is formed from industrial sulphur dioxide in the atmosphere.

Give *two* harmful effects of the sulphuric acid.

4 Certain lichens are never found in industrial areas, though they are abundant in non-industrial areas. Suggest one possible reason for this.

Describe an experiment which you could do to find out if your suggestion is correct.

5 There is much more pollution in the Mediterranean Sea than in the English Channel. Explain the possible reasons for this.

6 The following figures show the concentration of mercury in sea water and of various organisms expressed in parts per million:

sea water	0.00003	ppm
algae	0.03	ppm
fish	0.3	ppm
water birds	2.0	ppm

Suggest an explanation for these figures. What do they tell us about mercury as a pollutant?

Investigation 5

Measuring the noise level in different places

1 Using a sound meter, measure the maximum amount of noise in different places such as: a street corner, a railway station, an airport, a children's playground, a school dining hall, a reference library, a bridge over a motorway, a factory, a park.

2 Compare your results with Figure 7, and state whether the noise level in each place is quiet, normal, loud, very loud or dangerous.

How could you find the *average* noise level in a particular place.

*Tools
and techniques
of biology*

Testing hypotheses

Suppose we want to know why a certain type of plant grows well in place A but badly in place B.

The first thing to do is to put forward a *possible* reason. Scientists call this a **hypothesis**. A hypothesis explaining why plants grow better in one place than another might be that they differ in the amount of light they get.

The next step is to test this hypothesis to find out if it is true or not. This is done by carrying out an **experiment**. We grow the plants in different levels of illumination: some are brightly lit and others are poorly lit. If we find that the brightly lit ones grow better than the dimly lit ones, we conclude that the hypothesis is probably correct.

Scientists often investigate things by first thinking of a hypothesis, and then testing it by doing experiments. This procedure is sometimes called the **scientific method**.

Doing experiments

Suppose we want to test the hypothesis that light is needed for the leaves of a young plant to become green.

We obtain a plant and put it in the dark. If the green colour fails to develop, we will conclude that light is needed for it. However, there is something more that we must do: we must obtain a second plant and put it in the light. We need this second plant in order to provide a standard with which to compare the first plant. The second plant is called the **control**.

In carrying out this experiment it is essential that the two plants should be kept in exactly the same conditions, except for the light they receive. To put it in a general way: *we must keep all the variables constant except for the one whose effect we want to investigate.*

An experiment of this kind, in which the experimenter controls the conditions, is called a **controlled experiment**. Controlled experiments are frequently carried out by biologists, and you will probably do many such experiments during your biology course.

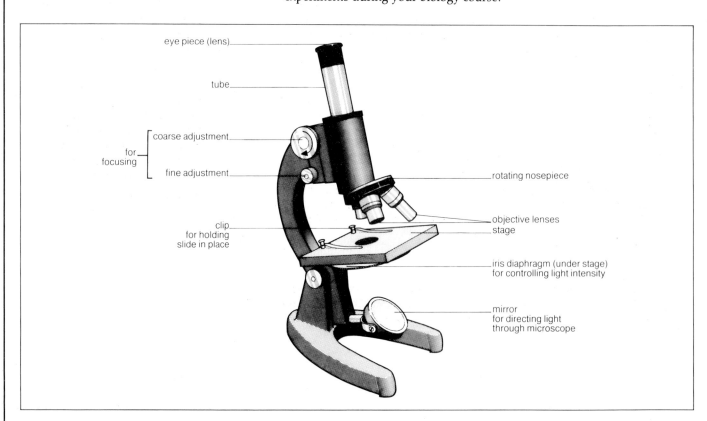

Investigation 1

Learning to use the microscope

There are many different kinds of light microscope, but they all work in the same way.
The one shown opposite is commonly used in schools.

1 Study your microscope carefully and compare it with the diagram. Yours may be slightly different. Make sure you understand it before you use it.

2 Objects to be viewed under the microscope are first placed on a glass slide and covered with a thin piece of glass called a coverslip. Your teacher will give you a specimen which has been mounted in this way.

3 Place the slide on the stage of your microscope: arrange it so the specimen is in the centre of the hole in the stage.

4 When you have positioned your slide, fix it in place with the two clips.

5 Rotate the nosepiece so the *small* objective lens is immediately above the specimen: the nosepiece should click into position.

6 Place a lamp in front of the microscope, and set the angle of the mirror so the light is directed up through the microscope.

7 Look down the microscope through the eye piece. Adjust the iris diaphragm so the field of vision is bright but not dazzling.

8 Look at the microscope from the side. Turn the coarse adjustment knob in the direction of the arrow in the diagram. This will make the tube move downwards.

9 Continue turning the knob until the tip of the objective lens is close to the slide.

10 Now look down the microscope again. Slowly turn the coarse adjustment knob in the other direction, so the tube gradually moves upwards. The specimen on the slide should eventually come into view.

11 Use the coarse and fine adjustment knobs to focus the object as sharply as possible.

12 If necessary re-adjust the iris diaphragm so the specimen is correctly illuminated. You will get a much better picture if you *don't* have too much light coming through the microscope.

You are now looking at the specimen under low power, i.e. at low magnification. To look at it under high power, i.e. at a greater magnification, proceed as follows:

13 Rotate the nosepiece so the *large* objective lens is immediately above the specimen. The nosepiece should click into position, as before.

14 If the specimen is not in focus, focus it with the fine adjustment knob. *Be careful that the tip of the objective lens does not touch the slide.*

15 Re-adjust the illumination if necessary.

You are now looking at the specimen under high power. Do you agree that it is now much more enlarged?

Always treat the microscope with the greatest care: it is an expensive precision instrument. Always carry it with both hands, and keep it covered when you are not using it. Make sure the lenses never get scratched or damaged: if they need cleaning tell your teacher.

Investigation 2

The magnifying power of the microscope

1 Place a transparent ruler on the stage of your microscope. Arrange it so that a line on the millimetre scale is immediately below the low power objective lens.

2 Focus onto the line.

   What does it look like?

3 Using the ruler, count how many millimetre divisions fit across the low power field of view.

   What is the diameter of the low power field of view   (a) in millimetres, and   (b) in micrometres?

4 Now rotate the nosepiece so the high power objective is immediately above a line on the ruler.

5 Focus onto the line.

   What does it look like now?

6 Using the ruler, count how many millimetre divisions fit across the high power field of view.

   What is the diameter of the high power field of view   (a) in millimetres, and   (b) in micrometres?

7 The magnifying power of your microscope is the magnifying power of the eye piece lens, multiplied by the magnifying power of the objective lens. The magnifying power of the eye piece and objectives is engraved on them. Find these figures.

   Work out the low and high power magnifications of your microscope.

Index

Page numbers in bold refer to the definitions of the entry.

Acknowledgements

The author and publishers are grateful to the following for permission to use their photographs:

Heather Angel: pages 4, 38, 41 top and bottom, 43, 46 left, 60, 62, 78, 80, 95, 121, 187 bottom, 201 top right, 203, 282, 284 bottom, 318, 320 top and bottom, 356, 358, 370 and 398; Flour Advisory Bureau, page 160 top; Aquilla/M. C. Wilkes: pages 56 and 214; Ardea Photographic: title page, half title page, 3, 12, 54, 81, 87, 90, 91, 96, 106, 217, 236 and 357 left; Barnaby's Picture Library: pages 1, 74 bottom, 94 top and bottom, 98, 99 top and bottom, 100, 191 centre, 212, 217 bottom and 286; BBC Hulton Picture Library: pages 293, 342 top, 344 left and right and 235; Bernsen's International Press Service Ltd.: page 349 top; Bibl. Ambrosiana Milan: page 117; Birds Eye: page 125; Blocpix U.S.A.: page 244; Dr G. Bond with permission from Intermediate Botany by L. J. F. Brimble (Macmillan): pages 180 and 181 top; Pat Brindley: page 181 bottom; British Museum (Natural History): page 61; British Tourist Authority: page 32 top; Camera Press Ltd.: 224, 246, 248, 264, 309, 310, 350, 367 left, centre and right and 394; Danish Agricultural Producers: page 378; Doctor: page 216; Douglas P. Wilson/Eric and David Hosking: page 22; Nick Evans: page 36 top; F. A. O. Photos: pages 2, 126, 191 bottom and 387; Forestry Commission: pages 48 and 52; Fox Photos Ltd.: page 395 bottom; M. Fraser, Chelsea College: page 198; Garden News: page 120 left; Gene Cox: pages 15 top and bottom, 204 and 210 right; Glasshouse Crop Research Institute: page 190 bottom left and right; Henry Grant: page 112; Harry Smith Horticultural Collection: pages 120 right, 122, 191 top, 284 top, 332 bottom, 334 and 349 bottom; Beverley Heath: pages 10 bottom left, 14 bottom, 24 right, 46 right, 103, 105, 147, 148, 149, 160 top, 186, 187 top, 200, 320 bottom, 326, 331, 332 top, 348 and 354; Henk Snoek Photography: page 230 top; Eric Hosking: pages 368 bottom and 382 bottom; Dr M. W. Jenison, Syracuse University New York, courtesy of the Society of American Bacteriologists: page 228 top; John Topham Picture Library: page 146; Dr H. B. D. Kettlewell, University of Oxford: page 362 left and right; King's College Hospital and Dental School: page 138; Lancashire Evening Post: page 239 bottom; Liverpool Hospital of Hygiene and Tropical Medicine: page 50; Mansell Collection: page 231 bottom; Middlesex Hospital: page 157; M.I.S.S. at the Royal College of Surgeons: pages 140, 152, 224 top, 238, 247, 268, 278, 364; Ken Moreman/Vision International: page 221; Mark Moylan: page 10 top; Musée Pasteur: page 376; National Coal Board: pages 277 and 397 left and right; Natural History Photographic Agency: page 360; Oxford Scientific Films: page 86; Pace: pages 170, 174, 217 top left and 252; Dr R. J. Pack, Department of Physiology, St. George's Medical School: page 231 top; Popperfoto Ltd.: pages 162, 190 top and 234; Dr K. R. Porter, Harvard University: page 16 right; Press Association Ltd.: page 129; Chris Ridgers V. R. U.: pages 16 left courtesy of Neurovirology Unit, Rayne Research Institute, St Thomas' Hospital, London, 32 top and 113 top; Rothamsted Experimental Station: pages 36 bottom, 70, 72, 74 top and 182; S. G. B. Group Ltd.: page 394 top left and top right; J. Sainsbury Ltd., Public Relations Department: page 124; Shell Photographic Service: page 77 bottom; Sporting Pictures (UK) Ltd.: pages 245 and 276 top; Syndication International (Daily Mirror): page 368 top; M. B. V. Roberts: page 357 right; U. K. Atomic Energy Authority: page 294; Upjohn Ltd.: page 32 bottom right; M. T. Walters and Associates Ltd.: page 161; C. James Webb: pages 77 top and 276 bottom; Wellcome Institute for the History of Medicine: pages 14 top, 32 bottom left, 113 middle and bottom, 116, 118 top and bottom, 179, 227, 228 bottom and 229; Harrington Wells: page 210 left; World Health organisation: pages 28, 37, 130 and 233.

Cover photograph: Stan Wayman/Life © Time Inc. 1965 from Colorific.

Certain illustrations are based upon already published sources as follows:

Page 51 Fig. 3:	R. Buchsbaum, Animals without Backbones. 2nd Edition. University of Chicago Press 1976 p. 145
Page 52 Fig. 4:	R. Buchsbaum, Animals without Backbones. 2nd Edition. University of Chicago Press 1976 p. 132
Page 83 Fig. 10:	A. Dale, Patterns of Life. 2nd Edition. Heinemann 1951 p. 64 Fig. 37
Page 83 Fig. 9:	Based on N. Tinbergen, Study of Instinct. Oxford University Press 1975
Page 86 Fig. 3:	J. Gray, How Animals Move. Cambridge University Press 1953 Plate VI facing p. 79
Page 97 Assign 8:	Data Based on G. V. T. Mathews 1948
Page 114 Fig. 6:	Adapted from King, et al Nutrition for Developing Countries. Oxford University Press 1972 Fig. 3.5
Page 119 Assign 6:	Data from J. Marks, The Vitamins in Health and Disease. Churchill 1968
Page 125 Fig. 3:	Adapted from Gwen Conacher, Food Freezing at Home. Electricity Council (30 Millbank, S.W.1) p. 5
Page 128 Tables 1 & 2:	Data from Ministry of Agriculture, Fisheries and Food
Page 128 Table 1:	Data from, Manual of Nutrition Ministry of Agriculture, Fisheries & Food. HMSO 1970
Page 130 Fig. 5:	Based on United Nations data (FAO)
Page 131 Assign 3 & 7:	Data from Ministry of Agriculture, Fisheries and Food
Page 157 Fig. 4:	Data extracted from Report of the Royal College of Physicians of London 1971, Smoking and Health Now. Published by Pitman
Page 167 Assign 6:	Data from F. G. Hall
Page 175 Fig. 6:	Based on, The Sunday Times Book of Body Maintenance. Michael Joseph 1978
Page 182 Fig. 5 and Page 183 Assign 7:	Data from Rothamsted Experimental Station Report 1968
Page 252 Fig. 1:	Based on W. Keble Martin, The Concise British Flora in Colour Ebury Press/Michael Joseph 1965
Page 253 Fig. 4:	Based on K. Mellanby
Page 253 Fig. 3:	Based on G. J. Tortora, Principles of Human Anatomy. Canfield Press 1977
Page 276 Fig. 2:	Based on Tortora as above
Page 289 Fig. 3:	Based on B. S. Beckett, Biology, A Modern Introduction. Oxford University Press 1976
Page 306 Fig. 1:	Based on R. J. Demarest and J. J. Sciarra, Conception, Birth and Contraception. Hodder & Stoughton 1969
Page 308 Fig. 3:	From C. Wood, Sex and Fertility. Thames and Hudson 1969
Page 319 Fig. 5:	Based on L. J. E. Brimble, Intermediate Botany. Macmillan 1977
Page 366 Fig. 4:	Adapted from F. G. W. Knowles, Diagrams of Human Biology. Harrap
Page 372 Fig. 4:	Based on E. G. Neal, Woodland Ecology. Heinemann 1958
Page 395 Fig. 4:	Based on J. W. Kimball, Man and Nature. Addison-Wesley 1975 p. 106